Modern Analytic Geometry

William Wooton
Los Angeles Pierce College

Edwin F. Beckenbach
University of California, Los Angeles

Frank J. Fleming
Los Angeles Pierce College

Modern Analytic Geometry

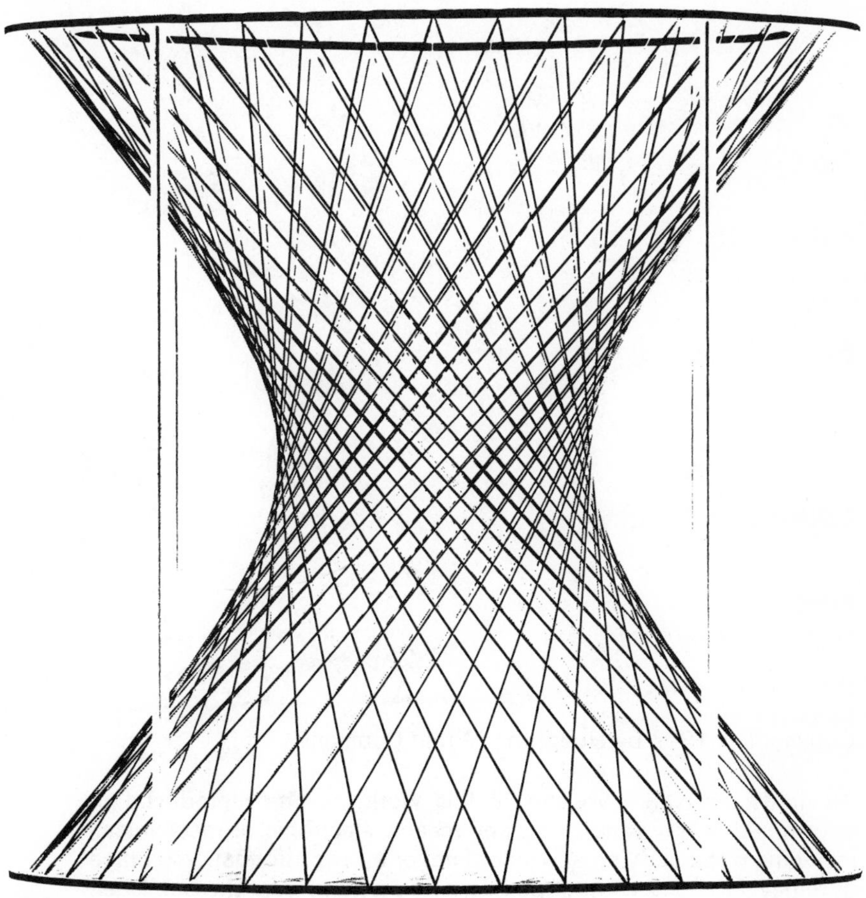

Houghton Mifflin Company · Boston
New York · Atlanta · Geneva, Illinois · Dallas · Palo Alto

Printed in the United States of America.

Library of Congress Catalog Card Number: 73–126910

ISBN 0–395–03743–3

Preface

We have written this book with the aim of providing a modern course in analytic geometry for today's students. The past decade has witnessed important changes in the textbooks used for the study of algebra, geometry, and trigonometry. It has become evident that the traditional course in analytic geometry should also be changed to reflect in spirit, as well as in content, the changes that have taken place in the study of mathematics.

This text is excellent preparation for the study of calculus and linear algebra. It is also an interesting course of and by itself. The material allows for a considerable amount of flexibility. Planned as a semester course, it might also be used in special cases as the basis for a full-year course. Also, several different semester courses are possible, depending on the preparation and aims of the students. The large number of exercises allows ample opportunity for adjustments in assignments, and the attention given to proof may be much or little. Asterisks signal those exercises which are particularly demanding.

Interest is secured at the outset by a study of the algebra and geometry of two-dimensional vectors, a topic new to many of the students. With this basis, vector methods can then be used in developing the concepts and techniques of analytic geometry. The student who continues his study of mathematics will find it very advantageous to have a working knowledge of vectors, and the material is inherently interesting even if one does not plan on future courses in mathematics.

An understanding of the traditional Cartesian methods, however, is still highly important to today's students and, accordingly, we have been very careful to include a thorough treatment of these methods in tandem with the vector approach. We believe the course presented in this book gives students an unusual opportunity to see how two apparently diverse branches of mathematics are, in reality, closely related, and to appreciate the value of having more than one method available for solving a given problem.

Study aids include such valuable lists as key trigonometric identities, fundamental properties of real numbers, important symbols, and a summary of formulas from analytic geometry. Color is used functionally to call attention to key concepts and processes and to clarify illustrative examples and diagrams.

Finally, students will enjoy the series of illustrated essays appearing at the close of chapters. These essays were included to deepen and extend the student's interest in the subject and its applications.

William Wooton
Edwin F. Beckenbach
Frank J. Fleming

Credits

Page	
iii, 359	Model courtesy of Massachusetts Institute of Technology; photograph by John T. Urban
43	Courtesy of *Scripta Mathematica*
45	Courtesy of Yale University
208	Courtesy of the Australian News and Information Bureau
210	Courtesy of the French Embassy Press and Information Division
303	Courtesy of General Dynamics
359	*top*—Model courtesy of John Greenfield; photograph by John T. Urban
385	Courtesy of Mrs. Oswald Veblen

Contents

Chapter 1 Vectors in the Plane

Chapter 2 Lines in the Plane

Chapter 3 Applications of Lines

Chapter 4 Conic Sections

Chapter 5 Transformation of Coordinates

Chapter 6 Curve Sketching

Chapter 7 Polar Coordinates

Chapter 8 Vectors in Space

Chapter 9 Lines and Planes in Space

Chapter 10 Surfaces and Transformations of Coordinates in Space

Symbols

Symbol	Meaning	Page
$A \times B$	Cartesian product	1
$\mathcal{R}$	set of real numbers	1
$\mathcal{R}^2$	$\mathcal{R} \times \mathcal{R}$	1
$d(\mathbf{A}, \mathbf{B})$	distance between points **A** and **B**	3
$\lvert x \rvert$	absolute value of x	4
$\mathbf{v}$, $\underset{\sim}{\mathrm{v}}$, $\vec{\mathrm{v}}$, (x, y)	vectors	6
u	scalar	6
$\lVert \mathbf{v} \rVert$	the norm of vector **v**	13
$m^\circ(\theta)$	degree measure of angle θ	14
$m^{\mathrm{R}}(\theta)$	radian measure of angle θ	14
cos	the cosine function	14
sin	the sine function	14
tan	the tangent function	15
$\doteq$	equals approximately	15
1°	1 degree	15
$1'$	1 minute	15
$\overline{\mathbf{ST}}$	the line segment with endpoints **S** and **T**	17
$\theta_{\mathbf{v}}$	the direction angle of vector **v**	18
$\mathbf{u} \cdot \mathbf{v}$	the inner product of the vectors **u** and **v**	27
$\mathbf{v_p}$	the vector $(-y, x)$, where $\mathbf{v} = (x, y)$	27
$\mathbf{Comp_u\, v}$	the vector component of **v** parallel to **u**	36
$\mathbf{Comp_{u_p}\, v}$	the vector component of **v** perpendicular to **u**	36
$\mathrm{Comp}_{\mathbf{t}}\, \mathbf{v}$	the scalar component of **v** parallel to **t**	37
$d(\mathbf{S}, \mathcal{L})$	the distance between a point **S** and a line $\mathcal{L}$	89
$l(\mathbf{S}, \mathbf{T})$	the directed distance from a point **S** to a point **T**	90
$\emptyset$	empty set or null set	97
$\begin{bmatrix} a_1 & b_1 \\ a_2 & b_2 \end{bmatrix}$	matrix	103
$\begin{vmatrix} a_1 & b_1 \\ a_2 & b_2 \end{vmatrix}$	determinant	103
D, det M	determinant of matrix M	103
e	eccentricity	154
Δ	discriminant of general quadratic equation	191
$f(x)$	the value of a function f at x (read "f of x")	213

$\frac{f}{g}$	the quotient of the functions f and g	220
$\log_{10}$	the logarithmic function to the base 10	230
$(r, m(\theta))$ or (r, θ)	polar coordinates of a point	243
sec	the secant function	256
csc	the cosecant function	256
cot	the cotangent function	256
T^{-1}	the inverse transformation *or* inverse of a matrix T	269, 302
$A \times B \times C$	Cartesian product of sets A, B, and C	273
$\mathcal{R}^3$	$\mathcal{R} \times \mathcal{R} \times \mathcal{R}$	273
$\mathbf{u} \times \mathbf{v}$	the cross product of the vectors **u** and **v**	294
I	the identity matrix	302
$(r, m(\theta), z)$ or (r, θ, z)	cylindrical coordinates of a point in space	370
$(\rho, m(\theta), m(\phi))$ or (ρ, θ, ϕ)	spherical coordinates of a point in space	372

Greek Alphabet

A	α	Alpha	I	ι	Iota	P	ρ	Rho
B	β	Beta	K	κ	Kappa	Σ	σ ς	Sigma
Γ	γ	Gamma	Λ	λ	Lambda	T	τ	Tau
Δ	δ	Delta	M	μ	Mu	Υ	υ	Upsilon
E	ϵ	Epsilon	N	ν	Nu	Φ	ϕ	Phi
Z	ζ	Zeta	Ξ	ξ	Xi	X	χ	Chi
H	η	Eta	O	o	Omicron	Ψ	ψ	Psi
Θ	θ	Theta	Π	π	Pi	Ω	ω	Omega

Script Alphabet

$\mathcal{A}$	$\mathcal{E}$	$\mathcal{I}$	$\mathcal{M}$	$\mathcal{Q}$	$\mathcal{U}$	$\mathcal{X}$
$\mathcal{B}$	$\mathcal{F}$	$\mathcal{J}$	$\mathcal{N}$	$\mathcal{R}$	$\mathcal{V}$	$\mathcal{Y}$
$\mathcal{C}$	$\mathcal{G}$	$\mathcal{K}$	$\mathcal{O}$	$\mathcal{S}$	$\mathcal{W}$	$\mathcal{Z}$
$\mathcal{D}$	$\mathcal{H}$	$\mathcal{L}$	$\mathcal{P}$	$\mathcal{T}$		

Chapter 1

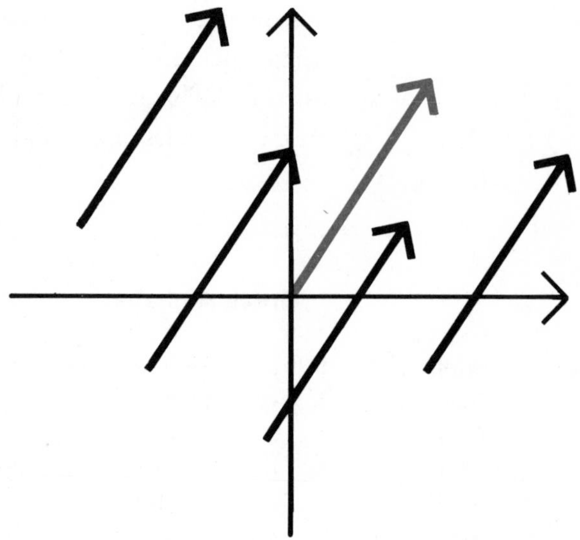

In this chapter, the algebra and geometry of two-dimensional vectors are introduced. The concepts and techniques developed here are then used in succeeding chapters to solve problems in analytic geometry.

Vectors in the Plane

Geometry of Ordered Pairs

1–1 Cartesian Coordinates

In your earlier studies of mathematics you learned that the Cartesian product $A \times B$ (read "A cross B") of the sets A and B is defined to be the set of all ordered pairs (x, y) in which the **first component**, x, is a member of A and the **second component**, y, is a member of B. Thus, if $A = \{1, 2, 3\}$ and $B = \{5, 6, 7\}$, then

$$A \times B = \{(1, 5), (1, 6), (1, 7), (2, 5), (2, 6), (2, 7), (3, 5), (3, 6), (3, 7)\}.$$

A set of ordered pairs such as $A \times B$ can be visualized as a point lattice, as shown in Figure 1–1.

Figure 1–1

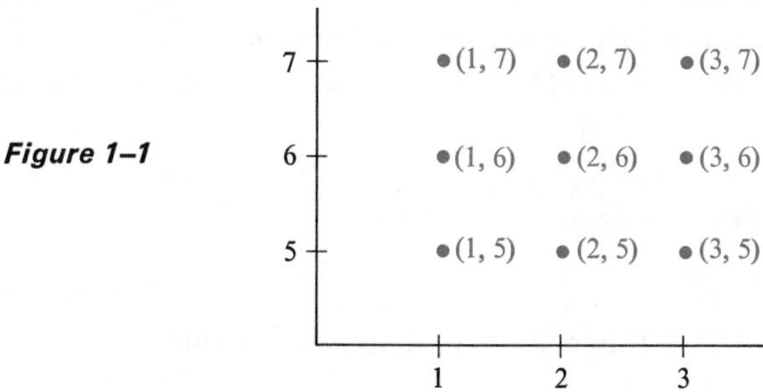

The Cartesian product you will encounter most often in this book is $\mathcal{R} \times \mathcal{R}$, or as we shall denote it, $\mathcal{R}^2$, where $\mathcal{R}$ is the set of real numbers. By the definition of Cartesian product,

$$\mathcal{R}^2 = \{(x, y): x \in \mathcal{R} \text{ and } y \in \mathcal{R}\}.$$

Recall that each ordered pair in $\mathcal{R}^2$ can be uniquely assigned to a point **S** in the plane by means of a **rectangular Cartesian**, or **Cartesian, coordinate system**.

The perpendicular number lines shown in Figure 1–2 are called the **axes** of such a system, and their point of intersection, **O**, is called the **origin**. (It is traditional to direct the number lines horizontally to the right and vertically up the page, as illustrated; this choice is by no means necessary, however, and in applications the perpendicular number lines should be chosen in whatever directions seem most convenient.) The four regions into which the axes partition the plane are called **quadrants**, and they are numbered I, II, III, and IV as shown.

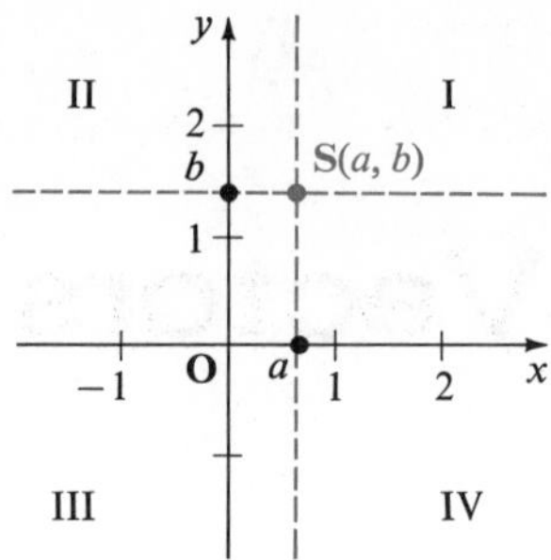

Figure 1–2

The assignment of a particular ordered pair (a, b) to a point **S** is made as follows:

1. Through the point corresponding to the number a on the horizontal axis (the **x-axis**), construct a line parallel to the vertical axis.
2. Through the point corresponding to the number b on the vertical axis (the **y-axis**), construct a line parallel to the horizontal axis.
3. The point **S** of intersection of these lines is then assigned the **coordinates** (a, b). **S** is called "the **graph** of (a, b)" or sometimes simply "the point (a, b)."

The first component, a, of (a, b) is sometimes called the **abscissa** of **S**, and the second component, b, is called the **ordinate** of **S**.

Notice that for each point **S** in the plane, if lines are drawn through **S** parallel to the axes, they will intersect the axes at unique points. The numbers a and b assigned to these points of intersection form one and only one ordered pair (a, b), so that the assignment of a given point **S** to an ordered pair (a, b) is also unique. Thus, there is a *one-to-one correspondence* between the set of points in the plane and the members of $\mathcal{R}^2$, and the graph of $\mathcal{R}^2$ is the entire *coordinate plane*.

Because of this one-to-one correspondence, if two ordered pairs correspond to the same point, they must be equal. Thus:

■ Two ordered pairs (a, b) and (c, d) in $\mathcal{R}^2$ correspond to the same point if and only if they are equal, that is, if and only if

$$a = c \quad \text{and} \quad b = d.$$

Example 1. For what real values of x and y is $(x + y, x - y) = (5, 3)$?

Solution: Since the given ordered pairs are equal, $x + y = 5$ and $x - y = 3$. You can solve the system

$$\begin{aligned} x + y &= 5 \\ x - y &= 3 \end{aligned} \qquad (1)$$

to find the required values for x and y. Adding the corresponding members of the two equations, you obtain

$$2x = 8,$$

or

$$x = 4.$$

Replacing x with 4 in the equation $x + y = 5$ produces

$$4 + y = 5,$$
$$y = 1.$$

You can check that the values $x = 4$ and $y = 1$ do actually satisfy Equations (1). Therefore, the required values for x and y are 4 and 1, respectively.

An important property you should recall about the coordinate plane is that if the same scale is used on both axes, then the distance between any two points $\mathbf{A}(x_1, y_1)$ and $\mathbf{B}(x_2, y_2)$ in the plane is given by the **distance formula,**

$$d(\mathbf{A}, \mathbf{B}) = \sqrt{(x_2 - x_1)^2 + (y_2 - y_1)^2}.$$

This is an immediate consequence of the familiar Pythagorean theorem

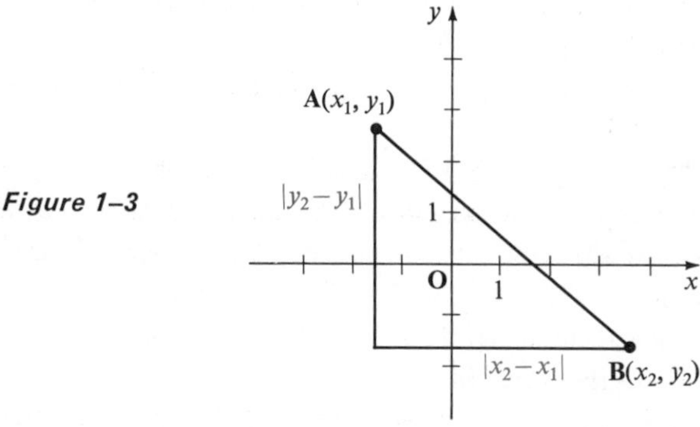

Figure 1–3

(Figure 1–3). In applying the distance formula, it does not matter which of the two given points is considered to be $\mathbf{A}(x_1, y_1)$ and which $\mathbf{B}(x_2, y_2)$, since

$$\sqrt{(x_1 - x_2)^2 + (y_1 - y_2)^2} = \sqrt{(x_2 - x_1)^2 + (y_2 - y_1)^2}.$$

Example 2. Find the distance between the points $\mathbf{A}(-4, 1)$ and $\mathbf{B}(3, 2)$.

Solution: Using $(x_1, y_1) = (-4, 1)$ and $(x_2, y_2) = (3, 2)$, you obtain

$$d(\mathbf{A}, \mathbf{B}) = \sqrt{[3 - (-4)]^2 + (2 - 1)^2}$$
$$= \sqrt{7^2 + 1^2} = \sqrt{50} = 5\sqrt{2}.$$

Exercises 1–1

In Exercises 1–8, find real-number values, if there are any, for which the equation is true. If no such real values exist, so state.

1. $(x + 3, 5) = (-1, 9 + x)$
2. $(x - 4, 2) = (3, x - 5)$
3. $(2x - 7, x + 2) = (-5, 3)$
4. $(3x + 2, 2x - 3) = (8, 1)$
5. $(x - 2y, 2x + y) = (-1, 3)$
6. $(2x + 3y, x + 4y) = (3, -1)$
7. $(x^2 - 2x, x^2 - x) = (3, 6)$
8. $(x^2 + 2x, 2x^2 + 3x) = (-1, -1)$

In Exercises 9–14, find the distance between the given points **S** and **T**. Express results in simplest radical form.

9. $\mathbf{S}(1, 3), \mathbf{T}(-2, 6)$
10. $\mathbf{S}(1, -6), \mathbf{T}(6, 6)$
11. $\mathbf{S}(\sqrt{2}, \sqrt{2}), \mathbf{T}(-\sqrt{2}, -\sqrt{2})$
12. $\mathbf{S}(\sqrt{3}, -\sqrt{3}), \mathbf{T}(-\sqrt{3}, \sqrt{3})$
13. $\mathbf{S}(4, \sqrt{3}), \mathbf{T}(2, -1)$
14. $\mathbf{S}(\sqrt{2}, -\sqrt{3}), \mathbf{T}(1, 2)$

15. Show that the triangle with vertices at the points $\mathbf{R}(0, 1)$, $\mathbf{S}(8, -7)$, and $\mathbf{T}(1, -6)$ is isosceles.
16. Show that the points $\mathbf{R}(-4, 4)$, $\mathbf{S}(-2, -4)$, and $\mathbf{T}(6, -2)$ are the vertices of an isosceles triangle.
17. Show that the point $\mathbf{Q}(1, -2)$ is equidistant from the points $\mathbf{R}(-11, 3)$, $\mathbf{S}(6, 10)$, and $\mathbf{T}(1, 11)$.
18. Show that the point $\mathbf{Q}(2, -3)$ is equidistant from the points $\mathbf{R}(6, 0)$, $\mathbf{S}(-2, -6)$, and $\mathbf{T}(-1, 1)$.
19. The points $\mathbf{Q}(1, 1)$, $\mathbf{R}(2, 5)$, $\mathbf{S}(6, 8)$, and $\mathbf{T}(5, 4)$ are the vertices of a quadrilateral. Show that the opposite sides of the quadrilateral are equal in length.
20. Are the opposite sides of the quadrilateral whose vertices are the points $\mathbf{Q}(-2, 3)$, $\mathbf{R}(5, 2)$, $\mathbf{S}(7, -4)$, and $\mathbf{T}(0, -2)$ equal in length?
21. Use the distance formula to show that the points $\mathbf{R}(-2, -5)$, $\mathbf{S}(1, -1)$, and $\mathbf{T}(4, 3)$ lie on the same line.
22. Show that the points $\mathbf{R}(-3, 3)$, $\mathbf{S}(2, 1)$, and $\mathbf{T}(7, -1)$ lie on the same line.
23. Show that $\mathbf{R}(1, 5)$ is the midpoint of the line segment with endpoints $\mathbf{S}(-2, 3)$ and $\mathbf{T}(4, 7)$.
24. Show that $\mathbf{M}\left(\frac{a + c}{2}, \frac{b + d}{2}\right)$ is the midpoint of the line segment with endpoints $\mathbf{S}(a, b)$ and $\mathbf{T}(c, d)$.
25. Find the midpoint of the line segment with endpoints $\mathbf{S}(-2, 9)$ and $\mathbf{T}(8, -1)$.
26. Find the midpoint of the line segment with endpoints $\mathbf{S}(-3, 5)$ and $\mathbf{T}(3, 2)$.
27. Show that for points $\mathbf{A}(x_1, 0)$ and $\mathbf{B}(x_2, 0)$, $d(\mathbf{A}, \mathbf{B}) = |x_2 - x_1|$.
28. Show that for points $\mathbf{C}(0, y_1)$ and $\mathbf{D}(0, y_2)$, $d(\mathbf{C}, \mathbf{D}) = |y_2 - y_1|$.

1–2 Points and Vectors

You have seen that you can associate each point in the coordinate plane with an ordered pair (x, y). For example, you can associate the point **R** in Figure 1–4 with the ordered pair $(2, -1)$. You can also associate a **displacement**, or **translation,** in the plane with the same ordered pair (x, y). For example, consider a particle which moves from a point **S** in the plane to another point, **T**, in the

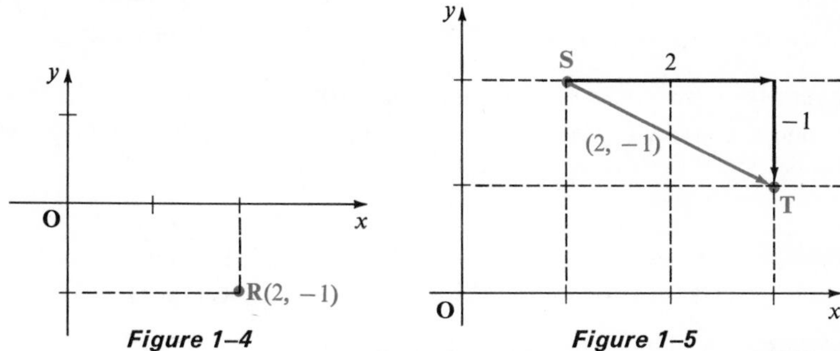

Figure 1–4 ***Figure 1–5***

plane along a straight line. If the particle moves 2 units to the right and 1 unit down, then this translation, or displacement, can be described by the ordered pair $(2, -1)$. Such a translation is pictured in Figure 1–5.

Since an ordered pair (x, y) of real numbers can be used to specify a translation in the plane, such an ordered pair is often called a **vector** (from the Latin word for *carrier*). The geometric representation of a vector (x, y) is an **arrow**, or **directed line segment**, in the plane; the arrow is called a **geometric vector**.

Figure 1–6

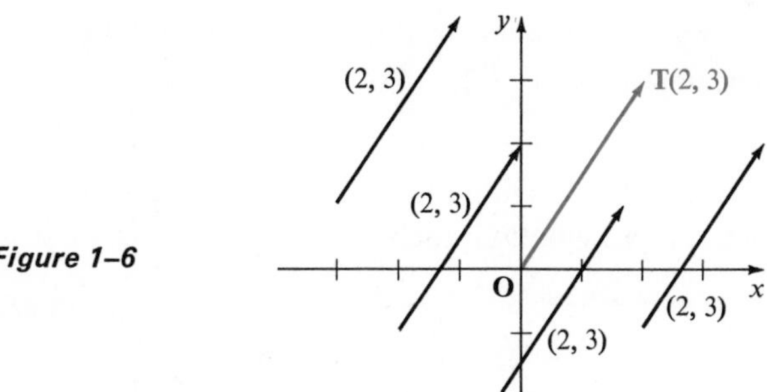

Since a translation can be visualized as having its **initial point** at any point **S** in the plane and its **terminal point** at a related point **T**, each vector (x, y) has infinitely many geometric representations in the plane. The particular arrow associated with (x, y) that has *initial point at the origin* is called the **standard representation** of (x, y), and the arrow is said to be in **standard position**. Clearly, the standard representation of the vector (x, y) has as terminal point the point **T** associated with (x, y). Figure 1–6 shows several geometric representations of $(2, 3)$; the arrow shown in color is its standard representation.

If an arrow with initial point $\mathbf{S}(a, b)$ is a geometric representation for the vector (x, y), as indicated in Figure 1–7, then the terminal point of the geometric vector is the point $\mathbf{T}(c, d)$ whose coordinates satisfy

$$(c, d) = (a + x, b + y).$$

Figure 1–7

Conversely, if an arrow with initial point $\mathbf{S}(a, b)$ has $\mathbf{T}(c, d)$ as its terminal point, then the arrow is a geometric representation for the vector (x, y), where

$$(x, y) = (c - a, d - b).$$

Example 1. What are the coordinates of the terminal point $\mathbf{T}$ of the geometric vector corresponding to $(3, -2)$ which has initial point $\mathbf{S}(-1, 4)$?

Solution: A sketch such as the one shown at the right helps you to visualize the situation. If (c, d) is used to denote the coordinates of the terminal point $\mathbf{T}$ of the arrow, then

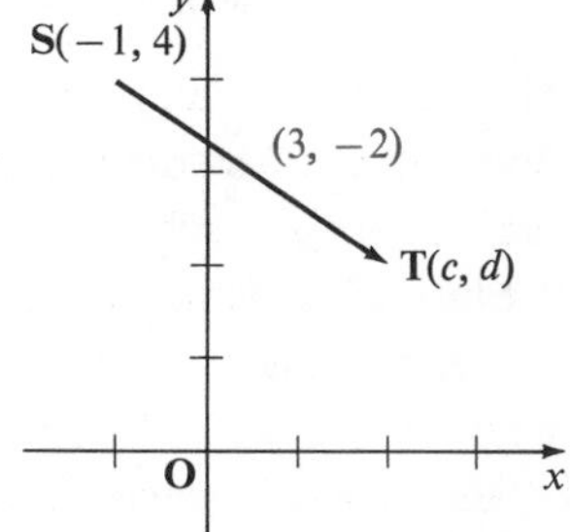

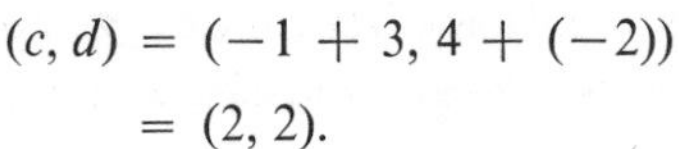

$$\begin{aligned}(c, d) &= (-1 + 3, 4 + (-2)) \\ &= (2, 2).\end{aligned}$$

Example 2. What vector corresponds to the arrow with initial point $\mathbf{S}(3, 0)$ and terminal point $\mathbf{T}(-2, -5)$?

Solution: Again, a sketch is helpful. If (x, y) denotes the vector with the given geometric representation, then

$$\begin{aligned}(x, y) &= (-2 - 3, -5 - 0) \\ &= (-5, -5).\end{aligned}$$

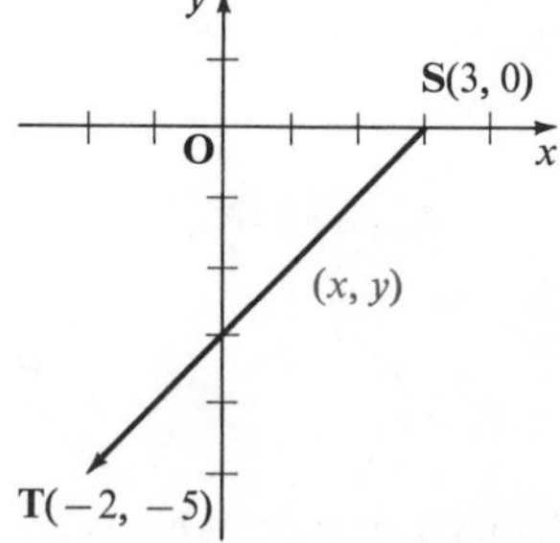

Notice that we ordinarily use such notation as $\mathbf{S}(x, y)$ to denote a point and its coordinates, and (x, y) to denote a vector or its geometric representation. We shall also use boldface letters, such as $\mathbf{v}$, $\mathbf{u}$, and $\mathbf{t}$, to denote vectors in $\mathcal{R}^2$. Italic letters, such as v, u, and t, will generally be used to represent real numbers. In this book, real numbers will often be referred to as **scalars**. In handwritten work, it is often convenient to represent a vector by a letter with a small arrow over it, like $\vec{\mathrm{v}}$, $\vec{\mathrm{t}}$, and $\vec{\mathrm{s}}$. Another way to denote a vector is to place a wavy line beneath a letter, as in $\underset{\sim}{\mathrm{v}}$, $\underset{\sim}{\mathrm{t}}$, and $\underset{\sim}{\mathrm{s}}$.

Exercises 1–2

In Exercises 1–10, sketch the arrow representation of the given vector which has initial point (a) the origin **O**(0, 0); (b) **Q**(3, 2); (c) **R**(2, −1); (d) **S**(−3, −2); and (e) **T**(−2, 0).

Example. (4, 2)

Solution:

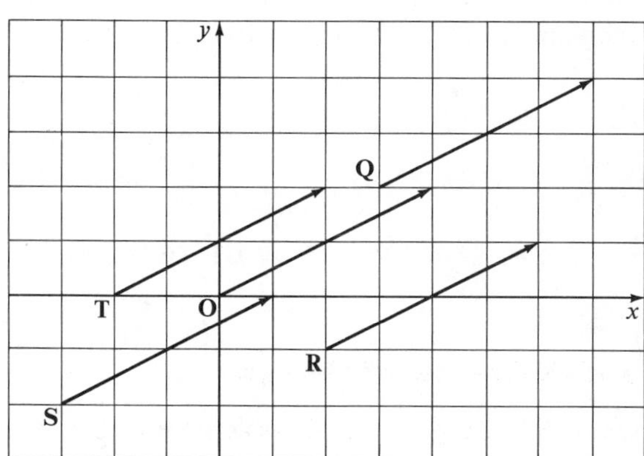

1. (1, 1)	**3.** (4, −1)	**5.** (−2, 1)	**7.** (−6, −1)	**9.** (3, 0)
2. (3, 4)	**4.** (2, −2)	**6.** (−3, 4)	**8.** (−2, −4)	**10.** (0, −5)

In Exercises 11–20, find the coordinates of the terminal point of the geometric representation of the given vector if its initial point is (a) **O**(0, 0); (b) **Q**(3, 2); (c) **R**(4, −1); (d) **S**(−3, 7); and (e) **T**(−6, −5).

11. (1, 6)
12. (2, 4)
13. (−3, 5)
14. (−4, 1)
15. (7, −8)
16. (3, −4)
17. (−3, −4)
18. (−5, −9)
19. (−2, 0)
20. (0, 4)

In Exercises 21–28, the coordinates of two points, **S** and **T**, are given. **S** is the initial point of the geometric representation of a vector (x, y), and **T** is the terminal point. Name the vector.

21. **S**(1, 2), **T**(5, 4)
22. **S**(3, 5), **T**(6, 8)
23. **S**(2, −1), **T**(−3, 2)
24. **S**(6, −4), **T**(−1, 5)
25. **S**(7, 12), **T**(−8, 3)
26. **S**(4, −2), **T**(−3, 6)
27. **S**(−6, −4), **T**(0, −4)
28. **S**(−5, −2), **T**(−5, 6)

29. Find the ordered pair (x, y) such that the arrow from **S**(x, y) to **T**(7, 3) represents the same vector as the arrow from **O**(0, 0) to **S**.

30. Find the ordered pair (x, y) such that the arrow from **S**(x, y) to **T**(9, −3) represents the same vector as the arrow from **O**(0, 0) to **S**.

31. The arrow from **R**(3, 5) to **S**(x, y) represents the same vector as the arrow from **S**(x, y) to **T**(8, 1). Find (x, y).

32. The arrow from **R**$(-4, -7)$ to **S**(x, y) represents the same vector as the arrow from **S**(x, y) to **T**(4, 11). Find (x, y).

33. Find the ordered pair (x, y) such that the arrow from **S**(x, y) to **T**$(6, -2)$ represents the same vector as the arrow from **Q**$(4, -7)$ to **R**(3, 1).

34. Find the ordered pair (x, y) such that the arrow from **S**$(-1, -2)$ to **T**(x, y) represents the same vector as the arrow from **Q**(2, 4) to **R**$(8, -2)$.

Basic Operations with Vectors

1–3 Sums and Differences of Vectors

The fact that any vector (x_1, y_1) can be viewed as representing a translation of x_1 units parallel to the x-axis followed by a translation of y_1 units parallel to the y-axis suggests that (x_1, y_1) can be considered as the "sum" of the vectors $(x_1, 0)$ and $(0, y_1)$. (See Figure 1–8.) More generally (see Figure 1–9), a trans-

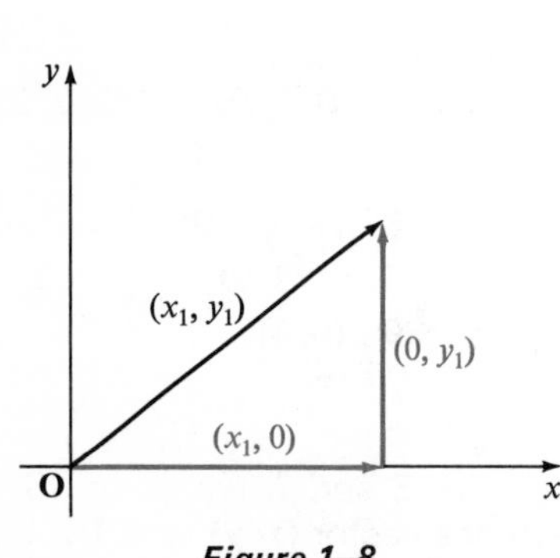

Figure 1–8

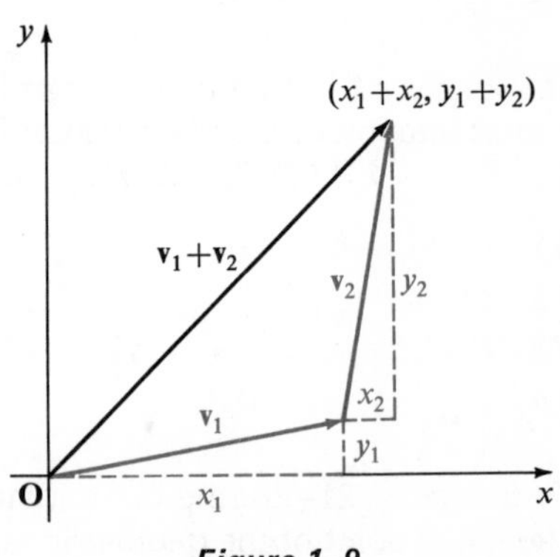

Figure 1–9

lation along any arrow corresponding to the vector $\mathbf{v}_1 = (x_1, y_1)$ followed by a translation from the terminal point of this arrow along the arrow representing the vector $\mathbf{v}_2 = (x_2, y_2)$ results in a net translation corresponding to the vector $(x_1 + x_2, y_1 + y_2)$. This fact leads us to define the **sum** of two vectors as follows:

■ If $\mathbf{v}_1 = (x_1, y_1) \in \mathcal{R}^2$ and $\mathbf{v}_2 = (x_2, y_2) \in \mathcal{R}^2$, then

$$\mathbf{v}_1 + \mathbf{v}_2 = (x_1, y_1) + (x_2, y_2) = (x_1 + x_2, y_1 + y_2).$$

Example 1. Find r and s so that

$$(r, s) = (3, 1) + (-2, 3),$$

and show the standard representation of each of the vectors (r, s), $(3, 1)$, and $(-2, 3)$.

Solution: You have

$$(r, s) = (3, 1) + (-2, 3) = (3 + (-2), 1 + 3) = (1, 4).$$

The standard representation of each vector is shown at the right.

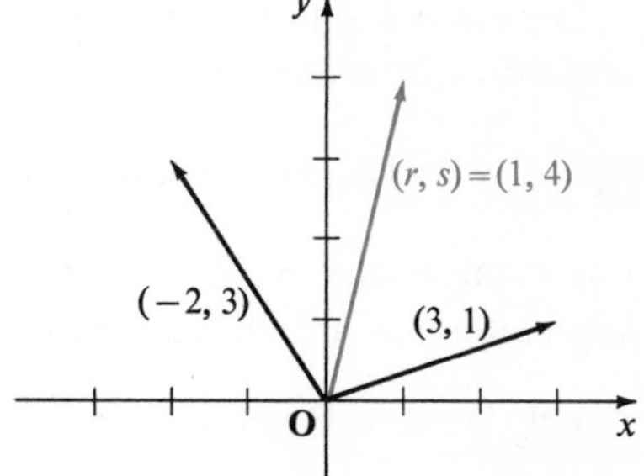

If dashed lines are inserted in the diagram above to complete a parallelogram having the (standard) geometric representations of $(-2, 3)$ and $(3, 1)$ as adjacent sides, then the diagonal of the parallelogram is the (standard) geometric representation of the sum $(1, 4)$, as shown in Figure 1–10. For this reason, vector addition is sometimes referred to as addition according to the **parallelogram rule**.

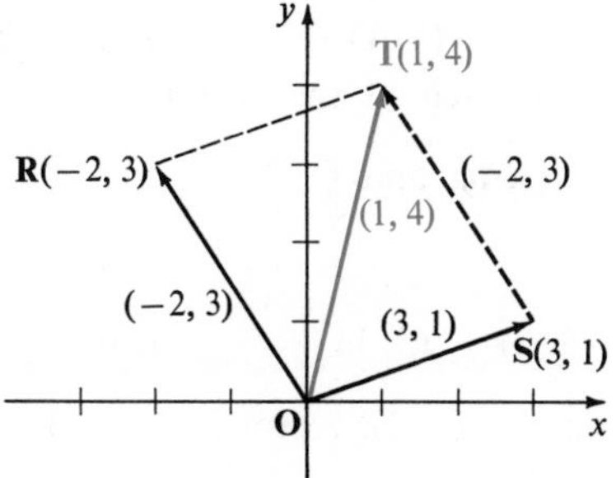

Figure 1–10

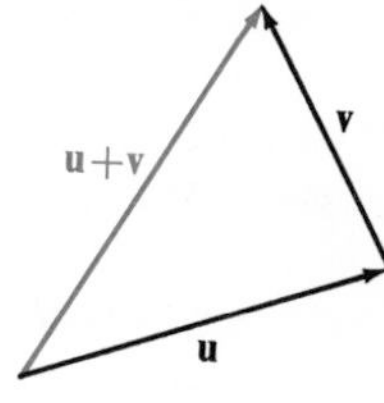

Figure 1–11

Also, as Figure 1–10 shows, there is an arrow representation of $(-2, 3)$ which has initial point at $\mathbf{S}(3, 1)$ and terminal point at $\mathbf{T}(1, 4)$. This illustrates the **triangle rule** for picturing a vector sum: If an arrow representing a vector $\mathbf{v}$ is drawn with initial point at the terminal point of an arrow representing a vector $\mathbf{u}$, then the arrow joining the initial point of $\mathbf{u}$ with the terminal point of $\mathbf{v}$ represents the vector sum $\mathbf{u} + \mathbf{v}$ (Figure 1–11).

Since for any vector $(x, y) \in \mathcal{R}^2$, you have

$$(x, y) + (0, 0) = (x + 0, y + 0) = (x, y),$$

and

$$(0, 0) + (x, y) = (0 + x, 0 + y) = (x, y),$$

the vector (0, 0), called the **zero vector**, serves as the *additive identity* element for the addition of vectors in $\mathcal{R}^2$. The zero vector is usually represented by the symbol $\mathbf{0}$; its geometric representation is simply a point.

The symbol $-\mathbf{v} = -(x, y)$ represents the vector whose components are the negatives of the components of $\mathbf{v} = (x, y)$. That is:

■ If $\mathbf{v} = (x, y) \in \mathcal{R}^2$, then $-\mathbf{v} = -(x, y) = (-x, -y)$.

For example, if $\mathbf{v} = (2, -4)$, then $-\mathbf{v} = (-2, -(-4)) = (-2, 4)$. Since for any vector $\mathbf{v} = (x, y)$, it is true that

$$\mathbf{v} + (-\mathbf{v}) = (x, y) + (-x, -y) = (x + (-x), y + (-y)) = (0, 0) = \mathbf{0},$$

$-\mathbf{v}$ is called the **additive inverse**, or **negative**, of $\mathbf{v}$. This leads us to a natural definition for the **difference** of two vectors:

■ If $\mathbf{v}_1 = (x_1, y_1) \in \mathcal{R}^2$ and $\mathbf{v}_2 = (x_2, y_2) \in \mathcal{R}^2$, then

$$\begin{aligned} \mathbf{v}_1 - \mathbf{v}_2 &= \mathbf{v}_1 + (-\mathbf{v}_2) \\ &= (x_1, y_1) + (-x_2, -y_2) \\ &= (x_1 - x_2, y_1 - y_2). \end{aligned}$$

Example 2. Find r and s so that $(r, s) = (3, 1) - (-2, 3)$, and show the standard representation of each of the vectors

$$(r, s),\ (3, 1),\ \text{and}\ (-2, 3).$$

Solution: You have

$$(r, s) = (3, 1) - (-2, 3) = (3, 1) + (2, -3) = (5, -2).$$

The standard representation of each vector is shown below.

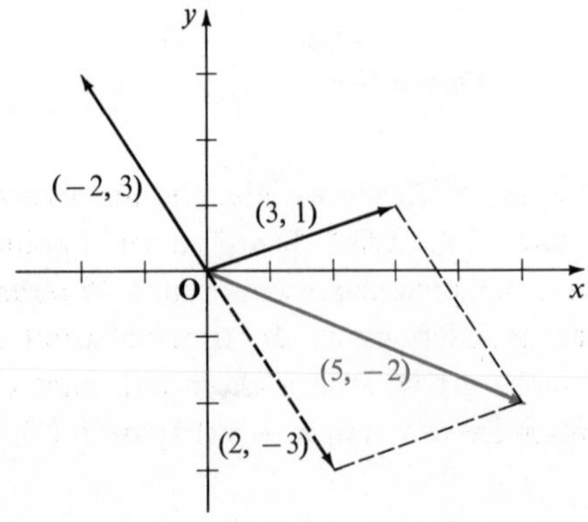

In the diagram for Example 2, notice that (2, −3), the additive inverse of (−2, 3), has a standard geometric representation collinear with, and of the same length as, that of (−2, 3), but with opposite direction. Notice also that the standard geometric representation of the *difference* of (3, 1) and (−2, 3), that is, (5, −2), is the standard geometric representation of the *sum* of (3, 1) and (2, −3).

In general, the standard geometric representation of $-\mathbf{v}$ is collinear with, and of the same length as, that of $\mathbf{v}$, but with opposite direction. A geometric representation of the difference $\mathbf{u} - \mathbf{v}$ can therefore be obtained by applying the parallelogram rule or the triangle rule to the sum $\mathbf{u} + (-\mathbf{v})$, as illustrated in Figure 1–13(a) below.

There is another convenient way of picturing a vector difference. If, for example, the difference vector

$$(5, -2) = (3, 1) - (-2, 3)$$

of Example 2 is represented by an arrow with initial point $\mathbf{S}(-2, 3)$, then its terminal point $\mathbf{T}$ has coordinates

$$(-2, 3) + (5, -2), \quad \text{or} \quad (3, 1),$$

as shown in Figure 1–12. This illustrates the following fact: If $\mathbf{u}$ and $\mathbf{v}$ are any vectors in $\mathcal{R}^2$, then since the difference $\mathbf{u} - \mathbf{v}$ satisfies

$$\mathbf{v} + (\mathbf{u} - \mathbf{v}) = \mathbf{u},$$

there are arrow representations for $\mathbf{u}$, $\mathbf{v}$, and $\mathbf{u} - \mathbf{v}$ that form a triangle in the plane, with arrows directed as illustrated in Figure 1–13(b). This, in turn, suggests why $\mathbf{u} - \mathbf{v}$ is often referred to as "the vector from $\mathbf{v}$ to $\mathbf{u}$."

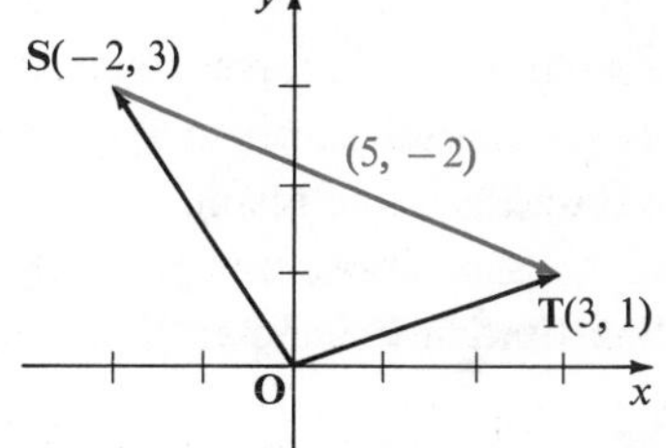

Figure 1–12

(*a*)

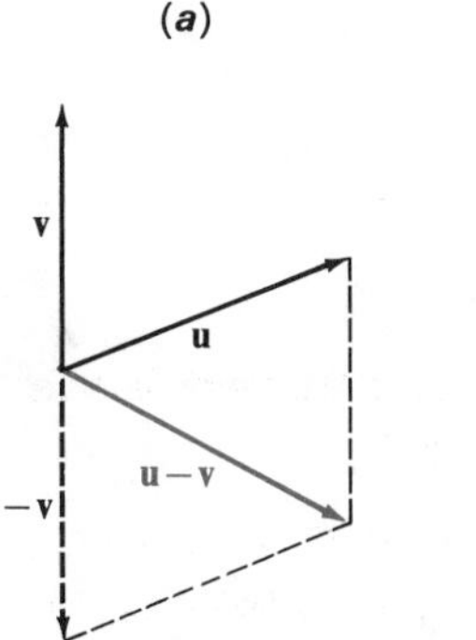

(*b*)

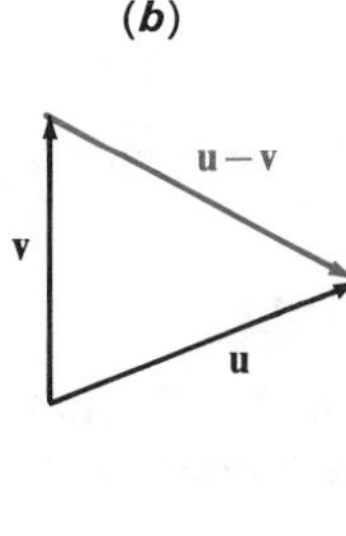

Figure 1–13

The following list summarizes the basic properties of vector addition. Proofs of these properties are left as exercises (page 13).

Properties of Vector Addition

If $\mathbf{u}$, $\mathbf{v}$, and $\mathbf{s}$ are vectors in $\mathcal{R}^2$, then

1. $\mathbf{u} + \mathbf{v} \in \mathcal{R}^2$	Closure
2. $\mathbf{u} + \mathbf{v} = \mathbf{v} + \mathbf{u}$	Commutativity
3. $(\mathbf{u} + \mathbf{v}) + \mathbf{s} = \mathbf{u} + (\mathbf{v} + \mathbf{s})$	Associativity
4. $\mathbf{u} + \mathbf{0} = \mathbf{u}$ and $\mathbf{0} + \mathbf{u} = \mathbf{u}$	Additive identity property
5. $\mathbf{u} + (-\mathbf{u}) = \mathbf{0}$ and $(-\mathbf{u}) + \mathbf{u} = \mathbf{0}$	Additive inverse property

Notice that these properties have the same formal appearance as the properties of real-number addition (see page 392). Notice in particular that, like the addition of real numbers, the addition of vectors is basically a binary operation; the sum of three or more vectors is defined successively by

$$\mathbf{u} + \mathbf{v} + \mathbf{s} = (\mathbf{u} + \mathbf{v}) + \mathbf{s},$$
$$\mathbf{u} + \mathbf{v} + \mathbf{s} + \mathbf{t} = (\mathbf{u} + \mathbf{v} + \mathbf{s}) + \mathbf{t},$$

and so on. Of course, since addition of vectors is associative and commutative, you may add vectors in any convenient groups of two, and in any order, and still obtain the same result.

Another important property of both vectors and real numbers is the general **Substitution Principle**:

■ Changing the symbol by which a mathematical object is named in a mathematical expression does not change the meaning of the expression.

With regard to vector addition, this principle can be restated as follows:

If $\mathbf{u} = \mathbf{v}$ and $\mathbf{s} = \mathbf{t}$, then $\mathbf{u} + \mathbf{s} = \mathbf{v} + \mathbf{t}$.

Exercises 1–3

In Exercises 1–12, find r and s so that the given statement is true. Sketch the standard representation of each vector involved.

1. $(r, s) = (4, 1) + (2, -3)$

2. $(r, s) = (3, 4) + (-1, -5)$

3. $(r, s) + (3, -1) = (2, 5)$

4. $(r, s) + (6, -2) = (7, 3)$

5. $(5, -3) + (0, 0) = (r, s)$

6. $(2, -3) + (4, 6) = (r, s)$

7. $(r, s) = (7, -2) - (3, -5)$

8. $(r, s) = (6, 5) - (2, 3)$

9. $(7, -3) - (r, s) = (5, 1)$

10. $(-2, 4) - (r, s) = (3, -2)$

11. $(r, s) - (1, 0) = (5, 1)$

12. $(r, s) - (3, -2) = (4, 5)$

In Exercises 13–18, $\mathbf{s} = (-1, 4)$, $\mathbf{t} = (3, 2)$, and $\mathbf{u} = (-4, -1)$. Find a vector $\mathbf{v}$ for which the given equation is true.

13. $\mathbf{s} - \mathbf{v} = \mathbf{t}$

14. $\mathbf{v} - \mathbf{s} = \mathbf{u}$

15. $\mathbf{t} + \mathbf{s} = \mathbf{v} - \mathbf{u}$

16. $\mathbf{u} + \mathbf{t} = \mathbf{v} - \mathbf{s}$

17. $\mathbf{t} + \mathbf{s} = \mathbf{t} - \mathbf{v}$

18. $\mathbf{u} - \mathbf{s} = \mathbf{t} + \mathbf{v}$

In Exercises 19–29, if $\mathbf{a}$, $\mathbf{u}$, $\mathbf{v}$, $\mathbf{s}$, and $\mathbf{t}$ are vectors in $\mathcal{R}^2$, prove each statement.

Example. $\mathbf{u} + \mathbf{v} \in \mathcal{R}^2$.

Solution: Let $\mathbf{u} = (x_1, y_1)$ and $\mathbf{v} = (x_2, y_2)$. Then

$$\mathbf{u} + \mathbf{v} = (x_1, y_1) + (x_2, y_2) = (x_1 + x_2, y_1 + y_2).$$

Since addition is closed in $\mathcal{R}$, $x_1 + x_2 \in \mathcal{R}$ and $y_1 + y_2 \in \mathcal{R}$. Therefore, $(x_1 + x_2, y_1 + y_2) \in \mathcal{R}^2$.

19. $\mathbf{u} + \mathbf{v} = \mathbf{v} + \mathbf{u}$

20. $(\mathbf{u} + \mathbf{v}) + \mathbf{s} = \mathbf{u} + (\mathbf{v} + \mathbf{s})$

21. $\mathbf{v} + (\mathbf{u} - \mathbf{v}) = \mathbf{u}$

22. If $\mathbf{u} + \mathbf{v} = \mathbf{0}$, then $\mathbf{u} = -\mathbf{v}$.

23. $\mathbf{u} + \mathbf{0} = \mathbf{u}$

24. $\mathbf{u} + (-\mathbf{u}) = \mathbf{0}$

25. $\mathbf{s} + \mathbf{t} \in \mathcal{R}^2$

26. $-\mathbf{u} - \mathbf{v} = -(\mathbf{u} + \mathbf{v})$

27. If $\mathbf{u} + \mathbf{v} = \mathbf{t}$, then $\mathbf{u} = \mathbf{t} - \mathbf{v}$.

28. If $\mathbf{a} + \mathbf{u} = \mathbf{u}$, then $\mathbf{a} = \mathbf{0}$ (uniqueness of additive identity).

29. If $\mathbf{a} + \mathbf{u} = \mathbf{0}$, then $\mathbf{a} = -\mathbf{u}$ (uniqueness of additive inverse).

1–4 Norm and Direction of a Vector

Corresponding to each vector $\mathbf{v} = (x, y)$ in $\mathcal{R}^2$, there is a unique scalar called the **norm**, or **magnitude**, of $\mathbf{v}$. The norm of $\mathbf{v}$ is denoted by the symbol $\|\mathbf{v}\|$ and is defined by

$$\|\mathbf{v}\| = \sqrt{x^2 + y^2}.$$

In accordance with the customary use of the radical symbol $\sqrt{\ }$, *the norm of each vector is nonnegative.*

Note, in particular, that if $\mathbf{v} = \mathbf{0} = (0, 0)$, then $\|\mathbf{v}\| = \sqrt{0} = 0$. Conversely, if $\|\mathbf{v}\| = 0$, then $\mathbf{v} = (0, 0) = \mathbf{0}$, since $\sqrt{x^2 + y^2}$ is 0 only if both x and y are 0. Thus $\|\mathbf{v}\| = 0$ if and only if $\mathbf{v}$ is the zero vector.

Geometrically, the norm of a vector can be interpreted as the length of any one of its arrow representations. For example, if an arrow representation of

$\mathbf{v} = (x, y)$ has initial point $\mathbf{A}(a, b)$, then $\mathbf{v}$ has terminal point $\mathbf{B}(a + x, b + y)$, and, by the distance formula,

$$d(\mathbf{A}, \mathbf{B}) = \sqrt{[(a + x) - a]^2 + [(b + y) - b]^2} = \sqrt{x^2 + y^2} = \|\mathbf{v}\|.$$

Example 1. Find the norm of $(3, -2)$.

Solution: $\|(3, -2)\| = \sqrt{3^2 + (-2)^2} = \sqrt{9 + 4} = \sqrt{13}.$

Each nonzero vector $\mathbf{v}$ in $\mathbb{R}^2$ can be assigned a **direction** as follows:

■ If $\mathbf{v} = (x, y) \in \mathbb{R}^2$ and $\mathbf{v} \neq \mathbf{0}$, then the direction of $\mathbf{v}$ is the measure of the angle θ for which

$$\sin\theta = \frac{y}{\|\mathbf{v}\|} = \frac{y}{\sqrt{x^2 + y^2}} \quad \text{and} \quad \cos\theta = \frac{x}{\|\mathbf{v}\|} = \frac{x}{\sqrt{x^2 + y^2}} \tag{1}$$

and $0 \leq m^\circ(\theta) < 360$ if θ is measured in degrees, or $0 \leq m^R(\theta) < 2\pi$ if θ is measured in radians.

In the rest of the chapter, direction angles will always be measured in degrees.

It follows at once from Equations (1) that

$$\mathbf{v} = (x, y) = (\|\mathbf{v}\| \cos\theta, \|\mathbf{v}\| \sin\theta). \tag{2}$$

Thus you can see that a vector is determined by its magnitude and its direction.

Since the direction of a vector $\mathbf{v}$ is defined to be the measure of an angle θ, it is meaningful to call the angle θ the **direction angle** of $\mathbf{v}$. Geometrically, θ is determined by an arrow representation of $\mathbf{v}$ and the ray which has the same

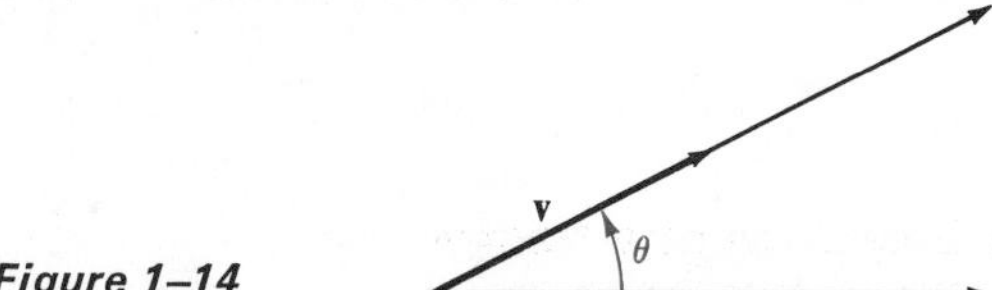

Figure 1–14

initial point as this arrow and which is parallel to the x-axis and oriented in the direction of the positive x-axis (Figure 1–14). The ray parallel to the x-axis is assumed to be the initial side of θ. The ray which contains the arrow for $\mathbf{v}$, and which has the same initial point as this arrow, is then the terminal side of θ.

Example 2. Find the norm and direction of the vector $(4, -3)$.

Solution: First, note that $\|(4, -3)\| = \sqrt{16 + 9} = 5.$

Then, by definition, the direction $m^\circ(\theta)$ of $(4, -3)$ is determined by

$$\sin\theta = \frac{-3}{5} \quad \text{and} \quad \cos\theta = \frac{4}{5}\cdot$$

Observe from a sketch, or from the fact that $\sin\theta < 0$ and $\cos\theta > 0$, that the terminal side of θ lies in Quadrant IV. Thus, from Table 2 on page 387, you obtain $m°(\theta) \doteq 323° \, 10'$.

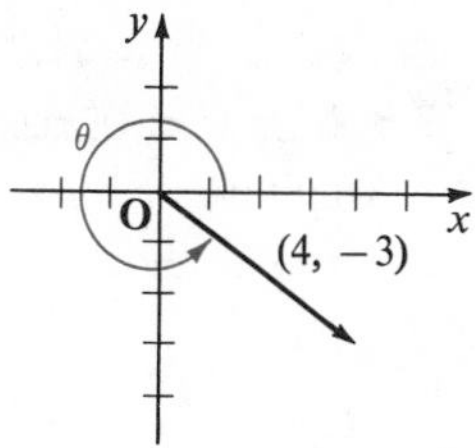

If, for $x \neq 0$, careful attention is paid to whether $x > 0$ or $x < 0$, and to whether $y \geq 0$ or $y < 0$, the trigonometric identity

$$\tan\theta = \frac{\sin\theta}{\cos\theta} = \frac{y}{x} \tag{3}$$

provides another means of obtaining the direction $m°(\theta)$ of $\mathbf{v} = (x, y)$. If $y = 0$, then $m°(\theta) = 0$ or $m°(\theta) = 180$ according as $x > 0$ or $x < 0$, respectively. If $y \neq 0$, then you can use Equation (3) to find a **reference angle** α, with $0 < m°(\alpha) < 90$, for which

$$\tan\alpha = \left|\frac{y}{x}\right|.$$

Then, taking account of quadrants (see Figure 1–15), you have

$m°(\theta) = m°(\alpha)$	if $x > 0, y > 0$	(Quadrant I),
$m°(\theta) = 180 - m°(\alpha)$	if $x < 0, y > 0$	(Quadrant II),
$m°(\theta) = 180 + m°(\alpha)$	if $x < 0, y < 0$	(Quadrant III),
$m°(\theta) = 360 - m°(\alpha)$	if $x > 0, y < 0$	(Quadrant IV).

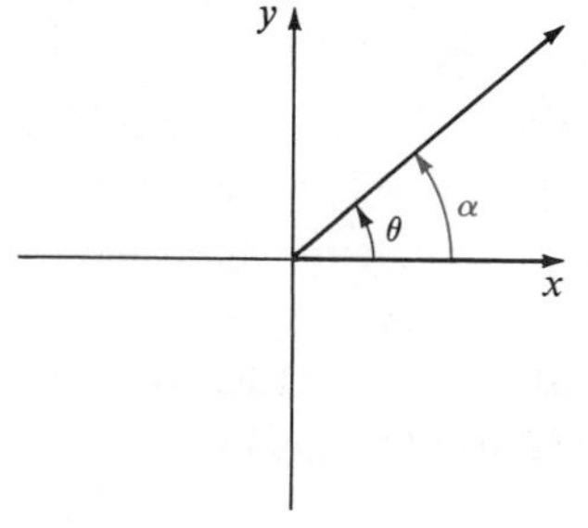

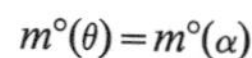
$m°(\theta) = m°(\alpha)$

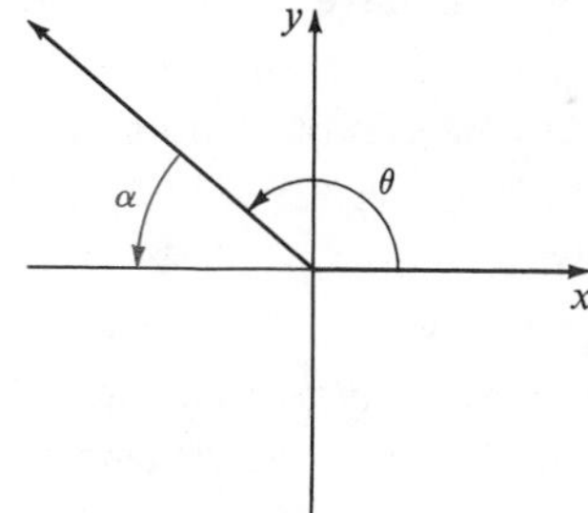

$m°(\theta) = 180 - m°(\alpha)$

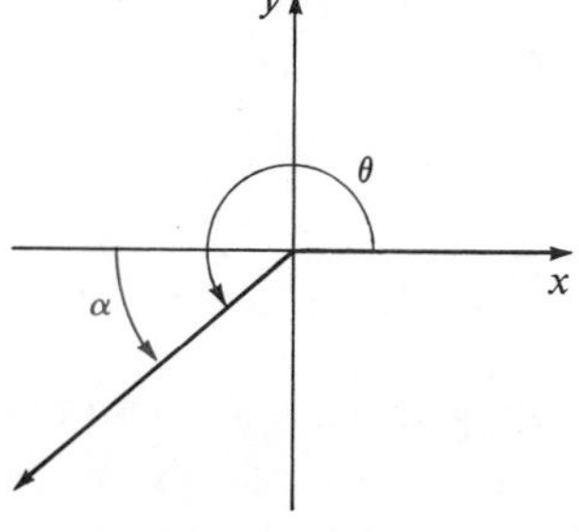

$m°(\theta) = 180 + m°(\alpha)$

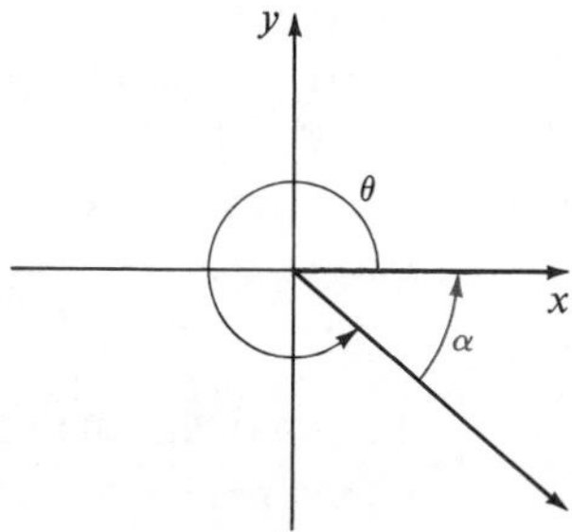

$m°(\theta) = 360 - m°(\alpha)$

Figure 1–15

Of course, if $x = 0$ but $y \neq 0$, then $m°(\theta) = 90$ or $m°(\theta) = 270$ according as $y > 0$ or $y < 0$, respectively.

Applying this method to the vector $(4, -3)$ in Example 2 above, for instance, you would have

$$\tan \theta = \frac{-3}{4} \quad \text{and} \quad \tan \alpha = \left|\frac{-3}{4}\right| = \frac{3}{4}.$$

Then, from Table 2, page 387, $m°(\alpha) \doteq 36° \, 50'$. Since $x > 0$ and $y < 0$,

$$m°(\theta) = 360 - m°(\alpha) \doteq 323° \, 10'.$$

Notice that for the zero vector $\mathbf{0}$, Equations (1) on page 14 would give you

$$\sin \theta = \frac{0}{\|\mathbf{0}\|} = \frac{0}{0} \quad \text{and} \quad \cos \theta = \frac{0}{\|\mathbf{0}\|} = \frac{0}{0}.$$

Since these expressions are meaningless, the direction of $\mathbf{0}$ is undetermined. In certain situations, however, it is useful to be able to think of the zero vector as having a particular direction (see Example 3 below). Therefore, we shall agree that we may assign any direction we wish to the zero vector. This agreement seems geometrically reasonable when you consider that the standard representation of the zero vector (the origin) is on *every* ray having the origin as initial point.

If two vectors have the same direction, then their standard representations lie on the same ray with initial point at the origin. In Section 1–3, we observed that the standard representation of the vector $-\mathbf{v}$ is collinear with, and of the same length as, that of $\mathbf{v}$, but with *opposite* direction; this will be demonstrated more formally in Example 4 below. Vectors having the same or opposite directions, that is, vectors whose directions either are equal or differ by $\pm 180°$, are defined to be **parallel vectors.** Vectors whose standard representations are at right angles to each other, that is, whose directions differ by $\pm 90°$ or $\pm 270°$, are said to be **perpendicular**, or **orthogonal, vectors.**

Example 3. Prove that the vectors $\mathbf{v} = (a, b)$ and $\mathbf{u} = (-b, a)$ have equal norms and are perpendicular vectors.

Solution: The norms are equal, since

$$\|\mathbf{u}\| = \sqrt{(-b)^2 + a^2} = \sqrt{a^2 + b^2} = \|\mathbf{v}\|.$$

To show that the vectors are perpendicular, suppose first that $\mathbf{v}$ is not the zero vector. You can then establish the result by showing that the points $\mathbf{O}(0, 0)$, $\mathbf{S}(a, b)$, and $\mathbf{T}(-b, a)$ are the vertices of a right triangle with hypotenuse $\overline{\mathbf{ST}}$ (see diagram below). You have

$$\begin{aligned}[d(\mathbf{S}, \mathbf{T})]^2 &= (-b - a)^2 + (a - b)^2 \\ &= (-b)^2 + 2ab + a^2 + a^2 - 2ab + b^2 \\ &= [(-b)^2 + a^2] + [a^2 + b^2] \\ &= [d(\mathbf{O}, \mathbf{T})]^2 + [d(\mathbf{O}, \mathbf{S})]^2,\end{aligned}$$

that is,

$$\|\mathbf{u} - \mathbf{v}\|^2 = \|\mathbf{u}\|^2 + \|\mathbf{v}\|^2.$$

Therefore, by the converse of the Pythagorean theorem, $\angle\mathbf{TOS}$

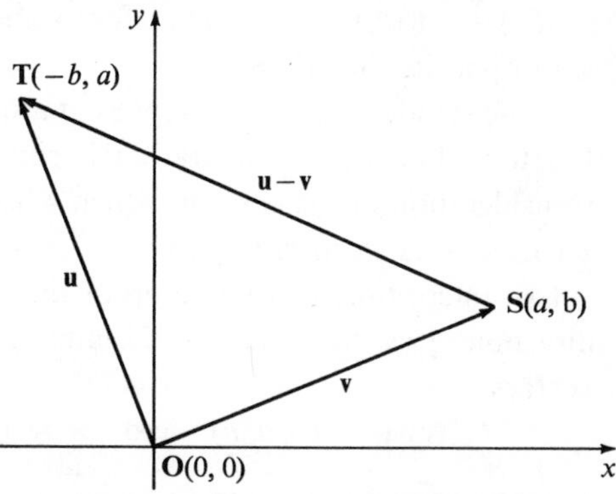

is a right angle and **u** and **v** are perpendicular vectors.

If **v** is the zero vector, then **u** and **v** are perpendicular vectors since, by agreement, any direction may be assigned to the zero vector. This completes the proof.

If $\mathbf{v} = (a, b)$ and $\mathbf{u} = (-b, a)$, a consideration of the possible signs of a and b shows (see Exercise 39, page 20) that if the standard representation of **v** is in Quadrant I, II, III, or IV, then the standard representation of **u** is in Quadrant II, III, IV, or I, respectively. That is, the standard representation of **u** is one quadrant "ahead" of the standard representation of **v**, in the sense of counterclockwise rotation.

Example 4. Prove that the vectors $\mathbf{v} = (a, b)$ and $\mathbf{w} = (-a, -b)$ have equal norms and are parallel vectors with opposite directions.

Solution: Notice first that

$$\|\mathbf{w}\| = \sqrt{(-a)^2 + (-b)^2} = \sqrt{a^2 + b^2} = \|\mathbf{v}\|.$$

(*Solution continued*)

Next observe that if $\theta_{\mathbf{v}}$ and $\theta_{\mathbf{w}}$ are the direction angles of **v** and **w**, respectively, and if $a \neq 0$, then

$$\tan \theta_{\mathbf{v}} = \frac{b}{a} \quad \text{and} \quad \tan \theta_{\mathbf{w}} = \frac{-b}{-a} = \frac{b}{a}.$$

The method discussed on page 15 can then be used to determine the directions of **v** and **w**. For example, suppose that the standard geometric representation of **v** lies in Quadrant II. Then $a < 0$ and $b > 0$. Hence $-a > 0$ and $-b < 0$, and the standard geometric representation of **w** lies in Quadrant IV. Since $\tan \theta_{\mathbf{v}} = \tan \theta_{\mathbf{w}}$, $\theta_{\mathbf{v}}$ and $\theta_{\mathbf{w}}$ have the same reference angle α. Thus you have $m^\circ(\theta_{\mathbf{v}}) = 180 - m^\circ(\alpha)$ and $m^\circ(\theta_{\mathbf{w}}) = 360 - m^\circ(\alpha)$, so that $m^\circ(\theta_{\mathbf{w}}) - m^\circ(\theta_{\mathbf{v}}) = 180$. Therefore **v** and **w** are parallel vectors with opposite directions.

If the standard geometric representation of **v** lies in Quadrants I, III, or IV, the demonstration is similar and is left to you. Consideration of the cases in which **v** lies on an axis or **v** is the zero vector is also left to you.

(An interesting alternative proof that **v** and **w** have opposite directions uses the result of Example 3 and the fact that the vectors

$$\mathbf{u} = (-b, a) \quad \text{and} \quad \mathbf{w} = (-a, -b)$$

bear the same relationship to each other as do the vectors

$$\mathbf{v} = (a, b) \quad \text{and} \quad \mathbf{u} = (-b, a),$$

with $-b$ and a in place of a and b, respectively. Thus, if $\mathbf{v} \neq \mathbf{0}$, **u** is perpendicular to **v** and one quadrant ahead of **v**, and **w** is perpendicular to **u** and one quadrant ahead of **u**.)

Exercises 1–4

In Exercises 1–16, find the norm and direction of the given vector. Use Table 2, page 387, as needed to approximate angle measures to the nearest ten minutes.

1. $(3, 3)$
2. $(-5, -5)$
3. $(\sqrt{3}, 1)$
4. $(-1, \sqrt{3})$
5. $(-3, 4)$
6. $(5, -12)$
7. $(-2, 0)$
8. $(0, -4)$
9. $(-6, -8)$
10. $(12, -5)$
11. $(3, 2) + (0, -6)$
12. $(-2, 4) + (-3, 8)$
13. $(6, 5) + (-2, -3)$
14. $(-3, 4) + (6, -1)$
15. $(5, 1) - (2, -2)$
16. $(7, -3) - (-5, 2)$

In Exercises 17–24, approximate each vector $\mathbf{v} = (x, y) \in \mathcal{R}^2$ by using the formula

$$\mathbf{v} = (x, y) = (\|\mathbf{v}\| \cos \theta, \|\mathbf{v}\| \sin \theta).$$

Use Table 2, page 387, to find values of $\cos \theta$ and $\sin \theta$ to two decimal places.

17. $\|\mathbf{v}\| = 5;\ m^\circ(\theta) = 30$

18. $\|\mathbf{v}\| = 3;\ m^\circ(\theta) = 45$

19. $\|\mathbf{v}\| = 6;\ m^\circ(\theta) = 90$

20. $\|\mathbf{v}\| = 10;\ m^\circ(\theta) = 120$

21. $\|\mathbf{v}\| = 2;\ m^\circ(\theta) = 300$

22. $\|\mathbf{v}\| = 4;\ m^\circ(\theta) = 285$

23. $\|\mathbf{v}\| = 7;\ m^\circ(\theta) = 210$

24. $\|\mathbf{v}\| = 8;\ m^\circ(\theta) = 180$

In Exercises 25–30, find the value of x or y so that the first vector is (a) parallel and (b) perpendicular to the second vector.

Example. $(x, 3), (2, 5)$

Solution: **(a)** Let α be the direction angle for $(x, 3)$ and θ be the direction angle for $(2, 5)$. Then

$$\tan \alpha = \frac{3}{x} \quad \text{and} \quad \tan \theta = \frac{5}{2}.$$

If the tangents of the direction angles of two vectors are equal, then the vectors have the same or opposite directions and, hence, are parallel. Thus $(x, 3)$ and $(2, 5)$ are parallel if

$$\frac{3}{x} = \frac{5}{2}, \quad \text{or} \quad x = \frac{6}{5}.$$

(b) From Example 3, page 16, you know that $(-5, 2)$ is perpendicular to $(2, 5)$. Thus $(x, 3)$ and $(2, 5)$ are perpendicular if $(x, 3)$ and $(-5, 2)$ are parallel, that is, if

$$\frac{3}{x} = \frac{2}{-5}, \quad \text{or} \quad x = -\frac{15}{2}.$$

25. $(x, 4), (6, 8)$

26. $(x, 5), (2, 9)$

27. $(3, y), (-2, 3)$

28. $(5, y), (7, -1)$

29. $(x, -2), (-3, 5)$

30. $(-3, y), (4, -1)$

In Exercises 31–36, find the specified vectors, leaving answers in radical form. (*Hint:* Use Equations (1) and (2), page 14.)

31. The vector (x, y) with norm 5 that has the same direction as $(3, 7)$.

32. The vector (x, y) with norm 2 that has the same direction as $(2, -3)$.

33. A vector (x, y) with norm 3 that is perpendicular to $(5, 2)$.

34. A vector (x, y) with norm 6 that is perpendicular to $(-3, 4)$.

35. The vector (x, y) with norm equal to that of $(4, -3)$ and direction the same as that of $(1, \sqrt{3})$.

36. The vector (x, y) with norm equal to that of $(-5, 12)$ and direction the same as that of $(-2, 2)$.

* **37.** Prove that for all vectors $\mathbf{u}$ and $\mathbf{v} \in \mathbb{R}^2$, $\|\mathbf{u} + \mathbf{v}\| \leq \|\mathbf{u}\| + \|\mathbf{v}\|$.

* **38.** Prove that for all vectors $\mathbf{u}$ and $\mathbf{v} \in \mathbb{R}^2$, $\|\mathbf{u} - \mathbf{v}\| \leq \|\mathbf{u}\| + \|\mathbf{v}\|$.

* **39.** Prove that if x and y are both different from 0, and (x, y) is in Quadrant I, II, III, or IV, then $(-y, x)$ is in Quadrant II, III, IV, or I, respectively.

* **40.** Prove that if x and y are both different from 0 and (x, y) is in Quadrant I, II, III, or IV, then $(y, -x)$ is in Quadrant IV, I, II, or III, respectively.

1–5 Product of a Scalar and a Vector

If $\mathbf{v} = (x, y) \in \mathbb{R}^2$, then

$$\begin{aligned}\mathbf{v} + \mathbf{v} + \mathbf{v} &= (\mathbf{v} + \mathbf{v}) + \mathbf{v}\\ &= (x + x, y + y) + (x, y)\\ &= (x + x + x, y + y + y)\\ &= (3x, 3y).\end{aligned}$$

Figure 1–16

By analogy with real numbers, this result suggests that $3\mathbf{v}$ might be defined by

$$3\mathbf{v} = 3(x, y) = (3x, 3y).$$

A geometric representation of the vector $\mathbf{v} + \mathbf{v} + \mathbf{v} = 3\mathbf{v}$ is shown in Figure 1–16. In general, let us agree that:

■ If $\mathbf{v} = (x, y) \in \mathbb{R}^2$ and if $r \in \mathbb{R}$, then

$$r\mathbf{v} = r(x, y) = (rx, ry).$$

The vector $r\mathbf{v}$ is called a **scalar multiple** of $\mathbf{v}$. Observe that *the product of a scalar and a vector is a vector.*

If $\mathbf{v} = (x, y)$, then the norm of the vector $r\mathbf{v}$ is

$$\|r(x, y)\| = \|(rx, ry)\| = \sqrt{r^2x^2 + r^2y^2} = \sqrt{r^2(x^2 + y^2)} = |r|\sqrt{x^2 + y^2}.$$

Thus:

■ $$\|r\mathbf{v}\| = |r|\,\|\mathbf{v}\|.$$

The direction of the vector $r\mathbf{v}$ is either the same as or the opposite of that of $\mathbf{v}$. That is,

the vectors $\mathbf{v}$ and $r\mathbf{v}$ are parallel.

This can be demonstrated as follows: Suppose that θ is the direction angle of

$\mathbf{v} = (x, y)$ and α is the direction angle of $r\mathbf{v} = (rx, ry)$. Then if $r \neq 0$ and $x \neq 0$, you have

$$\tan \theta = \frac{y}{x},$$

and

$$\tan \alpha = \frac{ry}{rx} = \frac{y}{x}.$$

Since $\tan \theta = \tan \alpha$, $\mathbf{v}$ and $r\mathbf{v}$ are parallel.

If $r \neq 0$ and $y \neq 0$, but $x = 0$, then $rx = 0$ and $ry \neq 0$. Hence $m°(\theta) = 90$ or 270 and $m°(\alpha) = 90$ or 270. Again, $\mathbf{v}$ and $r\mathbf{v}$ are parallel.

Finally, if either $r = 0$ or $\mathbf{v} = \mathbf{0}$, then $r\mathbf{v} = \mathbf{0}$. Since we have agreed to assign any convenient direction to $\mathbf{0}$, we can consider $r\mathbf{v}$ to be parallel to $\mathbf{v}$ in either of these cases.

Actually, by a consideration of signs, it can be shown that if $r > 0$, then $r\mathbf{v}$ has the same direction as $\mathbf{v}$, and if $r < 0$, then $r\mathbf{v}$ has the opposite direction. These results are illustrated in Figure 1–17.

Figure 1–17

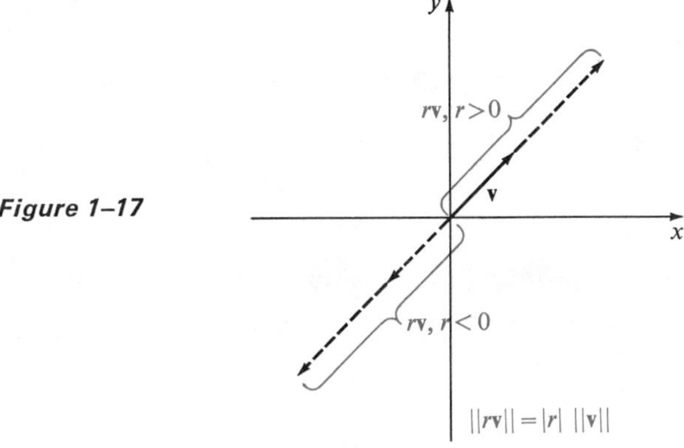

You have seen that for every $\mathbf{v} \in \mathbb{R}^2$ and every $r \in \mathbb{R}$, the vectors $\mathbf{v}$ and $r\mathbf{v}$ are parallel. If $\mathbf{v} \neq \mathbf{0}$, then the converse result also holds, as Example 1 below shows. Therefore:

■ If $\mathbf{u}$ and $\mathbf{v} \in \mathbb{R}^2$ and $\mathbf{v} \neq \mathbf{0}$, then $\mathbf{u}$ and $\mathbf{v}$ are parallel vectors if and only if there is a scalar r for which $\mathbf{u} = r\mathbf{v}$.

Example 1. Prove that if $\mathbf{u}$ and $\mathbf{v}$ are parallel vectors and $\mathbf{v} \neq \mathbf{0}$, then there exists a scalar r for which $\mathbf{u} = r\mathbf{v}$.

Solution: If $\mathbf{u} = \mathbf{0}$, then $\mathbf{u} = 0\mathbf{v}$ and $r = 0$. Otherwise, let $\mathbf{u} = (x_1, y_1)$ and $\mathbf{v} = (x_2, y_2)$, and let $\theta_{\mathbf{u}}$ and $\theta_{\mathbf{v}}$ be the direction angles of $\mathbf{u}$ and $\mathbf{v}$, respectively.

(Solution continued)

Then, you have

$$\sin \theta_{\mathbf{u}} = \frac{y_1}{\|\mathbf{u}\|}, \quad \cos \theta_{\mathbf{u}} = \frac{x_1}{\|\mathbf{u}\|},$$

and

$$\sin \theta_{\mathbf{v}} = \frac{y_2}{\|\mathbf{v}\|}, \quad \cos \theta_{\mathbf{v}} = \frac{x_2}{\|\mathbf{v}\|}.$$

Since, by hypothesis, **u** and **v** are parallel, either

$$m^\circ(\theta_{\mathbf{u}}) = m^\circ(\theta_{\mathbf{v}}) \quad \text{or} \quad m^\circ(\theta_{\mathbf{u}}) = m^\circ(\theta_{\mathbf{v}}) \pm 180.$$

If $m^\circ(\theta_{\mathbf{u}}) = m^\circ(\theta_{\mathbf{v}}) \pm 180$, for example, then it follows that

$$\frac{y_1}{\|\mathbf{u}\|} = \frac{-y_2}{\|\mathbf{v}\|} \quad \text{and} \quad \frac{x_1}{\|\mathbf{u}\|} = \frac{-x_2}{\|\mathbf{v}\|}.$$

These equations lead, respectively, to

$$y_1 = -\frac{\|\mathbf{u}\|}{\|\mathbf{v}\|} y_2 \quad \text{and} \quad x_1 = -\frac{\|\mathbf{u}\|}{\|\mathbf{v}\|} x_2.$$

By hypothesis, $\|\mathbf{v}\| \neq 0$, so that $-\dfrac{\|\mathbf{u}\|}{\|\mathbf{v}\|}$ is a real number r. Thus

$$y_1 = ry_2 \quad \text{and} \quad x_1 = rx_2,$$

and

$$(x_1, y_1) = r(x_2, y_2), \text{ or } \mathbf{u} = r\mathbf{v}, \text{ as desired.}$$

The proof for the case $m^\circ(\theta_{\mathbf{u}}) = m^\circ(\theta_{\mathbf{v}})$ is very similar and is left to you.

By using the properties of real numbers (page 392), together with the definitions in this chapter, you can prove that products of scalars and vectors obey the following laws:

Properties of Multiplication of a Vector by a Scalar

If **u** and **v** are vectors in $\mathcal{R}^2$ and r and s are scalars, then

1. $r\mathbf{u} \in \mathcal{R}^2$	Closure
2. $(rs)\mathbf{u} = r(s\mathbf{u})$	Associativity
3. $1\mathbf{u} = \mathbf{u}$	Multiplicative identity
4. $r\mathbf{u} = \mathbf{0}$ if and only if $r = 0$ or $\mathbf{u} = \mathbf{0}$.	Zero product
5. $-1\mathbf{u} = -\mathbf{u}$	Additive inverse
6. $r(\mathbf{u} + \mathbf{v}) = r\mathbf{u} + r\mathbf{v}$ and $(r + s)\mathbf{u} = r\mathbf{u} + s\mathbf{u}$	Distributivity
7. $\|r\mathbf{v}\| = \|r\| \, \|\mathbf{v}\|$	Norm property of scalar multiples

The first part of Property 6 is established in Example 2 below. Property 7 was proved on pages 20–21. Proofs of the other properties are left as exercises (Exercises 35–41, pages 25–26).

Example 2. Prove that if $r \in \mathcal{R}$, $\mathbf{v} = (x_1, y_1) \in \mathcal{R}^2$, and $\mathbf{t} = (x_2, y_2) \in \mathcal{R}^2$, then

$$r(\mathbf{v} + \mathbf{t}) = r\mathbf{v} + r\mathbf{t}.$$

Solution: You have

$$r(\mathbf{v} + \mathbf{t}) = r[(x_1, y_1) + (x_2, y_2)] = r(x_1 + x_2, y_1 + y_2).$$

By the definition of scalar multiple,

$$r(x_1 + x_2, y_1 + y_2) = \big(r(x_1 + x_2), r(y_1 + y_2)\big).$$

By the distributive law for real numbers,

$$\big(r(x_1 + x_2), r(y_1 + y_2)\big) = (rx_1 + rx_2, ry_1 + ry_2).$$

From the definition of the sum of two vectors,

$$(rx_1 + rx_2, ry_1 + ry_2) = (rx_1, ry_1) + (rx_2, ry_2).$$

Using the definition of the product of a scalar and a vector again, you have

$$(rx_1, ry_1) + (rx_2, ry_2) = r(x_1, y_1) + r(x_2, y_2) = r\mathbf{v} + r\mathbf{t},$$

and the demonstration is complete.

A vector whose norm is 1 is called a **unit vector**. For example, $(\frac{3}{5}, \frac{4}{5})$ is a unit vector, because

$$\|(\tfrac{3}{5}, \tfrac{4}{5})\| = \sqrt{(\tfrac{3}{5})^2 + (\tfrac{4}{5})^2} = \sqrt{\tfrac{9}{25} + \tfrac{16}{25}} = \sqrt{\tfrac{25}{25}} = 1.$$

You can represent *any* vector (x, y) in terms of a product of a scalar and a unit vector in the same direction as (x, y). For the zero vector this is trivial, of course, since for any unit vector $\mathbf{u}$, $0\mathbf{u} = \mathbf{0}$. For a nonzero vector (x, y), first observe that

$$\sqrt{x^2 + y^2}\left(\frac{x}{\sqrt{x^2 + y^2}}, \frac{y}{\sqrt{x^2 + y^2}}\right) = (x, y). \tag{1}$$

The norm of $\left(\frac{x}{\sqrt{x^2 + y^2}}, \frac{y}{\sqrt{x^2 + y^2}}\right)$ is given by

$$\left\|\left(\frac{x}{\sqrt{x^2 + y^2}}, \frac{y}{\sqrt{x^2 + y^2}}\right)\right\| = \sqrt{\frac{x^2 + y^2}{x^2 + y^2}} = 1.$$

Since (x, y) is a scalar multiple of $\left(\frac{x}{\sqrt{x^2 + y^2}}, \frac{y}{\sqrt{x^2 + y^2}}\right)$, and $\sqrt{x^2 + y^2}$ is positive, (x, y) has the same direction as $\left(\frac{x}{\sqrt{x^2 + y^2}}, \frac{y}{\sqrt{x^2 + y^2}}\right)$. Thus

$$\left(\frac{x}{\sqrt{x^2 + y^2}}, \frac{y}{\sqrt{x^2 + y^2}}\right)$$

is the unit vector with the same direction as (x, y), and the desired product appears in the left-hand member of Equation (1), page 23. A briefer version of this equation is

$$\|\mathbf{v}\| \left(\frac{\mathbf{v}}{\|\mathbf{v}\|}\right) = \mathbf{v}. \tag{2}$$

Example 3. Find the unit vector with the same direction as $(-4, 1)$.

Solution: You have

$$\|(-4, 1)\| = \sqrt{(-4)^2 + 1^2} = \sqrt{17}.$$

Thus the unit vector with the same direction as $(-4, 1)$ is $\left(\frac{-4}{\sqrt{17}}, \frac{1}{\sqrt{17}}\right)$.

Since the components of $\left(\frac{x}{\sqrt{x^2 + y^2}}, \frac{y}{\sqrt{x^2 + y^2}}\right)$ are just $\cos \theta$ and $\sin \theta$, respectively, where θ is the direction angle of (x, y), Equation (1), page 23, can also be written as

$$\|\mathbf{v}\|(\cos \theta, \sin \theta) = \mathbf{v}. \tag{3}$$

This is the factored form of Equation (2), page 14, giving $\mathbf{v}$ in terms of its norm and its direction angle.

Example 4. Express the vector $(\sqrt{2}, \sqrt{2})$ in terms of its norm and its direction angle θ.

Solution: You have

$$\|(\sqrt{2}, \sqrt{2})\| = \sqrt{2 + 2} = \sqrt{4} = 2.$$

Next, you find that

$$\cos \theta = \frac{\sqrt{2}}{2} \quad \text{and} \quad \sin \theta = \frac{\sqrt{2}}{2},$$

from which you have $m^\circ(\theta) = 45$. Therefore, from Equation (3),

$$(\sqrt{2}, \sqrt{2}) = 2(\cos 45^\circ, \sin 45^\circ).$$

Exercises 1–5

In Exercises 1–8, let $r = 3$, $s = -2$, $\mathbf{u} = (1, 3)$, $\mathbf{v} = (-2, 5)$, $\mathbf{m} = (-1, -3)$, and $\mathbf{n} = (0, -1)$. Express each of the following either as an ordered pair or as an appropriate scalar.

1. $r\mathbf{u} + s\mathbf{v}$
2. $r(\mathbf{u} + \mathbf{v})$
3. $(r + s)\mathbf{m}$
4. $(r + s)(\mathbf{u} + \mathbf{v})$
5. $\|r\mathbf{m} - s\mathbf{v}\|$
6. $\|r\mathbf{m}\| - \|s\mathbf{v}\|$
7. $\|s^2\mathbf{v}\| - \|s^2\mathbf{n}\|$
8. $\|rs\mathbf{u}\| + \|rs\mathbf{v}\|$

In Exercises 9–14, find the unit vector with the same direction as the given vector.

9. $(-1, 2)$
10. $(2, 5)$
11. $(3, 6)$
12. $(4, -1)$
13. $(-4, -2)$
14. $(-3, 5)$

In Exercises 15–20, determine whether or not the two given vectors are parallel.

15. $(3, 4), (15, 20)$
16. $(-9, 6), (3, -2)$
17. $(2, -3), (18, -24)$
18. $(9, 7), (-81, -56)$
19. $(4, -1), (-12, 3)$
20. $(5, 3), (-1, -0.6)$

In Exercises 21–26, express each vector in terms of its norm and direction angle in accordance with Equation (3), page 24. Use Table 2, page 387, as necessary to obtain approximations of $m^\circ(\theta)$ to the nearest 10 minutes.

21. $(\sqrt{3}, 1)$
22. $(1, -2\sqrt{2})$
23. $(3, -5)$
24. $(\sqrt{7}, 3)$
25. $(-2, -7)$
26. $(\sqrt{10}, -\sqrt{15})$

In Exercises 27–34, find a unit vector perpendicular to the given vector.

Example. $(3, -2)$

Solution: From Example 3 on page 16, a vector perpendicular to $(3, -2)$ is $(2, 3)$. The unit vector in the direction of $(2, 3)$ is

$$\left(\frac{2}{\sqrt{2^2 + 3^2}}, \frac{3}{\sqrt{2^2 + 3^2}}\right), \quad \text{or} \quad \left(\frac{2}{\sqrt{13}}, \frac{3}{\sqrt{13}}\right).$$

27. $(2, 2)$
28. $(-3, 3)$
29. $(3, 7)$
30. $(2, -2\sqrt{3})$
31. $(3, 0)$
32. $(0, -4)$
33. (a, b)
34. $(a, 2a)$

For r and $s \in \mathcal{R}$, $\mathbf{u} = (x_1, y_1) \in \mathcal{R}^2$, and $\mathbf{v} = (x_2, y_2) \in \mathcal{R}^2$, show that each of the statements in Exercises 35–41 is true.

35. $r\mathbf{u} \in \mathcal{R}^2$
36. $(rs)\mathbf{v} = r(s\mathbf{v})$
37. $1\mathbf{v} = \mathbf{v}$
38. $-1\mathbf{v} = -\mathbf{v}$

39. $(r + s)\mathbf{v} = r\mathbf{v} + s\mathbf{v}$

40. $r\mathbf{u} = \mathbf{0}$ if and only if $r = 0$ or $\mathbf{u} = \mathbf{0}$.

41. $(r + s)(\mathbf{u} + \mathbf{v}) = r\mathbf{u} + r\mathbf{v} + s\mathbf{u} + s\mathbf{v}$

* **42.** Show that if $\|\mathbf{u} + \mathbf{v}\| = \|\mathbf{u}\| + \|\mathbf{v}\|$, then **u** and **v** have the same direction.

* **43.** Show that if **u** and **v** have the same direction, then $\|\mathbf{u} + \mathbf{v}\| = \|\mathbf{u}\| + \|\mathbf{v}\|$.

1–6 Inner Products of Vectors

If **u** and **v** are nonzero vectors and **u** is perpendicular to **v**, then **u**, **v**, and $\mathbf{u} - \mathbf{v}$ (the vector from **v** to **u**) have geometric representations that form a right tri-

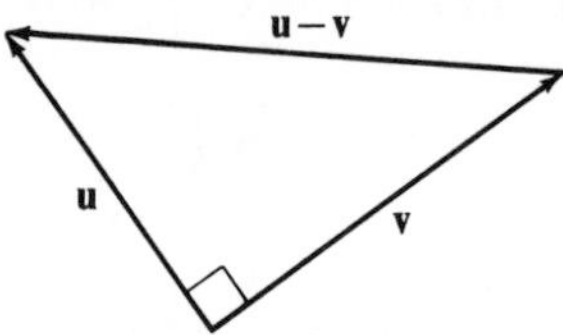

Figure 1–18

angle (Figure 1–18). Applying the Pythagorean theorem to this triangle, you have

$$\|\mathbf{u} - \mathbf{v}\|^2 = \|\mathbf{u}\|^2 + \|\mathbf{v}\|^2. \tag{1}$$

Now, with $\mathbf{u} = (x_1, y_1)$ and $\mathbf{v} = (x_2, y_2)$, Equation (1) is equivalent to

$$(x_1 - x_2)^2 + (y_1 - y_2)^2 = (x_1^2 + y_1^2) + (x_2^2 + y_2^2). \tag{2}$$

If the left-hand member of Equation (2) is expanded, you find that

$$x_1^2 - 2x_1x_2 + x_2^2 + y_1^2 - 2y_1y_2 + y_2^2 = (x_1^2 + y_1^2) + (x_2^2 + y_2^2),$$

or, after rearranging terms in the left-hand member, that

$$(x_1^2 + y_1^2) + (x_2^2 + y_2^2) - 2(x_1x_2 + y_1y_2) = (x_1^2 + y_1^2) + (x_2^2 + y_2^2). \tag{3}$$

By adding $-(x_1^2 + y_1^2) - (x_2^2 + y_2^2)$ to each member of Equation (3), you have

$$-2(x_1x_2 + y_1y_2) = 0,$$

from which you obtain

$$x_1x_2 + y_1y_2 = 0. \tag{4}$$

Each step in this process is reversible, so that Equation (4) implies Equation (1); and by the converse of the Pythagorean theorem, Equation (1) implies that **u** is perpendicular to **v**. Further, if either **u** or **v** is the zero vector, then **u** and **v** can be considered to be perpendicular, and of course Equation (4) is satisfied. We are accordingly led to the conclusion that

$$\mathbf{u} = (x_1, y_1) \text{ and } \mathbf{v} = (x_2, y_2) \text{ are perpendicular vectors if and only if } x_1x_2 + y_1y_2 = 0. \tag{5}$$

The expression $x_1x_2 + y_1y_2$ occurs so often in working with vectors (not just perpendicular vectors) that it is given a name and a special symbol. It is called the **inner product**, or **dot product**, or, sometimes, the **scalar product**, of the vectors (x_1, y_1) and (x_2, y_2), and this product is represented by the symbol $(x_1, y_1) \cdot (x_2, y_2)$. That is, you have the definition:

■ If $\mathbf{u} = (x_1, y_1) \in \mathcal{R}^2$ and $\mathbf{v} = (x_2, y_2) \in \mathcal{R}^2$, then

$$\mathbf{u} \cdot \mathbf{v} = (x_1, y_1) \cdot (x_2, y_2) = x_1x_2 + y_1y_2.$$

Observe that *the inner product of two vectors is a scalar*, the real number $x_1x_2 + y_1y_2$. For example,

$$(1, 3) \cdot (-1, 2) = (1)(-1) + (3)(2) = -1 + 6 = 5,$$

$$(2, 3) \cdot (7, -5) = (2)(7) + (3)(-5) = 14 + (-15) = -1,$$

and

$$(0, 0) \cdot (1, 3) = (0)(1) + (0)(3) = 0 + 0 = 0.$$

Statement (5) on page 26 can be reworded in terms of the inner product of **u** and **v** as follows:

■ If $\mathbf{u} \in \mathcal{R}^2$ and $\mathbf{v} \in \mathcal{R}^2$, then **u** and **v** are perpendicular vectors if and only if $\mathbf{u} \cdot \mathbf{v} = 0$.

Example 1. Show that $(3, 6)$ and $(-2, 1)$ are perpendicular vectors.

Solution: You have

$$(3, 6) \cdot (-2, 1) = (3)(-2) + (6)(1) = -6 + 6 = 0.$$

Therefore, $(3, 6)$ and $(-2, 1)$ are perpendicular vectors.

You saw in Example 3, page 16, that vectors of the form (x, y) and $(-y, x)$ are perpendicular. You can now reconfirm this easily by observing that

$$(x, y) \cdot (-y, x) = (x)(-y) + (y)(x) = -xy + xy = 0.$$

Because $(-y, x)$ is used so frequently in working with vectors, we give it a name. Thus, we define a vector $\mathbf{v_p}$ as follows:

■ If $\mathbf{v} = (x, y)$, then $\mathbf{v_p} = (-y, x)$.

Clearly, then, $\mathbf{v} \cdot \mathbf{v_p} = (x, y) \cdot (-y, x) = 0$.

Example 2. Find a unit vector perpendicular to the vector $(-2, 4)$.

Solution: If $\mathbf{v} = (-2, 4)$, then $\mathbf{v_P} = (-4, -2)$ is perpendicular to $\mathbf{v}$. Therefore, a unit vector perpendicular to $\mathbf{v}$ is

$$\frac{\mathbf{v_P}}{\|\mathbf{v_P}\|} = \frac{(-4, -2)}{\sqrt{(-4)^2 + (-2)^2}}$$

$$= \frac{(-4, -2)}{\sqrt{16 + 4}} = \frac{(-4, -2)}{\sqrt{20}} = \frac{(-4, -2)}{2\sqrt{5}}$$

$$= \left(\frac{-2}{\sqrt{5}}, \frac{-1}{\sqrt{5}}\right).$$

It is not difficult to verify that inner products of vectors have the following properties:

Properties of the Inner Product of Vectors

If $\mathbf{u}$, $\mathbf{v}$, and $\mathbf{s}$ are vectors in $\mathcal{R}^2$ and r is a scalar, then

1. $\mathbf{u} \cdot \mathbf{v} = \mathbf{v} \cdot \mathbf{u}$	Commutativity
2. $r(\mathbf{u} \cdot \mathbf{v}) = (r\mathbf{u}) \cdot \mathbf{v}$	Scalar associativity
3. $\mathbf{s} \cdot (\mathbf{u} + \mathbf{v}) = \mathbf{s} \cdot \mathbf{u} + \mathbf{s} \cdot \mathbf{v}$, and $(\mathbf{u} + \mathbf{v}) \cdot \mathbf{s} = \mathbf{u} \cdot \mathbf{s} + \mathbf{v} \cdot \mathbf{s}$	Distributivity
4. $\mathbf{v} \cdot \mathbf{v} = \|\mathbf{v}\|^2$	Norm property of inner products

Property 4 is established in Example 3 below. Proofs of the other properties are left as exercises (Exercises 31–34, page 29).

Example 3. Show that $\mathbf{v} \cdot \mathbf{v} = \|\mathbf{v}\|^2$.

Solution: Let $\mathbf{v} = (x, y)$. Then

$$\mathbf{v} \cdot \mathbf{v} = (x, y) \cdot (x, y) = x^2 + y^2 = \|\mathbf{v}\|^2.$$

Example 4. Prove that $\|\mathbf{u} + \mathbf{v}\|^2 = \|\mathbf{u}\|^2 + \|\mathbf{v}\|^2 + 2\mathbf{u} \cdot \mathbf{v}$.

Solution: Using the properties of inner products, you have:

$\|\mathbf{u} + \mathbf{v}\|^2 = (\mathbf{u} + \mathbf{v}) \cdot (\mathbf{u} + \mathbf{v})$	Norm property
$= \mathbf{u} \cdot (\mathbf{u} + \mathbf{v}) + \mathbf{v} \cdot (\mathbf{u} + \mathbf{v})$	Distributivity
$= \mathbf{u} \cdot \mathbf{u} + \mathbf{u} \cdot \mathbf{v} + \mathbf{v} \cdot \mathbf{u} + \mathbf{v} \cdot \mathbf{v}$	Distributivity
$= \mathbf{u} \cdot \mathbf{u} + \mathbf{v} \cdot \mathbf{v} + 2\mathbf{u} \cdot \mathbf{v}$	Commutativity and Associativity
$= \|\mathbf{u}\|^2 + \|\mathbf{v}\|^2 + 2\mathbf{u} \cdot \mathbf{v}$	Norm property

By observing that $\mathbf{u} - \mathbf{v} = \mathbf{u} + (-\mathbf{v})$, that $\mathbf{u} \cdot (-\mathbf{v}) = -(\mathbf{u} \cdot \mathbf{v})$, and that $\|-\mathbf{v}\| = \|\mathbf{v}\|$, you can use the result of Example 4 to show that

$$\|\mathbf{u} - \mathbf{v}\|^2 = \|\mathbf{u}\|^2 + \|\mathbf{v}\|^2 - 2\mathbf{u} \cdot \mathbf{v}.$$

From this equation, it follows that

$$2\mathbf{u} \cdot \mathbf{v} = \|\mathbf{u}\|^2 + \|\mathbf{v}\|^2 - \|\mathbf{u} - \mathbf{v}\|^2. \tag{6}$$

Therefore,

$$\mathbf{u} \cdot \mathbf{v} = \frac{\|\mathbf{u}\|^2 + \|\mathbf{v}\|^2 - \|\mathbf{u} - \mathbf{v}\|^2}{2}.$$

Thus, the inner product of the vectors **u** and **v** can be expressed in terms of their norms and the norm of the vector from **v** to **u**.

Exercises 1–6

In Exercises 1–8, find the inner product of the given vectors.

1. $(-1, 2), (3, 1)$
2. $(2, 4), (-4, 2)$
3. $(0, 4), (-2, 0)$
4. $(2, 0), (-3, 0)$
5. $(\sqrt{2}, 3), (-1, 5)$
6. $(4, \sqrt{3}), (-3, 2\sqrt{3})$
7. $(\sqrt{3}, 2), (2\sqrt{3}, 5)$
8. $(-\sqrt{3}, \sqrt{5}), (\sqrt{2}, \sqrt{6})$

In Exercises 9–16, find a unit vector perpendicular to the given vector.

9. $(3, 4)$
10. $(2, -\sqrt{5})$
11. $(-3, \sqrt{7})$
12. $(-2\sqrt{3}, \sqrt{3})$
13. $(3, -1)$
14. $(-5, -4)$
15. $(-3, 0)$
16. $(0, 5)$

In Exercises 17–24, find a value for x or y so that the first vector is perpendicular to the second.

17. $(x, 5), (3, 1)$
18. $(x, -2), (3, -4)$
19. $(-4, y), (3, 2)$
20. $(3, y), (0, 2)$
21. $(-x, 2), (2, 3)$
22. $(-x, -2), (-1, -5)$
23. $(-4, -y), (-3, -8)$
24. $(7, -y), (3, -21)$

In Exercises 25–30, find $\mathbf{u} \cdot \mathbf{v}$, if θ_1 is the direction angle of **u** and θ_2 is the direction angle of **v**. Use exact values for $\cos\theta$ and $\sin\theta$.

25. $\|\mathbf{u}\| = 2, m°(\theta_1) = 30; \|\mathbf{v}\| = 4, m°(\theta_2) = 60$
26. $\|\mathbf{u}\| = 3, m°(\theta_1) = 45; \|\mathbf{v}\| = 5, m°(\theta_2) = 30$
27. $\|\mathbf{u}\| = 4, m°(\theta_1) = 60; \|\mathbf{v}\| = 4, m°(\theta_2) = 150$
28. $\|\mathbf{u}\| = 7, m°(\theta_1) = 120; \|\mathbf{v}\| = 2, m°(\theta_2) = 135$
29. $\|\mathbf{u}\| = 3, m°(\theta_1) = 0; \|\mathbf{v}\| = 4, m°(\theta_2) = 330$
30. $\|\mathbf{u}\| = 6, m°(\theta_1) = 180; \|\mathbf{v}\| = 6, m°(\theta_2) = 270$

In Exercises 31–34, for $r \in \mathcal{R}$, $\mathbf{u} = (x_1, y_1) \in \mathcal{R}^2$, $\mathbf{v} = (x_2, y_2) \in \mathcal{R}^2$, and $\mathbf{s} = (x_3, y_3) \in \mathcal{R}^2$, show that the given statement is true.

31. $\mathbf{u} \cdot \mathbf{v} = \mathbf{v} \cdot \mathbf{u}$
32. $r(\mathbf{u} \cdot \mathbf{v}) = (r\mathbf{u}) \cdot \mathbf{v}$
33. $\mathbf{s} \cdot (\mathbf{u} + \mathbf{v}) = \mathbf{s} \cdot \mathbf{u} + \mathbf{s} \cdot \mathbf{v}$
34. $(\mathbf{u} + \mathbf{v}) \cdot \mathbf{s} = \mathbf{u} \cdot \mathbf{s} + \mathbf{v} \cdot \mathbf{s}$

35. Show that if $\mathbf{u}$ is perpendicular to $\mathbf{v}$, then $-\mathbf{u}$ is perpendicular to $\mathbf{v}$.
36. Show that if $\mathbf{u}$ is perpendicular to $\mathbf{v}$, then $r\mathbf{u}$ is perpendicular to $\mathbf{v}$.
37. Show that $\|\mathbf{u_p}\| = \|\mathbf{u}\|$.
38. Show that $(\mathbf{u_p})_\mathbf{p} = -\mathbf{u}$.
39. Show by counterexample that $\mathbf{u} \cdot \mathbf{x} = \mathbf{u} \cdot \mathbf{y}$ does not imply that either $\mathbf{x} = \mathbf{y}$ or $\mathbf{u} = \mathbf{0}$.
40. Show by counterexample that $\mathbf{u} \cdot \mathbf{x} = 0$ does not imply that either $\mathbf{u} = \mathbf{0}$ or $\mathbf{x} = \mathbf{0}$.
* 41. Show that if $\|\mathbf{u} + \mathbf{v}\| = \|\mathbf{u} - \mathbf{v}\|$, then $\mathbf{u}$ and $\mathbf{v}$ are perpendicular.
* 42. Show that $(\mathbf{u} - \mathbf{v}) \cdot (\mathbf{u} + \mathbf{v}) = \|\mathbf{u}\|^2 - \|\mathbf{v}\|^2$.

Relationships between Vectors

1–7 The Angle between Two Vectors

In Section 1–5, you saw that you can use the idea of *scalar multiple* to determine whether or not two vectors are parallel. There is another method for determining whether or not two vectors are parallel which is simple and which applies to all vectors, zero and nonzero. This method involves the inner product.

Let $\mathbf{u} = (x_1, y_1)$ and $\mathbf{v} = (x_2, y_2)$ be any two vectors in $\mathcal{R}^2$. Notice that

$$\mathbf{u} \cdot \mathbf{v_p} = (x_1, y_1) \cdot (-y_2, x_2) = -x_1y_2 + y_1x_2,$$

and (Exercise 23, page 33) that $-x_1y_2 + y_1x_2 = 0$ if and only if one of $\mathbf{u}$ and $\mathbf{v}$ is a scalar multiple of the other, that is, if and only if $\mathbf{u}$ and $\mathbf{v}$ are parallel. Therefore:

■ If $\mathbf{u} = (x_1, y_1) \in \mathcal{R}^2$ and $\mathbf{v} = (x_2, y_2) \in \mathcal{R}^2$, then $\mathbf{u}$ and $\mathbf{v}$ are parallel vectors if and only if $\mathbf{u} \cdot \mathbf{v_p} = 0$, that is, if and only if

$$(x_1, y_1) \cdot (-y_2, x_2) = 0.$$

Figure 1–19 illustrates the geometric interpretation of this result for nonzero vectors $\mathbf{u}$ and $\mathbf{v}$.

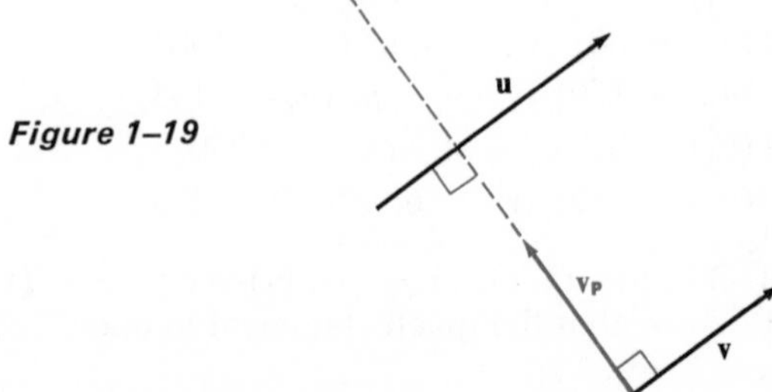

Figure 1–19

Example 1. Show that the vectors $(-1, 4)$ and $(3, -12)$ are parallel.

Solution: Let $\mathbf{u} = (-1, 4)$ and $\mathbf{v} = (3, -12)$. Then $\mathbf{v_p} = (12, 3)$, and $\mathbf{u} \cdot \mathbf{v_p} = (-1, 4) \cdot (12, 3) = (-1)(12) + (4)(3) = -12 + 12 = 0$. Therefore, **u** and **v** are parallel.

Recall from trigonometry that the Law of Cosines guarantees that in a triangle **ABC**, where a, b, and c are the lengths of the sides opposite angles A, B, and C, respectively,

$$a^2 = b^2 + c^2 - 2bc \cos A. \tag{1}$$

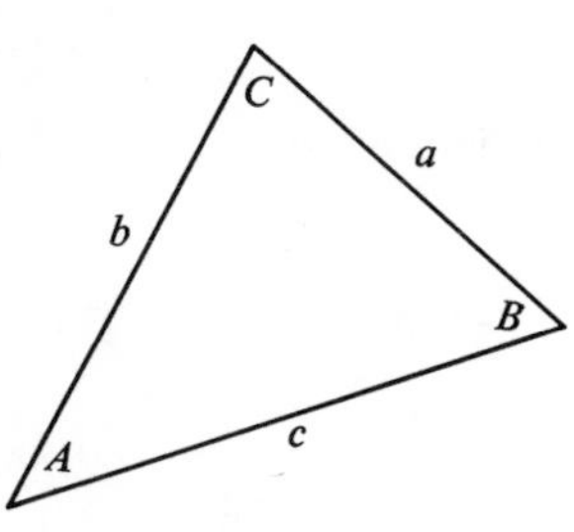

Figure 1–20

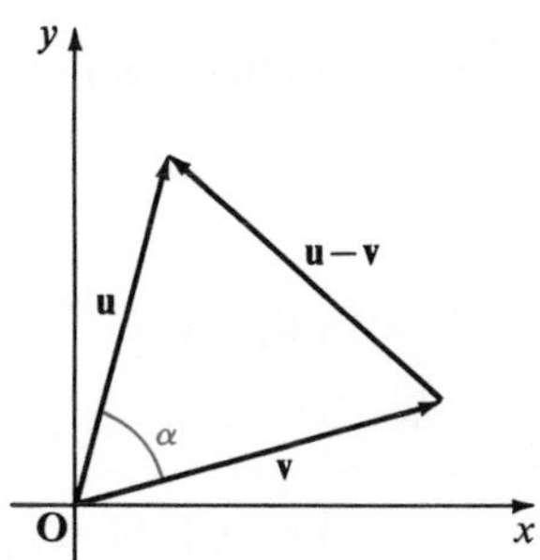

Figure 1–21

Now let **u** and **v** be nonzero, nonparallel vectors, and compare △**ABC** in Figure 1–20 with the triangle formed by geometric representations of the vectors **u**, **v**, and **u** − **v** in Figure 1–21. In this figure, α is the angle between the standard geometric representations of the vectors **u** and **v**; α is called the **angle between** the vectors **u** and **v**. Since α is an angle of a triangle, $m°(\alpha)$ lies between 0 and 180. If, in Equation (1), you replace a with $\|\mathbf{u} - \mathbf{v}\|$, b with $\|\mathbf{u}\|$, c with $\|\mathbf{v}\|$, and A with α, you have

$$\|\mathbf{u} - \mathbf{v}\|^2 = \|\mathbf{u}\|^2 + \|\mathbf{v}\|^2 - 2\|\mathbf{u}\|\,\|\mathbf{v}\| \cos \alpha, \tag{2}$$

which you can write equivalently as

$$2\|\mathbf{u}\|\,\|\mathbf{v}\| \cos \alpha = \|\mathbf{u}\|^2 + \|\mathbf{v}\|^2 - \|\mathbf{u} - \mathbf{v}\|^2. \tag{3}$$

Then, by Equation (6) on page 29, you can replace the right-hand member of Equation (3) with $2\mathbf{u} \cdot \mathbf{v}$ to obtain

$$2\|\mathbf{u}\|\,\|\mathbf{v}\| \cos \alpha = 2\mathbf{u} \cdot \mathbf{v}, \tag{4}$$

from which you have

$$\cos \alpha = \frac{\mathbf{u} \cdot \mathbf{v}}{\|\mathbf{u}\|\,\|\mathbf{v}\|}. \tag{5}$$

Equation (5) also holds (Exercises 21 and 22, page 33) if the nonzero vectors **u** and **v** are parallel. Thus, given nonzero vectors **u** and **v**, you can always use

Equation (5) to identify the measure of the angle α, with $0 \leq m°(\alpha) \leq 180$, between their standard geometric representations. (If **u** or **v** or both are the zero vector, then an angle α of any measure we please can be assigned as the angle between **u** and **v**; nevertheless, for brevity, we shall refer to *the* angle α between **u** and **v** even in this case, meaning "the assigned angle α.")

Example 2. Find an approximation for the measure of the angle between the vectors $(2, 3)$ and $(-1, 4)$.

Solution: Letting $\mathbf{u} = (2, 3)$ and $\mathbf{v} = (-1, 4)$, you have

$$\|\mathbf{u}\| = \sqrt{2^2 + 3^2} = \sqrt{13} \quad \text{and} \quad \|\mathbf{v}\| = \sqrt{(-1)^2 + 4^2} = \sqrt{17},$$

so that, from Equation (5),

$$\cos \alpha = \frac{(2, 3) \cdot (-1, 4)}{\sqrt{13}\sqrt{17}} = \frac{10}{\sqrt{221}}.$$

From Table 1, page 386, $\sqrt{221} \doteq 15$. Therefore,

$$\cos \alpha \doteq \tfrac{10}{15} \doteq 0.67.$$

From Table 2, page 387, you then get

$$m°(\alpha) \doteq 48.$$

If $m°(\alpha) = 90$, then Equation (5) on page 31 becomes

$$\cos 90° = \frac{\mathbf{u} \cdot \mathbf{v}}{\|\mathbf{u}\| \, \|\mathbf{v}\|},$$

or, since $\cos 90° = 0$,

$$\frac{\mathbf{u} \cdot \mathbf{v}}{\|\mathbf{u}\| \, \|\mathbf{v}\|} = 0. \tag{6}$$

Since $\|\mathbf{u}\| \neq 0$ and $\|\mathbf{v}\| \neq 0$, Equation (6) is equivalent to

$$\mathbf{u} \cdot \mathbf{v} = 0.$$

This result is consistent with the test (page 27) for the vectors **u** and **v** to be perpendicular.

You can also use Equation (5) on page 31 to determine whether or not two nonzero vectors **u** and **v** are parallel. You have learned that the measures of the direction angles of two parallel vectors either are equal or differ by $\pm 180°$. Hence, the measure of the angle between two parallel vectors is either 0° or 180°. Therefore, since $\cos 0° = 1$ and $\cos 180° = -1$, the vectors **u** and **v** are parallel if and only if

$$\frac{\mathbf{u} \cdot \mathbf{v}}{\|\mathbf{u}\| \, \|\mathbf{v}\|} = 1 \quad \text{or} \quad -1. \tag{7}$$

Exercises 1–7

In Exercises 1–12, determine whether the given pairs of vectors are parallel, or perpendicular, or neither parallel nor perpendicular. If they are neither parallel nor perpendicular, find the measure of the angle between them to the nearest degree.

1. $(-1, 2), (2, 1)$
2. $(2, 3), (6, -4)$
3. $(-3, 7), (6, -14)$
4. $(1, 5), (-2, -10)$
5. $(-4, 3), (1, 0)$
6. $(0, 1), (12, 5)$
7. $(\sqrt{3}, 1), (1, \sqrt{3})$
8. $(\sqrt{2}, \sqrt{2}), (0, 5)$
9. $(4, -1), (-1, -4)$
10. $(\sqrt{2}, \sqrt{3}), (-\sqrt{15}, -\sqrt{10})$
11. $(\sqrt{3}, 3), (2, \sqrt{2})$
12. $(-2, \sqrt{5}), (\sqrt{2}, -5)$

In Exercises 13–16, find, to the nearest degree, the measures of the angles of the triangle having the given points **A**, **B**, and **C** as vertices.

13. $\mathbf{A}(0, 0), \mathbf{B}(6, 0), \mathbf{C}(3, -3\sqrt{3})$
14. $\mathbf{A}(0, 0), \mathbf{B}(-3, 4), \mathbf{C}(4, 0)$
15. $\mathbf{A}(1, 1), \mathbf{B}(1, 4), \mathbf{C}(5, 1)$
16. $\mathbf{A}(-1, 2), \mathbf{B}(3, 2), \mathbf{C}(1, 5)$

In Exercises 17–22, for nonzero vectors **u**, **v**, and **t** in $\mathbb{R}^2$, show that the given statement is true.

* **17.** If $\mathbf{u} \parallel \mathbf{v}$ and $\mathbf{v} \parallel \mathbf{t}$, then $\mathbf{u} \parallel \mathbf{t}$.
* **18.** If $\mathbf{u} \perp \mathbf{v}$ and $\mathbf{v} \perp \mathbf{t}$, then $\mathbf{u} \parallel \mathbf{t}$.
* **19.** If $\mathbf{u} \perp \mathbf{v}$, then $\mathbf{u} \perp -\mathbf{v}$.
* **20.** If $\mathbf{u} \parallel \mathbf{v}$, then the angle between **u** and **t** is equal either to the angle between **v** and **t** or to the supplement of that angle.
* **21.** If $\mathbf{u} = r\mathbf{v}$, $r > 0$, then

$$\frac{\mathbf{u} \cdot \mathbf{v}}{\|\mathbf{u}\| \, \|\mathbf{v}\|} = 1.$$

* **22.** If $\mathbf{u} = r\mathbf{v}$, $r < 0$, then

$$\frac{\mathbf{u} \cdot \mathbf{v}}{\|\mathbf{u}\| \, \|\mathbf{v}\|} = -1.$$

* **23.** Show that if $\mathbf{u} = (x_1, y_1)$ and $\mathbf{v} = (x_2, y_2)$, then $-x_1y_2 + y_1x_2 = 0$ if and only if one of **u** and **v** is a scalar multiple of the other. *Hint:* For the "only if" part of the proof, consider the following three cases:

$$(1) \quad x_2 \neq 0,$$

$$(2) \quad x_2 = 0 \text{ but } y_2 \neq 0,$$

$$(3) \quad (x_2, y_2) = (0, 0).$$

1–8 Resolution of Vectors

The sum $\mathbf{s} + \mathbf{t} = \mathbf{v}$ of two vectors **s** and **t** is often called the **resultant** of **s** and **t**, and **s** and **t** are called **vector components** of **v**. Many times it is useful to be able to express a given vector **v** as the sum of two vector components with specified (nonparallel) directions. Expressing a vector **v** as the sum of vector components which are scalar multiples of given nonparallel vectors **s** and **t** is called **resolving** the vector **v** into components parallel to **s** and **t**, or expressing **v** as a **linear combination** of **s** and **t** (Figure 1–22).

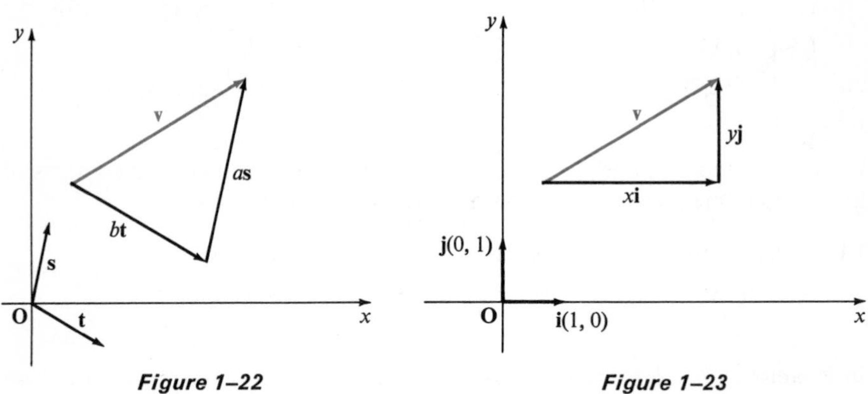

Figure 1–22 ***Figure 1–23***

It is possible to resolve any given vector in $\mathcal{R}^2$ into vector components which are scalar multiples of *any* two nonzero, nonparallel vectors **s** and **t**; in fact, this can always be done simply by solving a pair of simultaneous linear equations. The vectors **s** and **t** are then called a **basis** for the set, or **space**, of vectors in $\mathcal{R}^2$.

In the present section, we shall consider only the case in which the basis vectors are perpendicular (orthogonal) to each other. A particularly important pair of orthogonal *unit* vectors is the pair **i**, **j**, where

$$\mathbf{i} = (1, 0) \quad \text{and} \quad \mathbf{j} = (0, 1).$$

Figure 1–23 shows the standard geometric representations of **i** and **j**.

It is easy to express a vector $\mathbf{v} = (x, y)$ as the sum of scalar multiples of **i** and **j**. Since

$$(x, y) = (x, 0) + (0, y),$$

and since

$$(x, 0) = x(1, 0) \quad \text{and} \quad (0, y) = y(0, 1),$$

you have

$$(x, y) = x\mathbf{i} + y\mathbf{j}. \tag{1}$$

In this expression of (x, y) as a linear combination of $\mathbf{i}$ and $\mathbf{j}$, the scalars x and y are called **scalar components** of (x, y) parallel to $\mathbf{i}$ and $\mathbf{j}$, and the vectors $x\mathbf{i}$ and $y\mathbf{j}$ are the vector components of (x, y) parallel to $\mathbf{i}$ and $\mathbf{j}$ (see Figure 1–23).

Any vector $\mathbf{v}$ in $\mathcal{R}^2$ can be written in one and only one way as a linear combination of given orthogonal unit vectors $\mathbf{u} = (u_1, u_2)$ and $\mathbf{u_p} = (-u_2, u_1)$. That

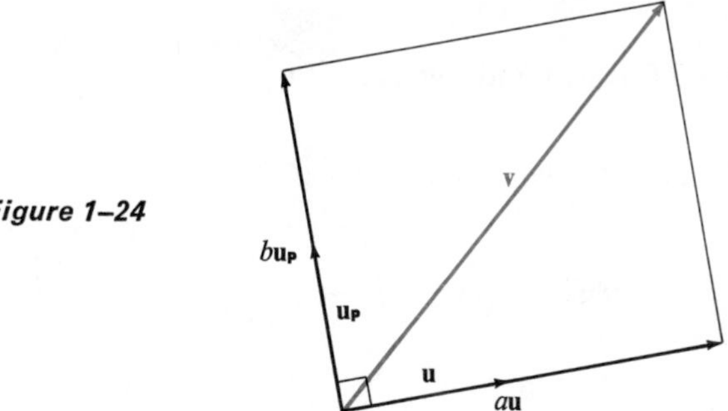

Figure 1–24

is (see Figure 1–24), there is one and only one pair of scalars, a and b, such that

$$\mathbf{v} = a\mathbf{u} + b\mathbf{u_p}. \tag{2}$$

To determine a and b, first take the scalar product of each member of Equation (2) with $\mathbf{u}$. You obtain

$$\begin{aligned} \mathbf{u} \cdot \mathbf{v} &= \mathbf{u} \cdot (a\mathbf{u} + b\mathbf{u_p}) \\ &= a(\mathbf{u} \cdot \mathbf{u}) + b(\mathbf{u} \cdot \mathbf{u_p}). \end{aligned} \tag{3}$$

Since $\mathbf{u} \cdot \mathbf{u} = \|\mathbf{u}\|^2 = 1$ and $\mathbf{u} \cdot \mathbf{u_p} = 0$, Equation (3) is equivalent to

$$\mathbf{u} \cdot \mathbf{v} = a. \tag{4}$$

Next take the scalar product of each member of Equation (2) with $\mathbf{u_p}$. You obtain

$$\begin{aligned} \mathbf{u_p} \cdot \mathbf{v} &= \mathbf{u_p} \cdot (a\mathbf{u} + b\mathbf{u_p}) \\ &= a(\mathbf{u_p} \cdot \mathbf{u}) + b(\mathbf{u_p} \cdot \mathbf{u_p}) \\ &= a(0) + b(1), \end{aligned}$$

or

$$\mathbf{u_p} \cdot \mathbf{v} = b. \tag{5}$$

Thus the only real numbers that can play the roles of a and b are $\mathbf{u} \cdot \mathbf{v}$ and $\mathbf{u_p} \cdot \mathbf{v}$.

You can verify that the values $\mathbf{u} \cdot \mathbf{v} = a$ and $\mathbf{u_p} \cdot \mathbf{v} = b$ do satisfy Equation (2) by expanding the expression $(\mathbf{u} \cdot \mathbf{v})\mathbf{u} + (\mathbf{u_p} \cdot \mathbf{v})\mathbf{u_p}$ (see Exercise 30, page 40). Thus the equation

■ $$\mathbf{v} = (\mathbf{u} \cdot \mathbf{v})\mathbf{u} + (\mathbf{u_p} \cdot \mathbf{v})\mathbf{u_p} \tag{6}$$

provides the unique resolution of any vector $\mathbf{v}$ in $\mathcal{R}^2$ into components that are scalar multiples of the orthogonal unit vectors $\mathbf{u}$ and $\mathbf{u_p}$.

We shall denote the vector components of $\mathbf{v}$ parallel to $\mathbf{u}$ and to $\mathbf{u_p}$ by $\mathbf{Comp_u\ v}$ and $\mathbf{Comp_{u_p}\ v}$, respectively. Thus,

$$\begin{aligned} \mathbf{v} &= \mathbf{Comp_u\ v} + \mathbf{Comp_{u_p}\ v} \\ &= (\mathbf{u} \cdot \mathbf{v})\mathbf{u} + (\mathbf{u_p} \cdot \mathbf{v})\mathbf{u_p}. \end{aligned}$$

Example 1. Find $\mathbf{Comp_u\ v}$ and $\mathbf{Comp_{u_p}\ v}$ if $\mathbf{u} = \left(\frac{1}{\sqrt{2}}, \frac{1}{\sqrt{2}}\right)$ and $\mathbf{v} = (2, 1)$.

Solution: Notice first that $\mathbf{u}$ is a unit vector:

$$\|\mathbf{u}\| = \sqrt{\left(\frac{1}{\sqrt{2}}\right)^2 + \left(\frac{1}{\sqrt{2}}\right)^2} = \sqrt{\frac{1}{2} + \frac{1}{2}} = 1.$$

Since $\mathbf{u} = \left(\frac{1}{\sqrt{2}}, \frac{1}{\sqrt{2}}\right)$, you have $\mathbf{u_p} = \left(-\frac{1}{\sqrt{2}}, \frac{1}{\sqrt{2}}\right)$. You can use Equation (6), page 35, to express $\mathbf{v}$ as a sum of components parallel to $\mathbf{u}$ and $\mathbf{u_p}$. You have

$$\begin{aligned} (2, 1) &= \left[\left(\frac{1}{\sqrt{2}}, \frac{1}{\sqrt{2}}\right) \cdot (2, 1)\right]\mathbf{u} + \left[\left(-\frac{1}{\sqrt{2}}, \frac{1}{\sqrt{2}}\right) \cdot (2, 1)\right]\mathbf{u_p} \\ &= \frac{3}{\sqrt{2}}\mathbf{u} - \frac{1}{\sqrt{2}}\mathbf{u_p}. \end{aligned}$$

Then

$$\mathbf{Comp_u\ v} = \frac{3}{\sqrt{2}}\left(\frac{1}{\sqrt{2}}, \frac{1}{\sqrt{2}}\right) = \left(\frac{3}{2}, \frac{3}{2}\right),$$

and

$$\mathbf{Comp_{u_p}\ v} = -\frac{1}{\sqrt{2}}\left(-\frac{1}{\sqrt{2}}, \frac{1}{\sqrt{2}}\right) = \left(\frac{1}{2}, -\frac{1}{2}\right).$$

Check: $\left(\frac{3}{2}, \frac{3}{2}\right) + \left(\frac{1}{2}, -\frac{1}{2}\right) = (2, 1)$.

You can express a vector $\mathbf{v}$ as the sum of scalar multiples of (nonzero) orthogonal vectors that are *not* unit vectors by modifying Equation (6), page 35. Suppose, for example, that you wish to express a vector $\mathbf{v}$ as the sum of a scalar multiple of a nonzero vector $\mathbf{t}$ and a scalar multiple of $\mathbf{t_p}$. Recall (page 24) that $\frac{\mathbf{t}}{\|\mathbf{t}\|}$ is the unit vector with the same direction as $\mathbf{t}$. Then note (Exercises 27 and 28, page 39) that $\left(\frac{\mathbf{t}}{\|\mathbf{t}\|}\right)_\mathbf{p} = \frac{\mathbf{t_p}}{\|\mathbf{t}\|}$ and that $\frac{\mathbf{t_p}}{\|\mathbf{t}\|}$ is the unit vector with the same direction as $\mathbf{t_p}$. Thus $\dfrac{\mathbf{t}}{\|\mathbf{t}\|}$ can be substituted for $\mathbf{u}$ and $\dfrac{\mathbf{t_p}}{\|\mathbf{t}\|}$ for $\mathbf{u_p}$ in Equation (6), page 35, to obtain

$$\mathbf{v} = \left(\frac{\mathbf{t}}{\|\mathbf{t}\|} \cdot \mathbf{v}\right)\frac{\mathbf{t}}{\|\mathbf{t}\|} + \left(\frac{\mathbf{t_p}}{\|\mathbf{t}\|} \cdot \mathbf{v}\right)\frac{\mathbf{t_p}}{\|\mathbf{t}\|}, \tag{7}$$

or

$$\mathbf{v} = \left(\frac{\mathbf{t} \cdot \mathbf{v}}{\|\mathbf{t}\|^2}\right)\mathbf{t} + \left(\frac{\mathbf{t_p} \cdot \mathbf{v}}{\|\mathbf{t}\|^2}\right)\mathbf{t_p}, \tag{8}$$

which is the required sum.

The scalar coefficients $\dfrac{\mathbf{t} \cdot \mathbf{v}}{\|\mathbf{t}\|}$ and $\dfrac{\mathbf{t_p} \cdot \mathbf{v}}{\|\mathbf{t}\|}$ of the unit vectors $\dfrac{\mathbf{t}}{\|\mathbf{t}\|}$ and $\dfrac{\mathbf{t_p}}{\|\mathbf{t}\|}$ in Equation (7) are the scalar components of $\mathbf{v}$ parallel to $\mathbf{t}$ and to $\mathbf{t_p}$, respectively. We write $\mathrm{Comp}_{\mathbf{t}}\, \mathbf{v} = \dfrac{\mathbf{t} \cdot \mathbf{v}}{\|\mathbf{t}\|}$ and $\mathrm{Comp}_{\mathbf{t_p}}\, \mathbf{v} = \dfrac{\mathbf{t_p} \cdot \mathbf{v}}{\|\mathbf{t}\|}$. (Note that in the notation for *vector* components, "**Comp**" is in boldface type, whereas in the notation for *scalar* components, "Comp" is in regular type.)

Example 2. Find the scalar components of $\mathbf{v} = (-2, 3)$ that are parallel to $\mathbf{t} = (1, 1)$ and to $\mathbf{t_p}$.

Solution: You first find that $\|\mathbf{t}\| = \sqrt{2}$ and that $\mathbf{t_p} = (-1, 1)$. Then, using Equation (7) above, you have

$$\mathbf{v} = \left(\frac{(1, 1)}{\sqrt{2}} \cdot (-2, 3)\right)\frac{\mathbf{t}}{\|\mathbf{t}\|} + \left(\frac{(-1, 1)}{\sqrt{2}} \cdot (-2, 3)\right)\frac{\mathbf{t_p}}{\|\mathbf{t}\|}$$

$$= \left(\frac{-2 + 3}{\sqrt{2}}\right)\frac{\mathbf{t}}{\|\mathbf{t}\|} + \left(\frac{2 + 3}{\sqrt{2}}\right)\frac{\mathbf{t_p}}{\|\mathbf{t}\|},$$

or

$$\mathbf{v} = \frac{1}{\sqrt{2}}\frac{\mathbf{t}}{\|\mathbf{t}\|} + \frac{5}{\sqrt{2}}\frac{\mathbf{t_p}}{\|\mathbf{t}\|}.$$

Therefore,

$$\mathrm{Comp}_{\mathbf{t}}\, \mathbf{v} = \frac{1}{\sqrt{2}} \quad \text{and} \quad \mathrm{Comp}_{\mathbf{t_p}}\, \mathbf{v} = \frac{5}{\sqrt{2}}.$$

Figure 1–25 illustrates the fact that

$$\mathbf{Comp_t\, v} = \left(\frac{\mathbf{t} \cdot \mathbf{v}}{\|\mathbf{t}\|^2}\right)\mathbf{t} \quad \text{and} \quad \mathbf{Comp_{t_p}\, v} = \left(\frac{\mathbf{t_p} \cdot \mathbf{v}}{\|\mathbf{t}\|^2}\right)\mathbf{t_p}$$

are *vectors* parallel to (but not necessarily in the same direction as) $\mathbf{t}$ and $\mathbf{t_p}$,

Figure 1–25

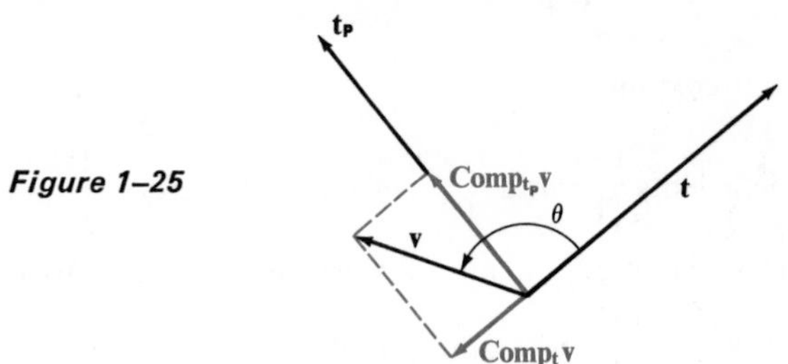

respectively, and that the sum of these vectors is $\mathbf{v}$. The *scalar*

$$\begin{aligned}\text{Comp}_{\mathbf{t}}\, \mathbf{v} &= \frac{\mathbf{t} \cdot \mathbf{v}}{\|\mathbf{t}\|} \\ &= \frac{\mathbf{t} \cdot \mathbf{v}}{\|\mathbf{t}\|\,\|\mathbf{v}\|}\,\|\mathbf{v}\| \\ &= \|\mathbf{v}\| \cos \alpha \quad \text{(Equation 5, page 31)}\end{aligned}$$

is either the norm or the negative of the norm of $\mathbf{Comp_t\, v}$,

$$\text{Comp}_{\mathbf{t}}\, \mathbf{v} = \pm\|\mathbf{Comp_t\, v}\|,$$

according as $\mathbf{t}$ and $\mathbf{Comp_t\, v}$ have the same or opposite directions (Exercise 29, page 39). Similarly,

$$\text{Comp}_{\mathbf{t_p}}\, \mathbf{v} = \pm\|\mathbf{Comp_{t_p}\, v}\|.$$

For the vectors illustrated in Figure 1–25, you have $\text{Comp}_{\mathbf{t}}\, \mathbf{v} = -\|\mathbf{Comp_t\, v}\|$ and $\text{Comp}_{\mathbf{t_p}}\, \mathbf{v} = \|\mathbf{Comp_{t_p}\, v}\|$.

Notice in Figure 1–25 that the geometric representation of the vector $\mathbf{Comp_t\, v}$ is the *perpendicular projection* of the geometric representation of $\mathbf{v}$ on the line containing the geometric representation of $\mathbf{t}$. For this reason, $\mathbf{Comp_t\, v}$ is sometimes called the **vector projection** of $\mathbf{v}$ on $\mathbf{t}$, and $\text{Comp}_{\mathbf{t}}\, \mathbf{v}$ is called the **scalar projection** of $\mathbf{v}$ on $\mathbf{t}$.

Exercises 1–8

In Exercises 1–8, express each vector as a linear combination of (a) **i** and **j**; (b) **u** and $\mathbf{u_p}$, where $\mathbf{u} = \left(\frac{1}{\sqrt{2}}, \frac{1}{\sqrt{2}}\right)$; and (c) **t** and $\mathbf{t_p}$, where $\mathbf{t} = (3, 4)$.

1. $(1, 2)$ **3.** $(-1, 0)$ **5.** $(-3, -4)$ **7.** $(5, -6)$
2. $(3, 5)$ **4.** $(-2, 3)$ **6.** $(-5, -12)$ **8.** $(4, -3)$

In Exercises 9–12, find the measure of the angle between the given vectors to the nearest degree. Use Table 2, page 387, as necessary.

9. $\mathbf{u} = -3\mathbf{i} + 4\mathbf{j}$; $\mathbf{v} = 4\mathbf{i} + 3\mathbf{j}$ **11.** $\mathbf{u} = 3\mathbf{i} - 4\mathbf{j}$; $\mathbf{v} = 5\mathbf{i} + 12\mathbf{j}$
10. $\mathbf{u} = 2\mathbf{i} + 3\mathbf{j}$; $\mathbf{v} = 2\mathbf{i} - 3\mathbf{j}$ **12.** $\mathbf{u} = 2\mathbf{i} - 3\mathbf{j}$; $\mathbf{v} = -4\mathbf{i} + 6\mathbf{j}$

In Exercises 13–16, find the vector projection and the scalar projection of **v** on **u**.

13. $\mathbf{u} = (3, 2)$, $\mathbf{v} = (1, -2)$ **15.** $\mathbf{u} = (4, -2)$, $\mathbf{v} = (-3, -4)$
14. $\mathbf{u} = (-4, 3)$, $\mathbf{v} = (2, -1)$ **16.** $\mathbf{u} = (-2, -7)$, $\mathbf{v} = (-5, -12)$

17. Find the scalar projection of **v** on **u** if **u** and **v** are perpendicular.

18. Find the scalar projection of **v** on **u** if **u** and **v** (**a**) have the same direction and (**b**) have opposite directions.

In Exercises 19–22, find the unit vector **u** for which the given statement is true.

19. $(-5, 10) = 5\mathbf{u} + 10\mathbf{u_p}$ **21.** $(0, 2\sqrt{5}) = 4\mathbf{u} + 2\mathbf{u_p}$
20. $(-\sqrt{2}, 5\sqrt{2}) = 4\mathbf{u} + 6\mathbf{u_p}$ **22.** $(-31, 27) = 13\mathbf{u} + 39\mathbf{u_p}$

In Exercises 23–26, show that the given statement is true.

23. $\mathbf{i} \cdot \mathbf{i} = 1$ **24.** $\mathbf{j} \cdot \mathbf{j} = 1$ **25.** $\mathbf{i} \cdot \mathbf{j} = 0$

26. $(x_1\mathbf{i} + y_1\mathbf{j}) \cdot (x_2\mathbf{i} + y_2\mathbf{j}) = x_1x_2 + y_1y_2$

* **27.** Show that for any nonzero vector $\mathbf{t} \in \mathcal{R}^2$,

$$\left(\frac{\mathbf{t}}{\|\mathbf{t}\|}\right)_{\mathbf{p}} = \frac{\mathbf{t_p}}{\|\mathbf{t}\|}.$$

* **28.** Show that $\frac{\mathbf{t_p}}{\|\mathbf{t}\|}$ is the unit vector with the same direction as $\mathbf{t_p}$.

* **29.** Show that for any vector **v** and any nonzero vector $\mathbf{t} \in \mathcal{R}^2$,

$$\mathbf{Comp_t\, v} = \frac{\mathbf{t}}{\|\mathbf{t}\|}\,\mathrm{Comp}_{\mathbf{t}}\,\mathbf{v},$$

and, therefore, that $\mathrm{Comp}_{\mathbf{t}}\,\mathbf{v} = \pm\|\mathbf{Comp_t\, v}\|$ according as **t** and $\mathbf{Comp_t\, v}$ have the same or opposite directions.

* **30.** Show that for any vector $\mathbf{v} \in \mathbb{R}^2$ and any unit vector $\mathbf{u} \in \mathbb{R}^2$,

$$\mathbf{v} = (\mathbf{u} \cdot \mathbf{v})\mathbf{u} + (\mathbf{u_p} \cdot \mathbf{v})\mathbf{u_p}.$$

[*Hint:* Let $\mathbf{u} = (u_1, u_2)$ and $\mathbf{v} = (v_1, v_2)$.]

* **31.** Show that for any vectors $\mathbf{v}$ and $\mathbf{t} \in \mathbb{R}^2$,

$$\|\mathbf{t}\|^2 \mathbf{v} = (\mathbf{t} \cdot \mathbf{v})\mathbf{t} + (\mathbf{t_p} \cdot \mathbf{v})\mathbf{t_p}.$$

* **32.** Show that for any vectors $\mathbf{v}$ and $\mathbf{t} \in \mathbb{R}^2$,

$$\|\mathbf{t}\|^2 \|\mathbf{v}\|^2 = (\mathbf{t} \cdot \mathbf{v})^2 + (\mathbf{t_p} \cdot \mathbf{v})^2.$$

* **33.** Use the result in Exercise 32 to show that $\|\mathbf{u}\|^2\|\mathbf{v}\|^2 \geq (\mathbf{u} \cdot \mathbf{v})^2$.

Chapter Summary

1. There is a one-to-one correspondence between the set of points in the plane and the set $\mathbb{R} \times \mathbb{R}$, or $\mathbb{R}^2$, of ordered pairs of real numbers.
2. As a consequence of the one-to-one correspondence mentioned above, two ordered pairs (a, b) and (c, d) represent the same point in the plane if and only if the two pairs are equal, that is, if and only if $a = c$ and $b = d$.
3. The distance between two points $\mathbf{A}(x_1, y_1)$ and $\mathbf{B}(x_2, y_2)$ in a coordinate plane in which the same scale is used on both axes is given by the **distance formula,**

$$d(\mathbf{A}, \mathbf{B}) = \sqrt{(x_2 - x_1)^2 + (y_2 - y_1)^2}.$$

4. A **vector** in $\mathbb{R}^2$ can be interpreted as a translation, or displacement, described by an ordered pair of real numbers. The first component denotes a displacement parallel to the x-axis; the second component denotes a displacement parallel to the y-axis.
5. A geometric representation of a vector $\mathbf{v}$ in $\mathbb{R}^2$, called a **geometric vector,** is an arrow, or directed line segment, in the plane. Every vector has infinitely many geometric representations, one starting at each point in the plane. The geometric representation of $\mathbf{v}$ having the origin as initial point is called the **standard representation** of $\mathbf{v}$ and is said to be in **standard position.**
6. If (a, b) and (c, d) are the initial and terminal points, respectively, of a geometric vector representing (x, y), then

$$(c, d) = (a + x, b + y) \qquad \text{and} \qquad (x, y) = (c - a, d - b).$$

7. If $\mathbf{v}_1 = (x_1, y_1)$ and $\mathbf{v}_2 = (x_2, y_2)$, then $\mathbf{v}_1 + \mathbf{v}_2 = (x_1 + x_2, y_1 + y_2)$.
8. Either the **parallelogram rule** or the **triangle rule** can be used to find an arrow representation of a vector sum.

9. The vector (0, 0), denoted **0**, is called the **zero vector**; its geometric representation is a point.

10. If $\mathbf{v} = (x, y)$, the vector $-\mathbf{v} = (-x, -y)$ is called the **additive inverse**, or **negative**, of **v**. The **difference** of the vectors $\mathbf{v}_1$ and $\mathbf{v}_2$ is

$$\mathbf{v}_1 - \mathbf{v}_2 = \mathbf{v}_1 + (-\mathbf{v}_2).$$

The standard geometric representation of $-\mathbf{v}$ is collinear with that of **v**, but has the opposite direction. There are arrow representations of the vectors **u**, **v**, and $\mathbf{u} - \mathbf{v}$ that form a triangle. The arrow representation of $\mathbf{u} - \mathbf{v}$ in this triangle can be described as "the vector from **v** to **u**"; it is found by applying a form of the triangle rule.

11. The **norm** $\|\mathbf{v}\|$ of a vector $\mathbf{v} = (x, y)$ is a unique scalar defined by

$$\|\mathbf{v}\| = \sqrt{x^2 + y^2}.$$

The norm of each vector is nonnegative and can be interpreted as the length of the corresponding geometric vector.

12. The **direction** of a nonzero vector $\mathbf{v} = (x, y)$ is the measure of the angle θ for which

$$\sin\theta = \frac{y}{\|\mathbf{v}\|} \quad \text{and} \quad \cos\theta = \frac{x}{\|\mathbf{v}\|}$$

and $0 \leq m^\circ(\theta) < 360$ if θ is measured in degrees, or $0 \leq m^{\mathrm{R}}(\theta) < 2\pi$ if θ is measured in radians. Geometrically, θ is determined by a ray R_2 which contains a geometric representation of **v** and has the same initial point **S** as **v** and a ray R_1, also having initial point **S**, which has the same direction as the positive ray of the x-axis. R_1 is the initial side of θ and R_2 the terminal side. If the signs of x and y are taken into account, and $x \neq 0$, then the measure of θ can be found from the formula $\tan\theta = \dfrac{y}{x}$.

13. By agreement, any direction may be assigned to the zero vector **0**.

14. Two vectors are **parallel** if and only if they have the same or opposite directions. Two nonzero vectors are **perpendicular (orthogonal)** if and only if their standard geometric representations are at right angles to each other.

15. The vectors (x, y) and $(-y, x)$ are perpendicular.

16. If r is a scalar and $\mathbf{v} = (x, y) \in \mathbb{R}^2$, then

$$r\mathbf{v} = (rx, ry).$$

The vector $r\mathbf{v}$ is called a **scalar multiple** of **v**. If $r > 0$, $r\mathbf{v}$ and **v** have the same direction; if $r < 0$, they have opposite directions. Vectors **u** and **v**, $\mathbf{v} \neq \mathbf{0}$, are parallel if and only if **u** is a scalar multiple of **v**.

17. A vector whose norm is 1 is called a **unit vector**.

18. Any vector can be represented as the product of a scalar and a unit vector having the same direction as the given vector.

19. Any vector can be represented in terms of its norm and direction angle. (For the zero vector, the representation is not unique.)

20. If $\mathbf{u} = (x_1, y_1)$ and $\mathbf{v} = (x_2, y_2)$, the scalar $x_1x_2 + y_1y_2$ is called the **inner product**, or **dot product**, or **scalar product**, of $\mathbf{u}$ and $\mathbf{v}$. This product is denoted by $\mathbf{u} \cdot \mathbf{v}$.

21. The vectors $\mathbf{u}$ and $\mathbf{v}$ are perpendicular if and only if $\mathbf{u} \cdot \mathbf{v} = 0$.

22. If α is the angle between two nonzero vectors $\mathbf{u}$ and $\mathbf{v}$, then

$$\cos \alpha = \frac{\mathbf{u} \cdot \mathbf{v}}{\|\mathbf{u}\| \, \|\mathbf{v}\|}.$$

23. A vector can be expressed as a sum of two vector components which are scalar multiples of any two given nonzero, nonparallel vectors. In this chapter, the given vectors are taken to be perpendicular to each other. The vectors $\mathbf{i} = (1, 0)$ and $\mathbf{j} = (0, 1)$ are often used. Thus, if $\mathbf{v} = (x, y)$, then $\mathbf{v} = x\mathbf{i} + y\mathbf{j}$. If $\mathbf{u} = (u_1, u_2)$ is a unit vector, then $\mathbf{u_p} = (-u_2, u_1)$ is a unit vector perpendicular to $\mathbf{u}$, and

$$\mathbf{v} = a\mathbf{u} + b\mathbf{u_p},$$

where a and b are scalars determined by $a = \mathbf{u} \cdot \mathbf{v}$ and $b = \mathbf{u_p} \cdot \mathbf{v}$.

24. The **scalar projection** of the vector $\mathbf{v}$ on the vector $\mathbf{u}$ is

$$\text{Comp}_{\mathbf{u}}\, \mathbf{v} = \frac{\mathbf{u} \cdot \mathbf{v}}{\|\mathbf{u}\|} = \|\mathbf{v}\| \cos \alpha,$$

where α is the angle between the vectors.

Chapter Review Exercises

1. Find values for x and y so that $(x + y, 2x - y) = (6, 3)$.

2. Find the distance between the points $\mathbf{A}(7, -5)$ and $\mathbf{B}(3, -2)$.

3. What are the coordinates of the terminal point $\mathbf{T}$ of the geometric representation of the vector $(2, -4)$ with initial point $\mathbf{S}(0, 5)$?

4. If $\mathbf{u} = (1, 2)$ and $\mathbf{v} = (5, 4)$, find **(a)** $\mathbf{u} + \mathbf{v}$ and **(b)** $2\mathbf{v} - 3\mathbf{u}$.

5. Find the norm and the tangent of the direction angle of the vector $(\sqrt{3}, 3) + (0, -2)$.

6. Find the unit vector with the same direction as $(-1, \sqrt{3})$.

7. Find a value for a so that $(3, 4)$ and $(8, a)$ are perpendicular vectors.

8. Find the cosine of the angle between the vectors $(3, -4)$ and $(12, 5)$.

9. Express $\mathbf{v} = (2, 5)$ in the form $a\mathbf{u} + b\mathbf{u_p}$, where $\mathbf{u}$ is the unit vector in the direction of $(3, 4)$.

10. Find the scalar projection of $\mathbf{v}$ on $\mathbf{u}$, where $\mathbf{v} = (2, 3)$ and $\mathbf{u} = (3, -1)$.

Historical Beginnings

The decisive steps in the initiation of modern analytic geometry were taken by two French mathematicians: by Pierre de Fermat (1601–1655) in 1629, and by René Descartes (1596–1650) in 1637.

Descartes was a highly respected philosopher and mathematician, and his works were widely read and much discussed. He published his *La Géométrie* (*Geometry*) in the form of three books, as an appendix to another work. It was in the second of these three books, entitled *De la nature des lignes courbes* (*On the nature of curved lines*), that he dealt with the methods of analytic geometry. These methods systematically use numbers to represent points and reduce the study of geometric properties of plane curves to an analysis of equations representing the curves; conversely, they open the way to a geometrical interpretation of the mathematical analysis of equations.

Descartes was well aware of the importance of what he had done. In the same year, 1637, he wrote that "What I have given in the second book on the nature and properties of curved lines, and the method of examining them, is, it seems to me, as far beyond the treatment in the ordinary geometry as the rhetoric of Cicero is beyond the *a, b, c* of children." This immodest appraisal, incidentally, was later to be concurred in by the great French mathematician Jacques Hadamard (1865–1963).

Fermat, unlike Descartes, was little interested in publishing his results. As a lawyer and amateur mathematician, he did his scientific work for the sheer

René Descartes

Pierre de Fermat

joy of it. Fortunately for us, though, he often wrote letters about what he had done. His greatest work was in number theory, although he shared with Blaise Pascal (1623–1662) the creation of probability theory, with Sir Isaac Newton (1642–1727) and Gottfried Wilhelm Leibniz (1646–1716) the creation of differential calculus, and with Descartes the creation of analytic geometry.

Fermat's work actually was more systematic than that of Descartes, but it was not communicated to others until 1636, and it had no effect on Descartes' geometrical investigations. By 1629, Fermat had found the general linear equation of a straight line, and also equations for circles, parabolas, ellipses, and hyperbolas.

In addition, Fermat had developed an analytical method for determining an equation of the line tangent to a given curve at a given point on the curve. In 1638 he wrote to Descartes describing this method, since it was vastly superior to the one Descartes had given in 1637. Although Descartes did not accept the superiority of Fermat's method, Newton, who was not yet born, later acknowledged that he had himself been influenced in his development of differential calculus by "Fermat's way of drawing tangents."

Fermat and Descartes were by no means the first to use coordinate systems. Astronomers, for example, had used longitude and latitude as coordinates for centuries, at least back to the time of Hipparchus (2nd century B.C.) and probably earlier.

Nor were they the first to use analytic methods to study curves. The locus of a point that moves at a constant rate outward along a ray, while the ray rotates at a constant rate about its endpoint, is now called the spiral of Archimedes. Archimedes (287–212 B.C.), the greatest mathematician and scientist of the ancient world, found how to determine the tangent line to this curve at any point of the curve by using geometrical equivalents of coordinate equations. Archimedes also devised a method for computing the area bounded by a curve, such as a circle. Since, among other things, differential calculus is concerned with tangents, and integral calculus with areas, Archimedes might be said to have anticipated the development of both differential and integral calculus by some 2000 years.

The term *Cartesian coordinate* is derived from the Latinized form, Cartesius, of Descartes' name. It is interesting to note that Descartes *did not use a second axis* in studying the properties of plane curves. Rather, he thought of a segment of the appropriate length as being erected, either perpendicularly or in a prescribed oblique direction (Descartes had a very modern grasp of the notion of a function), at each point of a single axis.

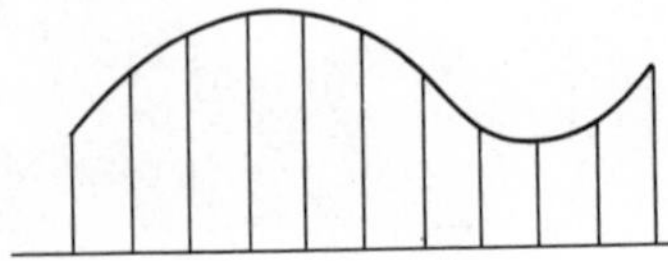

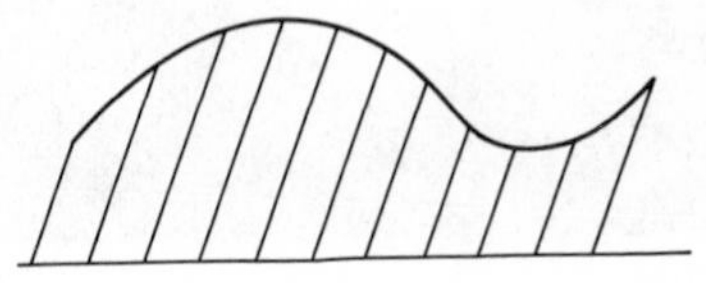

Also, Descartes considered only positive values. Thus, in modern terminology, he worked *only in the first quadrant*. Newton apparently was the first to use negative coordinates, and Leibniz introduced the word "coordinate" into analytic geometry.

Using Cartesian coordinates, Newton and Leibniz developed differential and integral calculus, and many mathematicians such as Carl Friedrich Gauss (German; 1777–1855), Georg Friedrich Bernhard Riemann (German; 1826–1866), and Hermann Minkowski (Russian; 1864–1909) developed geometry in diverse directions, giving a much deeper understanding of space and of the interlocking branches of mathematics, and furnishing one of the major cornerstones on which all physical sciences are based. The relativity theory of Albert Einstein (1879–1955) can be traced, with an admixture of genius here and there along the way, directly back to these early beginnings.

Vector analysis was somewhat slower in being developed. The theory of quaternions of Sir William Rowan Hamilton (British; 1805–1865) and the algebraic analysis of Hermann Grassmann (German; 1809–1877), which were developed in the first half of the nineteenth century, were molded, in a setting of analytic geometry, into modern vector analysis several decades later by Josiah Willard Gibbs (American; 1839–1903) and Oliver Heaviside (English; 1850–1925). In particular, Gibbs [who, according to Max Planck (German; 1858–1947), was "among the most renowned theoretical physicists of all time"] in a famous treatise dated 1881 introduced the symbols for scalar products and vector products that are used today.

Fermat and Descartes are considered to be the first modern mathematicians. In inventing analytic geometry they furnished the thrust that impelled a wave of mathematical creativity that continues today.

Josiah Willard Gibbs

Chapter 2

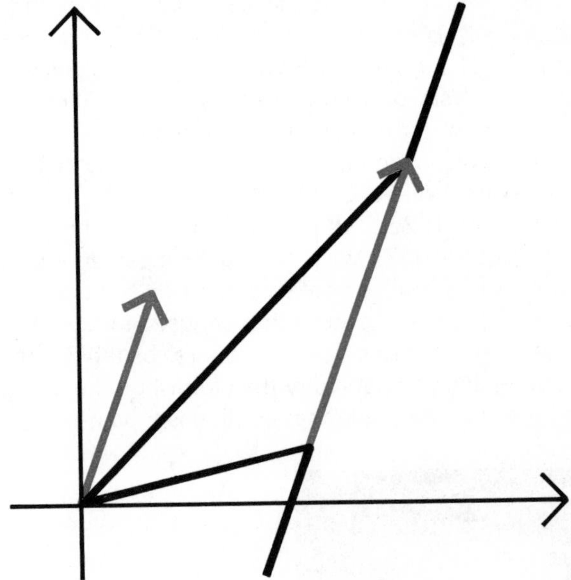

Lines and line segments in the plane can be specified both by vector equations and by Cartesian equations. In this chapter, a number of the most useful forms of these equations are presented and their interrelationship discussed.

Lines in the Plane

Vector Equations for Lines

2–1 Lines and Line Segments in the Plane

In plane geometry, you learned that a line is a set of points in the plane. In algebra, you found that such a set of points is the graph of the solution set in $\mathcal{R}^2$ of a *linear equation* in two variables, for which the **standard Cartesian form** is

$$Ax + By + C = 0,$$

where A and B are not both 0. (The condition "A and B are not both 0" is often expressed equivalently as "$A^2 + B^2 \neq 0$.") Later in this chapter we shall discuss such equations in detail, but first we shall take advantage of the relationship between points and vectors that we developed in Chapter 1 to show that a line can also be specified by a *vector equation*.

In discussing points and vectors in the plane, if a point is denoted by the capital letter **S**, for example, then it is convenient to denote the geometric vector from the origin to that point by the lower-case letter **s**.

You have learned that two points determine a line. Now you will see how that fact can be used to derive a vector equation for the line. Consider Figure 2–1, which shows the points **S**(4, 2) and **T**(5, 4) and the line $\mathcal{L}$ containing these points. If the standard geometric representations of the vectors $\mathbf{s} = (4, 2)$ and

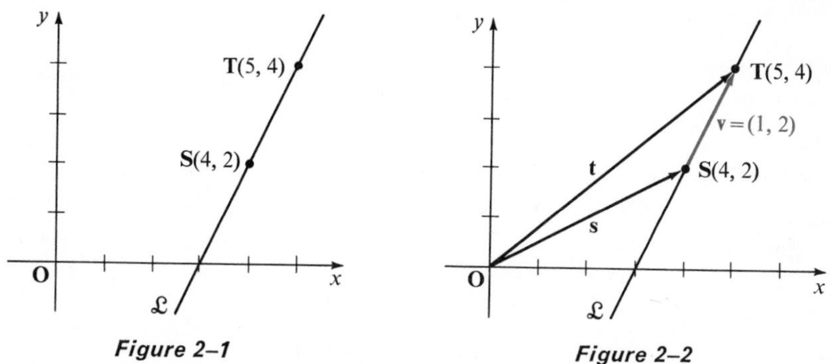

Figure 2–1 *Figure 2–2*

$\mathbf{t} = (5, 4)$ are added to the picture, the result is as shown in Figure 2–2. Notice in Figure 2–2 that the vector $\mathbf{v} = \mathbf{t} - \mathbf{s} = (5, 4) - (4, 2) = (1, 2)$ has a geometric representation lying on $\mathcal{L}$.

Now look at Figure 2–3, which shows the same situation as in Figure 2–2 except that the point $\mathbf{U}(x, y)$ on line $\mathcal{L}$ and the corresponding vector $\mathbf{u}$ have been added (and, for clarity, the vector $\mathbf{t}$ has been omitted). From the figure you can see that the vector $\mathbf{w} = \mathbf{u} - \mathbf{s}$ also has a geometric representation lying on $\mathcal{L}$ and hence (see Exercise 27, page 52) that $\mathbf{w}$ is parallel to $\mathbf{v}$.

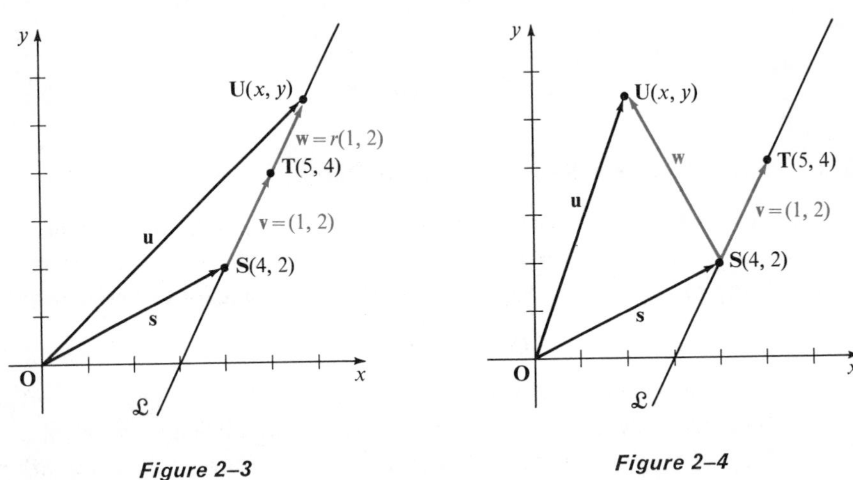

Figure 2–3

Figure 2–4

Figure 2–4 shows essentially the same situation as in Figure 2–3 except that this time the point $\mathbf{U}(x, y)$ does *not* lie on $\mathcal{L}$. In this case you can see that the geometric representation of $\mathbf{w} = \mathbf{u} - \mathbf{s}$ does *not* lie on $\mathcal{L}$ and hence (see Exercise 28, page 52) that $\mathbf{w}$ is *not* parallel to $\mathbf{v}$.

Thus, $\mathbf{U}(x, y)$ lies on $\mathcal{L}$ if and only if $\mathbf{w}$ is parallel to $\mathbf{v}$, that is, if and only if $\mathbf{u} - \mathbf{s}$ is parallel to $\mathbf{t} - \mathbf{s}$. From Section 1–5, however, you know that $\mathbf{w}$ is parallel to $\mathbf{v}$ if and only if $\mathbf{w} = r\mathbf{v}$, where r is a scalar. Thus $\mathbf{U}$ lies on $\mathcal{L}$ if and only if

$$\mathbf{w} = r\mathbf{v},$$

or

$$\mathbf{u} - \mathbf{s} = r(\mathbf{t} - \mathbf{s}).$$

Since $\mathbf{u} = (x, y)$, $\mathbf{s} = (4, 2)$, and $\mathbf{t} = (5, 4)$, $\mathbf{w} = \mathbf{u} - \mathbf{s} = (x - 4, y - 2)$ and, as noted earlier, $\mathbf{v} = \mathbf{t} - \mathbf{s} = (1, 2)$. Therefore, this last equation is equivalent to

$$(x - 4, y - 2) = r(1, 2),$$

$$(x, y) - (4, 2) = r(1, 2),$$

or

$$(x, y) = (4, 2) + r(1, 2).$$

In general, by similar reasoning you can conclude that if $\mathbf{S}(x_1, y_1)$ and $\mathbf{T}(x_2, y_2)$ are any two points in the plane (Figure 2–5), then the point $\mathbf{U}(x, y)$ lies on the line through **S** and **T** if and only if

$$\mathbf{u} - \mathbf{s} = r(\mathbf{t} - \mathbf{s}),$$

or

$$\mathbf{u} = \mathbf{s} + r(\mathbf{t} - \mathbf{s}), \qquad r \in \mathcal{R}. \tag{1}$$

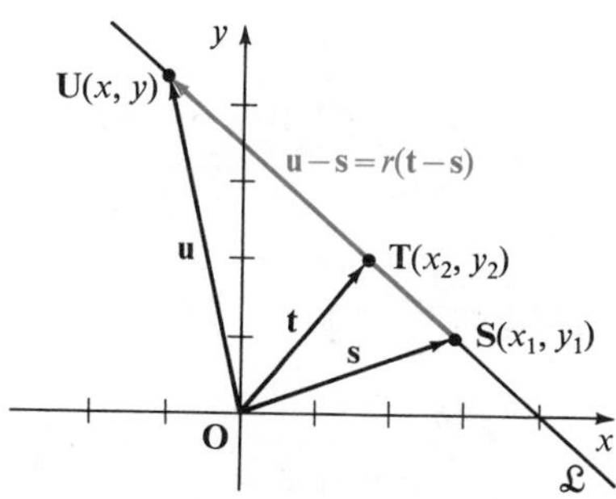

Figure 2–5

The set of points lying on $\mathcal{L}$ can be specified by

$$\{\mathbf{U}\colon \mathbf{u} = \mathbf{s} + r(\mathbf{t} - \mathbf{s}), \quad r \in \mathcal{R}\},$$

where $\mathbf{t} \neq \mathbf{s}$ since $\mathbf{T} \neq \mathbf{S}$.

Since **U** lies on $\mathcal{L}$ *if and only if* Equation (1) is satisfied, you say that Equation (1) is *an equation of* $\mathcal{L}$ and that $\mathcal{L}$ is *the graph of* Equation (1). In Equation (1), the scalar variable r is called a **parameter**, and Equation (1) is called the **standard parametric vector equation of the line through S and T.**

Example 1. Write a parametric vector equation for the line $\mathcal{L}$ through $\mathbf{S}(3, -2)$ and $\mathbf{T}(4, 4)$.

Solution: Draw a sketch. You have $\mathbf{s} = (3, -2)$ and $\mathbf{t} = (4, 4)$. Then

$$\mathbf{t} - \mathbf{s} = (4, 4) - (3, -2)$$
$$= (1, 6).$$

Accordingly, from Equation (1), a parametric vector equation for $\mathcal{L}$ is

$$\mathbf{u} = (3, -2) + r(1, 6).$$

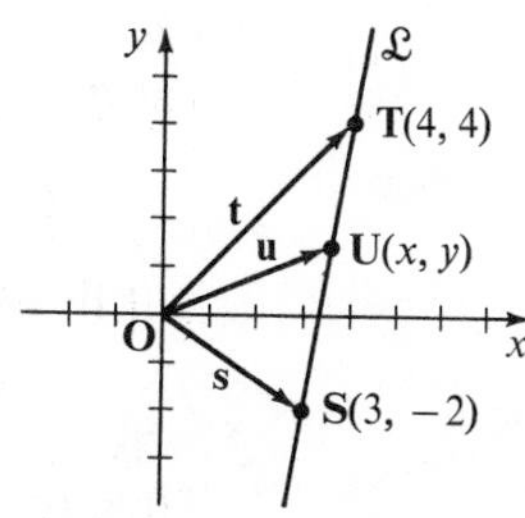

If the replacement set for r is restricted to a closed interval, $\{r: a \leq r \leq b\}$, then the graph of Equation (1) is a *line segment*. Note, in particular, that if $r = 0$ in Equation (1), then $\mathbf{u} = \mathbf{s}$ and $\mathbf{U}(x, y) = \mathbf{S}(x_1, y_1)$. Also, if $r = 1$, then $\mathbf{u} = \mathbf{t}$ and $\mathbf{U}(x, y) = \mathbf{T}(x_2, y_2)$. Thus, as suggested in Figure 2–6, as r goes through the interval $\{r: 0 \leq r \leq 1\}$, the point $\mathbf{U}(x, y)$ traces out the line

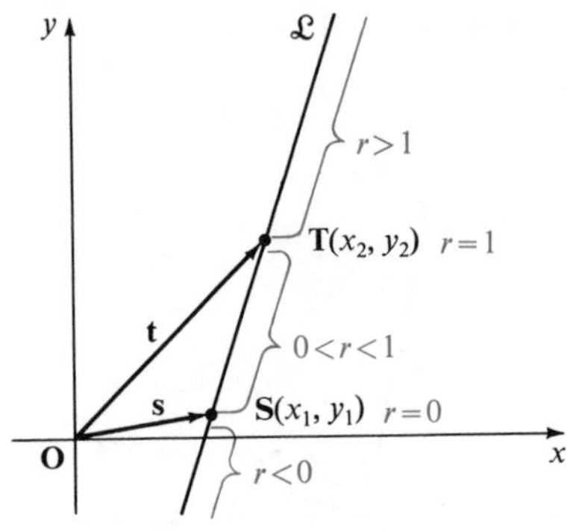

Figure 2–6

segment from $\mathbf{S}(x_1, y_1)$ to $\mathbf{T}(x_2, y_2)$. The remaining points on the line correspond to values of r satisfying $r < 0$ and $r > 1$.

You can use Equation (1), page 49, to find the coordinates of a point on the line segment $\overline{\mathbf{ST}}$ that lies a specified part of the way from $\mathbf{S}$ to $\mathbf{T}$. For example, to find the coordinates of the midpoint, you would take $r = \frac{1}{2}$.

Example 2. Find the coordinates of the points that trisect the line segment with endpoints $\mathbf{S}(3, -4)$ and $\mathbf{T}(6, 2)$.

Solution: You have $\mathbf{s} = (3, -4)$ and $\mathbf{t} = (6, 2)$. Therefore, the vector from $\mathbf{S}$ to $\mathbf{T}$ is

$$\mathbf{t} - \mathbf{s} = (6, 2) - (3, -4) = (3, 6).$$

Then the points of the segment $\overline{\mathbf{ST}}$ are given by

$$\mathbf{u} = (3, -4) + r(3, 6), \qquad 0 \leq r \leq 1.$$

The coordinates of the point one-third of the way from $\mathbf{S}$ to $\mathbf{T}$ are

$$(3, -4) + \tfrac{1}{3}(3, 6) = (3, -4) + (1, 2) = (4, -2),$$

and those of the point two-thirds of the way from $\mathbf{S}$ to $\mathbf{T}$ are

$$(3, -4) + \tfrac{2}{3}(3, 6) = (3, -4) + (2, 4) = (5, 0).$$

Thus the coordinates of the trisection points are $(4, -2)$ and $(5, 0)$.

If Equation (1), page 49, is written equivalently in terms of the parameter r and the coordinates of **S**, **T**, and **U**, you have

$$\begin{aligned}(x, y) &= (x_1, y_1) + r[(x_2, y_2) - (x_1, y_1)] \\ &= (x_1, y_1) + r(x_2 - x_1, y_2 - y_1),\end{aligned}$$

or

$$(x, y) = \big(x_1 + r(x_2 - x_1), y_1 + r(y_2 - y_1)\big).$$

This vector equation is equivalent to the two equations

$$x = x_1 + r(x_2 - x_1) \quad \text{and} \quad y = y_1 + r(y_2 - y_1), \qquad r \in \mathcal{R}. \tag{2}$$

These are called a system of **parametric Cartesian equations** for the line through **S** and **T**. In Section 2–5 you will see how such a system is related to the standard Cartesian form of an equation of a line.

Example 3. Write a system of parametric Cartesian equations for the line through the points **S**$(2, -1)$ and **T**$(5, 3)$.

Solution: If in Equations (2) you replace x_1, y_1, x_2, and y_2 with 2, -1, 5, and 3, respectively, you obtain

$$x = 2 + r(5 - 2) \quad \text{and} \quad y = -1 + r[3 - (-1)],$$

or

$$x = 2 + 3r \quad \text{and} \quad y = -1 + 4r.$$

Exercises 2–1

In Exercises 1–10, write a parametric vector equation and a system of parametric Cartesian equations for the line containing the given points **S** and **T**.

1. **S**$(2, 1)$, **T**$(0, 0)$
2. **S**$(3, 2)$, **T**$(1, 1)$
3. **S**$(4, -2)$, **T**$(4, 3)$
4. **S**$(5, -6)$, **T**$(2, -6)$
5. **S**$(-7, 2)$, **T**$(-3, -1)$
6. **S**$(-3, 1)$, **T**$(4, -2)$
7. **S**$(-6, -3)$, **T**$(-4, -2)$
8. **S**$(-1, -7)$, **T**$(-7, -1)$
9. **S**(a, b), **T**(b, a)
10. **S**$(2a, b)$, **T**$(3a, 2b)$

In Exercises 11–18, find the coordinates of (a) the midpoint and (b) the points of trisection of the segment whose endpoints **S** and **T** are given.

11. **S**$(5, -1)$, **T**$(-4, 2)$
12. **S**$(6, -2)$, **T**$(1, 7)$
13. **S**$(-3, -5)$, **T**$(3, 10)$
14. **S**$(4, 7)$, **T**$(-5, 3)$
15. **S**$(2, 5)$, **T**$(-10, -1)$
16. **S**$(-5, 3)$, **T**$(7, 21)$
17. **S**$(-3, 7)$, **T**$(4, 1)$
18. **S**$(5, -2)$, **T**$(12, -5)$

In Exercises 19–22, find a parametric vector equation for the segment joining:

19. $\mathbf{R}(2, 5)$ and the midpoint of the segment with endpoints $\mathbf{S}(5, 1)$ and $\mathbf{T}(7, -3)$.

20. $\mathbf{R}(-2, 6)$ and the midpoint of the segment with endpoints $\mathbf{S}(0, 3)$ and $\mathbf{T}(4, 0)$.

21. The midpoint of the segment with endpoints $\mathbf{Q}(-5, 2)$ and $\mathbf{R}(1, 6)$ and the point one-third of the way from $\mathbf{S}(-2, 6)$ to $\mathbf{T}(1, 9)$.

22. The point two-thirds of the way from $\mathbf{Q}(8, -2)$ to $\mathbf{R}(2, 7)$ and the point one-fourth of the way from $\mathbf{S}(1, 6)$ to $\mathbf{T}(9, 10)$.

* **23.** Show that the coordinates (x_0, y_0) of the midpoint of the segment with endpoints $\mathbf{S}(x_1, y_1)$ and $\mathbf{T}(x_2, y_2)$ are given by

$$x_0 = \frac{x_1 + x_2}{2}, \quad y_0 = \frac{y_1 + y_2}{2}.$$

* **24.** Show that the coordinates (x', y') and (x'', y'') of the points of trisection of the segment with endpoints $\mathbf{S}(x_1, y_1)$ and $\mathbf{T}(x_2, y_2)$ are given by

$$x' = \frac{2x_1 + x_2}{3}, \quad y' = \frac{2y_1 + y_2}{3}.$$

and

$$x'' = \frac{x_1 + 2x_2}{3}, \quad y'' = \frac{y_1 + 2y_2}{3}.$$

* **25.** Show that the medians of the triangle with vertices $\mathbf{R}(6, 1)$, $\mathbf{S}(-2, 3)$, and $\mathbf{T}(2, -7)$ meet in a common point located two-thirds of the way from each vertex to the opposite side. [*Hint:* Determine the coordinates of the point located two-thirds of the way from each vertex to the midpoint of the opposite side.]

* **26.** Repeat Exercise 25 for the triangle with vertices $\mathbf{O}(0, 0)$, $\mathbf{S}(x_1, y_1)$, and $\mathbf{T}(x_2, y_2)$.

* **27.** Prove that if the line $\mathcal{L}$ is the graph of the linear Cartesian equation $Ax + By + C = 0$, $A^2 + B^2 \neq 0$, then all vectors having geometric representations with initial point and terminal point on $\mathcal{L}$ are parallel. [*Hint:* Let $\mathbf{S}(x_1, y_1)$ and $\mathbf{T}(x_2, y_2)$ be any two points on $\mathcal{L}$. Show that $A(x_2 - x_1) + B(y_2 - y_1) = 0$, and accordingly (Section 1–6) that the vector $(x_2 - x_1, y_2 - y_1)$ is perpendicular to the vector (A, B).]

* **28.** Prove that if the line $\mathcal{L}$ is the graph of the linear Cartesian equation $Ax + By + C = 0$, $A^2 + B^2 \neq 0$, and $\mathbf{v}$ is a vector having a geometric representation with one endpoint on $\mathcal{L}$ and the other not on $\mathcal{L}$, then $\mathbf{v}$ is not parallel to any nonzero vector having a geometric representation with both endpoints on $\mathcal{L}$. [*Hint:* Let $\mathbf{S}(x_1, y_1)$ be any point on $\mathcal{L}$ and $\mathbf{T}(x_2, y_2)$ be any point *not* on $\mathcal{L}$. Show that $A(x_2 - x_1) + B(y_2 - y_1) \neq 0$, and accordingly that the vector $(x_2 - x_1, y_2 - y_1)$ is not perpendicular to the vector (A, B).]

2–2 Points on Lines

In Section 2–1, we used the following facts, which intuitively seem quite obvious, in deriving a parametric vector equation of a line: If a vector **v** has a geometric representation with its initial point and also its terminal point on a given line $\mathcal{L}$, then **v** is parallel to every other vector having such a geometric representation; but, unless $\mathbf{v} = \mathbf{0}$, **v** is not parallel to any vector having a geometric representation with one endpoint on $\mathcal{L}$ and the other endpoint *not* on $\mathcal{L}$. (For an outline of the proof of these statements, see Exercises 27 and 28, page 52.) We say that the line $\mathcal{L}$ is **parallel** to each vector **v** having a geometric representation with initial and terminal points on $\mathcal{L}$ and that each such vector **v** is parallel to $\mathcal{L}$. A *nonzero* vector **v** parallel to $\mathcal{L}$ is called a **direction vector** of $\mathcal{L}$.

You saw in Section 2–1 that a parametric vector equation, or a system of parametric Cartesian equations, of a line $\mathcal{L}$ can be found if the coordinates of two points on $\mathcal{L}$ are known. Such equations can also be found if the coordinates of a single point on $\mathcal{L}$ and a direction vector of $\mathcal{L}$ are known. Consider the line

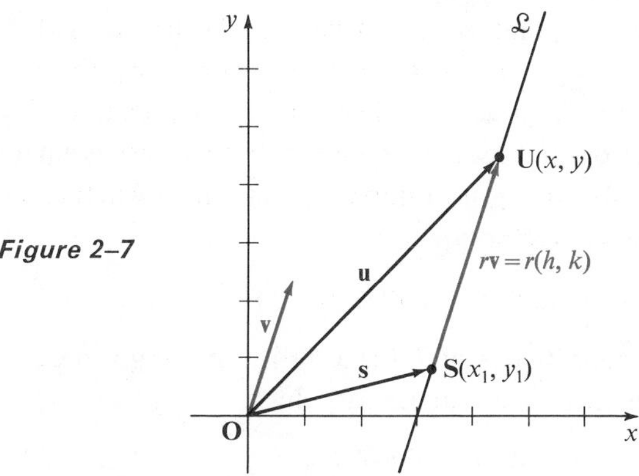

Figure 2–7

$\mathcal{L}$ through the point $\mathbf{S}(x_1, y_1)$ parallel to the nonzero vector $\mathbf{v} = (h, k)$. (See Figure 2–7.) From the remarks above, a point $\mathbf{U}(x, y)$ is on $\mathcal{L}$ if and only if

$$\mathbf{u} - \mathbf{s} = r\mathbf{v},$$

or

$$\mathbf{u} = \mathbf{s} + r\mathbf{v}, \qquad (1)$$

where r is a scalar. Equation (1) is called the **standard parametric vector equation of the line through S parallel to v.** Since Equation (1) can be written equivalently as

$$(x, y) = (x_1, y_1) + r(h, k),$$

the corresponding system of parametric Cartesian equations for $\mathcal{L}$ is

$$x = x_1 + rh, \qquad y = y_1 + rk, \qquad r \in \mathcal{R}. \qquad (2)$$

Example 1. Write a parametric vector equation and a system of parametric Cartesian equations for the line through the point $\mathbf{S}(7, -1)$ parallel to the vector $\mathbf{v} = (-2, 3)$.

Solution: Make a sketch. Directly from Equation (1), you have

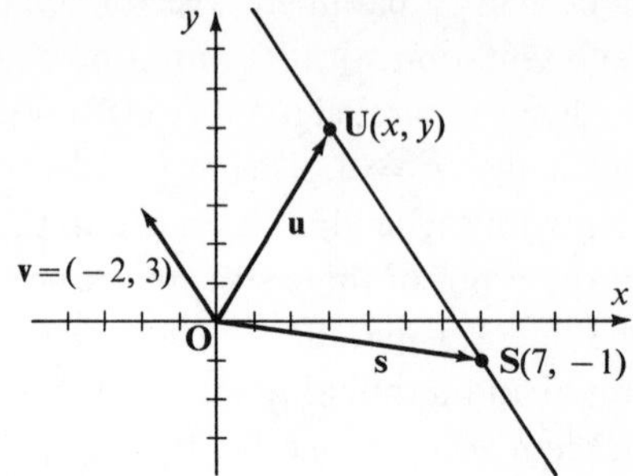

$$\mathbf{u} = (7, -1) + r(-2, 3).$$

Similarly, from Equation (2), you have

$$x = 7 - 2r, \; y = -1 + 3r.$$

Of course, any line has infinitely many direction vectors, since any of the infinitely many nonzero vectors parallel to the line is a direction vector of the line. In particular, if **P** and **Q** are any two points on the line $\mathcal{L}$, then both $\mathbf{p} - \mathbf{q}$ and $\mathbf{q} - \mathbf{p}$ are direction vectors of the line.

To determine whether a point $\mathbf{S}(x_1, y_1)$ lies on the line $\mathcal{L}$ with Cartesian equation $Ax + By + C = 0$, $A^2 + B^2 \neq 0$, you substitute the coordinates x_1 and y_1 in the equation and see whether a true statement results. For a parametric vector equation or a system of parametric Cartesian equations, the problem is somewhat different. For example, to determine whether $\mathbf{T}(3, 6)$ lies on the line with parametric vector equation

$$(x, y) = (1, 3) + r(1, 1),$$

you can substitute the coordinates of **T** for x and y in the given equation and then determine whether there is a scalar r for which

$$(3, 6) = (1, 3) + r(1, 1); \tag{3}$$

that is, for which

$$(3, 6) = (1 + r, 3 + r),$$

or

$$3 = 1 + r \quad \text{and} \quad 6 = 3 + r,$$

or

$$r = 2 \quad \text{and} \quad r = 3.$$

Since $2 \neq 3$, there is no real number r for which Equation (3) is true, and you can conclude that $\mathbf{T}(3, 6)$ does not lie on $\mathcal{L}$.

There is, however, an easier way to arrive at this conclusion. Note first that the result expressed by Equation (1) on page 53 can be restated as follows:

■ If **v** is a direction vector of a line $\mathcal{L}$ containing a point **S**, then a point **U** lies on $\mathcal{L}$ if and only if $\mathbf{u} - \mathbf{s}$ is parallel to **v**.

Next, recall (page 30) that the vectors $\mathbf{u} = (x_1, y_1)$ and $\mathbf{v} = (x_2, y_2)$ are parallel if and only if $\mathbf{u} \cdot \mathbf{v_p} = 0$, where $\mathbf{v_p} = (-y_2, x_2)$. These two results can be combined to obtain the following simple test for determining whether or not a point **U** lies on a line $\mathcal{L}$:

■ If $\mathbf{v}$ is a direction vector of a line $\mathcal{L}$ containing a point **S**, then a point **U** lies on $\mathcal{L}$ if and only if

$$(\mathbf{u} - \mathbf{s}) \cdot \mathbf{v_p} = 0. \tag{4}$$

In particular, if you know that $\mathcal{L}$ contains the two points **S** and **T**, then a point **U** is on $\mathcal{L}$ if and only if

$$(\mathbf{u} - \mathbf{s}) \cdot (\mathbf{t} - \mathbf{s})_\mathbf{p} = 0. \tag{5}$$

Notice that although Equations (4) and (5) involve vectors, they are actually scalar equations for $\mathcal{L}$ because of the scalar products involved. We shall return to this observation in Section 2–4.

Example 2. Show that if $\mathcal{L}$ is the line with parametric vector equation

$$\mathbf{u} = (3, 2) + r(-1, 2),$$

then (**a**) the point with coordinates (6, 1) does not lie on $\mathcal{L}$ and (**b**) the point with coordinates (5, −2) lies on $\mathcal{L}$.

Solution: By inspection of the given equation for $\mathcal{L}$, you can see that $\mathcal{L}$ passes through the point **S**(3, 2) and that $\mathbf{v} = (-1, 2)$ is a direction vector of $\mathcal{L}$. You then have $\mathbf{v_p} = (-2, -1)$.

(**a**) For $\mathbf{u} = (6, 1)$, you have

$$\mathbf{u} - \mathbf{s} = (6, 1) - (3, 2) = (3, -1),$$

and

$$\begin{aligned}(\mathbf{u} - \mathbf{s}) \cdot \mathbf{v_p} &= (3, -1) \cdot (-2, -1) \\ &= -6 + 1 = -5 \neq 0.\end{aligned}$$

Therefore $\mathbf{u} - \mathbf{s}$ is not parallel to $\mathbf{v}$, and hence the point with coordinates (6, 1) does not lie on $\mathcal{L}$.

(**b**) For $\mathbf{u} = (5, -2)$, you have

$$\mathbf{u} - \mathbf{s} = (5, -2) - (3, 2) = (2, -4),$$

and

$$(\mathbf{u} - \mathbf{s}) \cdot \mathbf{v_p} = (2, -4) \cdot (-2, -1) = -4 + 4 = 0.$$

Thus $\mathbf{u} - \mathbf{s}$ is parallel to $\mathbf{v}$, and the point with coordinates (5, −2) does lie on $\mathcal{L}$.

If the direction vector $\mathbf{v}$ in the equation

$$\mathbf{u} = \mathbf{s} + r\mathbf{v}$$

is a unit vector, then for any point $\mathbf{U}$ on the graph of the equation, $|r|$ is the distance between $\mathbf{S}$ and $\mathbf{U}$ (Figure 2–8). This follows from the fact that

$$\begin{aligned} d(\mathbf{S}, \mathbf{U}) = \|\mathbf{u} - \mathbf{s}\| &= \|r\mathbf{v}\| \\ &= |r|\,\|\mathbf{v}\| \\ &= |r|(1) \\ &= |r|. \end{aligned}$$

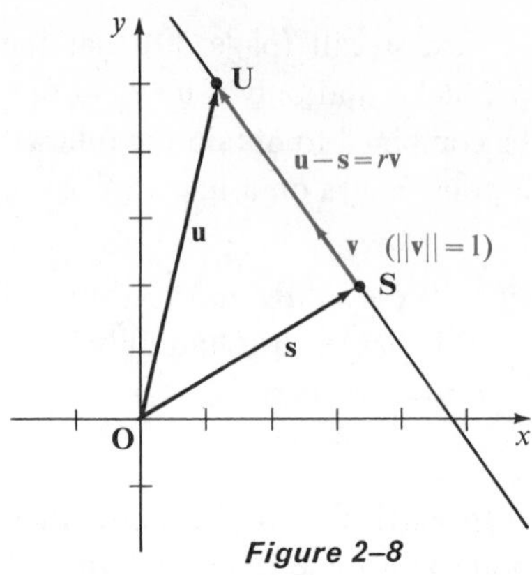

Figure 2–8

Example 3. If $\mathcal{L}$ is the line with equation $\mathbf{u} = (1, 6) + r(3, 4)$, find the coordinates of the points on $\mathcal{L}$ that are 10 units from $\mathbf{S}(1, 6)$.

Solution: First find the unit vector in the same direction as (3, 4). This vector (see page 24) is

$$\left(\frac{3}{\sqrt{9 + 16}}, \frac{4}{\sqrt{9 + 16}}\right) = \left(\frac{3}{5}, \frac{4}{5}\right).$$

Then another equation for $\mathcal{L}$ is

$$\mathbf{u} = (1, 6) + r(\tfrac{3}{5}, \tfrac{4}{5}).$$

You want to find the coordinates of the points $\mathbf{U}(x, y)$ for which $|r| = 10$, that is, for which $r = 10$ or $r = -10$. Thus you have:

$r = 10$	$r = -10$
$(x_1, y_1) = (1, 6) + 10(\tfrac{3}{5}, \tfrac{4}{5})$	$(x_2, y_2) = (1, 6) - 10(\tfrac{3}{5}, \tfrac{4}{5})$
$= (1, 6) + (6, 8)$	$= (1, 6) - (6, 8)$
$= (7, 14)$	$= (-5, -2)$

Therefore $\mathbf{U}_1(7, 14)$ and $\mathbf{U}_2(-5, -2)$ are the desired points.

Exercises 2–2

In Exercises 1–8, write a parametric vector equation and a system of parametric Cartesian equations for the line $\mathcal{L}$ through the given point $\mathbf{S}$ parallel to the given vector $\mathbf{v}$.

1. $\mathbf{S}(3, 2)$; $\mathbf{v} = (1, 1)$
2. $\mathbf{S}(5, 7)$; $\mathbf{v} = (2, 3)$
3. $\mathbf{S}(4, -3)$; $\mathbf{v} = (-1, 2)$
4. $\mathbf{S}(1, -5)$; $\mathbf{v} = (3, -1)$
5. $\mathbf{S}(-2, 4)$; $\mathbf{v} = (-1, -2)$
6. $\mathbf{S}(-5, 2)$; $\mathbf{v} = (-3, 1)$
7. $\mathbf{S}(-6, -4)$; $\mathbf{v} = (-3, -2)$
8. $\mathbf{S}(-3, -7)$; $\mathbf{v} = (4, -2)$

In Exercises 9–16, state whether or not the given point **S** lies on the line with the given parametric vector equation.

9. $\mathbf{S}(2, -1)$; $\mathbf{u} = (1, 2) + r(-1, 3)$
10. $\mathbf{S}(-8, 6)$; $\mathbf{u} = (-2, 3) + r(2, -1)$
11. $\mathbf{S}(3, 2)$; $\mathbf{u} = (1, 1) + r(2, -3)$
12. $\mathbf{S}(-1, 2)$; $\mathbf{u} = (4, 7) + r(3, 3)$
13. $\mathbf{S}(-1, 1)$; $\mathbf{u} = (-2, -3) + r(1, 4)$
14. $\mathbf{S}(2, -7)$; $\mathbf{u} = (-3, 2) + r(2, 1)$
15. $\mathbf{S}(0, -1)$; $\mathbf{u} = (3, -4) + r(-6, 2)$
16. $\mathbf{S}(-4, 0)$; $\mathbf{u} = (-1, -2) + r(3, -2)$

In Exercises 17–24, find a parametric vector equation for the line containing the given point **S** and having the given vector **v** as a direction vector.

17. $\mathbf{S}(2, 3)$; $\mathbf{v} = 3\mathbf{i} + 4\mathbf{j}$
18. $\mathbf{S}(4, 1)$; $\mathbf{v} = 2\mathbf{i} - 5\mathbf{j}$
19. $\mathbf{S}(3, -1)$; $\mathbf{v} = 5\mathbf{i} + 2\mathbf{j}$
20. $\mathbf{S}(1, -5)$; $\mathbf{v} = -\mathbf{i} - 3\mathbf{j}$
21. $\mathbf{S}(-6, 2)$; $\mathbf{v} = 3\mathbf{i} + \mathbf{j}$
22. $\mathbf{S}(-5, 3)$; $\mathbf{v} = 4\mathbf{i} - 2\mathbf{j}$
23. $\mathbf{S}(-1, -1)$; $\mathbf{v} = -2\mathbf{i} + 7\mathbf{j}$
24. $\mathbf{S}(-3, -5)$; $\mathbf{v} = -4\mathbf{i} - \mathbf{j}$

In Exercises 25–30, determine whether or not the two given parametric vector equations specify the same line.

25. $\mathbf{u} = (2, 1) + r(3, -1)$; $\mathbf{u} = (2, 1) + r(-3, 1)$
26. $\mathbf{u} = (3, 4) + r(2, -2)$; $\mathbf{u} = (3, 4) + r(-2, 2)$
27. $\mathbf{u} = (2, 3) + r(-1, 2)$; $\mathbf{u} = (1, 5) + r(2, -4)$
28. $\mathbf{u} = (-3, 1) + r(1, -3)$; $\mathbf{u} = (3, -1) + r(-1, 3)$
29. $\mathbf{u} = (-1, -2) + r(-2, 4)$; $\mathbf{u} = (1, 0) + r(1, -2)$
30. $\mathbf{u} = (0, 3) + r(-1, 5)$; $\mathbf{u} = (-1, -2) + r(2, -10)$

In Exercises 31–34, find the coordinates of the points $\mathbf{U}_1$ and $\mathbf{U}_2$ on the line with the given parametric vector equation that are located the given distance from the given point **S**.

31. On $\mathbf{u} = (4, -2) + r(1, 1)$; $3\sqrt{2}$ units from $\mathbf{S}(4, -2)$
32. On $\mathbf{u} = (-3, -1) + r(6, 8)$; 5 units from $\mathbf{S}(-3, -1)$
33. On $\mathbf{u} = (0, 4) + r(5, 12)$; 26 units from $\mathbf{S}(5, 16)$
34. On $\mathbf{u} = (-1, 6) + r(1, 4)$; $2\sqrt{17}$ units from $\mathbf{S}(1, 14)$

* **35.** Given that line $\mathcal{L}$ with parametric vector equation $\mathbf{u} = \mathbf{s} + r\mathbf{v}$ contains the point **T**, show that for any point **U** on $\mathcal{L}$, there is a scalar k such that $\mathbf{u} - \mathbf{t} = k\mathbf{v}$.

* **36.** Use the result of Exercise 35 to show that $(\mathbf{u} - \mathbf{t}) \cdot \mathbf{v_p} = 0$. What does this imply for the line through **T** with direction vector **v**?

2–3 Slope of a Line; Parallel and Perpendicular Lines

You may recall from earlier mathematics courses that the ratio of the rise of a line segment to its run is called the *slope of the segment*, and that the slope of a segment is ordinarily designated by the letter m. Thus you have

$$m = \frac{\text{rise}}{\text{run}}.$$

If $\mathbf{v} = (h, k)$ is a direction vector of a line $\mathcal{L}$ containing a point $\mathbf{S}$, then $\mathcal{L}$ has a parametric vector equation of the form

$$\mathbf{u} = \mathbf{s} + r(h, k), \qquad r \in \mathcal{R}. \tag{1}$$

If the value 1 is assigned to r, you can see that the coordinates of a second point, $\mathbf{Q}$, on $\mathcal{L}$ can be found by adding h to the first coordinate and k to the second coordinate of the given point $\mathbf{S}$. Thus, h and k are the run and the rise, respectively, of the segment $\overline{\mathbf{SQ}}$, and if $h \neq 0$, then $\frac{k}{h}$ is the slope of $\overline{\mathbf{SQ}}$ [Figure 2–9(a)].

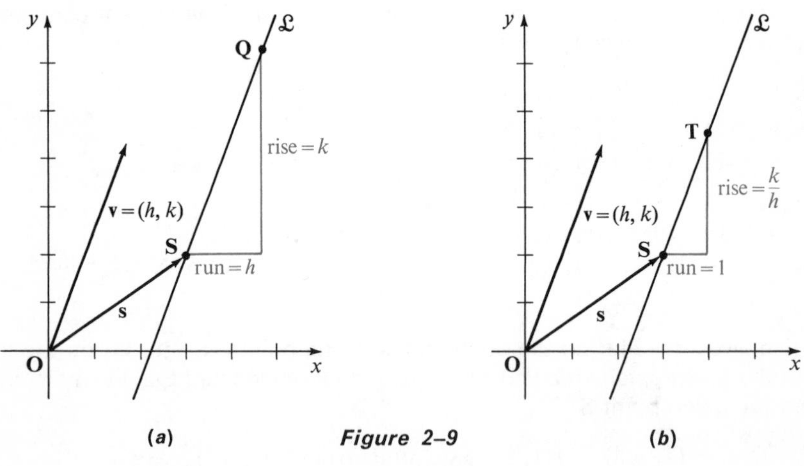

(*a*) *Figure 2–9* (*b*)

Again, for $h \neq 0$, if the value $\frac{1}{h}$ is assigned to r, you can see from Equation (1) that the coordinates of a second point, $\mathbf{T}$, on $\mathcal{L}$ can be found by adding $\frac{1}{h}(h, k) = \left(1, \frac{k}{h}\right)$ to $\mathbf{s}$, that is, by adding 1 to the first coordinate of $\mathbf{S}$ and $\frac{k}{h}$ to the second coordinate of $\mathbf{S}$. Since 1 is added to the first coordinate of $\mathbf{S}$ and $\frac{k}{h}$ is added to the second coordinate, you can think of $\frac{k}{h}$ as the change, along the line $\mathcal{L}$, in the vertical direction per unit change in the horizontal direction, as shown in Figure 2–9(b).

Notice that the number $\frac{k}{h}$ does not depend on the choice of the point **S** on $\mathcal{L}$ but only on the direction vector (h, k). Neither does $\frac{k}{h}$ depend on the *particular* direction vector (h, k) of $\mathcal{L}$, since any direction vector of $\mathcal{L}$ is of the form $c(h, k)$, or (ch, ck), where $c \neq 0$, and $\frac{ck}{ch} = \frac{k}{h}$. Consequently, you can define the *slope of a line* as follows:

■ If $\mathcal{L}$ is a line with direction vector (h, k), where $h \neq 0$, then the slope m of $\mathcal{L}$ is given by

$$m = \frac{k}{h}.$$

It follows from this definition that m is the slope of a line $\mathcal{L}$ if and only if $(1, m)$, or $\left(1, \frac{k}{h}\right)$, is a direction vector of $\mathcal{L}$. Thus, Equation (1), page 58, can be written as

$$\mathbf{u} = \mathbf{s} + r(1, m), \qquad r \in \mathcal{R}.$$

Example 1. Determine the slope m of the line containing the points $\mathbf{S}(5, 1)$ and $\mathbf{T}(3, -2)$, and write a parametric vector equation in the form $\mathbf{u} = \mathbf{s} + r(1, m)$ for the line.

Solution: A direction vector for the line is

$$\mathbf{t} - \mathbf{s} = (3, -2) - (5, 1) = (-2, -3).$$

Then, by definition, $m = \frac{-3}{-2} = \frac{3}{2}$. Since $\mathbf{S}(5, 1)$ is on the line, a parametric vector equation for the line is $\mathbf{u} = (5, 1) + r(1, \frac{3}{2})$.

The method of solution in Example 1 suggests that it would be useful to have an expression for the slope of a line through two given points in terms of the coordinates of the points. Since a direction vector for the line $\mathcal{L}$ through the points $\mathbf{S}(x_1, y_1)$ and $\mathbf{T}(x_2, y_2)$ is

$$\mathbf{v} = \mathbf{t} - \mathbf{s} = (x_2, y_2) - (x_1, y_1) = (x_2 - x_1, y_2 - y_1),$$

it follows from the definition of slope that if $x_2 \neq x_1$, then the slope of $\mathcal{L}$ is given by

■
$$m = \frac{y_2 - y_1}{x_2 - x_1}. \tag{2}$$

A line whose slope is *positive* "rises" from left to right; a line whose slope is *negative* "falls" from left to right. In Figure 2–10, $\mathcal{L}_1$ has positive slope and $\mathcal{L}_2$ has negative slope.

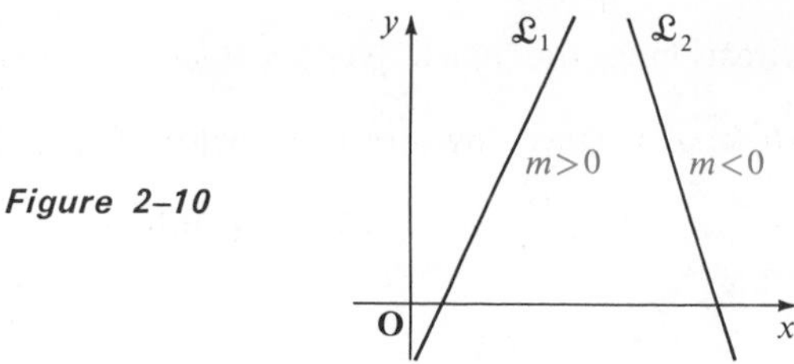

Figure 2–10

A line with a direction vector of the form $(h, 0)$, $h \neq 0$, is a **horizontal line** and has slope $\frac{0}{h} = 0$. Since the vector $(h, 0)$ has a geometric representation lying on the x-axis, a horizontal line is parallel to the x-axis (see definition below). A line with a direction vector of the form $(0, k)$, $k \neq 0$, is a **vertical line.** Vertical lines have no slope $\left(\text{since } \frac{k}{0} \text{ is not defined}\right)$ and are parallel to the y-axis.

You have been accustomed, in the study of Euclidean geometry, to thinking of parallel lines as lines which lie in the same plane and which have no point in common. Recall, however (page 16), that two vectors are said to be parallel if and only if they have the same or opposite directions. Also (page 53), a line $\mathcal{L}$ and a nonzero vector $\mathbf{v}$ are said to be parallel if and only if $\mathbf{v}$ has a geometric representation lying on $\mathcal{L}$, that is, if and only if $\mathbf{v}$ is a direction vector of $\mathcal{L}$. It is now convenient for us to give a new definition of parallel lines in terms of direction vectors. The new definition is equivalent (Exercises 45 and 46, page 63) to the familiar geometric definition that parallel lines are non-intersecting coplanar lines, except that, as you will see, every line in the plane will now be considered to be parallel to itself.

■ Lines $\mathcal{L}_1$ and $\mathcal{L}_2$ in the coordinate plane will be said to be **parallel** if and only if a direction vector of $\mathcal{L}_1$ is parallel to a direction vector of $\mathcal{L}_2$.

Of course, since all direction vectors of a given line are parallel (page 53), if any direction vector of $\mathcal{L}_1$ is parallel to any direction vector of $\mathcal{L}_2$, then *all* direction vectors of $\mathcal{L}_1$ are parallel to *all* direction vectors of $\mathcal{L}_2$. Also, since any nonzero vector parallel to a direction vector of a line is *itself* a direction vector of the line (page 53), any direction vector of $\mathcal{L}_1$ is also a direction vector of $\mathcal{L}_2$. Thus lines $\mathcal{L}_1$ and $\mathcal{L}_2$ are parallel if and only if they have a direction vector (h, k) in common. Recall, though, that a line with direction vector

(h, k) is either vertical (if $h = 0$) or has slope $\frac{k}{h}$ (if $h \neq 0$). Therefore lines $\mathcal{L}_1$ and $\mathcal{L}_2$ are parallel if and only if either (i) they are both vertical or (ii) they have the same slope, $\frac{k}{h}$.

Example 2. If $\mathcal{L}_1$ contains $\mathbf{S}(3, -1)$, $\mathcal{L}_2$ contains $\mathbf{T}(2, 5)$, and $\mathcal{L}_1$ and $\mathcal{L}_2$ have $\mathbf{v} = (1, 2)$ as a direction vector, are $\mathcal{L}_1$ and $\mathcal{L}_2$ coincident lines?

Solution: By definition, the lines $\mathcal{L}_1$ and $\mathcal{L}_2$ are parallel. They coincide if and only if $\mathbf{S}$ and $\mathbf{T}$ are on both. Thus (see pages 54 and 55), they coincide if and only if $\mathbf{t} - \mathbf{s}$ is parallel to both, that is, if and only if $(\mathbf{t} - \mathbf{s}) \cdot \mathbf{v_p} = 0$. You have $(\mathbf{t} - \mathbf{s}) \cdot \mathbf{v_p} = [(2, 5) - (3, -1)] \cdot (-2, 1) = (-1, 6) \cdot (-2, 1) = 8 \neq 0$. Therefore, $\mathcal{L}_1$ and $\mathcal{L}_2$ are not coincident.

You have just seen that parallel lines can be defined in terms of their direction vectors. You can similarly define perpendicular lines in terms of their direction vectors. Recall (page 16) that vectors whose directions differ by $\pm 90°$ or $\pm 270°$ are said to be perpendicular, or orthogonal, vectors.

■ Lines $\mathcal{L}_1$ and $\mathcal{L}_2$ in the coordinate plane will be said to be **perpendicular**, or **orthogonal**, if and only if a direction vector of $\mathcal{L}_1$ is perpendicular to a direction vector of $\mathcal{L}_2$.

This definition is equivalent to the familiar definition from Euclidean geometry that coplanar lines are perpendicular if and only if they intersect at right angles.

Of course, if any direction vector of $\mathcal{L}_1$ is perpendicular to any direction vector of $\mathcal{L}_2$, then *all* direction vectors of $\mathcal{L}_1$ are perpendicular to *all* direction vectors of $\mathcal{L}_2$. Thus, since two vectors are perpendicular if and only if their dot product is 0, if $\mathbf{v}_1$ is any direction vector of $\mathcal{L}_1$ and $\mathbf{v}_2$ is any direction vector of $\mathcal{L}_2$, then $\mathcal{L}_1$ and $\mathcal{L}_2$ are perpendicular if and only if $\mathbf{v}_1 \cdot \mathbf{v}_2 = 0$.

Example 3. Show that the line $\mathcal{L}_1$ with parametric vector equation $\mathbf{u} = (3, -1) + r(2, 3)$ is perpendicular to the line $\mathcal{L}_2$ with parametric vector equation $\mathbf{u} = (2, -1) + r(6, -4)$.

Solution: From the given equations, you can see that $(2, 3)$ is a direction vector of $\mathcal{L}_1$ and $(6, -4)$ is a direction vector of $\mathcal{L}_2$. Since $(2, 3) \cdot (6, -4) = 12 - 12 = 0$, $\mathcal{L}_1$ and $\mathcal{L}_2$ are perpendicular.

On page 59, you saw that every nonvertical line has a direction vector of the form $(1, m)$, where m is the slope of the line. Thus if $\mathcal{L}_1$ is a nonvertical line with slope m_1 and $\mathcal{L}_2$ is a nonvertical line with slope m_2, then $(1, m_1)$ and $(1, m_2)$ are direction vectors of $\mathcal{L}_1$ and $\mathcal{L}_2$, respectively. Now $\mathcal{L}_1$ and $\mathcal{L}_2$ are perpendicular if and only if $(1, m_1)$ and $(1, m_2)$ are perpendicular, that is, if and only if

$$\begin{aligned}(1, m_1) \cdot (1, m_2) &= 0,\\ 1 + m_1 m_2 &= 0,\\ m_1 m_2 &= -1,\end{aligned}$$

or

$$m_1 = -\frac{1}{m_2} \quad \text{and} \quad m_2 = -\frac{1}{m_1}.$$

Thus:

■ Two nonvertical lines are perpendicular if and only if their slopes are *negative reciprocals* of each other.

Vertical lines are perpendicular to horizontal lines. As discussed earlier, the former have no slope and the latter have slope 0. For these lines, the negative-reciprocal test does not apply.

Exercises 2–3

In Exercises 1–8, find the slope of the line containing the given points **S** and **T**, and write a parametric vector equation for the line in the form $\mathbf{u} = \mathbf{s} + r(1, m)$.

1. $\mathbf{S}(5, 1)$ and $\mathbf{T}(-4, 2)$
2. $\mathbf{S}(0, 7)$ and $\mathbf{T}(5, 0)$
3. $\mathbf{S}(3, -4)$ and $\mathbf{T}(-2, 1)$
4. $\mathbf{S}(2, 1)$ and $\mathbf{T}(1, -2)$
5. $\mathbf{S}(2, -3)$ and $\mathbf{T}(-4, -3)$
6. $\mathbf{S}(-1, -5)$ and $\mathbf{T}(4, -5)$
7. $\mathbf{S}(-2, -3)$ and $\mathbf{T}(-1, -7)$
8. $\mathbf{S}(-5, -4)$ and $\mathbf{T}(2, 5)$

In Exercises 9–16, write a parametric vector equation for the line through the given point **S** with the given slope m.

9. $\mathbf{S}(3, -4)$; $m = 2$
10. $\mathbf{S}(-2, 1)$; $m = -3$
11. $\mathbf{S}(0, -5)$; $m = 0$
12. $\mathbf{S}(2, -3)$; no slope
13. $\mathbf{S}(-2, -3)$; $m = \frac{2}{3}$
14. $\mathbf{S}(1, -5)$; $m = -\frac{3}{4}$
15. $\mathbf{S}(1, 0)$; no slope
16. $\mathbf{S}(-3, 4)$; $m = 0$

In Exercises 17–21, determine whether the lines with the given parametric vector equations are (a) parallel, (b) perpendicular, or (c) neither parallel nor perpendicular.

17. $\mathcal{L}_1$: $\mathbf{u} = (2, -1) + r(3, 2)$, $\mathcal{L}_2$: $\mathbf{u} = (-3, 1) + r(6, 4)$
18. $\mathcal{L}_1$: $\mathbf{u} = (4, 7) + r(-1, 3)$, $\mathcal{L}_2$: $\mathbf{u} = (-2, 5) + r(6, 2)$

19. $\mathcal{L}_1$: $\mathbf{u} = (3, -5) + r(2, -3)$, $\mathcal{L}_2$: $\mathbf{u} = (-1, 1) + r(-6, 9)$

20. $\mathcal{L}_1$: $\mathbf{u} = (1, -2) + r(-2, -3)$, $\mathcal{L}_2$: $\mathbf{u} = (4, 2) + r(4, -3)$

21. $\mathcal{L}_1$: $\mathbf{u} = (6, 1) + r(-3, 6)$, $\mathcal{L}_2$: $\mathbf{u} = (-2, 1) + r(4, 2)$

22. Determine which of the parallel lines in Exercises 17–21 coincide.

In Exercises 23–28, determine whether or not the three given points, **R**, **S**, and **T**, are collinear.

23. $\mathbf{R}(2, 1)$, $\mathbf{S}(-1, 3)$, $\mathbf{T}(5, -1)$

24. $\mathbf{R}(1, -2)$, $\mathbf{S}(7, 1)$, $\mathbf{T}(-3, -4)$

25. $\mathbf{R}(8, -3)$, $\mathbf{S}(-4, 5)$, $\mathbf{T}(2, 4)$

26. $\mathbf{R}(3, -1)$, $\mathbf{S}(-2, -3)$, $\mathbf{T}(-1, -3)$

27. $\mathbf{R}(1, 0)$, $\mathbf{S}(0, -3)$, $\mathbf{T}(3, 3)$

28. $\mathbf{R}(2, 0)$, $\mathbf{S}(0, -5)$, $\mathbf{T}(4, 5)$

In Exercises 29–36, write a parametric vector equation for the line through the given point **S** (a) parallel to and (b) perpendicular to the line with the given parametric vector equation.

29. $\mathbf{S}(-2, 1)$; $\mathbf{u} = (3, 0) + r(2, -1)$

30. $\mathbf{S}(1, 2)$; $\mathbf{u} = (-3, -2) + r(3, 4)$

31. $\mathbf{S}(3, 4)$; $\mathbf{u} = (1, 5) + r(5, -2)$

32. $\mathbf{S}(2, 1)$; $\mathbf{u} = (-1, 7) + r(1, 3)$

33. $\mathbf{S}(0, 4)$; $\mathbf{u} = (2, 2) + r(-3, 3)$

34. $\mathbf{S}(0, 0)$; $\mathbf{u} = (1, -1) + r(4, 3)$

35. $\mathbf{S}(-2, -3)$; $\mathbf{u} = (3, 3) + r(5, 6)$

36. $\mathbf{S}(4, -2)$; $\mathbf{u} = (1, 2) + r(3, -4)$

In Exercises 37–42, determine whether the line $\mathcal{L}_1$ containing the given points **Q** and **R** is (a) parallel, (b) perpendicular, or (c) neither parallel nor perpendicular, to the line $\mathcal{L}_2$ containing the given points **S** and **T**.

37. $\mathcal{L}_1$: $\mathbf{Q}(3, 1)$ and $\mathbf{R}(4, 3)$; $\mathcal{L}_2$: $\mathbf{S}(2, 4)$ and $\mathbf{T}(4, 8)$

38. $\mathcal{L}_1$: $\mathbf{Q}(2, -5)$ and $\mathbf{R}(1, 2)$; $\mathcal{L}_2$: $\mathbf{S}(6, 5)$ and $\mathbf{T}(-1, 4)$

39. $\mathcal{L}_1$: $\mathbf{Q}(-1, -2)$ and $\mathbf{R}(2, 2)$; $\mathcal{L}_2$: $\mathbf{S}(-5, 7)$ and $\mathbf{T}(3, 1)$

40. $\mathcal{L}_1$: $\mathbf{Q}(4, -5)$ and $\mathbf{R}(2, -1)$; $\mathcal{L}_2$: $\mathbf{S}(6, -2)$ and $\mathbf{T}(-2, 3)$

41. $\mathcal{L}_1$: $\mathbf{Q}(1, -1)$ and $\mathbf{R}(3, 0)$; $\mathcal{L}_2$: $\mathbf{S}(0, 3)$ and $\mathbf{T}(1, 5)$

42. $\mathcal{L}_1$: $\mathbf{Q}(2, 0)$ and $\mathbf{R}(0, -2)$; $\mathcal{L}_2$: $\mathbf{S}(-1, 3)$ and $\mathbf{T}(2, -4)$

43. Given that $\mathcal{L}$ is the graph of the parametric vector equation $\mathbf{u} = (a, b) + r(c, d)$, write a parametric vector equation for the line $\mathcal{M}$ that is parallel to $\mathcal{L}$ and contains the midpoint of the segment with endpoints $\mathbf{S}(x_1, y_1)$ and $\mathbf{T}(x_2, y_2)$.

44. In Exercise 43, write a parametric vector equation for the line $\mathcal{N}$ that is perpendicular to $\mathcal{L}$ and contains the midpoint of the given segment.

* **45.** Let $\mathcal{L}_1$ and $\mathcal{L}_2$ be noncoincident parallel lines with direction vector $\mathbf{v}$, and let **S** be a point on $\mathcal{L}_1$ but not on $\mathcal{L}_2$ and **T** be a point on $\mathcal{L}_2$ but not on $\mathcal{L}_1$. Show that if there were a point **R** on both $\mathcal{L}_1$ and $\mathcal{L}_2$, then you would have $(\mathbf{r} - \mathbf{s}) \cdot \mathbf{v_p} = 0$ and $(\mathbf{r} - \mathbf{t}) \cdot \mathbf{v_p} = 0$. Show then, since $(\mathbf{t} - \mathbf{s}) = (\mathbf{r} - \mathbf{s}) - (\mathbf{r} - \mathbf{t})$, that you would have $(\mathbf{t} - \mathbf{s}) \cdot \mathbf{v_p} = 0$ and accordingly that **T** would be on $\mathcal{L}_1$. What does this prove?

* **46.** Prove that if **A**, **B**, and **C** are collinear points, then $(\mathbf{b} - \mathbf{a}) \cdot (\mathbf{c} - \mathbf{a})_\mathbf{p} = 0$.

Cartesian Equations for Lines

2–4 Standard Cartesian Form of an Equation of a Line

The standard Cartesian form (page 47) of an equation of a line $\mathcal{L}$ in the coordinate plane,

$$Ax + By + C = 0, \qquad A^2 + B^2 \neq 0,$$

can be derived in a number of ways. One of these is outlined in Exercises 35 and 36 on page 68. In this section, you will see how the standard Cartesian form can be derived from a vector equation of $\mathcal{L}$.

Any nonzero vector perpendicular to a direction vector of a line $\mathcal{L}$ is said to be a **normal vector** of $\mathcal{L}$. Figure 2–11 shows a line $\mathcal{L}$ containing the point

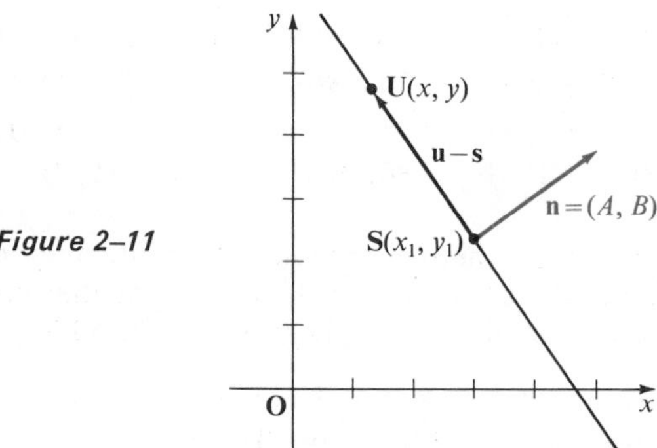

Figure 2–11

$\mathbf{S}(x_1, y_1)$, together with a normal vector $\mathbf{n} = (A, B)$ of $\mathcal{L}$, where A and B represent real numbers not both 0. If $\mathbf{U}(x, y)$ is any point in the coordinate plane, then $\mathbf{U}$ is on $\mathcal{L}$ if and only if $\mathbf{u} - \mathbf{s}$ is parallel to $\mathcal{L}$, that is, if and only if $\mathbf{u} - \mathbf{s}$ is perpendicular to $\mathbf{n}$. It follows (see page 55) that an equation of $\mathcal{L}$ is

$$(\mathbf{u} - \mathbf{s}) \cdot \mathbf{n} = 0,$$

$$\mathbf{u} \cdot \mathbf{n} - \mathbf{s} \cdot \mathbf{n} = 0,$$

or

$$\mathbf{u} \cdot \mathbf{n} = \mathbf{s} \cdot \mathbf{n}.$$

Since $\mathbf{u} = (x, y)$, $\mathbf{s} = (x_1, y_1)$, and $\mathbf{n} = (A, B)$, this last equation can be written as

$$(x, y) \cdot (A, B) = (x_1, y_1) \cdot (A, B),$$

or

$$Ax + By = Ax_1 + By_1.$$

Now, because x_1, y_1, A, and B are constants, the number $Ax_1 + By_1$ is also a constant, which, for convenience, can be denoted by $-C$. You then have, as an equation of $\mathcal{L}$,

$$Ax + By + C = 0, \qquad A^2 + B^2 \neq 0. \tag{1}$$

The Cartesian equation (1) is frequently called a **scalar equation** of $\mathcal{L}$ because it contains no vectors.

Example 1. Find a Cartesian equation in standard form for the line $\mathcal{L}$ containing the point $\mathbf{S}(1, 4)$ and having $\mathbf{v} = (3, -2)$ as a direction vector.

Solution (1): Since a direction vector of $\mathcal{L}$ is $(3, -2)$, a normal vector of $\mathcal{L}$ is $\mathbf{n} = (2, 3)$. Then, because $\mathbf{S}(1, 4)$ is on $\mathcal{L}$, if $\mathbf{U}(x, y)$ is any point on $\mathcal{L}$, you have

$$(\mathbf{u} - \mathbf{s}) \cdot \mathbf{n} = 0.$$

Since $\mathbf{u} = (x, y)$, $\mathbf{s} = (1, 4)$, and $\mathbf{n} = (2, 3)$, this equation is equivalent to

$$[(x, y) - (1, 4)] \cdot (2, 3) = 0,$$
$$(x, y) \cdot (2, 3) - (1, 4) \cdot (2, 3) = 0,$$

or

$$2x + 3y - 14 = 0,$$

which is the required Cartesian equation.

Solution (2): Note that since $(3, -2)$ is a direction vector of $\mathcal{L}$, a normal vector of $\mathcal{L}$ is $\mathbf{n} = (A, B) = (2, 3)$. Then, in Equation (1) above, you have $A = 2$ and $B = 3$, that is,

$$2x + 3y + C = 0.$$

Since $\mathbf{S}(1, 4)$ is on $\mathcal{L}$, you can substitute 1 for x and 4 for y in this equation to obtain

$$2(1) + 3(4) + C = 0,$$

or

$$C = -14.$$

Thus a Cartesian equation in standard form for $\mathcal{L}$ is

$$2x + 3y - 14 = 0.$$

Since the vectors (A, B) and $(-B, A)$ are perpendicular, if they are, respectively, normal vectors for lines $\mathcal{L}_1$ and $\mathcal{L}_2$ in the plane, then $\mathcal{L}_1$ and $\mathcal{L}_2$ must be perpendicular (Figure 2–12). It follows that equations of the form

$$\begin{aligned} Ax + By + C &= 0, \\ -Bx + Ay + C' &= 0, \end{aligned} \tag{2}$$

where $A^2 + B^2 \neq 0$, are Cartesian equations for perpendicular lines.

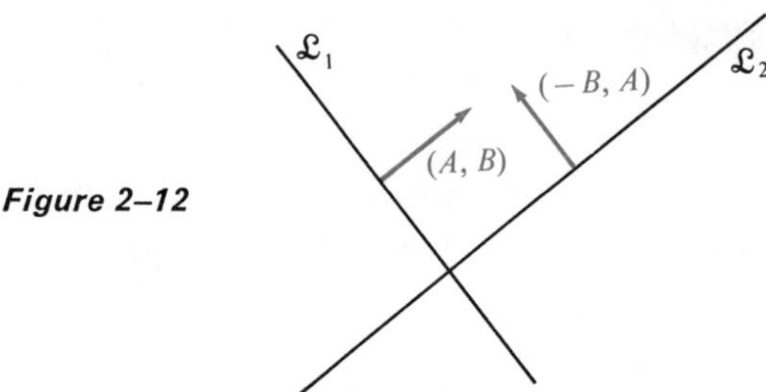

Figure 2–12

Example 2. Find a Cartesian equation in standard form for the line that contains $\mathbf{S}(5, -1)$ and is perpendicular to the line with Cartesian equation

$$3x - 4y + 6 = 0.$$

Solution: From Equations (2), the desired equation has the form

$$4x + 3y + C' = 0.$$

Since $\mathbf{S}(5, -1)$ is on the line, you can replace x with 5 and y with -1 to obtain

$$4(5) + 3(-1) + C' = 0,$$

or

$$C' = -17.$$

Therefore a Cartesian equation in standard form for the line is

$$4x + 3y - 17 = 0.$$

If (A, B) is a normal vector of a line $\mathcal{L}$, then it is also a normal vector of any line parallel to $\mathcal{L}$. Thus, the equations

$$\begin{aligned} Ax + By + C &= 0, \\ Ax + By + C' &= 0, \end{aligned} \tag{3}$$

where $A^2 + B^2 \neq 0$, are Cartesian equations for parallel lines.

Example 3. Find a Cartesian equation in standard form for the line that contains the point $\mathbf{S}(3, -2)$ and is parallel to the line with Cartesian equation $2x - 5y - 4 = 0$.

Solution: From Equations (3), the desired equation is of the form

$$2x - 5y + C' = 0.$$

Since $\mathbf{S}(3, -2)$ is on the line, you can substitute 3 for x and -2 for y to obtain

$$2(3) - 5(-2) + C' = 0,$$

or

$$C' = -16.$$

Therefore a Cartesian equation in standard form for the line is

$$2x - 5y - 16 = 0.$$

Notice that when an equation of a line $\mathcal{L}$ is given in the standard Cartesian form,

$$Ax + By + C = 0, \qquad A^2 + B^2 \neq 0,$$

then a normal vector of $\mathcal{L}$ is $\mathbf{n} = (A, B)$ and a direction vector of $\mathcal{L}$ is $\mathbf{v} = (-B, A)$. Thus, the slope of $\mathcal{L}$ is given by

$$m = -\frac{A}{B}, \quad \text{if } B \neq 0.$$

Exercises 2–4

In Exercises 1–8, find a Cartesian equation in standard form for the line containing the given point **S** and having the given normal vector **n**.

1. $\mathbf{S}(4, 2)$; $\mathbf{n} = (3, 5)$
2. $\mathbf{S}(3, 7)$; $\mathbf{n} = (2, 4)$
3. $\mathbf{S}(2, -1)$; $\mathbf{n} = (1, -7)$
4. $\mathbf{S}(5, -3)$; $\mathbf{n} = (-6, 1)$
5. $\mathbf{S}(-7, 3)$; $\mathbf{n} = (-1, -2)$
6. $\mathbf{S}(-6, 1)$; $\mathbf{n} = (-3, 3)$
7. $\mathbf{S}(-2, -2)$; $\mathbf{n} = (1, 1)$
8. $\mathbf{S}(-8, -3)$; $\mathbf{n} = (-4, 2)$

In Exercises 9–14, find a Cartesian equation in standard form for the line containing the given point **S** and having the given vector **v** as a direction vector.

9. $\mathbf{S}(3, 7)$; $\mathbf{v} = 5\mathbf{i} - 2\mathbf{j}$
10. $\mathbf{S}(2, -1)$; $\mathbf{v} = 4\mathbf{i} + 7\mathbf{j}$
11. $\mathbf{S}(-4, 5)$; $\mathbf{v} = -6\mathbf{i} + 3\mathbf{j}$
12. $\mathbf{S}(6, 3)$; $\mathbf{v} = -3\mathbf{i} + 5\mathbf{j}$
13. $\mathbf{S}(7, -5)$; $\mathbf{v} = 6\mathbf{i}$
14. $\mathbf{S}(0, 5)$; $\mathbf{v} = -3\mathbf{j}$

In Exercises 15–20, which pairs of lines with the given Cartesian equations are perpendicular, and which pairs are parallel?

15. $2x - 3y + 7 = 0$, $4x - 6y - 15 = 0$
16. $5x - 7y + 13 = 0$, $-15x + 21y + 25 = 0$
17. $4x + 5y - 9 = 0$, $10x - 8y + 17 = 0$
18. $6x - 8y + 19 = 0$, $4x + 3y + 5 = 0$
19. $3x - 6y + 8 = 0$, $-2x + 4y - 7 = 0$
20. $8x + 20y - 7 = 0$, $15x - 6y - 4 = 0$

In Exercises 21–28, write a Cartesian equation in standard form for the line which contains the given point **S** and is (a) perpendicular and (b) parallel to the line with the given Cartesian equation.

21. $\mathbf{S}(1, 3)$; $2x - 5y + 7 = 0$
22. $\mathbf{S}(5, 2)$; $3x + 4y - 8 = 0$
23. $\mathbf{S}(3, -2)$; $5x - 7y + 10 = 0$
24. $\mathbf{S}(2, -4)$; $3x - 4y + 12 = 0$
25. $\mathbf{S}(-6, 2)$; $5x + 6y - 9 = 0$
26. $\mathbf{S}(-3, 5)$; $6x - 3y + 18 = 0$
27. $\mathbf{S}(-1, -1)$; $x - 2y - 7 = 0$
28. $\mathbf{S}(-8, -3)$; $4x + 7y - 11 = 0$

In Exercises 29–34, find a Cartesian equation for the line containing the given points **S** and **T**.

29. $\mathbf{S}(3, 5)$ and $\mathbf{T}(2, 6)$
30. $\mathbf{S}(2, 7)$ and $\mathbf{T}(5, 1)$
31. $\mathbf{S}(4, -5)$ and $\mathbf{T}(-2, 3)$
32. $\mathbf{S}(-4, 2)$ and $\mathbf{T}(3, -7)$
33. $\mathbf{S}(-1, -1)$ and $\mathbf{T}(0, -1)$
34. $\mathbf{S}(3, -5)$ and $\mathbf{T}(3, 2)$

In Exercises 35 and 36, use the following fact from plane geometry: The line which is the perpendicular bisector of a segment $\overline{\mathbf{ST}}$ is the set of all points equidistant from **S** and **T**.

* **35.** Prove that every line in a coordinate plane is the graph of some linear Cartesian equation $Ax + By + C = 0$, $A^2 + B^2 \neq 0$. [*Hint:* Let $\mathcal{L}$ be a line in the plane and choose points $\mathbf{S}(x_1, y_1)$ and $\mathbf{T}(x_2, y_2)$ so that $\mathcal{L}$ is the perpendicular bisector of $\overline{\mathbf{ST}}$. Show that the following statements are equivalent:

$$\mathbf{U}(x, y) \text{ is on } \mathcal{L}$$

$$d(\mathbf{U}, \mathbf{S}) = d(\mathbf{U}, \mathbf{T}) \tag{1}$$

$$\sqrt{(x - x_1)^2 + (y - y_1)^2} = \sqrt{(x - x_2)^2 + (y - y_2)^2}$$

$$2(x_2 - x_1)x + 2(y_2 - y_1)y + (x_1^2 + y_1^2 - x_2^2 - y_2^2) = 0 \tag{2}$$

Is the last equation linear in the variables x and y? That is, is it of the form $Ax + By + C = 0$, $A^2 + B^2 \neq 0$?]

* **36.** Prove that the graph of every linear Cartesian equation in two variables is a line. [*Hint:* Let $Ax + By + C = 0$, $A^2 + B^2 \neq 0$, be a linear equation in x and y. Equation (2) of Exercise 35 suggests that we choose numbers x_1, x_2, y_1, and y_2, as coordinates of points **S** and **T**, such that

$$2(x_2 - x_1) = A,\ 2(y_2 - y_1) = B,\ x_1^2 + y_1^2 - x_2^2 - y_2^2 = C. \tag{3}$$

Show that this choice can be made by solving the first two equations in (3) for x_2 and y_2 and then eliminating x_2 and y_2 from the third equation in (3). The given equation $Ax + By + C = 0$ is now the *same* as (2). Use the fact that Equations (2) and (1) are equivalent to show that the graph of $Ax + By + C = 0$ is $\{\mathbf{U}\colon d(\mathbf{U}, \mathbf{S}) = d(\mathbf{U}, \mathbf{T})\}$.]

* **37.** Let $\mathcal{L}$ be the line with Cartesian equation $Ax + By + C = 0, A^2 + B^2 \neq 0$. Write a Cartesian equation for the line $\mathcal{M}$ that is parallel to $\mathcal{L}$ and contains the point $\mathbf{S}(a, b)$.

* **38.** Write a Cartesian equation for the line $\mathcal{N}$ that is perpendicular to $\mathcal{L}$ in Exercise 37 and contains the point $\mathbf{S}(a, b)$.

* **39.** Show that if the point $\mathbf{S}(h, k)$ lies on a line $\mathcal{L}$ parallel to the line with Cartesian equation $Ax + By + C = 0$, with $B \neq 0$, then $\mathcal{L}$ intersects the y-axis at $\mathbf{Q}\left(0, \frac{A}{B}h + k\right)$.

* **40.** Show that if the point $\mathbf{S}(h, k)$ lies on a line $\mathcal{L}$ parallel to the line with Cartesian equation $Ax + By + C = 0$, with $A \neq 0$, then $\mathcal{L}$ intersects the x-axis at $\mathbf{R}\left(h + \frac{B}{A}k, 0\right)$.

2–5 Point-Slope and Two-Point Forms of Equations of Lines

The standard Cartesian equation of a line

$$Ax + By + C = 0, \qquad A^2 + B^2 \neq 0,$$

can be written equivalently in a variety of different forms. Some of these are of special interest because they enable you to obtain an equation of a line directly from certain given geometric facts about the line, or they enable you to identify certain geometric characteristics of a line directly from its equation. Two of these useful forms will be discussed in this section.

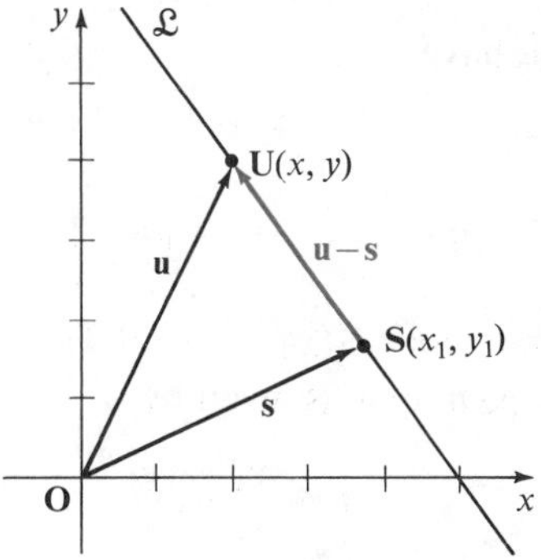

Figure 2–13

Figure 2–13 shows a line $\mathcal{L}$ through a given point $\mathbf{S}(x_1, y_1)$. If $\mathbf{U}(x, y)$ is any point on $\mathcal{L}$ other than $\mathbf{S}$, then a direction vector for $\mathcal{L}$ is

$$\begin{aligned} \mathbf{u} - \mathbf{s} &= (x, y) - (x_1, y_1) \\ &= (x - x_1, y - y_1). \end{aligned}$$

Now you saw on page 59 that if $x \neq x_1$, then the slope m of $\mathcal{L}$ is given by

$$\frac{y - y_1}{x - x_1} = m, \tag{1}$$

from which you have

$$y - y_1 = m(x - x_1). \tag{2}$$

This Cartesian equation for $\mathcal{L}$ is called the **point-slope form** of the equation for the line. Although Equation (1) is not meaningful for $x = x_1$, Equation (2) is satisfied by the coordinates of *every* point (x, y) on $\mathcal{L}$, since for $(x, y) = (x_1, y_1)$ you have

$$y_1 - y_1 = m(x_1 - x_1),$$

or

$$0 = 0.$$

If both the slope of a line and any point on it are specified, you can write an equation of the form (2) directly and then transform it into an equivalent Cartesian equation in standard form, as illustrated in the following example.

Example 1. Write a Cartesian equation in standard form for the line through the point $\mathbf{S}(7, -1)$ with slope 3.

Solution: If you set $x_1 = 7$, $y_1 = -1$, and $m = 3$ in Equation (2), you obtain

$$y - (-1) = 3(x - 7).$$

Then you have

$$y + 1 = 3x - 21,$$

or

$$3x - y - 22 = 0.$$

If a line $\mathcal{L}$ contains the points $\mathbf{S}(x_1, y_1)$ and $\mathbf{T}(x_2, y_2)$, with $x_2 \neq x_1$, then, as you have seen, the slope m of $\mathcal{L}$ is given by

$$m = \frac{y_2 - y_1}{x_2 - x_1}.$$

If you substitute this expression for m in Equation (2) above, you obtain the equivalent equation

$$y - y_1 = \frac{y_2 - y_1}{x_2 - x_1}(x - x_1). \tag{3}$$

This Cartesian equation is called the **two-point form** of the equation for $\mathcal{L}$.

Example 2. Write a Cartesian equation in standard form for the line containing the points **S**(2, 3) and **T**(4, −7).

Solution: If, in Equation (3), page 70, you replace x_1 and y_1 with the coordinates of **S**(2, 3), and x_2 and y_2 with the coordinates of **T**(4, −7), you obtain

$$y - 3 = \frac{-7 - 3}{4 - 2}(x - 2),$$

$$y - 3 = -5(x - 2),$$

$$y - 3 = -5x + 10,$$

or

$$5x + y - 13 = 0.$$

If the x-coordinates of two given points are equal, then the line containing these points is vertical (Figure 2–14), and the slope of the line is undefined. However, since every point on the line must have the same x-coordinate, say x_1, the line is completely determined by specifying this coordinate, and accordingly a Cartesian equation for the line is

$$x = x_1.$$

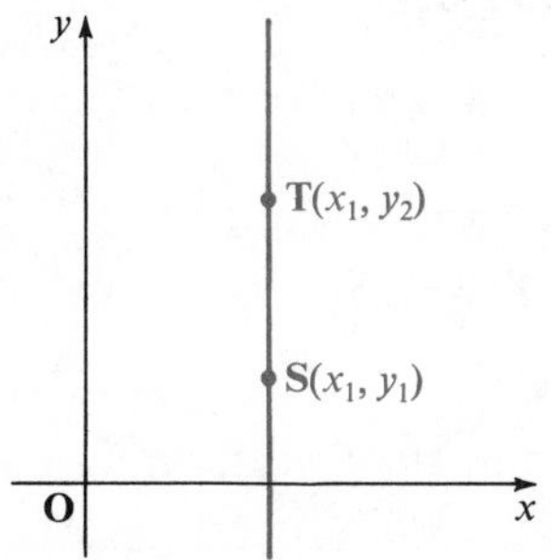

Figure 2–14

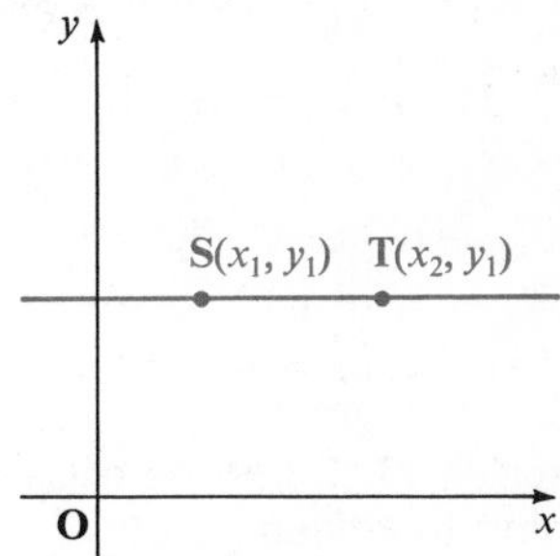

Figure 2–15

On the other hand, if the y-coordinates of two given points are equal, then the line containing these points is horizontal (Figure 2–15), and the line has slope 0. It follows immediately from Equation (2), page 70, that

$$y - y_1 = 0,$$

and therefore that

$$y = y_1$$

is a Cartesian equation for the line.

In Sections 2–1 and 2–2, you found two forms of a system of parametric equations for a line:

$$\begin{aligned} x &= x_1 + r(x_2 - x_1) \\ y &= y_1 + r(y_2 - y_1) \end{aligned} \quad \text{and} \quad \begin{aligned} x &= x_1 + rh \\ y &= y_1 + rk \end{aligned}$$

You can now see how these systems are related to the equations discussed in this section. In each case, solve the first equation for r:

$$r = \frac{x - x_1}{x_2 - x_1} \qquad r = \frac{x - x_1}{h}$$

Then substitute for r in the second equation:

$$y = y_1 + \frac{x - x_1}{x_2 - x_1}(y_2 - y_1) \qquad y = y_1 + \frac{x - x_1}{h}(k),$$

or

$$y - y_1 = \frac{y_2 - y_1}{x_2 - x_1}(x - x_1) \qquad y - y_1 = \frac{k}{h}(x - x_1). \tag{4}$$

Since $\frac{k}{h}$ is the slope of the line if $h \neq 0$, you can see that the resulting single equations (4) correspond to the forms developed in this section.

Exercises 2–5

In Exercises 1–10, write a Cartesian equation in standard form for the line containing the given point **S** and having the given slope *m*.

1. $\mathbf{S}(2, 5)$; $m = 2$
2. $\mathbf{S}(6, 1)$; $m = 4$
3. $\mathbf{S}(3, -4)$; $m = -1$
4. $\mathbf{S}(7, -2)$; $m = 5$
5. $\mathbf{S}(-1, 5)$; $m = -3$
6. $\mathbf{S}(-4, 6)$; $m = -2$
7. $\mathbf{S}(-1, -1)$; $m = \frac{1}{2}$
8. $\mathbf{S}(-3, -5)$; $m = -\frac{2}{3}$
9. $\mathbf{S}(2, -6)$; $m = -\frac{3}{5}$
10. $\mathbf{S}(-4, -1)$; $m = 0$

In Exercises 11–22, write a Cartesian equation in standard form for the line containing the given points **S** and **T**.

11. $\mathbf{S}(1, 5)$ and $\mathbf{T}(3, 2)$
12. $\mathbf{S}(4, 1)$ and $\mathbf{T}(2, 6)$
13. $\mathbf{S}(2, -4)$ and $\mathbf{T}(-1, 2)$
14. $\mathbf{S}(3, -7)$ and $\mathbf{T}(1, -4)$
15. $\mathbf{S}(-6, 2)$ and $\mathbf{T}(7, -2)$
16. $\mathbf{S}(-4, 3)$ and $\mathbf{T}(-2, 1)$
17. $\mathbf{S}(-3, 4)$ and $\mathbf{T}(-1, -1)$
18. $\mathbf{S}(-2, -3)$ and $\mathbf{T}(-5, -4)$
19. $\mathbf{S}(4, -3)$ and $\mathbf{T}(7, -3)$
20. $\mathbf{S}(-3, 2)$ and $\mathbf{T}(6, 2)$
21. $\mathbf{S}(3, 5)$ and $\mathbf{T}(3, -2)$
22. $\mathbf{S}(-4, 6)$ and $\mathbf{T}(-4, -3)$

In Exercises 23–26, the coordinates of the vertices of a triangle are given. In each exercise, find a Cartesian equation in standard form for the lines containing the sides of the triangle.

23. $\mathbf{A}(0, 0)$, $\mathbf{B}(4, 2)$, $\mathbf{C}(-2, 6)$
24. $\mathbf{A}(0, 0)$, $\mathbf{B}(-6, 2)$, $\mathbf{C}(2, 4)$
25. $\mathbf{A}(-1, 3)$, $\mathbf{B}(3, 7)$, $\mathbf{C}(7, 3)$
26. $\mathbf{A}(-3, -3)$, $\mathbf{B}(-3, 3)$, $\mathbf{C}(5, 0)$

27–30. Repeat Exercises 23–26 for the lines containing the medians of the triangle.

31–34. Repeat Exercises 23–26 for the lines containing the altitudes of the triangle.

For Exercises 35 and 36, recall from geometry that the line segment whose endpoints are the midpoints of two sides of a triangle is parallel to the third side and equal in length to one-half of the third side, and that this segment is bisected by a median of the triangle.

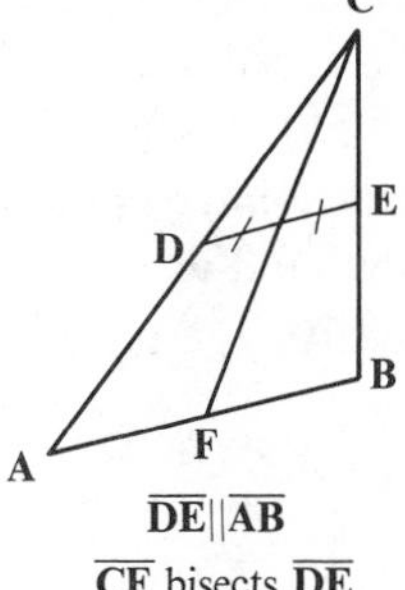

$\overline{\mathbf{DE}} \parallel \overline{\mathbf{AB}}$
$\overline{\mathbf{CF}}$ bisects $\overline{\mathbf{DE}}$

35. Find Cartesian equations in standard form for the lines containing the medians of the triangle having $\mathbf{S}(2, 6)$, $\mathbf{T}(5, 2)$, and $\mathbf{O}(0, 0)$ as midpoints of its sides.

36. Repeat Exercise 35 for the triangle having $\mathbf{R}(0, 5)$, $\mathbf{S}(2, 3)$, and $\mathbf{T}(-3, -3)$ as midpoints of its sides.

* **37.** Use the method of Example 3 on page 56 to find the coordinates of the vertices of the triangle specified in Exercise 35.

* **38.** Use the method of Example 3 on page 56 to find the coordinates of the vertices of the triangle specified in Exercise 36.

* **39.** Use the results of Exercise 37 to find Cartesian equations for the lines containing the altitudes of the triangle specified in Exercise 35.

* **40.** Use the results of Exercise 38 to find Cartesian equations for the lines containing the altitudes of the triangle specified in Exercise 36.

2–6 Slope-Intercept and Intercept Forms of Equations of Lines

You are now familiar with the forms

$$Ax + By + C = 0,$$

$$y - y_1 = m(x - x_1),$$

and

$$y - y_1 = \frac{y_2 - y_1}{x_2 - x_1}(x - x_1)$$

(where $A^2 + B^2 \neq 0$ and $x_2 \neq x_1$) for Cartesian equations of lines. Two other useful forms will be considered in this section.

As suggested by Figure 2–16, if a line $\mathcal{L}$ is not parallel to the y-axis, then it must intersect the y-axis at some point **T** (see Exercise 39, page 69). Since the x-coordinate of any point on the y-axis is 0, the coordinates of **T** must be of the form $(0, b)$, $b \in \mathfrak{R}$. The number b is called the ***y*-intercept** of $\mathcal{L}$. If you replace x_1 with 0 and y_1 with b in Equation (2) on page 70, you have

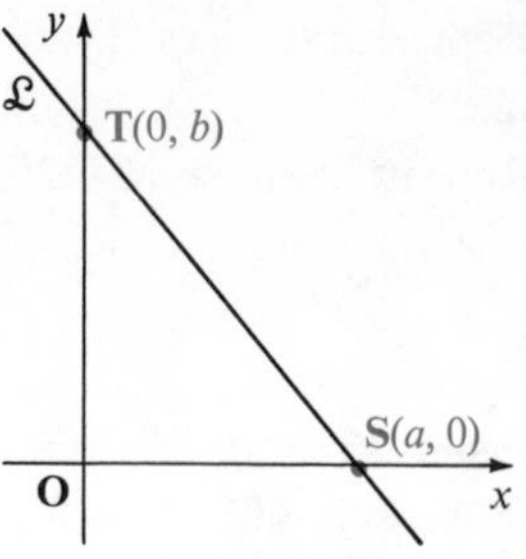

Figure 2–16

$$y - b = m(x - 0),$$

or

$$y = mx + b. \tag{1}$$

This form for the equation of $\mathcal{L}$ is called the **slope-intercept form**, because m is the slope of $\mathcal{L}$ and b is the y-intercept of $\mathcal{L}$.

If the standard Cartesian form for an equation of a nonvertical line $\mathcal{L}$ is solved for y in terms of x, you have

$$\begin{aligned} Ax + By + C &= 0 \quad (B \neq 0), \\ By &= -Ax - C, \\ y &= -\frac{A}{B}x - \frac{C}{B}. \end{aligned}$$

Comparing this equation with Equation (1) above, you can see that

$$m = -\frac{A}{B} \quad \text{and} \quad b = -\frac{C}{B}.$$

This expression for m, of course, agrees with that noted in Section 2–4.

Example 1. Determine the slope and y-intercept of the line with Cartesian equation $3x - 4y = 12$.

Solution: Write the equation equivalently in slope-intercept form. You have

$$\begin{aligned} 3x - 4y &= 12, \\ -4y &= -3x + 12, \\ y &= \tfrac{3}{4}x - 3. \end{aligned}$$

By inspection, the slope is $\frac{3}{4}$ and the y-intercept is -3.

If a line $\mathcal{L}$ is not parallel to the x-axis, then it must intersect the x-axis in some point. This point has coordinates of the form $(a, 0)$, $a \in \mathcal{R}$, because the y-coordinate of every point on the x-axis is 0. (See Figure 2–16.) The number a is called the ***x*-intercept** of $\mathcal{L}$.

If a line $\mathcal{L}$ has x-intercept a and y-intercept b, with $a \neq 0$ and $b \neq 0$, you can use the coordinates of the points of intersection of $\mathcal{L}$ and the coordinate axes, $(a, 0)$ and $(0, b)$, in the two-point form of the equation of a line (Equation (3), page 70) to obtain

$$y - 0 = \frac{b - 0}{0 - a}(x - a),$$

$$-ay = bx - ab,$$

$$bx + ay = ab,$$

or, multiplying both members by $\frac{1}{ab}$,

$$\frac{x}{a} + \frac{y}{b} = 1. \qquad (2)$$

This equation is called the **intercept form** of the equation of $\mathcal{L}$.

Example 2. Write a Cartesian equation in standard form for the line $\mathcal{L}$ with x-intercept 3 and y-intercept -5.

Solution: By using Equation (2) above, with $a = 3$ and $b = -5$, you have

$$\frac{x}{3} + \frac{y}{-5} = 1,$$

from which you obtain

$$-5x + 3y = -15,$$

or

$$5x - 3y - 15 = 0.$$

This is the required Cartesian equation for the line $\mathcal{L}$.

When the equation of a line $\mathcal{L}$ is given in the form $Ax + By + C = 0$, where A, B, and C are all different from 0, you can determine a and b, the x- and y-intercepts of $\mathcal{L}$, by simply substituting 0 for y and x, respectively. You have

$$a = -\frac{C}{A} \quad \text{and} \quad b = -\frac{C}{B}.$$

Exercises 2–6

In Exercises 1–8, write a Cartesian equation in standard form for the line with the given slope m and y-intercept b.

1. $m = 2;\ b = 5$
2. $m = -3;\ b = 2$
3. $m = 4;\ b = -4$
4. $m = -2;\ b = -1$
5. $m = \frac{2}{3};\ b = 0$
6. $m = -\frac{1}{2};\ b = 0$
7. $m = -\frac{4}{5};\ b = 4$
8. $m = 0;\ b = -5$

In Exercises 9–16, find the slope m and y-intercept b of the line with the given Cartesian equation.

9. $3x - 4y + 8 = 0$
10. $2x - 5y + 10 = 0$
11. $5x + 4y - 16 = 0$
12. $x + 3y + 9 = 0$
13. $7x - 2y - 4 = 0$
14. $3x + 4y + 5 = 0$
15. $3y + 6 = 0$
16. $5y - 4 = 0$

In Exercises 17–24, write a Cartesian equation in standard form for the line with the given x-intercept a and y-intercept b.

17. $a = 2,\ b = 3$
18. $a = 5,\ b = 1$
19. $a = -3,\ b = 2$
20. $a = -4,\ b = 4$
21. $a = 7,\ b = -2$
22. $a = 5,\ b = -3$
23. $a = -1,\ b = -2$
24. $a = -5,\ b = -4$

In Exercises 25–32, find the x-intercept a and the y-intercept b of the line with the given Cartesian equation.

25. $3x + 4y - 12 = 0$
26. $2x + 3y - 18 = 0$
27. $5x - 3y + 15 = 0$
28. $x - 7y - 14 = 0$
29. $5x - 2y + 10 = 0$
30. $8x + 3y - 12 = 0$
31. $3x - 5y + 8 = 0$
32. $4x + 6y + 7 = 0$

In Exercises 33–40, find a Cartesian equation in standard form for the line with the given characteristics:

33. Containing the origin, and perpendicular to the line with x- and y-intercepts 5 and $-\frac{2}{3}$, respectively.
34. Containing the point $\mathbf{S}(3, -2)$, and perpendicular to the line through $\mathbf{T}(1, 1)$ with slope $\frac{3}{2}$.
35. The sum of the intercepts is 7, and the slope is $-\frac{11}{3}$. (*Hint:* Let $\frac{b}{a} = -m$.)
36. The sum of the intercepts is 2, and the slope is $\frac{9}{5}$.
37. The sum of the intercepts is 0, and the line contains the point $\mathbf{S}(2, 4)$.

38. The sum of the intercepts is -1, and the line contains the point $S(2, 2)$. [Two solutions]

* **39.** The product of the intercepts is 12, and the line contains the point $S(3, 1)$.

* **40.** The product of the intercepts is 6, and the line contains the point $S(\frac{3}{2}, -3)$. [Two solutions]

* **41.** For what value of a are the lines with equations

$$\frac{x}{a} + \frac{y}{-2} = 1 \quad \text{and} \quad \frac{x}{-4} + \frac{y}{-2} = 1$$

perpendicular?

* **42.** For what value of b are the lines with equations

$$\frac{x}{2} + \frac{y}{b} = 1 \qquad \text{and} \qquad \frac{x}{2} - \frac{y}{4} = 1$$

perpendicular?

2–7 Symmetric Form of an Equation of a Line

The components h and k of a direction vector (h, k) for a line $\mathcal{L}$ are called **direction numbers** for $\mathcal{L}$. These numbers play an important role in another useful form of an equation of a line which is not parallel to either coordinate axis.

Consider a line $\mathcal{L}$ through $S(x_1, y_1)$ with a direction vector $\mathbf{v} = (h, k)$. A parametric vector equation for $\mathcal{L}$ is

$$(x, y) = (x_1, y_1) + r(h, k), \qquad r \in \mathcal{R},$$

from which you obtain the parametric Cartesian equations

$$x = x_1 + rh \quad \text{and} \quad y = y_1 + rk. \tag{1}$$

If $h \neq 0$ and $k \neq 0$, then you can solve each of these equations for r to obtain

$$\frac{x - x_1}{h} = r = \frac{y - y_1}{k},$$

from which you have

$$\frac{x - x_1}{h} = \frac{y - y_1}{k}. \tag{2}$$

Equation (2) is called the **symmetric form** of the equation for $\mathcal{L}$.

If either h or k is 0, then $\mathcal{L}$ is parallel to one of the coordinate axes, that is, $\mathcal{L}$ is either vertical or horizontal. In these cases, the symmetric form does not apply. However, from Equation (1) above you obtain $x = x_1$ or $y = y_1$, respectively, as an equation for $\mathcal{L}$.

A pair of direction numbers for $\mathcal{L}$ can be determined by identifying any two distinct points $\mathbf{S}(x_1, y_1)$ and $\mathbf{T}(x_2, y_2)$ on $\mathcal{L}$ and noting from Equation (1) that

$$x_2 = x_1 + ch, \qquad y_2 = y_1 + ck,$$

where c is a constant. Accordingly, you have

$$(x_2 - x_1, y_2 - y_1) = c(h, k),$$

and therefore $x_2 - x_1$ and $y_2 - y_1$ are a pair of direction numbers for $\mathcal{L}$.

Example 1. Write a Cartesian equation in symmetric form for the line $\mathcal{L}$ through the points $\mathbf{S}(3, -3)$ and $\mathbf{T}(6, 1)$.

Solution: A pair of direction numbers are

$$h = x_2 - x_1 = 6 - 3 = 3$$

and

$$k = y_2 - y_1 = 1 - (-3) = 4.$$

Then, replacing x_1 and y_1 in Equation (2), page 77, with the respective coordinates of $\mathbf{S}(3, -3)$, or of $\mathbf{T}(6, 1)$, you have

$$\frac{x-3}{3} = \frac{y+3}{4} \quad \text{or} \quad \frac{x-6}{3} = \frac{y-1}{4}.$$

You can verify that each of these equations actually represents the same line by reducing each equation to the standard Cartesian form.

Given a Cartesian equation in the standard form $Ax + By + C = 0$ for a line $\mathcal{L}$, if A and $B \neq 0$, you can write an equivalent equation in symmetric form by simply identifying a point $\mathbf{S}(x_1, y_1)$ on the graph of $Ax + By + C = 0$, and then observing that $\mathbf{v} = (-B, A)$ is a direction vector for the graph. Hence, in symmetric form, you have

$$\frac{x - x_1}{-B} = \frac{y - y_1}{A}$$

as an equation of $\mathcal{L}$.

Example 2. Write a Cartesian equation in symmetric form for the graph of

$$3x - 4y + 7 = 0.$$

Solution: First find a solution of $3x - 4y + 7 = 0$ by assigning a value, say 1, to x and solving for y. You have

$$\begin{aligned} 3(1) - 4y + 7 &= 0, \\ -4y &= -10, \\ y &= \tfrac{5}{2}. \end{aligned}$$

Thus $\mathbf{S}(1, \frac{5}{2})$ is a point on the graph of the given equation. Since $A = 3$ and $B = -4$, a direction vector for the graph is $\mathbf{v} = (4, 3)$. Therefore an equation in symmetric form is

$$\frac{x - 1}{4} = \frac{y - \frac{5}{2}}{3}.$$

A direction vector for a line can be used to specify the *direction angle* of the line. Recall that the direction angle of a vector $\mathbf{v} = (h, k)$ satisfies the conditions $0 \leq m^\circ(\theta) < 360$. If $\mathbf{v}$ is a direction vector of a line $\mathcal{L}$, then, as noted earlier, every nonzero scalar multiple of $\mathbf{v}$ is also a direction vector of $\mathcal{L}$. In particular, $-\mathbf{v}$ is a direction vector of $\mathcal{L}$, and $-\mathbf{v}$ has the direction opposite to that of $\mathbf{v}$. Accordingly, either the direction angle of $\mathbf{v}$ or the direction angle of $-\mathbf{v}$ satisfies the condition $0 \leq m^\circ(\theta) < 180$, and the angle satisfying this condition is taken to be the **direction angle** θ of the line $\mathcal{L}$. (See Figure 2–17.)

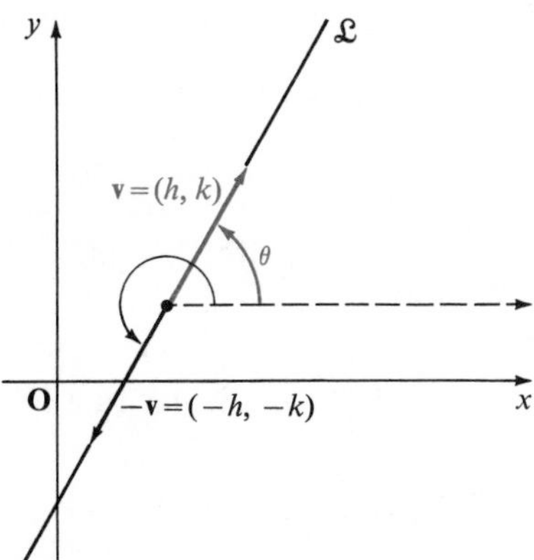

Figure 2–17

If $h = 0$, then $m^\circ(\theta) = 90$. Otherwise, since $\dfrac{-k}{-h} = \dfrac{k}{h}$, θ can be determined from the equation

$$\tan \theta = \frac{k}{h}, \qquad 0 \leq m^\circ(\theta) < 180.$$

Notice that $\tan \theta$ is simply the slope of $\mathcal{L}$.

A direction vector for the line $\mathcal{L}$ with equation $Ax + By + C = 0$, $A^2 + B^2 \neq 0$, is $(-B, A)$. Accordingly, if $B \neq 0$, then the direction angle θ of $\mathcal{L}$ is given by

$$\tan \theta = -\frac{A}{B}, \qquad 0 \leq m^\circ(\theta) < 180.$$

Exercises 2–7

In Exercises 1–12, write a Cartesian equation in symmetric form for the line $\mathcal{L}$ containing the given points **S** and **T**, and find the direction angle of $\mathcal{L}$ to the nearest degree.

1. $\mathbf{S}(5, 2)$ and $\mathbf{T}(3, 6)$
2. $\mathbf{S}(6, 4)$ and $\mathbf{T}(2, 3)$
3. $\mathbf{S}(5, 9)$ and $\mathbf{T}(3, -2)$
4. $\mathbf{S}(-1, 3)$ and $\mathbf{T}(4, -3)$
5. $\mathbf{S}(-3, 4)$ and $\mathbf{T}(-1, -3)$
6. $\mathbf{S}(2, -4)$ and $\mathbf{T}(-3, -2)$
7. $\mathbf{S}(4, -5)$ and $\mathbf{T}(2, -1)$
8. $\mathbf{S}(3, -8)$ and $\mathbf{T}(-3, -7)$
9. $\mathbf{S}(-1, -3)$ and $\mathbf{T}(2, 3)$
10. $\mathbf{S}(-6, -4)$ and $\mathbf{T}(4, -1)$
11. $\mathbf{S}(-2, -3)$ and $\mathbf{T}(-3, -5)$
12. $\mathbf{S}(-1, -7)$ and $\mathbf{T}(7, -1)$

In Exercises 13–20, find a Cartesian equation in symmetric form equivalent to the given equation.

13. $2x - y + 4 = 0$
14. $x + 3y - 6 = 0$
15. $2x + 5y - 10 = 0$
16. $3x + 4y - 24 = 0$
17. $2x + 3y - 7 = 0$
18. $7x + 3y - 5 = 0$
19. $2x - 7y + 3 = 0$
20. $5x - 9y + 45 = 0$

In Exercises 21–24, find a Cartesian equation in standard form for the line $\mathcal{L}$ with the given characteristics.

21. Containing $\mathbf{S}(-5, 3)$, and having a direction angle of 60°.
22. Containing $\mathbf{S}(2, 6)$, and having a direction angle of 45°.
23. Containing $\mathbf{S}(3, -2)$, and perpendicular to a line with a direction angle of 30°.
24. Containing $\mathbf{S}(-4, -3)$, and perpendicular to a line with a direction angle of 135°.

* 25. Show that the lines with symmetric equations

$$\frac{x - x_1}{h} = \frac{y - y_1}{k} \quad \text{and} \quad \frac{x - x_1}{k} = -\frac{y - y_1}{h}$$

are perpendicular.

* 26. Show that if the lines with symmetric equations

$$\frac{x - x_1}{h_1} = \frac{y - y_1}{k_1} \quad \text{and} \quad \frac{x - x_2}{h_2} = \frac{y - y_2}{k_2}$$

are parallel, then $h_1k_2 = h_2k_1$.

* 27. Show that the x- and y-intercepts of the graph of the symmetric equation

$$\frac{x - x_1}{h} = \frac{y - y_1}{k}$$

are

$$\frac{kx_1 - hy_1}{k} \quad \text{and} \quad \frac{hy_1 - kx_1}{h},$$

respectively.

Chapter Summary

1. Lines in the plane have both **vector** and **Cartesian equations**. The parametric vector equation

$$\mathbf{u} = \mathbf{s} + r(\mathbf{t} - \mathbf{s}), \qquad r \in \mathcal{R},$$

specifies the line in the plane containing the (distinct) points **S** and **T**. The variable r in the equation is a **parameter**. If the replacement set for r is a closed interval, then the graph of the equation is a line segment. A system of **parametric Cartesian equations** for the line containing the points $\mathbf{S}(x_1, y_1)$ and $\mathbf{T}(x_2, y_2)$ is

$$x = x_1 + r(x_2 - x_1) \quad \text{and} \quad y = y_1 + r(y_2 - y_1).$$

2. A point **U** lies on the line $\mathcal{L}$ containing the point $\mathbf{S}(x_1, y_1)$ and having direction vector $\mathbf{v} = (h, k)$ if and only if $\mathbf{u} - \mathbf{s}$ is parallel to $\mathbf{v}$. A parametric vector equation for $\mathcal{L}$ is $\mathbf{u} = \mathbf{s} + r\mathbf{v}$, and a system of parametric Cartesian equations for $\mathcal{L}$ is $x = x_1 + rh$, $y = y_1 + rk$. A point $\mathbf{U}(x, y)$ lies on $\mathcal{L}$ if and only if $(\mathbf{u} - \mathbf{s}) \cdot \mathbf{v_p} = 0$.
3. If (h, k), with $h \neq 0$, is a direction vector of a line $\mathcal{L}$, then the slope m of $\mathcal{L}$ is $\frac{k}{h}$. This number is also the ratio of the rise to the run of any segment on $\mathcal{L}$. For the line $\mathcal{L}$ containing the points $\mathbf{S}(x_1, y_1)$ and $\mathbf{T}(x_2, y_2)$, with $x_2 \neq x_1$, you have

$$m = \frac{y_2 - y_1}{x_2 - x_1}.$$

4. Nonvertical parallel lines have the same slope. The product of the slopes of perpendicular lines, neither of which is vertical, is -1.
5. A nonzero vector perpendicular to a direction vector **v** of a line $\mathcal{L}$ is a **normal vector** of $\mathcal{L}$. If (A, B), where $A^2 + B^2 \neq 0$, is a normal vector of $\mathcal{L}$, then a **Cartesian equation in standard form** for $\mathcal{L}$ is

$$Ax + By + C = 0,$$

where C is a suitable constant. A line with equation $Ax + By + C' = 0$ is parallel to $\mathcal{L}$, and a line with equation $-Bx + Ay + C' = 0$ is perpendicular to $\mathcal{L}$. The slope of $\mathcal{L}$ is $-\frac{A}{B}$ if $B \neq 0$.
6. The **point-slope form** of the equation of the line $\mathcal{L}$ containing the point $\mathbf{S}(x_1, y_1)$ and having slope m is

$$y - y_1 = m(x - x_1).$$

The **two-point form** of the equation of the nonvertical line $\mathcal{L}$ containing the points $\mathbf{S}(x_1, y_1)$ and $\mathbf{T}(x_2, y_2)$ is

$$y - y_1 = \frac{y_2 - y_1}{x_2 - x_1}(x - x_1).$$

7. The **slope-intercept form** of the equation of the line $\mathcal{L}$ having slope m and y-intercept b is

$$y = mx + b.$$

The **intercept form** of the equation of the line $\mathcal{L}$ with x- and y-intercepts a and b, respectively, with $a \neq 0$ and $b \neq 0$, is

$$\frac{x}{a} + \frac{y}{b} = 1.$$

8. If $\mathbf{v} = (h, k)$ is a direction vector of a line $\mathcal{L}$, then h and k are called **direction numbers** for $\mathcal{L}$. The direction angle θ of $\mathcal{L}$ satisfies $0 \leq m^\circ(\theta) < 180$; if $h \neq 0$, then $m^\circ(\theta)$ can be determined from the equation

$$\tan\theta = \frac{k}{h} = m, \qquad 0 \leq m^\circ(\theta) < 180,$$

where m is the slope of $\mathcal{L}$. If $h \neq 0$ and $k \neq 0$, and $\mathcal{L}$ contains the point $\mathbf{S}(x_1, y_1)$, then the **symmetric form** of the equation for $\mathcal{L}$ is

$$\frac{x - x_1}{h} = \frac{y - y_1}{k}.$$

Chapter Review Exercises

1. Find a parametric vector equation and a system of parametric Cartesian equations for the line containing the points $\mathbf{S}(3, -2)$ and $\mathbf{T}(-1, 5)$.
2. Find the coordinates of the midpoint of the segment with endpoints $\mathbf{S}(1, 7)$ and $\mathbf{T}(3, -9)$.
3. Write a parametric vector equation for the line containing $\mathbf{S}(5, -1)$ and having $\mathbf{v} = (3, 1)$ as a direction vector.
4. Determine whether or not the point $\mathbf{T}(1, -8)$ lies on the line with parametric vector equation $\mathbf{u} = (3, 2) + r(1, 5)$, $r \in \mathcal{R}$.
5. Write a parametric vector equation for the line containing the point $\mathbf{S}(2, -5)$ and having slope $-\frac{2}{3}$.
6. Write a parametric vector equation for the line which contains the point $\mathbf{S}(-5, 7)$ and is perpendicular to the line with parametric vector equation $\mathbf{u} = (1, 4) + r(-2, 3)$, $r \in \mathcal{R}$.
7. Find a Cartesian equation in standard form for the line containing the point $\mathbf{S}(4, -1)$ and having direction vector $\mathbf{v} = (1, 3)$.
8. Write a Cartesian equation in standard form for the line containing the points $\mathbf{S}(-5, 7)$ and $\mathbf{T}(0, 3)$.
9. Find the slope and y-intercept of the graph of $4x - 2y + 7 = 0$.
10. Write an equation in symmetric form for the line containing the points $\mathbf{S}(7, 1)$ and $\mathbf{T}(-1, 9)$.

Straightedge and Compass Constructions

The ancient Greek mathematicians asked and extensively investigated the question of determining how simple geometric figures can be constructed by means of straightedge and compass alone.

It should be understood, in the first place, just what the nature of this problem is. A straightedge is an instrument by means of which you can determine a line, or line segment of indefinite length, passing through any two given points in the plane—nothing more. Similarly, a compass is an instrument by means of which you can determine a circle, or arc of a circle, having center at any given point and radius the distance between any two

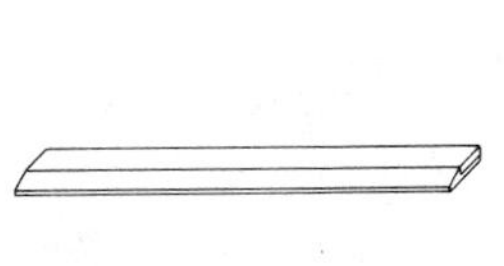

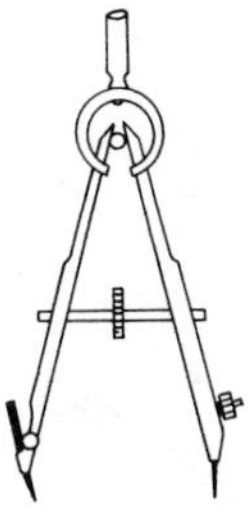

given points in the plane—nothing more. The "line" and "circle" are not the marks you make on a piece of paper, which are three-dimensional sets of molecules with their whirling electrons, but are the ideal line and circle of geometry.

The construction begins with a given *initial configuration* consisting of a finite number of points, lines, and circles. It must be completed in a *finite number of steps*, and it must be *mathematically exact*.

From the foregoing criteria, it might appear that physically the problem of straightedge-and-compass construction is not a particularly practical one. By means of a ruler and protractor, or other more sophisticated instrumenta-

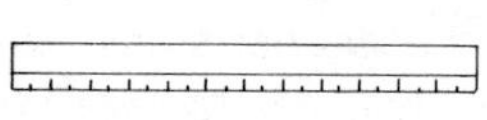

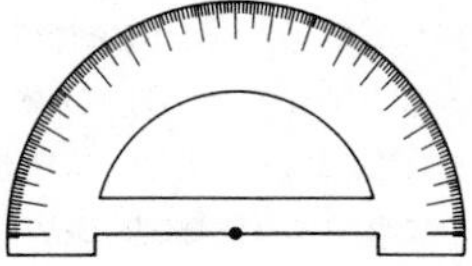

tion, you can determine the length of line segments and the measure of angles quite accurately. The problem nevertheless is historically an important one, for it helped mathematicians to focus their attention on basic questions. Knowledge of basic principles of mathematics and science is essential to sound practical scientific progress.

A few of the familiar simple straightedge-and-compass constructions are illustrated below.

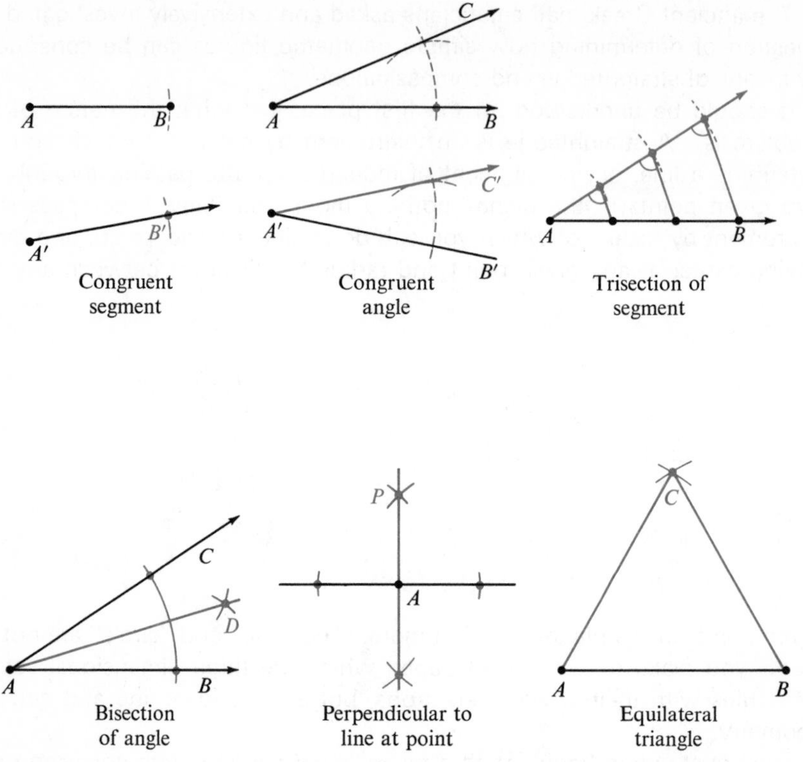

Congruent segment

Congruent angle

Trisection of segment

Bisection of angle

Perpendicular to line at point

Equilateral triangle

The initial configurations with which you will be especially concerned in this essay and in the one at the end of Chapter 3, in particular, the circle and the angle of measure 60°, can themselves be constructed by means of straightedge and compass from the simple configuration consisting of just two points. Thus a circle can be drawn when its center and one of its points are given; and an angle of measure 60° can be obtained by constructing an equilateral triangle, as illustrated above. Accordingly, any construction that can be made starting with one of these configurations can also be made starting with a configuration consisting of just two points. For brevity, it will be said that a configuration *can be constructed*, or that the configuration *is constructible*, if it can be constructed by means of straightedge and compass starting with just two points.

Suppose that, starting with a configuration consisting of just two points, A and B, and using $d(A, B)$ as the unit of length, you can specify a straightedge-and-compass construction of a segment of length p. Then you say that *the number p is* (straightedge-and-compass) *constructible*.

Each counting number is constructible, as indicated in (*a*) below. If the positive number p is constructible, then so are $\frac{1}{p}$ and $\sqrt{p}$, as illustrated in (*b*) and (*c*), respectively.

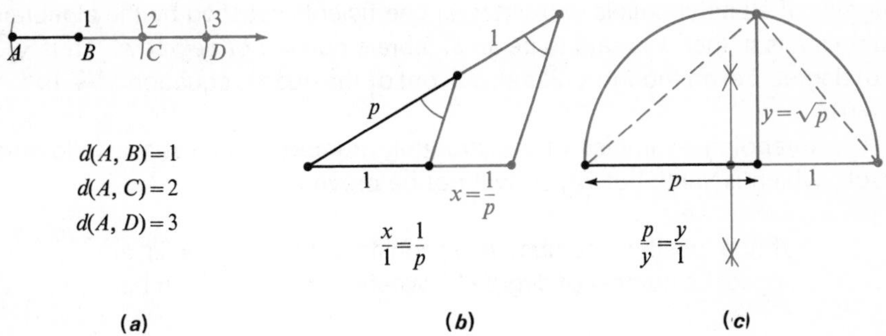

If the positive numbers p and q are constructible, then so are $p + q$, $p - q$ (if $p > q$), pq, and $\frac{p}{q}$:

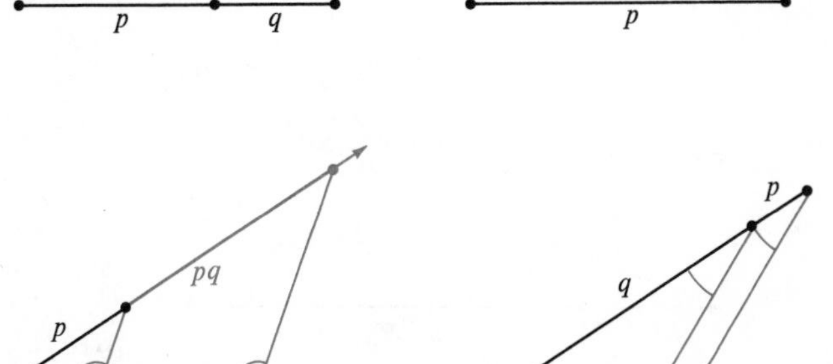

It follows that such numbers as

$$x = \frac{\sqrt{5} - 1}{2}, \ y = \sqrt{\sqrt{\tfrac{1}{7}} + 2}, \quad \text{and} \quad z = \sqrt{3 + \sqrt{9 - \sqrt{6}}}$$

are constructible.

Notice, for example, that the second of the constructible numbers given above satisfies

$$y^2 = \sqrt{\tfrac{1}{7}} + 2,$$
$$y^2 - 2 = \sqrt{\tfrac{1}{7}},$$
$$y^4 - 4y^2 + 4 = \tfrac{1}{7},$$
$$7y^4 - 28y^2 + 27 = 0.$$

If the number k is a root of a polynomial equation with integral coefficients, then k is said to be an *algebraic number*. For example, the rational number $\frac{2}{3}$ and the irrational number $\sqrt{3}$ are algebraic since they are roots of the equations $3x - 2 = 0$ and $x^2 - 3 = 0$, respectively. If the least of the degrees of all polynomials with integral coefficients satisfied by the algebraic number k is n, then k is said to be an algebraic number *of degree* n. Thus $\sqrt{3}$ is of degree 2 even though $\sqrt{3}$ is also a root of the quartic equation $x^4 - 6x^2 + 9 = 0$.

The foregoing examples of constructible numbers suggest the following result, which is valid, but which will not be proved here:

> If the positive number k is constructible, then k is an algebraic number of degree 2^n, where n is a whole number.

The number π is a *transcendental*, or nonalgebraic, number; that is, there is no polynomial equation with integral coefficients of which π is a root. This striking result was established only in 1882, by the German mathematician Carl Louis Ferdinand von Lindemann (1852–1939). It follows that π is not a constructible number. The diagram below suggests a valid (but hardly practical) construction of a close approximation to π, but not of π itself.

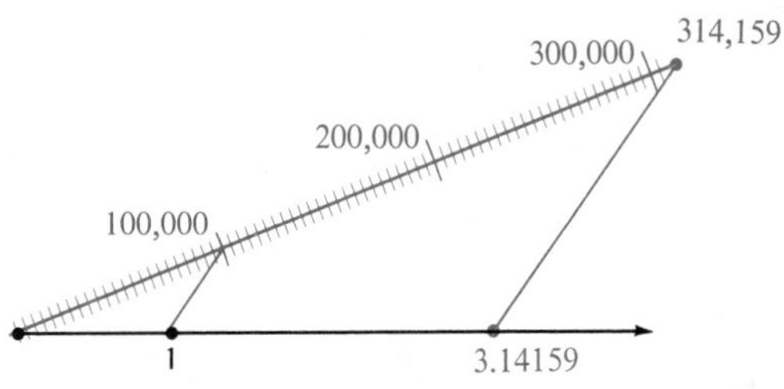

The result of Lindemann given above finally settled one of the three famous straightedge-and-compass construction problems of the ancient Greeks,

that of "squaring the circle," or constructing a square bounding a region having the same area as that of a given circular region. If the radius of the given circle is taken as the unit of length, then the area of the circular region is π, and the desired square must have sides of length $\sqrt{\pi}$. If $\sqrt{\pi}$ were constructible, then π would be constructible. You have seen, however, that this is not the case. Hence, there can be no straightedge-and-compass construction for squaring the circle.

Another of the three famous construction problems of the ancient Greeks was that of "duplicating the cube," or constructing a cube whose volume is

exactly twice that of a given cube. If the length of an edge of the cube is taken as the unit length, then the desired cube must have edges of length $\sqrt[3]{2}$. Since $\sqrt[3]{2}$ is a root of the equation $x^3 - 2 = 0$, $\sqrt[3]{2}$ is an algebraic number, and $\sqrt[3]{2}$ is of degree at most 3. It can be shown fairly easily that $\sqrt[3]{2}$ is not a root of a linear or quadratic equation with integral coefficients, that is, that $\sqrt[3]{2}$ is not of degree 1 or 2. Hence $\sqrt[3]{2}$ is an algebraic number of degree 3. Since 3 is not of the form 2^n, where n is a whole number, $\sqrt[3]{2}$ is not constructible. Therefore there can be no straightedge-and-compass construction for duplicating the cube.

The third famous straightedge-and-compass problem of antiquity was that of trisecting any angle that might be given. This problem will be discussed in the essay at the end of Chapter 3.

Chapter 3

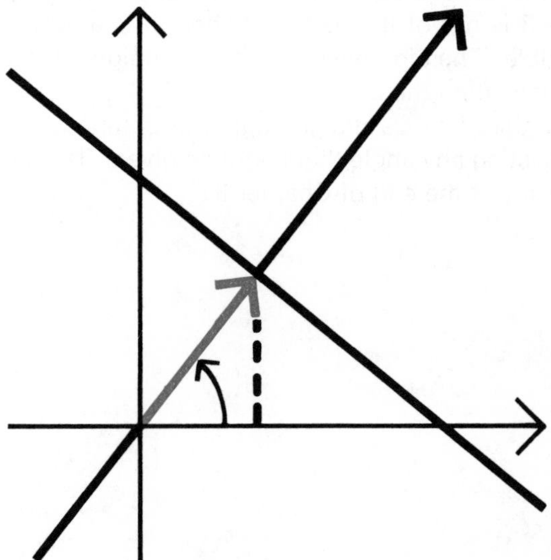

In this chapter, the analytic methods developed earlier are used to solve a variety of different types of problems. Normal and determinant forms for equations of lines are derived and analytic proofs of geometric theorems discussed.

Applications of Lines

Relationships involving Lines

3–1 Distance between a Point and a Line

In plane geometry, you learned that the distance between a point **S** and a line $\mathcal{L}$, denoted by $d(\mathbf{S}, \mathcal{L})$, is defined to be the length of the perpendicular segment between **S** and $\mathcal{L}$ (Figure 3–1).

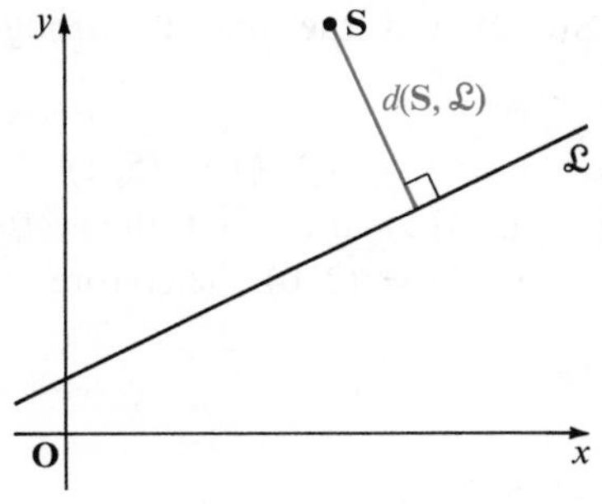

Figure 3–1

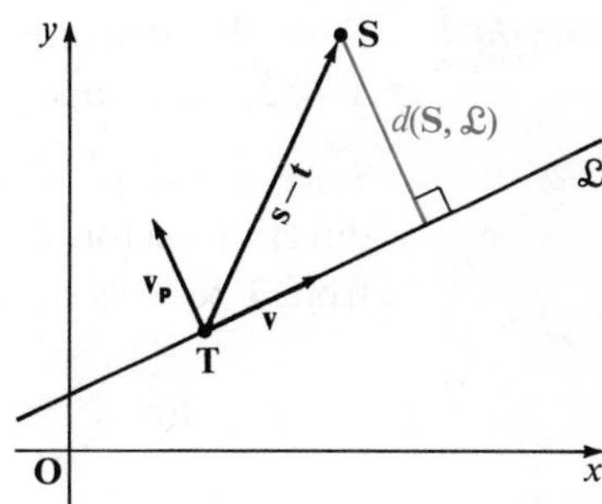

Figure 3–2

You can determine $d(\mathbf{S}, \mathcal{L})$ by vector methods provided you know the coordinates of **S**, the coordinates of any point **T** on $\mathcal{L}$, and a direction vector **v** of $\mathcal{L}$. To do this, notice from Figure 3–2 that the required distance is the absolute value of the scalar component of $\mathbf{s} - \mathbf{t}$ parallel to $\mathbf{v_p}$. That is,

$$d(\mathbf{S}, \mathcal{L}) = |\text{Comp}_{\mathbf{v_p}} (\mathbf{s} - \mathbf{t})|,$$

or (recall page 37 and Exercise 37, page 30)

$$d(\mathbf{S}, \mathcal{L}) = \frac{|(\mathbf{s} - \mathbf{t}) \cdot \mathbf{v_p}|}{\|\mathbf{v_p}\|}. \qquad (1)$$

The distance between **S** and $\mathfrak{L}$ does not depend on the selection of a particular point **T** on $\mathfrak{L}$. To see that this is true, consider any two points $\mathbf{T}_1$ and $\mathbf{T}_2$ on $\mathfrak{L}$. From Figure 3–3, you can see that

$$\mathbf{s} - \mathbf{t}_1 = (\mathbf{t}_2 - \mathbf{t}_1) + (\mathbf{s} - \mathbf{t}_2).$$

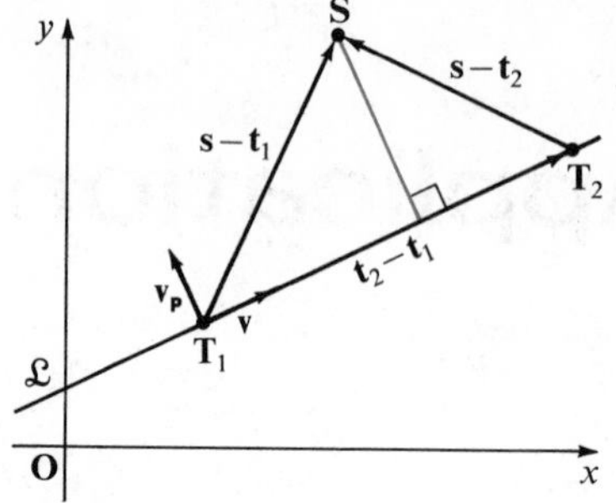

Figure 3–3

Then, taking the dot product of each member of this equation and $\mathbf{v_p}$, you obtain

$$(\mathbf{s} - \mathbf{t}_1)\cdot\mathbf{v_p} = (\mathbf{t}_2 - \mathbf{t}_1)\cdot\mathbf{v_p} + (\mathbf{s} - \mathbf{t}_2)\cdot\mathbf{v_p} = 0 + (\mathbf{s} - \mathbf{t}_2)\cdot\mathbf{v_p}.$$

Therefore,

$$\frac{|(\mathbf{s} - \mathbf{t}_1)\cdot\mathbf{v_p}|}{\|\mathbf{v_p}\|} = \frac{|(\mathbf{s} - \mathbf{t}_2)\cdot\mathbf{v_p}|}{\|\mathbf{v_p}\|}.$$

Example 1. Find the distance between **S**(6, 2) and the line $\mathfrak{L}$ through **T**(4, 2) with slope $\frac{1}{5}$.

Solution: Since the slope of $\mathfrak{L}$ is $\frac{1}{5}$, the vector $\mathbf{v} = 5(1, \frac{1}{5}) = (5, 1)$ is a direction vector of $\mathfrak{L}$. Then $\mathbf{v_p} = (-1, 5)$, and $\mathbf{s} - \mathbf{t}$, the vector from **T** to **S**, is equal to $(6, 2) - (4, 2) = (2, 0)$. Therefore

$$d(\mathbf{S}, \mathfrak{L}) = \frac{|(\mathbf{s} - \mathbf{t})\cdot\mathbf{v_p}|}{\|\mathbf{v_p}\|}$$

$$= \frac{|(2, 0)\cdot(-1, 5)|}{\sqrt{(-1)^2 + 5^2}} = \frac{|-2|}{\sqrt{26}} = \frac{2}{\sqrt{26}}.$$

To find $d(\mathbf{S}, \mathfrak{L})$ when the equation for $\mathfrak{L}$ is given in the standard Cartesian form,

$$Ax + By + C = 0, \qquad (2)$$

you can begin by transforming the equation into a special form that was systematically employed by the German geometer Ludwig Otto Hesse (1811–1874) and called by him the *normal form.* To derive the normal form, we shall use the concept of a *directed line:* If **v** is any direction vector of a line $\mathfrak{M}$ in the plane, then $\mathfrak{M}$ may be thought of as having either of two directions: that of **v** or that of $-\mathbf{v}$. When a direction is assigned to a line, then the line is said to be a **directed line.** If **S** and **T** are any two points on a directed line $\mathfrak{M}$, then $l(\mathbf{S}, \mathbf{T})$, the **directed distance** from **S** to **T**, is equal to $d(\mathbf{S}, \mathbf{T})$ or $-d(\mathbf{S}, \mathbf{T})$ according as $\mathfrak{M}$

and the geometric vector from **S** to **T** have the same or opposite directions. Thus the directed distance from **S** to **T** is the scalar component of the vector **t** − **s** parallel to $\mathfrak{M}$.

Now, for a given line $\mathcal{L}$, let $\mathfrak{M}$ be a *directed* line through the origin **O** perpendicular to $\mathcal{L}$; let α be the angle from the positive x-direction to the positive direction on $\mathfrak{M}$, $0 \leq m^\circ(\alpha) < 360$; let **P** be the point of intersection of $\mathcal{L}$ and $\mathfrak{M}$; and let p be the *directed distance* from **O** to **P**, as indicated in Figure 3–4.

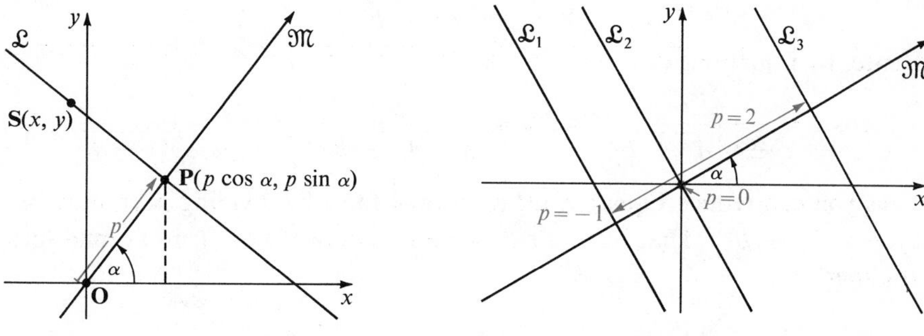

Figure 3–4 **Figure 3–5**

(Note that if $\mathfrak{M}$ and the vector from **O** to **P** have the same direction, then the directed distance from **O** to **P** is positive and the directed distance from **P** to **O** is negative.) From Figure 3–4, you can see that $\mathcal{L}$ is determined by α and p.

Figure 3–5 shows lines $\mathcal{L}_1$, $\mathcal{L}_2$, and $\mathcal{L}_3$ with $m^\circ(\alpha) = 30$ and $p = -1$, 0, and 2, respectively. For these lines, you could equally well choose $m^\circ(\alpha) = 210$ and $p = 1$, 0, and -2, respectively.

By the definition of $\cos \alpha$ and $\sin \alpha$, the coordinates of **P** in Figure 3–4 are $(p \cos \alpha, p \sin \alpha)$. Since the point **O** is also on $\mathfrak{M}$, the vector from **O** to **P**, $(p \cos \alpha, p \sin \alpha)$, is parallel to $\mathfrak{M}$; hence the vector $(\cos \alpha, \sin \alpha)$ is also parallel to $\mathfrak{M}$ and is thus a direction vector of $\mathfrak{M}$. Any point $\mathbf{S}(x, y)$ in the plane is on $\mathcal{L}$ if and only if the vector

$$\mathbf{s} - \mathbf{p} = (x - p \cos \alpha, y - p \sin \alpha)$$

is perpendicular to $\mathfrak{M}$; that is, if and only if

$$\begin{aligned}(x - p \cos \alpha, y - p \sin \alpha) \cdot (\cos \alpha, \sin \alpha) &= 0,\\ (x \cos \alpha - p \cos^2 \alpha) + (y \sin \alpha - p \sin^2 \alpha) &= 0,\\ x \cos \alpha + y \sin \alpha - p(\cos^2 \alpha + \sin^2 \alpha) &= 0,\end{aligned}$$

or

$$x \cos \alpha + y \sin \alpha - p = 0. \qquad (3)$$

Equation (3) is the **normal form** of the equation for $\mathcal{L}$.

If Equations (2) and (3) represent the same line $\mathcal{L}$, then the coefficients in the two equations must be proportional:

$$A = k \cos \alpha, \qquad B = k \sin \alpha, \qquad C = -kp, \tag{4}$$

where k is a constant. To determine k, notice that

$$A^2 + B^2 = k^2 \cos^2 \alpha + k^2 \sin^2 \alpha = k^2(\cos^2 \alpha + \sin^2 \alpha) = k^2,$$

so that

$$k = \pm\sqrt{A^2 + B^2}.$$

Hence, by Equations (4),

$$\cos \alpha = \frac{A}{\pm\sqrt{A^2 + B^2}}, \quad \sin \alpha = \frac{B}{\pm\sqrt{A^2 + B^2}}, \quad p = \frac{-C}{\pm\sqrt{A^2 + B^2}}. \tag{4'}$$

Thus you can reduce Equation (2) to normal form by dividing both members by $\pm\sqrt{A^2 + B^2}$. That is, you can write Equation (2) for $\mathcal{L}$ in normal form as either

$$\frac{A}{\sqrt{A^2 + B^2}}x + \frac{B}{\sqrt{A^2 + B^2}}y + \frac{C}{\sqrt{A^2 + B^2}} = 0 \tag{5}$$

or

$$\frac{A}{-\sqrt{A^2 + B^2}}x + \frac{B}{-\sqrt{A^2 + B^2}}y + \frac{C}{-\sqrt{A^2 + B^2}} = 0. \tag{5'}$$

Example 2. Write an equation in normal form for the line $\mathcal{L}$ with Cartesian equation

$$2x + 2y + 3 = 0,$$

and determine α, $0 \leq m^\circ(\alpha) < 360$, and p, first **(a)** with $p > 0$ and then **(b)** with $p < 0$. Draw a sketch in each case.

Solution: Since $A = 2$ and $B = 2$, you have $k = \pm\sqrt{2^2 + 2^2} = \pm 2\sqrt{2}$.

(a) Since $C = 3$, which is positive, and $C = -kp$, to get $p > 0$ you choose $k = -2\sqrt{2}$. Thus, the normal form is

$$\frac{2}{-2\sqrt{2}}x + \frac{2}{-2\sqrt{2}}y + \frac{3}{-2\sqrt{2}} = 0,$$

or

$$-\frac{1}{\sqrt{2}}x - \frac{1}{\sqrt{2}}y - \frac{3}{2\sqrt{2}} = 0.$$

Since $\cos \alpha = -\frac{1}{\sqrt{2}}$, $\sin \alpha = -\frac{1}{\sqrt{2}}$, and $-p = -\frac{3}{2\sqrt{2}}$,

you have $m^\circ(\alpha) = 225$ and $p = \frac{3}{2\sqrt{2}} = \frac{3}{4}\sqrt{2} \doteq 1.06.$

(b) To get $p < 0$, you choose $k = 2\sqrt{2}$ and have the normal form

$$\frac{1}{\sqrt{2}}x + \frac{1}{\sqrt{2}}y + \frac{3}{2\sqrt{2}} = 0.$$

Since $\cos\alpha = \dfrac{1}{\sqrt{2}}$, $\sin\alpha = \dfrac{1}{\sqrt{2}}$, and $-p = \dfrac{3}{2\sqrt{2}}$, you have $m°(\alpha) = 45$ and $p = -\dfrac{3}{2\sqrt{2}} = -\frac{3}{4}\sqrt{2} \doteq -1.06$.

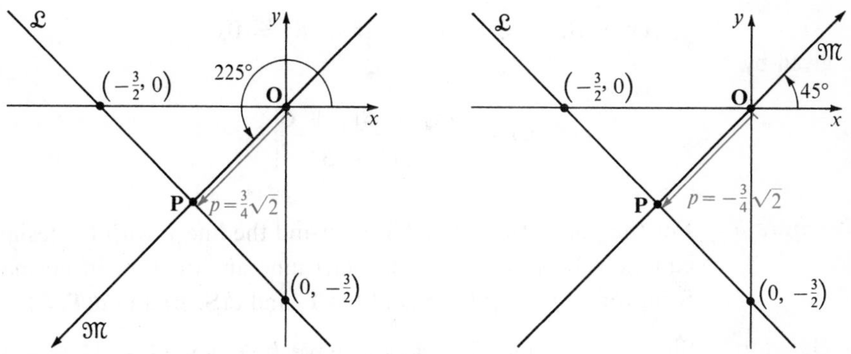

To determine the distance between the point $\mathbf{S}(x_1, y_1)$ and the line $\mathfrak{L}$ with normal equation

$$x\cos\alpha + y\sin\alpha - p = 0,$$

notice first that the line $\mathfrak{L}'$ that is parallel to $\mathfrak{L}$ and contains $\mathbf{S}$ has a normal equation of the form

$$x\cos\alpha + y\sin\alpha - p' = 0,$$

where p' is the directed distance from $\mathbf{O}$ to $\mathfrak{L}'$ (Figure 3–6). Since $\mathbf{S}(x_1, y_1)$ is on $\mathfrak{L}'$, you have

$$x_1\cos\alpha + y_1\sin\alpha - p' = 0,$$

and therefore

$$p' = x_1\cos\alpha + y_1\sin\alpha.$$

Figure 3–6

Now the directed distance from $\mathcal{L}$ to $\mathcal{L}'$ is $p' - p$, and this is also the directed distance from $\mathcal{L}$ to **S**. Therefore,

$$d(\mathbf{S}, \mathcal{L}) = |p' - p| = |x_1 \cos \alpha + y_1 \sin \alpha - p|. \tag{6}$$

Thus, the distance between **S** and $\mathcal{L}$ is determined by taking the absolute value of the expression obtained by substituting the coordinates of **S** in the left-hand member of a normal equation for $\mathcal{L}$.

It follows from Equations (4′) and (6) that the distance between the point $\mathbf{S}(x_1, y_1)$ and the line $\mathcal{L}$ with Cartesian equation

$$Ax + By + C = 0, \qquad A^2 + B^2 \neq 0,$$

is given by

$$d(\mathbf{S}, \mathcal{L}) = \left| \frac{Ax_1 + By_1 + C}{\sqrt{A^2 + B^2}} \right|.$$

Example 3. For the points $\mathbf{S}(1, 1)$ and $\mathbf{T}(1, 2)$ and the line $\mathcal{L}$ with Cartesian equation $3x + 4y - 8 = 0$, determine an equation in normal form for $\mathcal{L}$, $p' - p$ for **S** and for **T**, and $d(\mathbf{S}, \mathcal{L})$ and $d(\mathbf{T}, \mathcal{L})$.

Solution: Since $A = 3$ and $B = 4$, you have $k = \pm\sqrt{3^2 + 4^2} = \pm 5$. Suppose you choose $k = 5$. The normal form of the equation of $\mathcal{L}$ is then

$$\tfrac{3}{5}x + \tfrac{4}{5}y - \tfrac{8}{5} = 0.$$

For **S**, you have

$$p' - p = \tfrac{3}{5}(1) + \tfrac{4}{5}(1) - \tfrac{8}{5} = -\tfrac{1}{5},$$

and

$$d(\mathbf{S}, \mathcal{L}) = |-\tfrac{1}{5}| = \tfrac{1}{5}.$$

For **T**, you have

$$p' - p = \tfrac{3}{5}(1) + \tfrac{4}{5}(2) - \tfrac{8}{5} = \tfrac{3}{5},$$

and

$$d(\mathbf{T}, \mathcal{L}) = |\tfrac{3}{5}| = \tfrac{3}{5}.$$

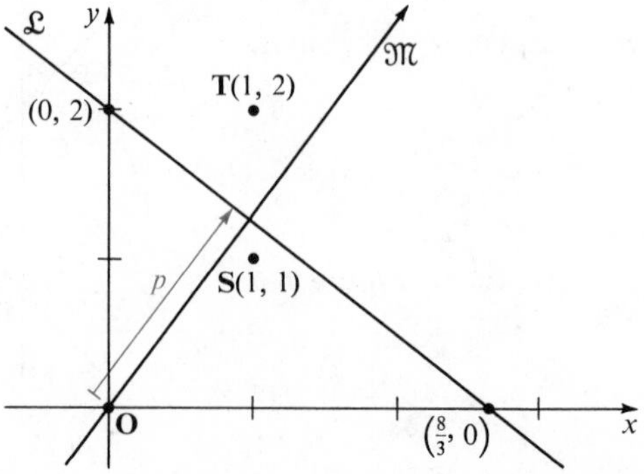

Note in the preceding solution that $p' - p < 0$ for **S** and $p' - p > 0$ for **T**. This indicates that **S** and **T** are on opposite sides of $\mathcal{L}$, and, in fact, since $p > 0$, that **S** and **O** are on the same side of $\mathcal{L}$.

Exercises 3–1

In Exercises 1–12, find the distance between the given point **S** and the line having the given direction vector **v** or slope m and passing through the given point **T**.

1. $\mathbf{S}(3, 4)$; $\mathbf{v} = (2, 1)$, $\mathbf{T}(4, 7)$
2. $\mathbf{S}(2, 5)$; $\mathbf{v} = (3, -1)$, $\mathbf{T}(1, 3)$
3. $\mathbf{S}(-1, 3)$; $\mathbf{v} = (2, 2)$, $\mathbf{T}(4, -2)$
4. $\mathbf{S}(4, -7)$; $\mathbf{v} = (-5, -6)$, $\mathbf{T}(1, 0)$
5. $\mathbf{S}(-5, 1)$; $\mathbf{v} = (4, -6)$, $\mathbf{T}(0, -1)$
6. $\mathbf{S}(-6, -2)$; $\mathbf{v} = (1, 5)$, $\mathbf{T}(0, 0)$
7. $\mathbf{S}(1, 6)$; $m = 2$, $\mathbf{T}(3, 3)$
8. $\mathbf{S}(5, 1)$; $m = -3$, $\mathbf{T}(2, -1)$
9. $\mathbf{S}(4, -2)$; $m = \frac{1}{2}$, $\mathbf{T}(5, -3)$
10. $\mathbf{S}(-3, 7)$; $m = -\frac{2}{3}$, $\mathbf{T}(-4, 6)$
11. $\mathbf{S}(-3, -4)$; $m = -\frac{1}{7}$, $\mathbf{T}(2, 3)$
12. $\mathbf{S}(-4, 0)$; $m = -4$, $\mathbf{T}(0, 4)$

In Exercises 13–20, find the distance between the point **S** and the line $\mathcal{L}$ with the given equation.

13. $\mathbf{S}(5, 7)$; $3x + 4y + 12 = 0$
14. $\mathbf{S}(3, 6)$; $2x - 3y + 6 = 0$
15. $\mathbf{S}(-2, 4)$; $5x - 4y - 10 = 0$
16. $\mathbf{S}(-3, 7)$; $6x - 5y - 15 = 0$
17. $\mathbf{S}(4, -3)$; $x - 8y + 5 = 0$
18. $\mathbf{S}(5, -8)$; $4x + y - 3 = 0$
19. $\mathbf{S}(-1, -5)$; $5x - 12y + 7 = 0$
20. $\mathbf{S}(-3, -6)$; $4x - 3y + 6 = 0$

In Exercises 21–24, find the lengths of the altitudes of the triangle whose vertices **R**, **S**, and **T** are given.

21. $\mathbf{R}(0, 4)$, $\mathbf{S}(4, -3)$, $\mathbf{T}(-3, 1)$
22. $\mathbf{R}(1, 0)$, $\mathbf{S}(2, 5)$, $\mathbf{T}(-2, 2)$
23. $\mathbf{R}(7, 0)$, $\mathbf{S}(-1, 0)$, $\mathbf{T}(1, -1)$
24. $\mathbf{R}(4, -1)$, $\mathbf{S}(1, 7)$, $\mathbf{T}(-3, 3)$

25–28. Find the area of the triangle whose vertices **R**, **S**, and **T** are given in Exercises 21–24.

In Exercises 29 and 30, find the distance between the parallel lines with equations as given.

29. $3x - y - 8 = 0$ and $3x - y - 15 = 0$
30. $x - 3y + 12 = 0$ and $x - 3y - 18 = 0$

31. Find k so that the point $(2, k)$ is equidistant from the lines with equations $x + y - 2 = 0$ and $x - 7y + 2 = 0$.
32. Find k so that the point $(k, 4)$ is equidistant from the lines with equations $13x - 9y - 10 = 0$ and $x + 3y - 6 = 0$.

In Exercises 33–36, find equations of the bisectors of the angles whose sides lie in the lines with the given equations.

Example. $x + 2y - 3 = 0$ and $2x + y - 5 = 0$

Solution: You use the fact that each point on the bisector of an angle is equidistant from the sides. Let (x', y') be the coordinates of a point equidistant from the given lines. Then you have

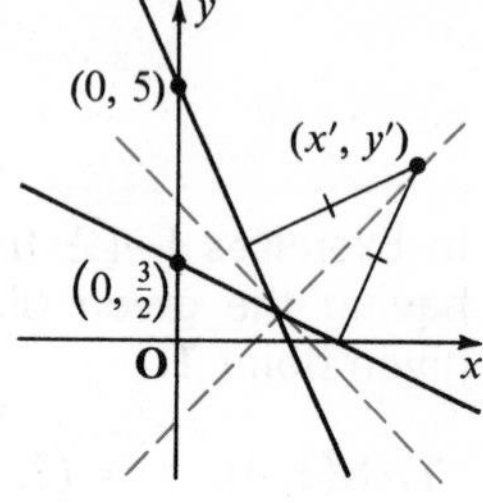

$$\frac{|x' + 2y' - 3|}{\sqrt{5}} = \frac{|2x' + y' - 5|}{\sqrt{5}}.$$

From the definition of absolute value, this equation is equivalent to

$$\frac{x' + 2y' - 3}{\sqrt{5}} = \frac{2x' + y' - 5}{\sqrt{5}} \quad or \quad \frac{x' + 2y' - 3}{\sqrt{5}} = -\frac{2x' + y' - 5}{\sqrt{5}},$$

$$x' + 2y' - 3 = 2x' + y' - 5 \quad or \quad x' + 2y' - 3 = -2x' - y' + 5,$$

$$x' - y' - 2 = 0 \quad or \quad 3x' + 3y' - 8 = 0.$$

Therefore the equations of the angle bisectors are

$$x - y - 2 = 0 \quad \text{and} \quad 3x + 3y - 8 = 0.$$

33. $3x + 4y - 2 = 0$ and $4x + 3y + 2 = 0$

34. $3x - 4y + 1 = 0$ and $5x + 12y - 2 = 0$

35. $x + 3y - 2 = 0$ and $2x - 6y + 5 = 0$

36. $x + y - 6 = 0$ and $3x - 3y + 5 = 0$

* **37.** Find equations of the angle bisectors of the triangle with vertices **R**(6, 2), **S**(−2, −4), and **T**($-\frac{42}{5}$, 8). (*Hint:* Use a sketch to help select the required equations.)

* **38.** Find equations of the angle bisectors of the triangle whose sides lie in the lines with equations $x + 2y - 4 = 0$, $x - 2y + 2 = 0$, and $2x - y - 8 = 0$.

* **39.** Find equations of the lines that are parallel to the line $\mathcal{L}$ with equation $3x - 4y + 10 = 0$ and are located a distance of 5 units from $\mathcal{L}$.

* **40.** Find equations of the lines that are parallel to the line $\mathcal{L}$ with equation $15x + 8y - 34 = 0$ and are located a distance of 4 units from $\mathcal{L}$.

* **41.** Find the distance between the point $\mathbf{S}(b, a)$ and the line with x- and y-intercepts a and b, respectively.

* **42.** Find the distance between the point $\mathbf{S}(b, -a)$ and the line with x- and y-intercepts a and b, respectively.

* **43.** Show that the area of the triangle with vertices $\mathbf{R}(x_1, y_1)$, $\mathbf{S}(x_2, y_2)$, and $\mathbf{T}(x_3, y_3)$ is

$$\tfrac{1}{2}|x_1(y_2 - y_3) + x_2(y_3 - y_1) + x_3(y_1 - y_2)|.$$

3–2 Intersection of Lines

In earlier mathematics courses, you learned that the intersection of sets A and B (denoted $A \cap B$) is defined to be the set of all elements common to both A and B. A special property of the sets of points we call *lines* is that in the plane the intersection of two *different* lines either is empty (if the lines are parallel) or else consists of a single point (if the lines are not parallel). If "$\mathcal{L}_1$" and "$\mathcal{L}_2$" denote the same line, then $\mathcal{L}_1$ and $\mathcal{L}_2$ are parallel and, of course, their intersection is the entire line.

You can determine whether or not two lines are parallel by simply comparing direction vectors, or, equivalently, normal vectors, of the lines. If the lines *are* parallel, then they coincide if and only if they have a point in common (see Exercise 45, page 63).

Example 1. If $\mathcal{L}_1$ and $\mathcal{L}_2$ are the lines specified by

$$\mathcal{L}_1 = \{(x, y)\colon 3x - 4y = 5\}$$

and

$$\mathcal{L}_2 = \{(x, y)\colon -6x + 8y + 9 = 0\},$$

find $\mathcal{L}_1 \cap \mathcal{L}_2$.

Solution: By inspection of the given equations, you see that $(3, -4)$ is a normal vector of $\mathcal{L}_1$ and $(-6, 8) = -2(3, -4)$ is a normal vector of $\mathcal{L}_2$. Therefore $(4, 3)$ is a direction vector of both $\mathcal{L}_1$ and $\mathcal{L}_2$, and thus $\mathcal{L}_1$ and $\mathcal{L}_2$ are parallel.

You can find the coordinates (x, y) of a point $\mathbf{S}$ on $\mathcal{L}_1$ by arbitrarily assigning a value to x and then solving the given equation of $\mathcal{L}_1$ for y. For example, if $x = 3$, then

$$y = -\tfrac{1}{4}(5 - 9) = 1,$$

and $\mathbf{S}(3, 1)$ is a point on $\mathcal{L}_1$. Substituting the coordinates of $\mathbf{S}$ in the equation for $\mathcal{L}_2$, you have $-6(3) + 8(1) + 9 = -18 + 8 + 9 \neq 0$. Thus $\mathbf{S}$ does not lie on $\mathcal{L}_2$, and $\mathcal{L}_1 \cap \mathcal{L}_2 = \emptyset$.

In algebra courses, you have seen that Cartesian methods can be used to show that nonparallel lines in the plane intersect in a unique point. You can also use vector methods to establish this result. Consider Figure 3–7, which shows the line $\mathcal{L}_1$ through the points **Q**$(-3, -2)$ and **R**$(0, 2)$ and the line $\mathcal{L}_2$ through the points **S**$(5, 0)$ and **T**$(6, -3)$; in this figure, $\mathcal{L}_1$ and $\mathcal{L}_2$ appear to intersect in a unique point. To prove that they do, you must show analytically (1) that there is only one point that can lie on both lines and (2) that this point actually does lie on both lines.

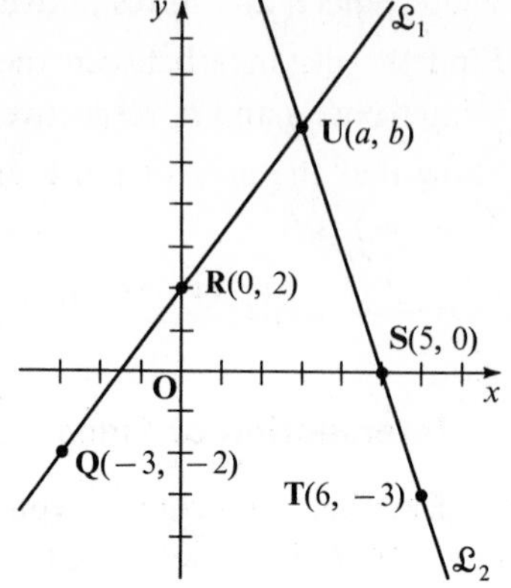

Figure 3–7

From the given information, a parametric vector equation for $\mathcal{L}_1$ is

$$(x, y) = (-3, -2) + r_1[(0, 2) - (-3, -2)], \qquad r_1 \in \mathcal{R},$$

or

$$(x, y) = (-3, -2) + r_1(3, 4);$$

and a parametric vector equation for $\mathcal{L}_2$ is

$$(x, y) = (5, 0) + r_2[(6, -3) - (5, 0)], \qquad r_2 \in \mathcal{R},$$

or

$$(x, y) = (5, 0) + r_2(1, -3).$$

Suppose that **U**(a, b) is a point lying on both lines. Then you must have

$$(a, b) = (-3, -2) + r_1(3, 4) \quad \textit{and} \quad (a, b) = (5, 0) + r_2(1, -3),$$

so that

$$(-3, -2) + r_1(3, 4) = (5, 0) + r_2(1, -3). \tag{1}$$

You can solve Equation (1) for r_1 and r_2 as follows: Since for any vector $\mathbf{v}$, $\mathbf{v} \cdot \mathbf{v_p} = 0$, you can eliminate r_1 by taking the inner product of each member of Equation (1) (recall the Distributive Property in Section 1–6) with the vector $(3, 4)_\mathbf{p} = (-4, 3)$ to obtain

$$\begin{aligned}(-3, -2) \cdot (-4, 3) + r_1(3, 4) \cdot (-4, 3) &= (5, 0) \cdot (-4, 3) + r_2(1, -3) \cdot (-4, 3) \\ 6 + r_1(0) &= -20 + r_2(-13), \\ 26 &= -13r_2, \\ r_2 &= -2.\end{aligned}$$

Similarly, you can eliminate r_2 by taking the inner product of each member of Equation (1) with $(1, -3)_\mathbf{p} = (3, 1)$. You obtain:

$$
\begin{aligned}
(-3, -2) \cdot (3, 1) + r_1(3, 4) \cdot (3, 1) &= (5, 0) \cdot (3, 1) + r_2(1, -3) \cdot (3, 1), \\
-11 + r_1(13) &= 15 + r_2(0), \\
13r_1 &= 26, \\
r_1 &= 2.
\end{aligned}
$$

Accordingly, if there is a point $\mathbf{U}(a, b)$ lying on both lines, then for it you must have $r_1 = 2$ and $r_2 = -2$. Substituting these values in the equations for $\mathcal{L}_1$ and $\mathcal{L}_2$, you get, respectively,

$$(a, b) = (-3, -2) + 2(3, 4) = (-3, -2) + (6, 8) = (3, 6)$$

and

$$(a, b) = (5, 0) - 2(1, -3) = (5, 0) + (-2, 6) = (3, 6).$$

Hence, $\mathcal{L}_1$ and $\mathcal{L}_2$ intersect in the unique point $\mathbf{U}(3, 6)$. Since this same method can be applied to *any* two given nonparallel lines, you can conclude that two nonparallel lines in the plane intersect in exactly one point.

If two lines in the plane are specified by a system of two linear equations in two variables, then their point of intersection can be determined by solving the two equations simultaneously. In algebra, you learned to solve a system of two linear equations in two variables by "substitution" and by "elimination of a variable," sometimes called the "linear-combination" method. Both of these techniques, as well as a vector method that parallels the "elimination" method, are illustrated in the solutions of the following example.

Example 2. Solve the system:

$$3x - 5y = 1 \quad (1)$$
$$4x + 3y = 11 \quad (2)$$

Solution by substitution:

Solve Equation (1) for y in terms of x:

$$y = \tfrac{3}{5}x - \tfrac{1}{5}. \quad (3)$$

Replace y in Equation (2) with $\frac{3}{5}x - \frac{1}{5}$ and solve for x:

$$
\begin{aligned}
4x + 3(\tfrac{3}{5}x - \tfrac{1}{5}) &= 11, \\
x &= 2.
\end{aligned}
$$

Replace x in Equation (3) with 2 and solve for y:

$$y = \tfrac{3}{5}(2) - \tfrac{1}{5} = 1.$$

Thus, the only possible solution for the system is (2, 1). To check that (2, 1) actually is a solution, you can substitute 2 for x and 1 for y in each of the original equations:

$$
\begin{aligned}
3(2) - 5(1) &= 1, \\
1 &= 1.
\end{aligned}
\qquad\Bigg|\qquad
\begin{aligned}
4(2) + 3(1) &= 11, \\
11 &= 11.
\end{aligned}
$$

Therefore, (2, 1) is the unique solution of the system.

Solution by elimination:

Multiply each member of Equation (1) by 3 and each member of Equation (2) by 5 to obtain the equivalent system:

$$\begin{aligned} 9x - 15y &= 3 \\ 20x + 15y &= 55 \end{aligned}$$

Add corresponding members of these equations to obtain

$$\begin{aligned} 29x &= 58, \\ x &= 2. \end{aligned}$$

You can now replace x with 2 in one of the equations and solve for y, as in the solution by substitution, or else return to the original equations and eliminate x as follows: Multiply each member of Equation (1) by -4 and each member of Equation (2) by 3 to obtain the equivalent system:

$$\begin{aligned} -12x + 20y &= -4 \\ 12x + 9y &= 33 \end{aligned}$$

Add corresponding members of these equations to obtain

$$\begin{aligned} 29y &= 29, \\ y &= 1. \end{aligned}$$

Thus, as before, the only possible solution is (2, 1), and you can check that (2, 1) actually is a solution by substituting in the original equations.

Solution using vectors:

The given system of equations, that is, the compound sentence

$$3x - 5y = 1 \quad \textit{and} \quad 4x + 3y = 11,$$

is equivalent to the vector equation

$$(3x - 5y, 4x + 3y) = (1, 11),$$

which, in turn, is equivalent to

$$x(3, 4) - y(5, -3) = (1, 11). \qquad (4)$$

You can solve this equation by the method used on page 98. That is, you can eliminate y by taking the inner product of each member of Equation (4) and $(5, -3)_{\mathbf{p}} = (3, 5)$. You have

$$\begin{aligned} x(3, 4) \cdot (3, 5) - y(5, -3) \cdot (3, 5) &= (1, 11) \cdot (3, 5), \\ 29x - 0y &= 58. \\ x &= 2. \end{aligned}$$

You can then eliminate x by taking the inner product of each member and $(3, 4)_p = (-4, 3)$. You have

$$x(3, 4) \cdot (-4, 3) - y(5, -3) \cdot (-4, 3) = (1, 11) \cdot (-4, 3),$$
$$0x + 29y = 29,$$
$$y = 1.$$

Thus, the only possible solution of the system is (2, 1), and, as before, you can check that (2, 1) is a solution by substituting in the original Equations (1) and (2).

If you compare the solution by elimination and the solution using vectors, you can readily see that they are algebraically equivalent. The single vector equation (4) is equivalent to the two Cartesian equations (1) and (2). To eliminate y, you multiply each member of Equation (1) by 3 and each member of (2) by 5 and add, or equivalently you form the inner product of each member of (4) with (3, 5). Similarly, to eliminate x, you multiply each member of (1) and (2) by -4 and 3, respectively, and add, or equivalently you form the inner product of each member of (4) with $(-4, 3)$.

Exercises 3–2

In Exercises 1–6, find $\mathcal{L}_1 \cap \mathcal{L}_2$ if $\mathcal{L}_1$ and $\mathcal{L}_2$ have the given equations. (The parameters r_1, r_2, t_1, t_2 represent real numbers.)

1. $\mathcal{L}_1$: $(x, y) = (3, 4) + r_1(1, 1)$; $\mathcal{L}_2$: $(x, y) = (1, 2) + r_2(1, -3)$

2. $\mathcal{L}_1$: $(x, y) = (2, -3) + r_1(4, -2)$; $\mathcal{L}_2$: $(x, y) = (-2, 1) + r_2(-1, -2)$

3. $\mathcal{L}_1$: $x - 2y = 3$; $\mathcal{L}_2$: $2x + y = 1$

4. $\mathcal{L}_1$: $3x - y = 4$; $\mathcal{L}_2$: $x + 3y = 8$

5. $\mathcal{L}_1$: $(x, y) = (3 - 2t_1, 1 - t_1)$; $\mathcal{L}_2$: $(x, y) = (4 + 5t_2, 2 + 3t_2)$

6. $\mathcal{L}_1$: $(x, y) = (6 + 3t_1, 3 - 2t_1)$; $\mathcal{L}_2$: $(x, y) = (3 - 3t_2, 5 + 2t_2)$

In Exercises 7–10, solve the given system by substitution.

7. $x + 3y = 7$
$2x + y = -1$

8. $3x - 2y = 19$
$2x + 5y = 0$

9. $x - 2y = 7$
$-2x + 4y = -14$

10. $4x - 3y = -2$
$3x + 2y = 7$

In Exercises 11–16, solve the given system by elimination.

11. $x + 3y = 7$
$x - y = 3$

12. $x - 3y = -8$
$2x + y = 5$

13. $2x - 5y = 3$
$3x + y = 13$

14. $3x + 2y = 4$
$2x - 7y = 11$

15. $3x - 2y = 8$
$6x - 4y = 3$

16. $5x + y = -9$
$x - 3y = -5$

17–26. Solve the systems in Exercises 7–16 using vectors.

In Exercises 27–30, find the vertices of the triangle whose sides are contained in the lines with the given equations.

27. $2x + y = 4,$
$x - 3y = -5,$
$4x - 5y = 8.$

28. $x - y = -1,$
$x + 3y = 7,$
$x + y = -1.$

29. $(x, y) = (3, 1) + r_1(1, -2),$
$(x, y) = (5, 7) + r_2(3, 4),$
$(x, y) = (-3, -7) + r_3(1, 3).$

30. $(x, y) = (2, 4) + r_1(0, 1),$
$(x, y) = (5, 4) + r_2(2, 1),$
$(x, y) = (3, 3) + r_3(-1, 1).$

In Exercises 31–36, find a Cartesian equation in standard form for the line with the given characteristics:

31. Containing the point $\mathbf{S}(4, \frac{8}{3})$ and the point of intersection of the lines with equations $3x - 4y - 2 = 0$ and $12x - 15y - 8 = 0$.

32. Containing the point $\mathbf{S}(1, 1)$ and the point of intersection of the lines with equations $2x - 5y + 9 = 0$ and $4x + y + 7 = 0$.

33. Containing the point of intersection of the lines with equations $3x + y - 16 = 0$ and $2x - 7y - 3 = 0$, and parallel to the line with equation $x - 3y + 2 = 0$.

34. Containing the point of intersection of the lines with equations $2x + y = 13$ and $7x - y = 2$, and parallel to the line with equation $x - y = 2$.

35. Containing the point of intersection of the lines with equations $x + 2y = 12$ and $3x - 4y = 26$, and perpendicular to the line with equation $x + y = 1$.

36. Containing the point of intersection of the lines with equations $3x + 4y + 10 = 0$ and $5x - 12y - 12 = 0$, and perpendicular to the line with equation $x - 2y + 6 = 0$.

37. Find equations of the lines containing the diagonals of the quadrilateral whose sides are contained in the lines with equations $x - 3y + 13 = 0$, $7x - y + 31 = 0$, $x - 3y - 7 = 0$, and $x + y - 11 = 0$.

38. The lines with equations $3x - 8y - 2 = 0$ and $3x - 8y - 44 = 0$ contain the opposite sides of a parallelogram, and one diagonal of the parallelogram lies in the line with equation $x + 2y + 4 = 0$. If the other diagonal of the parallelogram contains the point $(4, \frac{1}{5})$, find the vertices of the parallelogram.

39. Show that for $a_1b_2 - a_2b_1 \neq 0$, the system

$$a_1x + b_1y = c_1$$
$$a_2x + b_2y = c_2$$

has a unique solution given by

$$x = \frac{c_1b_2 - c_2b_1}{a_1b_2 - a_2b_1}, \qquad y = \frac{a_1c_2 - a_2c_1}{a_1b_2 - a_2b_1}.$$

40. Use the results of Exercise 39 to solve the system:

$$4x - 3y = 5$$
$$2x + 5y = 35$$

Analytic Methods

3–3 Determinants

Determinants are useful in the solution of systems of linear equations, and they also have many other applications in analytic geometry.

You will recall that a rectangular array of numerals such as that shown in color at the right represents a **matrix** (plural: **matrices**), just as a single numeral represents a number. Notice that the array of numerals is enclosed by brackets. Capital letters, such as M and N, are used to denote matrices.

$$\begin{array}{cc} & \text{column} \\ & \begin{array}{ccc} 1 & 2 & 3 \end{array} \\ \text{row} \begin{array}{c} 1 \\ 2 \end{array} & \begin{bmatrix} 3 & 2 & 5 \\ 1 & 0 & 6 \end{bmatrix} \end{array}$$

Each numeral in the array represents an **entry** of the matrix. The number of (horizontal) rows and the number of (vertical) columns of entries in the representation of a matrix determine its **dimensions**. For example, the matrix indicated above has two rows and three columns and accordingly is called a 2×3 (read "two by three") matrix. Notice that the number of rows is given *first* and *then* the number of columns. An entry of a matrix is referred to by giving its row and column numbers. In the matrix represented above, for example, the entry in the first row and third column is 5.

If a matrix has the same number n of rows as it has columns, then it is said to be a **square matrix of order n**. With each square matrix M having real-number entries there can be associated a particular real number, called the *determinant* of M and denoted by "det M" (read "determinant of M").

Determinants are sometimes represented in the same form as matrices, except that vertical bars are used instead of brackets. For example,

$$\det \begin{bmatrix} a_1 & b_1 \\ a_2 & b_2 \end{bmatrix} = \begin{vmatrix} a_1 & b_1 \\ a_2 & b_2 \end{vmatrix}.$$

The entries, rows, and columns of the matrix are then called the entries (or **elements**), rows, and columns of the determinant, and the order of the matrix is also said to be the order of the determinant.

The determinant of a 1×1 matrix $[a_1]$ is just the number a_1 itself. For example,

$$\det[-7] = -7.$$

(You should not use vertical bars in denoting a first-order determinant, since in this case the vertical-bar notation denotes absolute value.)

The determinant of a 2×2 matrix is defined as follows:

$$\begin{vmatrix} a_1 & b_1 \\ a_2 & b_2 \end{vmatrix} = a_1b_2 - a_2b_1. \tag{1}$$

For example, $\det\begin{bmatrix} 4 & 2 \\ -2 & 7 \end{bmatrix} = \begin{vmatrix} 4 & 2 \\ -2 & 7 \end{vmatrix} = 28 - (-4) = 32.$

The determinant of a 3×3 matrix is defined as follows:

$$\begin{vmatrix} a_1 & b_1 & c_1 \\ a_2 & b_2 & c_2 \\ a_3 & b_3 & c_3 \end{vmatrix} = a_1b_2c_3 + a_2b_3c_1 + a_3b_1c_2 - a_1b_3c_2 - a_2b_1c_3 - a_3b_2c_1. \tag{2}$$

The indicated sums of products in (1) and (2) above are called *expansions* of the determinants.

The determinant of a higher-order square matrix is defined in terms of determinants of next lower order by using *minors*. The **minor** of an entry in a determinant is defined to be the determinant obtained when you delete the row and column containing the entry. For example,

$$\text{the minor of 4 in } \begin{vmatrix} 4 & 2 & 3 \\ 1 & 0 & -7 \\ -6 & 5 & 3 \end{vmatrix} \text{ is } \begin{vmatrix} 0 & -7 \\ 5 & 3 \end{vmatrix}.$$

Similarly, the minor of 0 is $\begin{vmatrix} 4 & 3 \\ -6 & 3 \end{vmatrix}$, and the minor of -7 is $\begin{vmatrix} 4 & 2 \\ -6 & 5 \end{vmatrix}$.

In the right-hand member of Equation (2), you can factor the terms by pairs in several different ways. One way is this:

$$a_1(b_2c_3 - b_3c_2) - b_1(a_2c_3 - a_3c_2) + c_1(a_2b_3 - a_3b_2).$$

It follows that if you let A_1, B_1, and C_1 represent the minors of a_1, b_1, and c_1, respectively, then you can write

$$\begin{vmatrix} a_1 & b_1 & c_1 \\ a_2 & b_2 & c_2 \\ a_3 & b_3 & c_3 \end{vmatrix} = a_1A_1 - b_1B_1 + c_1C_1.$$

The right-hand member of this last equation is called the **expansion** of the determinant by minors of the entries in the first row.

By suitably arranging the terms in the right-hand member of Equation (2), you can show that a third-order determinant can be expanded by minors of the entries in any row or any column as follows:

1. Choose a row or column and form the product of each entry in the row or column with its minor.
2. Use the product obtained or its negative according as the sum of the number of the row and the number of the column containing the entry is even or odd.
3. The sum of the resulting numbers is the value of the determinant.

The same procedure can be applied to obtain expansions of fourth-order (and higher-order) determinants. For example, you have

$$\begin{vmatrix} a_1 & b_1 & c_1 & d_1 \\ a_2 & b_2 & c_2 & d_2 \\ a_3 & b_3 & c_3 & d_3 \\ a_4 & b_4 & c_4 & d_4 \end{vmatrix} = -b_1B_1 + b_2B_2 - b_3B_3 + b_4B_4,$$

where $B_1 = \begin{vmatrix} a_2 & c_2 & d_2 \\ a_3 & c_3 & d_3 \\ a_4 & c_4 & d_4 \end{vmatrix}$, $B_2 = \begin{vmatrix} a_1 & c_1 & d_1 \\ a_3 & c_3 & d_3 \\ a_4 & c_4 & d_4 \end{vmatrix}$, and so on.

Example 1. Expand $\begin{vmatrix} 1 & 3 & -2 \\ 2 & 0 & -3 \\ 4 & 5 & 1 \end{vmatrix}$ by minors of the second row.

Solution: The entries of the second row are 2, 0, and -3. The entry 2 is in the second row and first column; since $2 + 1 = 3$ (odd), you use the negative of its product with its minor. Similarly, you use the product of 0 and its minor and the negative of the product of -3 and its minor. Thus,

$$\begin{vmatrix} 1 & 3 & -2 \\ 2 & 0 & -3 \\ 4 & 5 & 1 \end{vmatrix} = -2\begin{vmatrix} 3 & -2 \\ 5 & 1 \end{vmatrix} + 0\begin{vmatrix} 1 & -2 \\ 4 & 1 \end{vmatrix} - (-3)\begin{vmatrix} 1 & 3 \\ 4 & 5 \end{vmatrix}$$
$$= -2(13) + 0(9) + 3(-7) = -47.$$

In Exercise 39, page 103, you saw that if $a_1b_2 - a_2b_1 \neq 0$ then the system of equations

$$\begin{aligned} a_1x + b_1y &= c_1 \\ a_2x + b_2y &= c_2 \end{aligned} \tag{3}$$

has a unique solution, given by

$$x = \frac{c_1b_2 - c_2b_1}{a_1b_2 - a_2b_1}, \qquad y = \frac{a_1c_2 - a_2c_1}{a_1b_2 - a_2b_1}.$$

This solution can be written by means of determinants as

$$x = \frac{\begin{vmatrix} c_1 & b_1 \\ c_2 & b_2 \end{vmatrix}}{\begin{vmatrix} a_1 & b_1 \\ a_2 & b_2 \end{vmatrix}}, \qquad y = \frac{\begin{vmatrix} a_1 & c_1 \\ a_2 & c_2 \end{vmatrix}}{\begin{vmatrix} a_1 & b_1 \\ a_2 & b_2 \end{vmatrix}}.$$

Now let M denote the matrix $\begin{bmatrix} a_1 & b_1 \\ a_2 & b_2 \end{bmatrix}$ of the coefficients of x and y, let $D = \det M$, and note that the numerator determinants, which we shall denote by D_x and D_y, respectively, are the determinants of matrices obtained from M by replacing the coefficients of the respective variables x and y in Equations (3) with the constant terms c_1 and c_2. Thus if $D \neq 0$, then the system (3) of equations has a unique solution, given by

$$x = \frac{D_x}{D}, \qquad y = \frac{D_y}{D}.$$

For any counting number n, the pattern can be extended to express the unique solution of any system of n linear equations in n variables for which the determinant D of the matrix M of coefficients of the variables is not 0. The expression of the solution by means of determinants is called **Cramer's Rule.**

In particular, for the system

$$\begin{aligned} a_1x + b_1y + c_1z &= d_1 \\ a_2x + b_2y + c_2z &= d_2 \\ a_3x + b_3y + c_3z &= d_3 \end{aligned}$$

you have

$$D = \begin{vmatrix} a_1 & b_1 & c_1 \\ a_2 & b_2 & c_2 \\ a_3 & b_3 & c_3 \end{vmatrix}, \qquad D_x = \begin{vmatrix} d_1 & b_1 & c_1 \\ d_2 & b_2 & c_2 \\ d_3 & b_3 & c_3 \end{vmatrix},$$

$$D_y = \begin{vmatrix} a_1 & d_1 & c_1 \\ a_2 & d_2 & c_2 \\ a_3 & d_3 & c_3 \end{vmatrix}, \qquad D_z = \begin{vmatrix} a_1 & b_1 & d_1 \\ a_2 & b_2 & d_2 \\ a_3 & b_3 & d_3 \end{vmatrix},$$

and if $D \neq 0$, then the solution is unique and is given by

$$x = \frac{D_x}{D}, \qquad y = \frac{D_y}{D}, \qquad z = \frac{D_z}{D}.$$

For applications to analytic geometry, the following property of determinants (see Exercises 12 and 13, page 107) is quite useful.

■ If two rows (or two columns) of a determinant have corresponding entries that are equal, then the determinant is equal to 0.

Example 2. Show that for any two (distinct) points $\mathbf{S}(x_1, y_1)$ and $\mathbf{T}(x_2, y_2)$ in the plane,

$$\begin{vmatrix} x & y & 1 \\ x_1 & y_1 & 1 \\ x_2 & y_2 & 1 \end{vmatrix} = 0 \tag{4}$$

is an equation for the line containing **S** and **T**.

Solution: If you replace x with x_1 and y with y_1 in Equation (4), you get

$$\begin{vmatrix} x_1 & y_1 & 1 \\ x_1 & y_1 & 1 \\ x_2 & y_2 & 1 \end{vmatrix} = 0,$$

which is true since the first two rows have corresponding entries that are equal. Therefore **S** is on the graph of the equation. Similarly, replacing x with x_2 and y with y_2, you see that **T** is on the graph of the equation. An expansion of the determinant in Equation (4) is

$$x(y_1 - y_2) - y(x_1 - x_2) + (x_1y_2 - x_2y_1).$$

Since $\mathbf{S} \neq \mathbf{T}$, the coefficients of x and y are not both 0. Therefore Equation (4) is a linear equation in x and y, and since its graph contains **S** and **T**, its graph is the line containing **S** and **T**.

Exercises 3–3

In Exercises 1–16, if M is the given matrix, find det M.

1. $\begin{bmatrix} 4 & 7 \\ 2 & -3 \end{bmatrix}$

2. $\begin{bmatrix} 5 & -1 \\ 6 & 2 \end{bmatrix}$

3. $\begin{bmatrix} 6 & 3 \\ 4 & 2 \end{bmatrix}$

4. $\begin{bmatrix} 2 & 6 \\ 1 & 3 \end{bmatrix}$

5. $\begin{bmatrix} 0 & 0 \\ 7 & -4 \end{bmatrix}$

6. $\begin{bmatrix} 2 & 0 \\ -5 & 0 \end{bmatrix}$

7. $\begin{bmatrix} \frac{3}{5} & \frac{4}{5} \\ -\frac{4}{5} & \frac{3}{5} \end{bmatrix}$

8. $\begin{bmatrix} \frac{5}{4} & \frac{3}{4} \\ \frac{3}{4} & \frac{5}{4} \end{bmatrix}$

9. $\begin{bmatrix} a & 1 \\ -b & 1 \end{bmatrix}$

10. $\begin{bmatrix} a & b \\ b & a \end{bmatrix}$

11. $\begin{bmatrix} a & b \\ -b & a \end{bmatrix}$

12. $\begin{bmatrix} a & b \\ a & b \end{bmatrix}$

13. $\begin{bmatrix} a & b & c \\ a & b & c \\ s & p & q \end{bmatrix}$

14. $\begin{bmatrix} 5 & 7 & 2 \\ 1 & 0 & 0 \\ 6 & 9 & 2 \end{bmatrix}$

15. $\begin{bmatrix} 4 & 5 & 6 \\ 2 & 4 & 3 \\ 0 & 7 & 0 \end{bmatrix}$

16. $\begin{bmatrix} 3 & -2 & 1 \\ -4 & 3 & 7 \\ 2 & 1 & 0 \end{bmatrix}$

In Exercises 17 and 18, evaluate the determinant by expanding it by elements of the second column.

17. $\begin{vmatrix} 1 & 3 & 2 \\ 4 & -2 & 3 \\ 5 & 0 & 6 \end{vmatrix}$

18. $\begin{vmatrix} 4 & -1 & 2 \\ 3 & 0 & 1 \\ 4 & 2 & 6 \end{vmatrix}$

19–20. Do Exercises 17 and 18 by expanding by elements of the third row.

In Exercises 21–24, find the solution set of the given system of linear equations by using Cramer's Rule.

21. $4x - 3y = 5$
$2x + y = 5$

22. $3x + 2y = 1$
$x + y = -1$

23. $x + 2y = 0$
$x + z = 1$
$3y - 2z = -3$

24. $2x - 3y + z = 1$
$6x - 6y - z = 0$
$4x + 6y - 2z = 2$

In Exercises 25 and 26, use a third-order determinant to write an equation of the line containing the given points **S** and **T**.

25. $\mathbf{S}(2, -1)$, $\mathbf{T}(3, 0)$

26. $\mathbf{S}(-3, 4)$, $\mathbf{T}(5, 2)$

* **27.** Show that the points (not necessarily distinct) $\mathbf{R}(x_1, y_1)$, $\mathbf{S}(x_2, y_2)$, and $\mathbf{T}(x_3, y_3)$ are collinear if and only if

$$\begin{vmatrix} x_1 & y_1 & 1 \\ x_2 & y_2 & 1 \\ x_3 & y_3 & 1 \end{vmatrix} = 0.$$

* **28.** Find an equation in determinant form for the line containing the point $\mathbf{S}(x_1, y_1)$ and having y-intercept b.

* **29.** Find an equation in determinant form for the line with x-intercept a and y-intercept b.

* **30.** Show that

$$\begin{vmatrix} x & y & 1 \\ x_1 & y_1 & 1 \\ 1 & m & 0 \end{vmatrix} = 0$$

is an equation for the line containing the point $\mathbf{S}(x_1, y_1)$ and having slope m.

* **31.** Show that the slope-intercept form of the equation of a line can be written in determinant form as

$$\begin{vmatrix} x & y & 1 \\ 0 & b & 1 \\ 1 & m & 0 \end{vmatrix} = 0.$$

* **32.** Show that the line passing through the origin and containing the point $\mathbf{S}(x_1, y_1)$ has an equation of the form

$$\begin{vmatrix} x & y \\ x_1 & y_1 \end{vmatrix} = 0.$$

* **33.** Show that the points $\mathbf{S}(x_1, y_1)$ and $\mathbf{T}(x_2, y_2)$ are collinear with the origin if and only if

$$\begin{vmatrix} x_1 & y_1 \\ x_2 & y_2 \end{vmatrix} = 0.$$

* **34.** Show that the area of the triangle with vertices $\mathbf{R}(x_1, y_1)$, $\mathbf{S}(x_2, y_2)$, and $\mathbf{T}(x_3, y_3)$ is

$$\pm\tfrac{1}{2}\begin{vmatrix} x_1 & y_1 & 1 \\ x_2 & y_2 & 1 \\ x_3 & y_3 & 1 \end{vmatrix}.$$

[*Hint:* See Exercise 43, page 97.]

* **35.** Show that the vectors $\mathbf{s} = (x_1, y_1)$ and $\mathbf{t} = (x_2, y_2)$ are parallel if and only if

$$\begin{vmatrix} x_1 & y_1 \\ x_2 & y_2 \end{vmatrix} = 0.$$

* **36.** Show that if the vectors $\mathbf{s} = (x_1, y_1)$ and $\mathbf{t} = (x_2, y_2)$ are not parallel, then any vector $\mathbf{v} = (x, y)$ can be expressed as a linear combination of $\mathbf{s}$ and $\mathbf{t}$,

$$\mathbf{v} = a\mathbf{s} + b\mathbf{t},$$

where a and b are given by

$$a = \frac{\begin{vmatrix} x & x_2 \\ y & y_2 \end{vmatrix}}{\begin{vmatrix} x_1 & x_2 \\ y_1 & y_2 \end{vmatrix}} \quad \text{and} \quad b = \frac{\begin{vmatrix} x_1 & x \\ y_1 & y \end{vmatrix}}{\begin{vmatrix} x_1 & x_2 \\ y_1 & y_2 \end{vmatrix}}.$$

[*Hint:* The vector equation $(x, y) = a(x_1, y_1) + b(x_2, y_2)$ is equivalent to a system of two Cartesian equations in the variables a and b.]

3–4 Analytic Proofs

Analytic methods can be used very effectively in proving theorems from Euclidean plane geometry. The proofs can be given in terms of Cartesian coordinates of points or, alternatively, in terms of vectors.

When coordinates are used to prove a theorem, you can sometimes make the proof easier if the coordinate axes are oriented in a particular way with respect to the figure involved; no loss of generality results, since the location of the coordinate axes in the plane is arbitrary. When vector methods are used, however, the location of the figure with respect to the coordinate axes ordinarily is not important; in fact, the precise location of the figure usually does not enter into the discussion.

Theorem. The lengths of the opposite sides of a parallelogram are equal.

Proof using coordinates:

For a parallelogram **ABCD** (see diagram below), the coordinate axes can be chosen with the origin at **A** and the positive ray of the x-axis

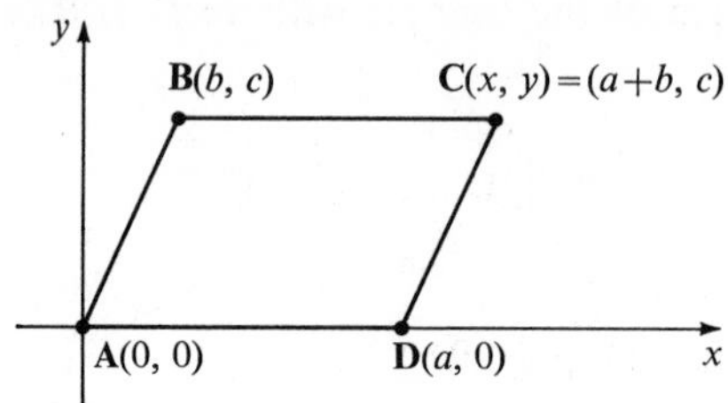

along $\overline{\mathbf{AD}}$. Then the coordinates of **A** are (0, 0) and the coordinates of **D** are of the form $(a, 0)$, with $a > 0$. Let **B** have coordinates (b, c), with $c \neq 0$, as shown. You can find (x, y), the coordinates of **C**, as follows:

Since $\overline{\mathbf{BC}} \parallel \overline{\mathbf{AD}}$, and the slope of $\overline{\mathbf{AD}}$ is 0, it follows that the slope of $\overline{\mathbf{BC}}$ is 0. Therefore you have $\dfrac{y - c}{x - b} = 0$, so that

$$y = c.$$

Since $\overline{\mathbf{CD}} \parallel \overline{\mathbf{AB}}$, slope $\overline{\mathbf{CD}}$ = slope $\overline{\mathbf{AB}}$. Thus, if $x \neq a$,

$$\frac{y - 0}{x - a} = \frac{c - 0}{b - 0}.$$

However, since $y = c$, as already determined, you have

$$\frac{c - 0}{x - a} = \frac{c - 0}{b - 0},$$

or

$$x = a + b.$$

If $x = a$, so that $\overline{\mathbf{CD}}$ is vertical, then $\overline{\mathbf{AB}}$ must also be vertical, and therefore $b = 0$; thus you have $x = a + b$ in this case also.

Once the coordinates $(a + b, c)$ of **C** are known, you can use the distance formula to find the lengths of the sides:

$$d(\overline{\mathbf{AB}}) = \sqrt{(b - 0)^2 + (c - 0)^2} = \sqrt{b^2 + c^2}$$
$$d(\overline{\mathbf{CD}}) = \sqrt{(a + b - a)^2 + (c - 0)^2} = \sqrt{b^2 + c^2}$$
$$d(\overline{\mathbf{BC}}) = \sqrt{(a + b - b)^2 + (c - c)^2} = \sqrt{a^2} = a$$
$$d(\overline{\mathbf{AD}}) = \sqrt{(a - 0)^2 + (0 - 0)^2} = \sqrt{a^2} = a$$

Therefore the lengths of the opposite sides are equal.

Proof using vectors:

Let **ABCD** be a parallelogram, and let boldface lower-case letters refer to vectors from the origin to the respective points.

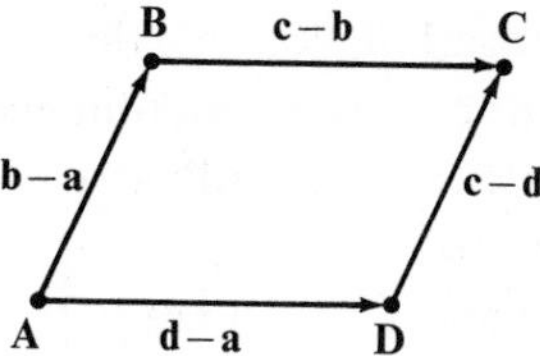

Since $\overline{\mathbf{CD}} \parallel \overline{\mathbf{AB}}$ and $\overline{\mathbf{BC}} \parallel \overline{\mathbf{AD}}$, you have

$$\mathbf{c} - \mathbf{d} = k_1(\mathbf{b} - \mathbf{a}) \quad \text{and} \quad \mathbf{c} - \mathbf{b} = k_2(\mathbf{d} - \mathbf{a}),\ k_1, k_2 \in \mathcal{R}. \tag{1}$$

Since

$$\mathbf{c} - \mathbf{a} = (\mathbf{c} - \mathbf{b}) + (\mathbf{b} - \mathbf{a})$$

and

$$\mathbf{c} - \mathbf{a} = (\mathbf{c} - \mathbf{d}) + (\mathbf{d} - \mathbf{a}),$$

you have

$$(\mathbf{c} - \mathbf{b}) + (\mathbf{b} - \mathbf{a}) = (\mathbf{c} - \mathbf{d}) + (\mathbf{d} - \mathbf{a}).$$

Therefore, from Equations (1),

$$k_2(\mathbf{d} - \mathbf{a}) + (\mathbf{b} - \mathbf{a}) = k_1(\mathbf{b} - \mathbf{a}) + (\mathbf{d} - \mathbf{a}),$$

or

$$(k_2 - 1)(\mathbf{d} - \mathbf{a}) = (k_1 - 1)(\mathbf{b} - \mathbf{a}). \tag{2}$$

By assumption, neither $\mathbf{d} - \mathbf{a}$ nor $\mathbf{b} - \mathbf{a}$ is the zero vector. Thus if $k_2 - 1 \neq 0$ and $k_1 - 1 \neq 0$, then Equation (2) implies that $\mathbf{d} - \mathbf{a}$ and $\mathbf{b} - \mathbf{a}$ are nonzero scalar multiples of each other and hence are parallel (page 21). But this is impossible, since the adjacent sides of a parallelogram are *not* parallel. Therefore you can conclude that $k_2 - 1 = 0$ and $k_1 - 1 = 0$, that is, that $k_2 = 1$ and $k_1 = 1$. Then from Equation (1) above, you have

$$\mathbf{c} - \mathbf{d} = \mathbf{b} - \mathbf{a} \quad \text{and} \quad \mathbf{c} - \mathbf{b} = \mathbf{d} - \mathbf{a}. \tag{3}$$

Accordingly,

$$\|\mathbf{c} - \mathbf{d}\| = \|\mathbf{b} - \mathbf{a}\|$$

and

$$\|\mathbf{c} - \mathbf{b}\| = \|\mathbf{d} - \mathbf{a}\|,$$

and the proof is complete.

For a further illustration of the two methods, here is another example:

Theorem. The diagonals of a parallelogram bisect each other.

Proof using coordinates:

In the proof on page 110, you saw that the vertices of the parallelogram **ABCD** can be given coordinates as shown below. By Exercise 23, page 52, the coordinates of the midpoint of the segment joining the points with coordinates (x_1, y_1) and (x_2, y_2) are $\left(\frac{x_1 + x_2}{2}, \frac{y_1 + y_2}{2}\right)$. Accordingly, the midpoint **M** of $\overline{\mathbf{AC}}$ has coordinates $\left(\frac{a + b}{2}, \frac{c}{2}\right)$ and the midpoint **N** of $\overline{\mathbf{BD}}$ also has coordinates $\left(\frac{a + b}{2}, \frac{c}{2}\right)$. Therefore, **M** = **N**. Since the midpoint of each diagonal is on the other diagonal, the diagonals bisect each other.

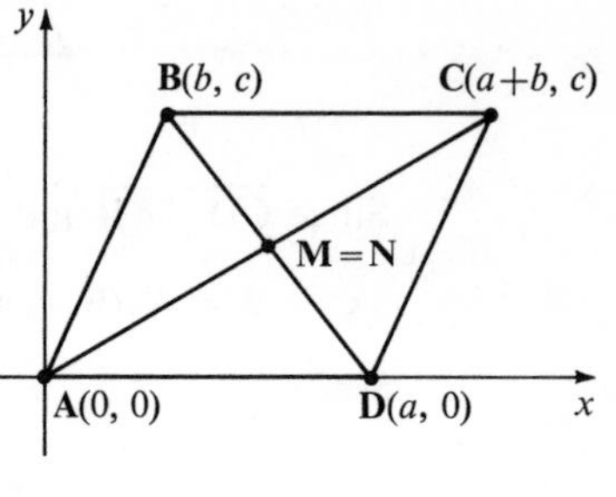

Proof using vectors:

Let **ABCD** be a parallelogram, with **M** and **N** the midpoints of the diagonals $\overline{\mathbf{AC}}$ and $\overline{\mathbf{BD}}$, respectively, as shown below. Then

$$\mathbf{m} - \mathbf{a} = \tfrac{1}{2}(\mathbf{c} - \mathbf{a}),$$

so that

$$\mathbf{m} = \tfrac{1}{2}(\mathbf{c} + \mathbf{a}).$$

Similarly, you have

$$\mathbf{n} = \tfrac{1}{2}(\mathbf{b} + \mathbf{d}).$$

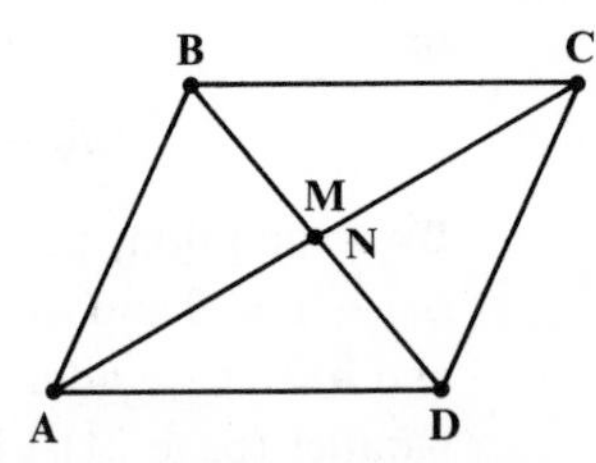

By Equation (3) on page 111, you have

$$\mathbf{c} - \mathbf{d} = \mathbf{b} - \mathbf{a}.$$

Then

$$\mathbf{c} - \mathbf{d} + (\mathbf{d} + \mathbf{a}) = \mathbf{b} - \mathbf{a} + (\mathbf{d} + \mathbf{a}),$$

$$\mathbf{c} + \mathbf{a} = \mathbf{b} + \mathbf{d},$$

$$\tfrac{1}{2}(\mathbf{c} + \mathbf{a}) = \tfrac{1}{2}(\mathbf{b} + \mathbf{d}).$$

Thus **m** = **n**, that is, **M** = **N**, and the theorem is proved.

Exercises 3–4

In Exercises 1–10, prove the given theorem using Cartesian coordinates.

1. The diagonals of a rectangle are equal in length.
2. The diagonals of a square are perpendicular to each other.
3. The diagonals of a rhombus are perpendicular to each other.
4. The midpoint of the hypotenuse of a right triangle is equidistant from the three vertices.
5. The bisectors of the angles of a triangle meet in a point.
6. The sum of the squares of the lengths of the four sides of a parallelogram is equal to the sum of the squares of the lengths of the diagonals.
7. The diagonals of an isosceles trapezoid are of equal length.
8. The diagonals of a trapezoid and the line joining the midpoints of the parallel sides meet in a point.
9. The line segment joining the midpoints of the nonparallel sides of a trapezoid is parallel to the bases, and its length is equal to one-half the sum of the lengths of the bases.
10. The line segment joining the midpoints of two sides of a triangle is parallel to the third side and its length is half the length of the third side.

In Exercises 11–16, prove the given theorem using vectors.

***11.** The medians to the two equal sides of an isosceles triangle are equal in length.
***12.** The perpendicular bisector of the base of an isosceles triangle passes through the vertex of the triangle.
***13.** The midpoints of the sides of a quadrilateral are the vertices of a parallelogram.
***14.** In an equilateral triangle, each median is an altitude.
***15.** The medians of a triangle meet in a point whose distance from each vertex is two-thirds the length of the median from that vertex.
***16.** The perpendicular bisectors of the sides of a triangle meet in a point.

***17–21.** Use vectors to prove the theorems stated in Exercises 2, 3, 6, 9, and 10.

***22–25.** Use Cartesian coordinates to prove the theorems stated in Exercises 12, 13, 15, and 16.

****26.** Prove that if the lines containing two opposite sides of a quadrilateral meet in a point **S**, and the lines containing the other two sides meet in a point **T**, then the midpoint of the segment $\overline{\mathbf{ST}}$ is collinear with the midpoints of the diagonals of the quadrilateral. [*Hint:* Let the origin be at one of the vertices of the quadrilateral.]

Chapter Summary

1. The distance between a point $\mathbf{S}(x_1, y_1)$ and a line $\mathcal{L}$ having direction vector $\mathbf{v}$ and containing a point $\mathbf{T}$ is given by

$$d(\mathbf{S}, \mathcal{L}) = \frac{|(\mathbf{s} - \mathbf{t}) \cdot \mathbf{v_P}|}{\|\mathbf{v_P}\|}.$$

If $\mathcal{L}$ has equation $Ax + By + C = 0$, then

$$d(\mathbf{S}, \mathcal{L}) = \frac{|Ax_1 + By_1 + C|}{\sqrt{A^2 + B^2}}.$$

2. The **intersection** of lines $\mathcal{L}_1$ and $\mathcal{L}_2$ is either a single point if they are not parallel, a line if $\mathcal{L}_1$ and $\mathcal{L}_2$ denote the same line, or the empty set if $\mathcal{L}_1$ and $\mathcal{L}_2$ are parallel and distinct.

3. $\det [a_1] = a_1$; $\det \begin{bmatrix} a_1 & b_1 \\ a_2 & b_2 \end{bmatrix} = \begin{vmatrix} a_1 & b_1 \\ a_2 & b_2 \end{vmatrix} = a_1b_2 - a_2b_1.$ Higher-order determinants can be expanded by minors, for example:

$$\begin{vmatrix} a_1 & b_1 & c_1 \\ a_2 & b_2 & c_2 \\ a_3 & b_3 & c_3 \end{vmatrix} = a_1A_1 - b_1B_1 + c_1C_1,$$

where $A_1 = \begin{vmatrix} b_2 & c_2 \\ b_3 & c_3 \end{vmatrix}$, $B_1 = \begin{vmatrix} a_2 & c_2 \\ a_3 & c_3 \end{vmatrix}$, and $C_1 = \begin{vmatrix} a_2 & b_2 \\ a_3 & b_3 \end{vmatrix}$.

4. Systems of linear equations can be solved by substitution or elimination. They can also be solved by a vector method or by Cramer's Rule.

5. For a system of two linear equations in two variables,

$$\begin{aligned} a_1x + b_1y &= c_1 \\ a_2x + b_2y &= c_2 \end{aligned}$$

in which $a_1b_2 - a_2b_1 \neq 0$, Cramer's Rule is that the unique solution is given by

$$x = \frac{D_x}{D}, \qquad y = \frac{D_y}{D},$$

where

$$D = \begin{vmatrix} a_1 & b_1 \\ a_2 & b_2 \end{vmatrix}, \qquad D_x = \begin{vmatrix} c_1 & b_1 \\ c_2 & b_2 \end{vmatrix}, \qquad D_y = \begin{vmatrix} a_1 & c_1 \\ a_2 & c_2 \end{vmatrix}.$$

6. Equations for lines can be written in determinant form. The two-point form is

$$\begin{vmatrix} x & y & 1 \\ x_1 & y_1 & 1 \\ x_2 & y_2 & 1 \end{vmatrix} = 0.$$

7. Proofs of geometric theorems can be formulated by analytic methods, either in terms of Cartesian coordinates of points or in terms of vectors.

Chapter Review Exercises

1. Find the distance between the point $\mathbf{S}(4, -1)$ and the line $\mathcal{L}$ with equation

$$5x + 12y - 2 = 0.$$

2. Find the distance between the parallel lines with equations

$$3x - 4y - 15 = 0 \quad \text{and} \quad 3x - 4y + 10 = 0.$$

3. Find the intersection of the lines $\mathcal{L}_1$ and $\mathcal{L}_2$ with the given parametric vector equations:

$$\mathcal{L}_1\colon (x, y) = (-2, -2) + r_1(2, 3), \ \mathcal{L}_2\colon (x, y) = (4, 2) + r_2(1, -1).$$

4. Solve the system of linear equations

$$\begin{aligned} x - 3y &= 4 \\ 2x + 5y &= -3 \end{aligned}$$

by elimination.

5. Solve the system of linear equations

$$\begin{aligned} 4x - 2y &= 3 \\ 3x + 5y &= 6 \end{aligned}$$

using vectors.

6. Solve the system of linear equations

$$\begin{aligned} 2x + 3y &= 7 \\ 3x - 2z &= 1 \\ y + 3z &= 2 \end{aligned}$$

by Cramer's Rule.

7. Evaluate the determinant

$$\begin{vmatrix} 2 & 1 & 3 \\ 0 & 4 & 2 \\ -3 & 1 & 5 \end{vmatrix}$$

by expanding it by minors of the second row.

8. Use a third-order determinant to write an equation of the line with x-intercept 7 and y-intercept -3.

9. Find the area of the triangle having vertices $\mathbf{O}(0, 0)$, $\mathbf{S}(3, 5)$, and $\mathbf{T}(6, 4)$.

10. Prove that the angles opposite the sides of equal length in an isosceles triangle are of equal measure.

Constructible and Nonconstructible Angles

There are many angles that can be trisected by means of straightedge-and-compass construction. For example, a right angle can be trisected by this means since, as you saw on page 84, an equilateral triangle can be constructed starting with a configuration of just two points, and, of course, any one of its 60° angles can be bisected.

A 60° angle, however, cannot be trisected by straightedge-and-compass construction; that is, a 20° angle cannot be constructed by this method. To see why this is so, notice first that if a 20° angle could be constructed, then so could cos 20°. Now recall the trigonometric identity

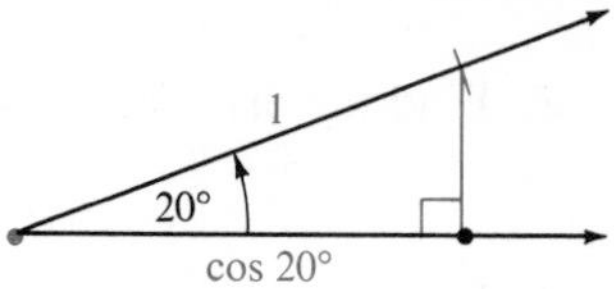

$$\cos 3\theta = 4\cos^3\theta - 3\cos\theta.$$

Since $\cos 60° = \frac{1}{2}$, it follows that cos 20° satisfies the equation

$$\frac{1}{2} = 4x^3 - 3x,$$

or

$$8x^3 - 6x - 1 = 0. \tag{1}$$

Hence cos 20° is an algebraic number, of degree at most 3. It can be shown that no solution of Equation (1) is also a solution of a linear or quadratic equation with integral coefficients. Therefore, cos 20° is an algebraic number of degree 3, and accordingly cos 20° is not constructible. Hence there can be no straightedge-and-compass construction for the trisection of an angle of 60°.

Many constructions have been advanced for the trisection of an arbitrary angle, but of course by the above discussion any such construction either does not satisfy the criteria for a straightedge-and-compass construction or is not exact for all angles.

A routine construction for an *approximate* trisection proceeds as follows: Since any angle can be bisected, and one of the resulting angles again bisected, for any angle θ you can construct angles of degree measure

$$\tfrac{1}{4}m°(\theta),\ (\tfrac{1}{4} + \tfrac{1}{16})m°(\theta),\ (\tfrac{1}{4} + \tfrac{1}{16} + \tfrac{1}{64})m°(\theta),$$

and so on. Since the geometric series

$$\frac{1}{4} + \frac{1}{16} + \frac{1}{64} + \cdots + \frac{1}{4^n} + \cdots$$

converges to $\frac{1}{3}$, by continuing this process you can in a finite number of steps construct an angle of degree measure arbitrarily near to $\frac{1}{3}m°(\theta)$; after the nth step, the error is precisely $\dfrac{1}{3 \cdot 4^n}m°(\theta)$.

Other "trisections" that have been offered cleverly obscure the small error in the construction. Just remember two things: (*a*) in any "trisection" the burden of *proof* is on the person claiming the solution, and (*b*) a valid straightedge-and-compass proof *cannot* be forthcoming.

A configuration due to Archimedes, which in fact is not a straightedge-and-compass construction, does involve the trisection of any given angle AOB. The configuration is drawn as follows:

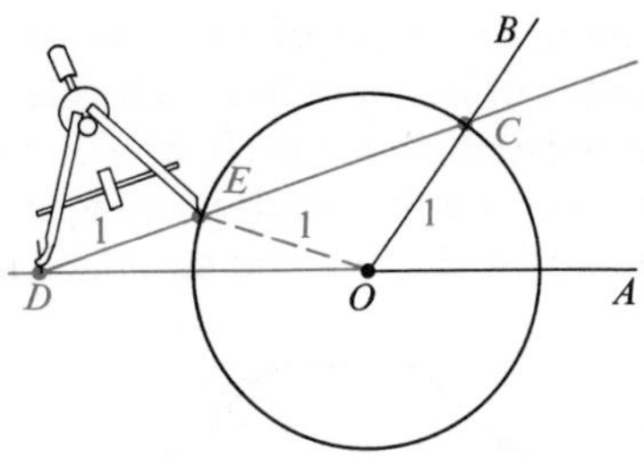

For the given angle AOB, draw the unit circle with center at O, let this circle cut the ray OB at the point C, and extend the ray OA backward from O. Hold one tip of a compass firmly at a point D of a straightedge, place this point on the extension of OA, and let the straightedge also pass through the point C. Now place the other tip of the compass at the point E on the straightedge at unit distance from D in the direction toward C, and adjust the straightedge in such a way the E falls on the circle, as shown. (Alternatively, you could keep the point E on the circle, let the straightedge also pass through the point C, and adjust the straightedge in such a way that the point D falls on the extension of OA.) Since triangles OCE and ODE are isosceles, you have

$$m^\circ(\angle EDO) = m^\circ(\angle EOD),$$

and

$$m^\circ(\angle CEO) = m^\circ(\angle OCE). \tag{2}$$

Then, because the measure of an exterior angle of a triangle is equal to the sum of the measures of the opposite interior angles,

$$m^\circ(\angle CEO) = 2m^\circ(\angle EDO), \tag{3}$$

and

$$m^\circ(\angle AOB) = m^\circ(\angle EDO) + m^\circ(\angle OCE). \tag{4}$$

Therefore, from Equations (2) and (3) above,

$$2m^\circ(\angle EDO) = m^\circ(\angle OCE).$$

Substituting this last result into Equation (4), you obtain

$$m^\circ(\angle AOB) = 3m^\circ(\angle EDO). \tag{5}$$

From the diagram above, you can see that points A, O, and D, and also points C, E, and D, are collinear, so that

$$m^\circ(\angle EDO) = m^\circ(\angle ADC),$$

and hence Equation (5) is equivalent to,

$$m^\circ(\angle ADC) = \tfrac{1}{3}m^\circ(\angle AOB).$$

It should be noted, however, that although you might satisfy yourself visually that physically the configuration of Archimedes has been drawn accurately, the proper adjustment of the straightedge actually cannot be made in accordance with the criteria for a straightedge-and-compass construction.

Since, as you have seen, an angle of measure 20° is not constructible, neither is a regular nonagon, or polygon having 9 sides as shown below. For the same reason, regular polygons having 18, 36, 72, and so on, sides are not constructible. Neither, in fact, are the regular heptagon (7 sides) or, therefore, the regular polygons having 14, 28, 56, and so on, sides constructible.

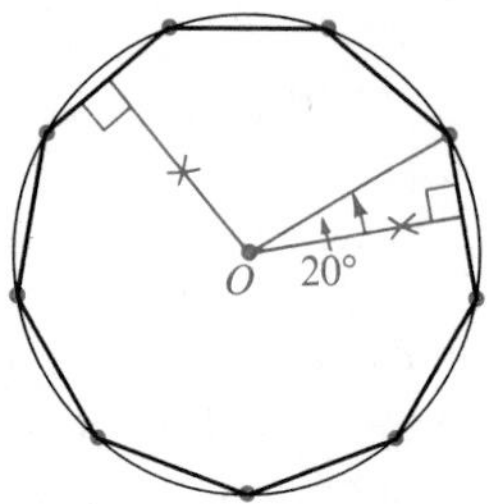

On the other hand, the equilateral triangle and the square are constructible, and therefore so are the regular hexagon and the regular octagon, and so on.

In 1796, when he was only eighteen years old, the great German mathematician Carl Friedrich Gauss (1777–1855) established the totally unanticipated result that a regular polygon having 17 sides is constructible.

Through the theory of algebraic numbers, it is now known that a regular polygon of k sides can be constructed if and only if k is a Fermat prime (that is, a prime number of the form $2^{2^n} + 1$), or a power of 2, or a product of different numbers of these sorts. There are only a few Fermat primes known. The first of these are 3, 5, 17, and 257. Among those regular polygons that can be constructed are the pentagon and the decagon, as will now be shown.

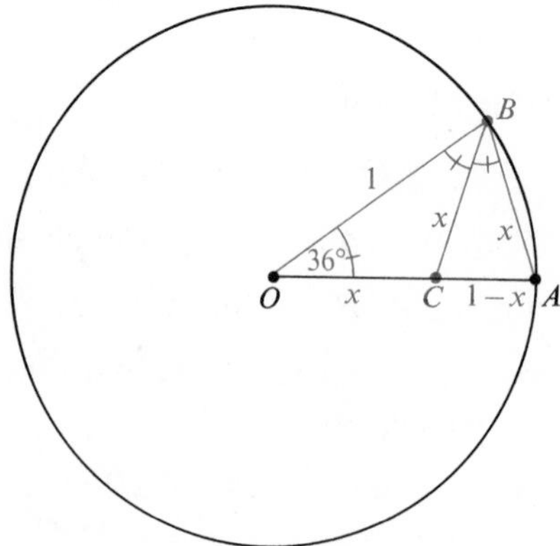

(After a regular decagon has been constructed, you can join alternate vertices to construct a regular pentagon.)

Starting with a circle of unit radius, you can construct an inscribed regular decagon if you can construct a line segment of length equal to the length x of a side of the decagon.

In the figure on page 118, isosceles triangles OAB and BAC are similar. Therefore, you have

$$\frac{1}{x} = \frac{x}{1 - x}, \tag{6}$$

$$x^2 + x - 1 = 0,$$

$$x = \tfrac{1}{2}(\sqrt{5} - 1).$$

Since x is an algebraic number of degree 2, x is constructible, as desired.

By Equation (6), the point on a unit segment at distance x, or $\frac{1}{2}(\sqrt{5} - 1)$, from one end of the segment divides the segment in "extreme and mean" proportion. The ratio $\frac{1}{x}$ is called the *golden ratio*. A rectangle the length of whose sides are in this ratio was called a *golden rectangle* by the ancient Greeks because of its aesthetically pleasing shape. If a square is removed

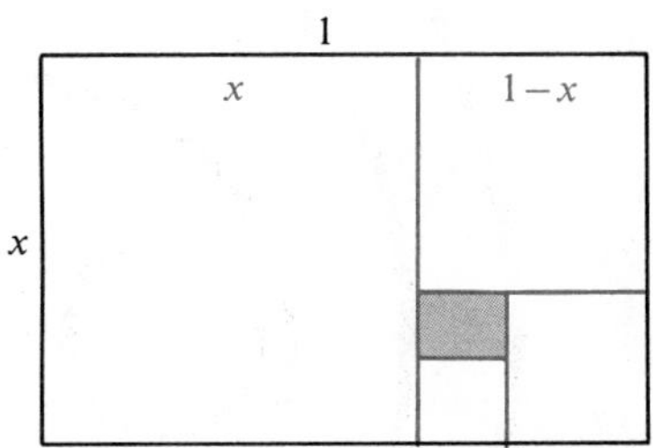

from one end of a golden rectangle, then the remaining rectangle is also golden. It is both interesting and amusing to note that this process of removing squares and leaving golden rectangles can theoretically be continued indefinitely.

Chapter 4

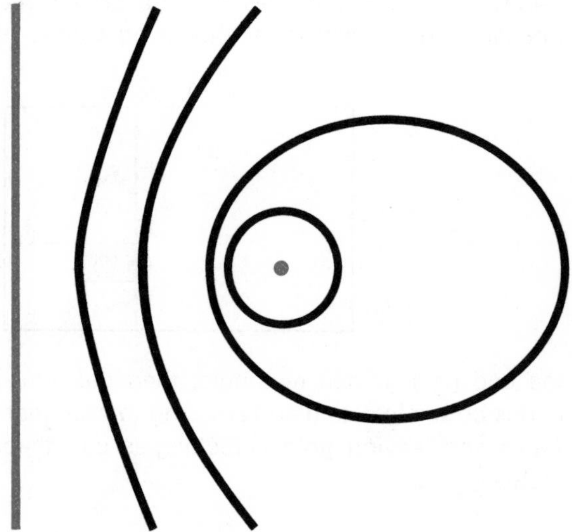

In this chapter, conic sections are defined, their equations derived, and their properties analyzed. The chapter concludes with a discussion of the focus-directrix property illustrated in the diagram above.

Conic Sections

Sets as Loci

4–1 Equation of a Locus

Any set of points, such as a line, can be called a **locus** (plural: **loci**). The word "locus" is ordinarily applied to the set of all points having some common geometric characteristic. For example (see Exercises 35 and 36, page 68), the locus of all points **U** in the plane that are equidistant from two fixed points **S** and **T** is a line, namely, the line $\mathcal{L}$ that is perpendicular to the segment $\overline{\mathbf{ST}}$ at its midpoint (Figure 4–1).

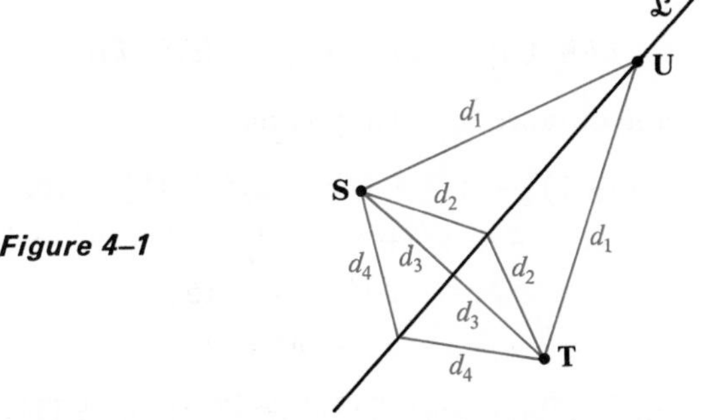

Figure 4–1

In Chapter 2, we considered the problem of determining a vector or a Cartesian equation of a specified line. This is a special instance of a common mathematical problem: finding an equation of a locus $\mathcal{S}$ when given a set of conditions that specifies $\mathcal{S}$. You say that the equation is **an equation of** $\mathcal{S}$, and that $\mathcal{S}$ is **the graph of** the equation, if and only if the equation is satisfied by the coordinates of every point of $\mathcal{S}$ *and* each point whose coordinates satisfy the equation is a point of $\mathcal{S}$. Ordinarily, in determining an equation for a locus, you try to find one that seems to be simple or in some standard form.

Example 1. Find a Cartesian equation for the locus $\mathcal{C}$ of all points such that the line segment from any point of $\mathcal{C}$ to $\mathbf{S}(-2, 0)$ is perpendicular to the line segment from the point to $\mathbf{T}(2, 0)$.

Solution 1 (using vectors):

Let $\mathbf{U}(x, y)$ represent any point on the locus. Since $\overline{\mathbf{US}} \perp \overline{\mathbf{UT}}$,

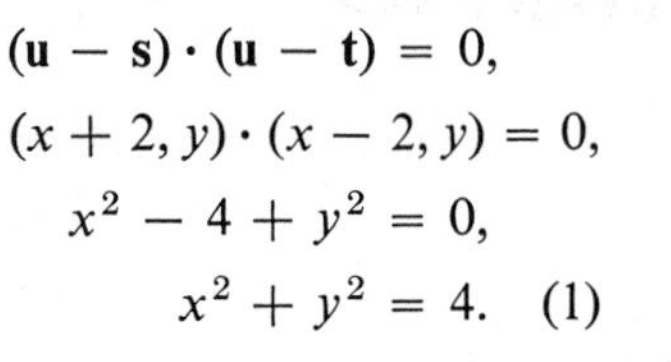

$$(\mathbf{u} - \mathbf{s}) \cdot (\mathbf{u} - \mathbf{t}) = 0,$$
$$(x + 2, y) \cdot (x - 2, y) = 0,$$
$$x^2 - 4 + y^2 = 0,$$
$$x^2 + y^2 = 4. \quad (1)$$

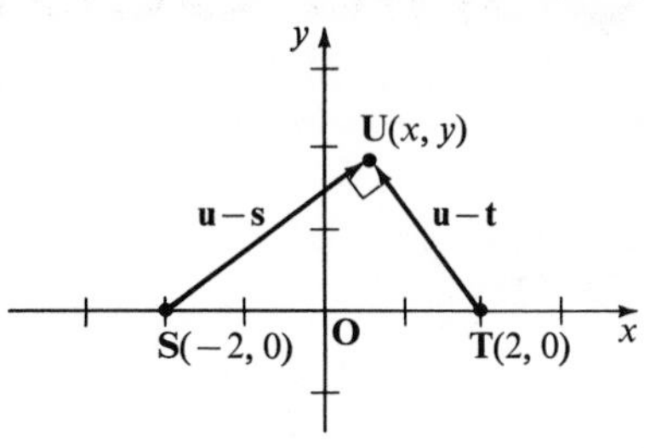

Thus the coordinates of any point on the locus satisfy Equation (1).

Conversely, since the algebraic steps are reversible, and since two vectors are perpendicular if their inner product is 0, any point whose coordinates satisfy Equation (1) is on the locus. Therefore Equation (1) is an equation of the locus.

Solution 2 (using a nonvector method):

If $\overline{\mathbf{US}} \perp \overline{\mathbf{UT}}$, then the triangle **STU** is a right triangle. Hence, by the Pythagorean theorem, for all points **U** on the locus, you have

$$[d(\mathbf{S}, \mathbf{U})]^2 + [d(\mathbf{T}, \mathbf{U})]^2 = [d(\mathbf{S}, \mathbf{T})]^2.$$

Then, by the distance formula, you have

$$[(x + 2)^2 + y^2] + [(x - 2)^2 + y^2] = 16,$$
$$x^2 + 4x + 4 + y^2 + x^2 - 4x + 4 + y^2 = 16,$$
$$2x^2 + 2y^2 + 8 = 16,$$

or

$$x^2 + y^2 = 4.$$

Again, the algebraic steps are reversible; and **STU** is a right triangle by the converse of the Pythagorean theorem. Therefore, Equation (1) is an equation of the locus.

The steps commonly used in deriving a standard or simple equation of a locus can be summarized as follows:

1. Let $\mathbf{U}(x, y)$ be any point on the locus, and write an equation satisfied by $\mathbf{u}$, or by x and y.
2. Perform the necessary transformations to simplify the equation.
3. Verify that any point whose coordinates satisfy the final equation is on the locus. (It often is convenient to do this by noting that the transformations in Step 2 are reversible.)

In deriving an equation of a locus, sometimes it is simpler to use vector methods, and at other times it is simpler to use Cartesian methods.

Example 2. Find a Cartesian equation of the locus $\mathcal{H}$ of all points such that the slope of the segment connecting each point of $\mathcal{H}$ with $\mathbf{S}(1, 2)$ is one-half the slope of the segment connecting the point with $\mathbf{T}(2, 5)$.

Solution: Let $\mathbf{U}(x, y)$ represent any point on the locus. Then, by the slope formula $m = \dfrac{y_2 - y_1}{x_2 - x_1}$, the slope of $\overline{\mathbf{US}}$ is $\dfrac{y-2}{x-1}$ and the slope of $\overline{\mathbf{UT}}$ is $\dfrac{y-5}{x-2}$. Hence, from the given condition, you have

$$\frac{y-2}{x-1} = \frac{1}{2}\left(\frac{y-5}{x-2}\right), \qquad (2)$$

from which, for $x \neq 1$ and $x \neq 2$, you obtain

$$\begin{aligned} 2(y-2)(x-2) &= (x-1)(y-5), \\ 2xy - 4x - 4y + 8 &= xy - y - 5x + 5, \\ xy + x - 3y + 3 &= 0. \end{aligned}$$

The left-hand member of Equation (2) is not defined for $x = 1$ and the right-hand member is not defined for $x = 2$; thus the locus does not include points with these x-coordinates.

Since the algebraic steps are reversible, you can conclude that

$$xy + x - 3y + 3 = 0, \quad x \neq 1, x \neq 2,$$

is an equation of $\mathcal{H}$.

Exercises 4–1

In Exercises 1–18, find an equation for the locus $\mathcal{C}$ of all points in the plane satisfying the stated conditions.

1. Each point of $\mathcal{C}$ is equidistant from the points $\mathbf{S}(1, 4)$ and $\mathbf{T}(3, 7)$.
2. Each point of $\mathcal{C}$ is equidistant from the points $\mathbf{S}(-3, 2)$ and $\mathbf{T}(2, -5)$.
3. Each point of $\mathcal{C}$ is 6 units from the point $\mathbf{S}(2, 4)$.
4. Each point of $\mathcal{C}$ is $\sqrt{7}$ units from the point $\mathbf{S}(-1, -3)$.
5. The line segment connecting each point of $\mathcal{C}$ with $\mathbf{S}(-3, 0)$ is perpendicular to the line segment connecting the point with $\mathbf{T}(3, 0)$.
6. Repeat Exercise 5 for $\mathbf{S}(1, 2)$ and $\mathbf{T}(5, -2)$.

7. The slope of the line segment connecting each point of $\mathcal{C}$ with $\mathbf{S}(1, 6)$ is twice the slope of the line segment connecting the point with $\mathbf{T}(3, 2)$.
8. The slope of the line segment connecting each point of $\mathcal{C}$ with $\mathbf{S}(2, -3)$ is two-thirds of the slope of the line segment connecting the point with $\mathbf{T}(-1, 4)$.
9. The slope of the line segment connecting each point of $\mathcal{C}$ with $\mathbf{S}(2, 5)$ is 2 more than the slope of the line segment connecting the point with $\mathbf{T}(-1, 2)$.
10. The slope of the line segment connecting each point of $\mathcal{C}$ with $\mathbf{S}(1, -2)$ is 3 less than the slope of the line segment connecting the point with $\mathbf{T}(-1, -2)$.
11. Each point of $\mathcal{C}$ is equidistant from $\mathbf{S}(6, 0)$ and the y-axis.
12. Each point of $\mathcal{C}$ is equidistant from $\mathbf{S}(0, 2)$ and the line with equation $y + 2 = 0$.
13. Each point of $\mathcal{C}$ is equidistant from $\mathbf{S}(1, 2)$ and the line with equation $x - y - 5 = 0$.
14. Each point of $\mathcal{C}$ is equidistant from $\mathbf{S}(3, -2)$ and the line with equation $x - y = 0$.
15. The distance from each point of $\mathcal{C}$ to $\mathbf{S}(4, 0)$ is one-half the distance from the point to the line with equation $x + 8 = 0$.
16. The distance from each point of $\mathcal{C}$ to $\mathbf{S}(4, 0)$ is twice its distance to the line with equation $x + 8 = 0$.

* 17. The sum of the distances from each point of $\mathcal{C}$ to $\mathbf{S}(-3, 0)$ and $\mathbf{T}(3, 0)$ is 10.

* 18. The absolute value of the difference of the distances from each point of $\mathcal{C}$ to $\mathbf{S}(-5, 0)$ and $\mathbf{T}(5, 0)$ is 8.

* 19. Let $\mathbf{U}$ be any point on the graph of $y = x^2$, and let $\mathbf{S}$ be the (perpendicular) projection of $\mathbf{U}$ on the x-axis. Find an equation for the locus of the midpoint $\mathbf{M}$ of $\overline{\mathbf{US}}$.

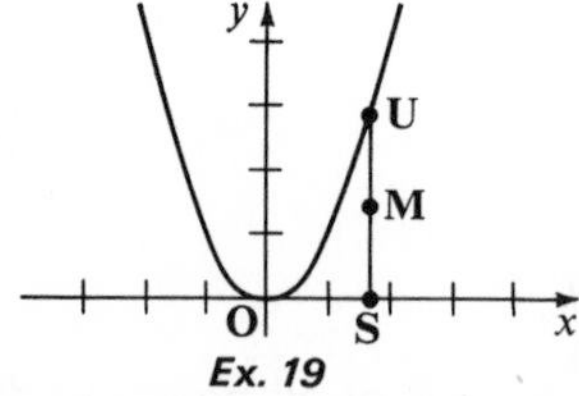

Ex. 19

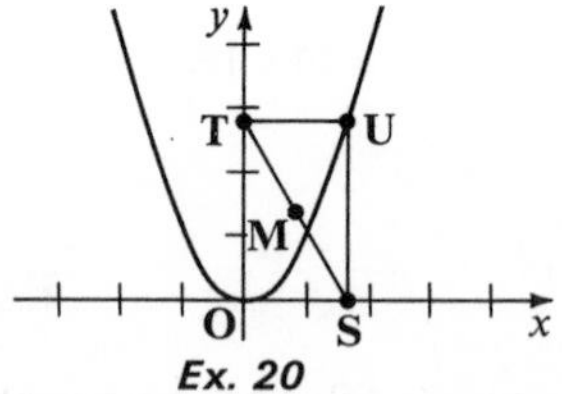

Ex. 20

* 20. Let $\mathbf{U}$ be any point on the graph of $y = x^2$, and let $\mathbf{S}$ and $\mathbf{T}$ be the (perpendicular) projections of $\mathbf{U}$ on the x- and y-axes, respectively. Find an equation for the locus of the midpoint $\mathbf{M}$ of $\overline{\mathbf{ST}}$.

* 21. Let $\mathbf{U}$ be any point on the graph of $y = 3x^2$, and let $\mathbf{S}$ and $\mathbf{T}$ be the (perpendicular) projections of $\mathbf{U}$ on the x- and y-axes, respectively. Find an equation for the locus of the point $\mathbf{Q}$ on $\overline{\mathbf{ST}}$ that lies one-third of the way from $\mathbf{S}$ to $\mathbf{T}$. [*Hint:* Draw a sketch similar to the one for Exercise 20.]

Conic Sections and their Properties

4–2 Circles

In the remainder of this chapter, and also in Chapter 5, we shall study the properties of an important class of loci called *conic sections*. Circles, parabolas, ellipses, and hyperbolas are all examples of conic sections. The derivation of the term "conic section" for these loci will be discussed in Chapter 5.

In studying plane geometry, you learned that a circle is the set (locus) of all points in the plane at a given distance (the radius) from a given point (the center). A segment with one endpoint at the center of a circle and the other endpoint on the circle is called a **radial segment** of the circle. Thus, the radius of a circle is the length of a radial segment.

As illustrated in Figure 4–2, a point $\mathbf{U}(x, y)$ in the plane is on the circle $\mathcal{C}$ of radius 4 with center $\mathbf{S}(2, -1)$ if and only if

$$\|\mathbf{u} - \mathbf{s}\| = 4. \qquad (1)$$

Notice, however, that

$$\begin{aligned} \|\mathbf{u} - \mathbf{s}\| &= \|(x, y) - (2, -1)\| \\ &= \|(x - 2, y + 1)\| \\ &= \sqrt{(x - 2)^2 + (y + 1)^2}. \end{aligned}$$

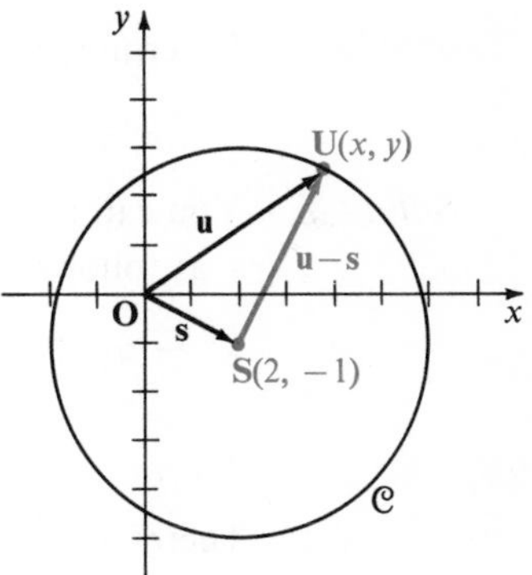

Figure 4–2

Hence, from Equation (1), you have

$$\sqrt{(x - 2)^2 + (y + 1)^2} = 4.$$

Because both members of this equation denote positive numbers, squaring both members produces an equivalent equation:

$$(x - 2)^2 + (y + 1)^2 = 16. \qquad (2)$$

Since the algebraic steps are reversible, Equation (2) is an equation of $\mathcal{C}$, and $\mathcal{C}$ is the graph, or locus, of Equation (2).

Using the same procedure, you can show that, in general, if a circle $\mathcal{C}$ has radius r and center $\mathbf{S}(h, k)$, then a Cartesian equation for $\mathcal{C}$ is

■ $$(x - h)^2 + (y - k)^2 = r^2. \qquad (3)$$

Example 1. Write a Cartesian equation for the circle of radius 6 with center $\mathbf{S}(4, -7)$.

Solution: Setting $h = 4$, $k = -7$, and $r = 6$, from Equation (3) you obtain $(x - 4)^2 + [y - (-7)]^2 = 6^2$, or

$$(x - 4)^2 + (y + 7)^2 = 36.$$

If you expand the left-hand member of Equation (3), page 125, you obtain

$$x^2 - 2hx + h^2 + y^2 - 2ky + k^2 = r^2,$$

or

$$x^2 + y^2 - 2hx - 2ky + h^2 + k^2 - r^2 = 0.$$

This equation, in turn, is of the general form

$$x^2 + y^2 + Dx + Ey + F = 0, \tag{4}$$

where D, E, and F are constants.

You can write an equation which is in the form (4) equivalently in the form

$$(x - h)^2 + (y - k)^2 = c \tag{5}$$

by completing the square in x and y.

Example 2. Determine the locus in $\mathcal{R}^2$ of the equation

$$x^2 + y^2 - 6x + 4y + 4 = 0.$$

Solution: You can rewrite the equation and complete the square in x and y as follows:

$$\begin{aligned} x^2 - 6x + (-3)^2 + y^2 + 4y + (2)^2 &= -4 + (-3)^2 + (2)^2, \\ (x - 3)^2 + (y + 2)^2 &= 9, \\ (x - 3)^2 + [y - (-2)]^2 &= 3^2. \end{aligned}$$

Therefore, the locus is a circle of radius 3 with center at the point $\mathbf{S}(3, -2)$.

Since the square of the radius of a circle must be a positive number, an equation of the form (4) will not necessarily have a locus which is a circle. If after the squares in x and y are completed and the equation is written in the form (5), the right-hand member, c, is 0, then the locus is a single point. If c is negative, then the locus is the empty set, $\emptyset$.

You can use Equation (4), above, to determine an equation of a circle $\mathcal{C}$ when the coordinates of three points on $\mathcal{C}$ are known.

Example 3. Find a Cartesian equation for the circle passing through the points $\mathbf{Q}(3, -2)$, $\mathbf{S}(-1, -4)$, and $\mathbf{T}(2, -5)$.

Solution: The constants D, E, and F must be such that the coordinates of each point satisfy

$$x^2 + y^2 + Dx + Ey + F = 0.$$

Therefore:

$$\begin{aligned} 3^2 + (-2)^2 + 3D - 2E + F &= 0 \\ (-1)^2 + (-4)^2 - D - 4E + F &= 0 \\ (2)^2 + (-5)^2 + 2D - 5E + F &= 0, \end{aligned} \quad \text{or}$$

$$\begin{aligned} 3D - 2E + F &= -13 \\ -D - 4E + F &= -17 \\ 2D - 5E + F &= -29 \end{aligned}$$

Solving this system for D, E, and F, you find that

$$D = -2, E = 6, \text{ and } F = 5.$$

Therefore, a Cartesian equation for the circle is

$$x^2 + y^2 - 2x + 6y + 5 = 0.$$

Because a line $\mathcal{L}$ which is tangent to a circle $\mathcal{C}$ is perpendicular to the radial segment containing the point of contact **T** (Figure 4–3), you know that if **v** is a direction vector of the line containing the radial segment $\overline{\mathbf{TS}}$, then $\mathbf{v_p}$ is a direction vector of $\mathcal{L}$. You can use this information to find an equation for $\mathcal{L}$ if you know the coordinates of **T** and an equation for $\mathcal{C}$.

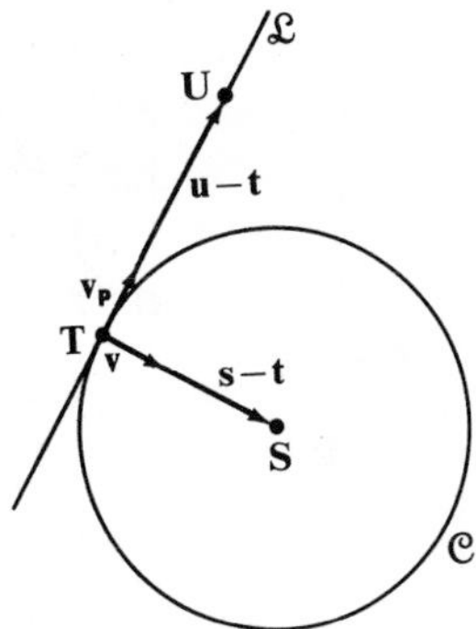

Figure 4–3

Example 4. Find a Cartesian equation for the line $\mathcal{L}$ that is tangent at the point **T**$(6, -4)$ to the circle $\mathcal{C}$ having equation

$$x^2 + y^2 - 4x + 2y - 20 = 0.$$

Solution: Notice first that **T** is on $\mathcal{C}$ since

$$6^2 + (-4)^2 - 4(6) + 2(-4) - 20 = 36 + 16 - 24 - 8 - 20 = 0.$$

Completing the squares in x and y in the equation for $\mathcal{C}$, you have

$$\begin{aligned} (x^2 - 4x + 4) + (y^2 + 2y + 1) &= 20 + 4 + 1, \\ (x - 2)^2 + (y + 1)^2 &= 25. \end{aligned}$$

Therefore $\mathcal{C}$ has center **S**$(2, -1)$, and a direction vector for the line containing the radial segment $\overline{\mathbf{TS}}$ is

$$\mathbf{s} - \mathbf{t} = (2, -1) - (6, -4) = (-4, 3).$$

Thus a point **U**(x, y) is on $\mathcal{L}$ if and only if

$$\begin{aligned} (\mathbf{u} - \mathbf{t}) \cdot (\mathbf{s} - \mathbf{t}) &= 0, \\ [(x, y) - (6, -4)] \cdot (-4, 3) &= 0, \\ (x - 6, y + 4) \cdot (-4, 3) &= 0, \\ -4x + 24 + 3y + 12 &= 0, \\ 4x - 3y - 36 &= 0. \end{aligned}$$

As suggested by Figure 4–3, the radius r of a circle $\mathcal{C}$ is equal to the (perpendicular) distance between the center $\mathbf{S}$ of $\mathcal{C}$ and any line $\mathcal{L}$ that is tangent to $\mathcal{C}$. You can use this fact to determine an equation for a circle given the coordinates (x_1, y_1) of its center and an equation $Ax + By + C = 0$ of a tangent line.

Example 5. Find a Cartesian equation for the circle $\mathcal{C}$ with center $\mathbf{S}(5, 4)$ if the line $\mathcal{L}$ with equation $x + y = 3$ is tangent to $\mathcal{C}$.

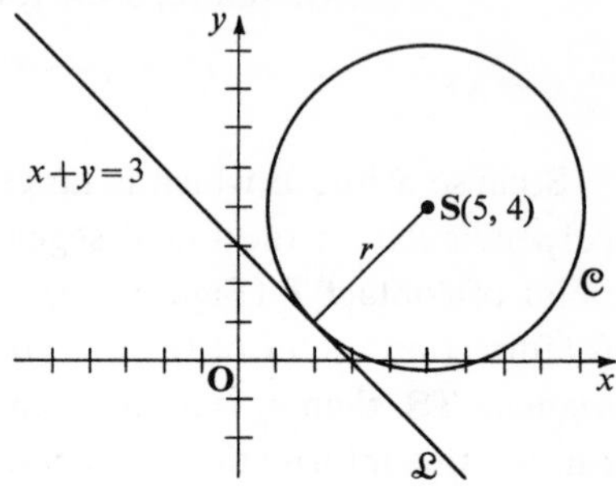

Solution: Make a sketch. In the equation $Ax + By + C = 0$ for $\mathcal{L}$, you have $A = 1$, $B = 1$, and $C = -3$. Also, you have $(x_1, y_1) = (5, 4)$.

Therefore (see page 94), the radius r of $\mathcal{C}$ is given by

$$r = d(\mathbf{S}, \mathcal{L})$$

$$= \left|\frac{Ax_1 + By_1 + C}{\sqrt{A^2 + B^2}}\right| = \left|\frac{5 + 4 - 3}{\sqrt{1^2 + 1^2}}\right| = \frac{6}{\sqrt{2}}.$$

Substituting (5, 4) for (h, k) and $\frac{6}{\sqrt{2}}$ for r in Equation (3) on page 125, you have

$$(x - 5)^2 + (y - 4)^2 = 18$$

as an equation for $\mathcal{C}$.

Exercises 4–2

In Exercises 1–6, find an equation of the form (3) on page 125 for the circle with the given radius r and the given center $\mathbf{S}$.

1. $r = 2$, $\mathbf{S}(3, 4)$
2. $r = 4$, $\mathbf{S}(1, 5)$
3. $r = 7$, $\mathbf{S}(2, -3)$
4. $r = 5$, $\mathbf{S}(-3, 4)$
5. $r = 3$, $\mathbf{S}(-2, -4)$
6. $r = 6$, $\mathbf{S}(-8, -1)$

In Exercises 7–14, find an equation of the form (3) on page 125 for the circle $\mathcal{C}$ with the given equation.

7. $x^2 + y^2 - 2x - 4y - 20 = 0$
8. $x^2 + y^2 - 6x - 8y = 0$
9. $x^2 + y^2 + 10x - 6y - 15 = 0$
10. $x^2 + y^2 - 8x + 12y - 29 = 0$
11. $x^2 + y^2 + 14x + 16y + 13 = 0$
12. $x^2 + y^2 + 12x + 10y - 3 = 0$
13. $4x^2 + 4y^2 + 12x - 20y - 66 = 0$
14. $2x^2 + 2y^2 + 2x + 2y - 1 = 0$

In Exercises 15–18, find an equation of the form (4) on page 126 for the circle $\mathcal{C}$ passing through the given points **Q**, **S**, and **T**.

15. $\mathbf{Q}(5, 4)$, $\mathbf{S}(4, -3)$, $\mathbf{T}(-2, 5)$

16. $\mathbf{Q}(5, 7)$, $\mathbf{S}(6, 0)$, $\mathbf{T}(-1, -1)$

17. $\mathbf{Q}(1, 9)$, $\mathbf{S}(8, 2)$, $\mathbf{T}(-9, 9)$

18. $\mathbf{Q}(3, -10)$, $\mathbf{S}(10, 7)$, $\mathbf{T}(-7, -10)$

In Exercises 19–24, find a Cartesian equation for the line $\mathcal{L}$ that is tangent at the given point **T** to the circle $\mathcal{C}$ having the given equation.

19. $\mathbf{T}(0, 0)$; $x^2 + y^2 - 8x + 6y = 0$

20. $\mathbf{T}(0, 0)$; $x^2 + y^2 - 10x - 4y = 0$

21. $\mathbf{T}(2, 1)$; $x^2 + y^2 - 12x + 14y + 5 = 0$

22. $\mathbf{T}(3, 2)$; $x^2 + y^2 - 12x + 8y + 7 = 0$

23. $\mathbf{T}(-3, 1)$; $x^2 + y^2 + 8x - 2y + 16 = 0$

24. $\mathbf{T}(2, -1)$; $x^2 + y^2 - 4x + 6y + 9 = 0$

In Exercises 25–30, find a Cartesian equation for the circle $\mathcal{C}$ that has the given center **S** and is tangent to the line with the given equation.

25. $\mathbf{S}(3, 4)$, $x = 8$

26. $\mathbf{S}(5, 2)$, $y = 5$

27. $\mathbf{S}(1, -5)$, $x + y = 5$

28. $\mathbf{S}(-2, 3)$, $x - y = 7$

29. $\mathbf{S}(-2, -4)$, $2x - y = 4$

30. $\mathbf{S}(-3, -5)$, $2x + 3y = 5$

In Exercises 31–38, find a Cartesian equation for the circle satisfying the given conditions.

31. Containing the points $\mathbf{S}(-1, -3)$ and $\mathbf{T}(-5, 3)$, with center on the line with equation $x - 2y + 2 = 0$.

32. Containing the points $\mathbf{S}(0, 0)$ and $\mathbf{T}(6, 2)$, with center on the line with equation $2x - y = 0$.

33. Tangent to the x-axis at the point $\mathbf{S}(4, 0)$, and containing the point $\mathbf{T}(7, 1)$.

34. Tangent to the line with equation $4x - 3y - 2 = 0$ at the point $\mathbf{S}(6, -1)$, and containing the point $\mathbf{T}(6, 1)$.

35. Center at $\mathbf{S}(-1, 4)$, and tangent to the line with equation $5x + 12y + 9 = 0$.

36. Center at $\mathbf{S}(2, 4)$, and tangent to the line with equation $x + y - 4 = 0$.

37. Tangent to the line with equation $2x - y + 6 = 0$ at the point $\mathbf{S}(-1, 4)$, and having radius $3\sqrt{5}$. (Two solutions.)

38. Tangent to the line with equation $x - 2y - 3 = 0$ at the point $\mathbf{S}(-1, -2)$, and having radius $\sqrt{5}$. (Two solutions.)

* **39.** Show that for three noncollinear points $\mathbf{R}(x_1, y_1)$, $\mathbf{S}(x_2, y_2)$, $\mathbf{T}(x_3, y_3)$,

$$\begin{vmatrix} x^2 + y^2 & x & y & 1 \\ x_1^2 + y_1^2 & x_1 & y_1 & 1 \\ x_2^2 + y_2^2 & x_2 & y_2 & 1 \\ x_3^2 + y_3^2 & x_3 & y_3 & 1 \end{vmatrix} = 0$$

is a Cartesian equation of the circle containing **R**, **S**, and **T**.

* **40.** Determine the graph of the equation in Exercise 39 in case **R**, **S**, and **T** are distinct collinear points.

4–3 Parabolas

The set $\mathcal{P}$ of points in the plane each of which is located the same distance from a fixed line $\mathcal{D}$ and a fixed point **F** not on $\mathcal{D}$ is called a **parabola** (Figure 4–4).

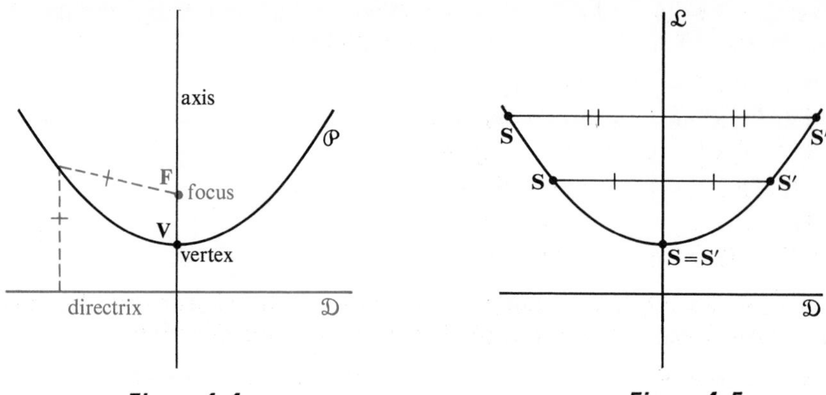

Figure 4–4 *Figure 4–5*

The point **F** is called the **focus** of the parabola, and the line $\mathcal{D}$ is called its **directrix**. The line that contains the focus of a parabola and is perpendicular to the directrix is called the **axis** (or **axis of symmetry**) of the parabola. The point **V** of intersection of a parabola with its axis is called the **vertex** of the parabola.

Two points **S** and **S′** are said to be *symmetric with respect to a line* $\mathcal{L}$ if $\mathcal{L}$ is the perpendicular bisector of the segment $\overline{\mathbf{SS'}}$. A *set* of points is said to be *symmetric with respect to a line* $\mathcal{L}$ if for every point **S** in the set there is a point **S′** in the set such that **S** and **S′** are symmetric points with respect to $\mathcal{L}$ (Figure 4–5). It follows directly from the definition of *parabola* that a parabola is symmetric with respect to its axis.

Figure 4–6 illustrates a simple mechanical construction of a portion of the parabola with focus **F** and directrix $\mathcal{D}$. Place a T-square (or other straightedge) at right angles to $\mathcal{D}$ at **A** and choose point **B** on the T-square. At **F** and **B** fasten the ends of a string of length $d(\mathbf{A}, \mathbf{B})$. The point of a pencil which holds the string taut against the edge of the T-square (at **U** in Figure 4–6) will describe the arc of a parabola as the T-square is moved always at right angles to $\mathcal{D}$, with **A** moving along $\mathcal{D}$.

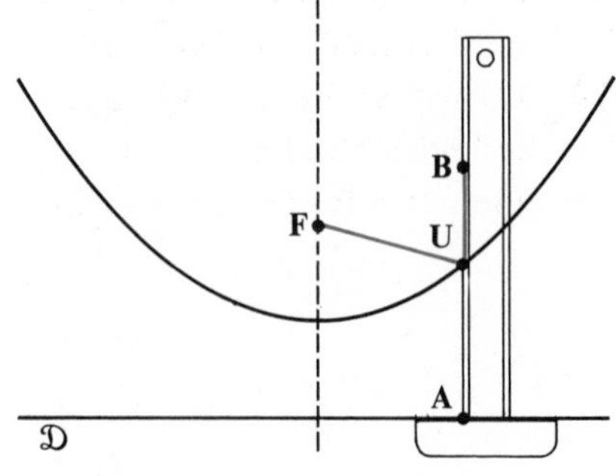

Figure 4–6

To obtain an equation for the parabola having the point **F**(0, 2) as its focus and the line $\mathfrak{D}$ with equation $y = -2$ as its directrix (Figure 4–7), you can begin

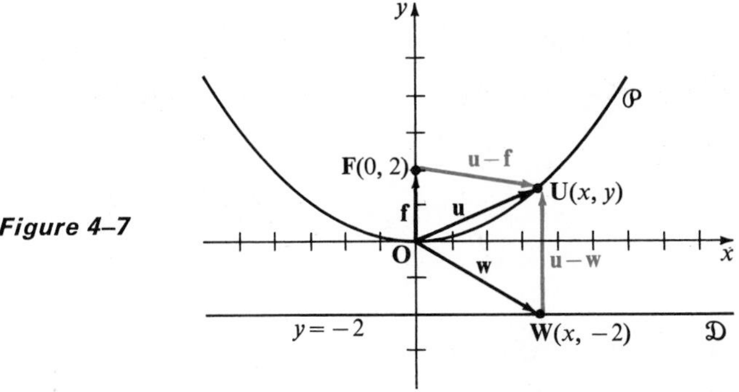

Figure 4–7

by observing that **U**(x, y) is a point of the locus if and only if

$$\|\mathbf{u} - \mathbf{f}\| = \|\mathbf{u} - \mathbf{w}\|,$$

where **W** is the (perpendicular) projection of **U** on $\mathfrak{D}$. Then, since **U** has coordinates (x, y), **F** has coordinates (0, 2), and **W** has coordinates $(x, -2)$, you have

$$\begin{aligned}
\|(x, y) - (0, 2)\| &= \|(x, y) - (x, -2)\|, \\
\|(x - 0, y - 2)\| &= \|(x - x, y + 2)\|, \\
\sqrt{x^2 + (y - 2)^2} &= \sqrt{0^2 + (y + 2)^2}, \\
x^2 + (y - 2)^2 &= (y + 2)^2, \\
x^2 + y^2 - 4y + 4 &= y^2 + 4y + 4,
\end{aligned}$$

or

$$x^2 = 8y.$$

Since the square roots involved are nonnegative, the steps are reversible, and the graph of $x^2 = 8y$ is the specified parabola.

By similar reasoning, you can show that an equation for the parabola with focus **F**$(0, p)$ and directrix the line $\mathfrak{D}$ with equation $y = -p$ is

$$x^2 = 4py, \tag{1}$$

and an equation for the parabola with focus **F**$(p, 0)$ and directrix the line $\mathfrak{D}$ with equation $x = -p$ is

$$y^2 = 4px. \tag{2}$$

Equations (1) and (2) are called **standard forms** of the equation for a parabola with vertex at the origin and focus on a coordinate axis. If $p > 0$, then the parabola opens upward or to the right; if $p < 0$, then the parabola opens downward or to the left. Figure 4–8 on the following page shows the possibilities.

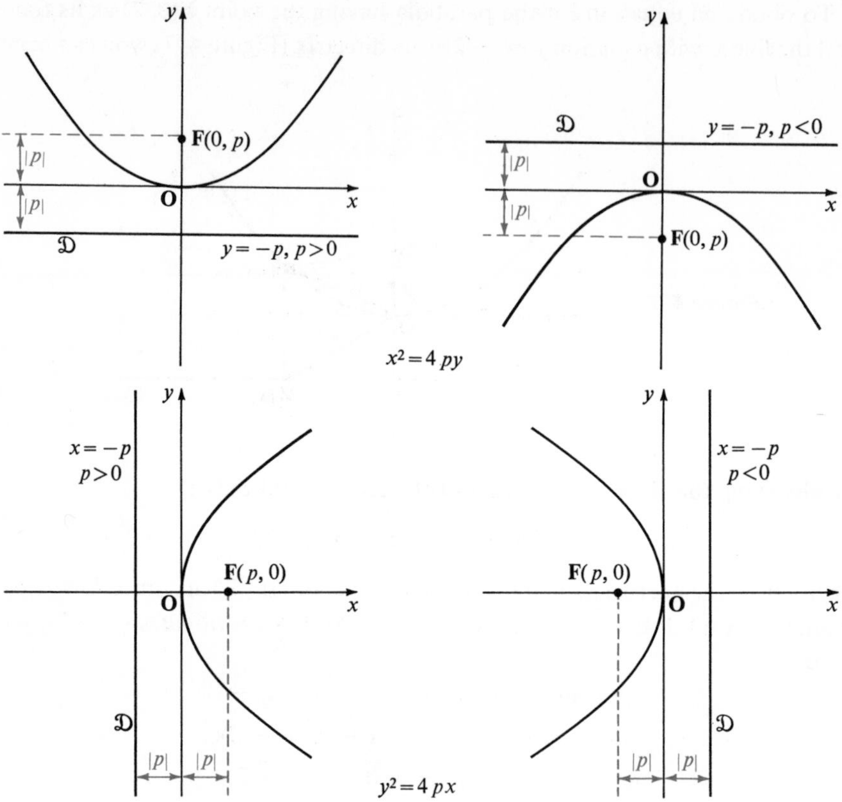

$x^2 = 4py$

$y^2 = 4px$

Figure 4–8

Notice that in each case the vertex is located $|p|$ units from both the focus and the directrix.

Example. Find the coordinates of the focus **F** and an equation for the directrix $\mathfrak{D}$ of the parabola with equation $y^2 = -8x$. Sketch the curve.

Solution: Comparing $y^2 = -8x$ with Equation (2) on page 131, you can see that $p = -2$. Hence, the coordinates of **F** are $(-2, 0)$ and $\mathfrak{D}$ has equation $x = 2$. Since y^2 is always nonnegative, and $y^2 = -8x$, x must be nonpositive for each point on the parabola, and the graph appears as shown.

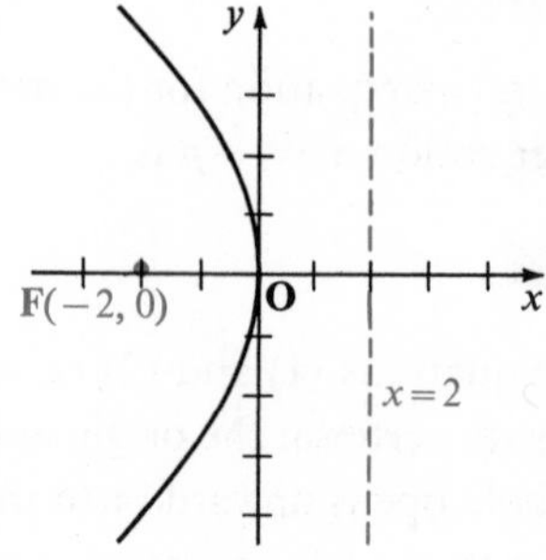

Exercises 4–3

In Exercises 1–8, find a Cartesian equation for the parabola with vertex at the origin and focus at the given point **F**. Also find a Cartesian equation for the directrix.

1. F(0, 3)
2. F(0, 5)
3. F(2, 0)
4. F(4, 0)
5. F(−1, 0)
6. F(0, −4)
7. F(0, −5)
8. F(−3, 0)

In Exercises 9–12, find the coordinates of the focus and a Cartesian equation for the directrix of the parabola having vertex at the origin and passing through the given points **S** and **T**.

Example 1. **S**(−3, 3) and **T**(3, 3)

Solution: Make a sketch. By inspection, you can see that **S**(−3, 3) and

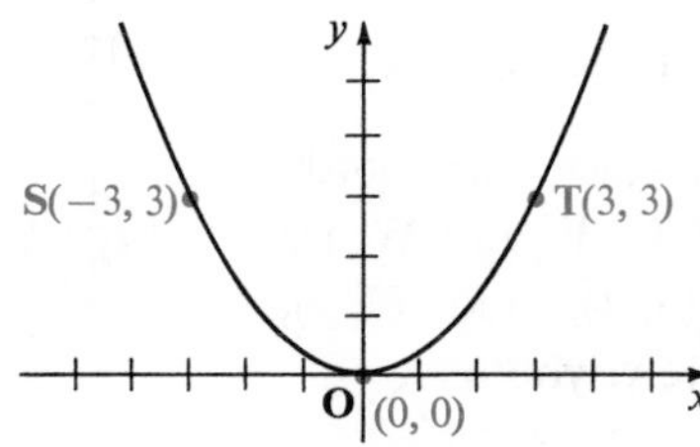

T(3, 3) are symmetric with respect to the y-axis. Then, since the vertex is at the origin, the y-axis is the axis of the parabola, and accordingly the parabola has an equation of the form $x^2 = 4py$. Substituting the coordinates of either **S** or **T** in this equation yields

$$9 = 12p,$$

or

$$p = \tfrac{3}{4}.$$

Therefore the coordinates of the focus are $(0, \tfrac{3}{4})$, and an equation for the directrix is $y = -\tfrac{3}{4}$.

9. **S**(−4, 2) and **T**(4, 2)
10. **S**(−5, 5) and **T**(−5, −5)
11. **S**(2, 4) and **T**(2, −4)
12. **S**(−6, −8) and **T**(6, −8)

In Exercises 13–16, find a Cartesian equation for the parabola satisfying the given conditions.

13. Vertex at the origin, axis along the x-axis, containing the point $\mathbf{S}(8, 8)$.
14. Vertex at the origin, axis along the x-axis, containing the point $\mathbf{S}(9, 6)$.
15. Vertex at the origin, axis along the y-axis, containing the point $\mathbf{S}(-2, 4)$.
16. Vertex at the origin, axis along the y-axis, containing the point $\mathbf{S}(-4, 48)$.

In Exercises 17–22, use the definition of a parabola to find a Cartesian equation for the parabola with the given focus, **F**, and the given directrix.

Example 2. $\mathbf{F}(3, 4)$;
equation of directrix: $x = 7$.

Solution: Sketch the curve as shown.

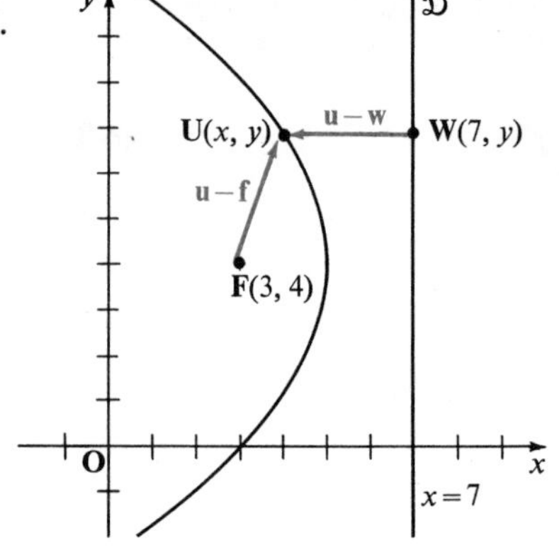

A point $\mathbf{U}(x, y)$ is on the parabola if and only if

$$\|\mathbf{u} - \mathbf{f}\| = \|\mathbf{u} - \mathbf{w}\|.$$

Then, since the coordinates of **U**, **F**, and **W** are (x, y), $(3, 4)$, and $(7, y)$, respectively, you have

$$\|(x, y) - (3, 4)\| = \|(x, y) - (7, y)\|,$$

or

$$\sqrt{(x - 3)^2 + (y - 4)^2} = \sqrt{(x - 7)^2 + 0^2}.$$

Squaring both members and simplifying, you have

$$(x - 3)^2 + (y - 4)^2 = (x - 7)^2,$$
$$x^2 - 6x + 9 + y^2 - 8y + 16 = x^2 - 14x + 49,$$
$$y^2 - 8y + 8x - 24 = 0.$$

Since the algebraic steps are reversible, this is an equation for the parabola.

17. $\mathbf{F}(5, 2)$; equation of directrix: $x = 1$.
18. $\mathbf{F}(-3, 2)$; equation of directrix: $x = -7$.
19. $\mathbf{F}(4, -5)$; equation of directrix: $y = 1$.
20. $\mathbf{F}(-2, -5)$; equation of directrix: $y = -4$.
21. $\mathbf{F}(8, 0)$; equation of directrix: $x = 1$.
22. $\mathbf{F}(0, -3)$; equation of directrix: $y = 10$.

The general form for a Cartesian equation of a parabola with axis parallel to the y-axis is

$$y = ax^2 + bx + c, \qquad a \neq 0,$$

with the parabola opening upward if $a > 0$ and downward if $a < 0$. In Exercises 23–28, find an equation of this form for the parabola $\mathcal{P}$ containing the given points **Q**, **S**, and **T**.

Example 3. **Q**(0, 1), **S**(−1, 4), **T**(2, 1)

Solution: Since each point is on the graph of $y = ax^2 + bx + c$, the coordinates of the points must satisfy this equation. Thus you have

$$\begin{aligned} 1 &= a(0)^2 + b(0) + c \\ 4 &= a(-1)^2 + b(-1) + c \\ 1 &= a(2)^2 + b(2) + c \end{aligned}$$

or

$$\begin{aligned} c &= 1 \\ a - b + c &= 4 \\ 4a + 2b + c &= 1. \end{aligned}$$

Solving this system for a, b, and c, you find that $a = 1$, $b = -2$, and $c = 1$. Therefore, an equation for $\mathcal{P}$ is

$$y = x^2 - 2x + 1.$$

23. **Q**(0, 1), **S**(1, 6), **T**(−1, 0)

24. **Q**(0, 2), **S**(1, −2), **T**(−1, 13)

25. **Q**(0, 3), **S**(1, 3), **T**(2, 1)

26. **Q**(0, 2), **S**(1, −1), **T**(−1, 1)

27. **Q**(0, −5), **S**(1, −2), **T**(−2, 7)

28. **Q**(0, 6), **S**(1, 1), **T**(−2, −14)

* **29.** Show that a Cartesian equation for the parabola $\mathcal{P}$ with focus **F**$(0, p)$ and directrix the line $\mathcal{D}$ with equation $y = -p$ is $x^2 = 4py$.

* **30.** Show that a Cartesian equation for the parabola $\mathcal{P}$ with focus **F**$(p, 0)$ and directrix the line $\mathcal{D}$ with equation $x = -p$ is $y^2 = 4px$.

The length of the segment which has midpoint at the focus of a parabola, is perpendicular to the axis of the parabola, and has endpoints lying on the parabola is called the *focal width* of the parabola. The segment itself is called the *latus rectum* of the parabola.

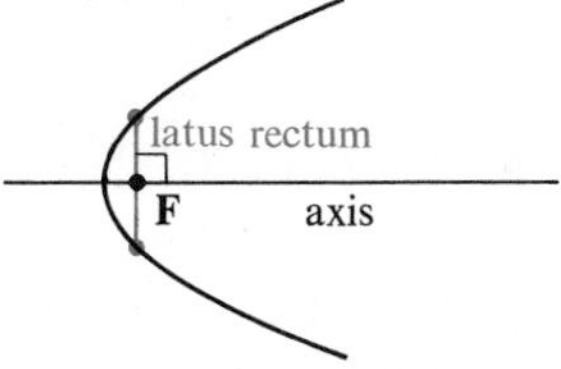

* **31.** Show that the focal width of the parabola with equation $x^2 = 4py$ is $|4p|$.

* **32.** Show that the focal width of the parabola with equation $y^2 = 4px$ is $|4p|$.

* **33.** Find a Cartesian equation for the circle containing the vertex and the two endpoints of the latus rectum of the parabola with equation $x^2 = 4py$.

* **34.** Show that the circle $\mathcal{C}$ having the latus rectum of the parabola $\mathcal{P}$ with equation $x^2 = 4py$ as a diameter is tangent to the directrix of $\mathcal{P}$.

*** 35.** Show that for any three noncollinear points $\mathbf{R}(x_1, y_1)$, $\mathbf{S}(x_2, y_2)$, and $\mathbf{T}(x_3, y_3)$, with x_1, x_2, and x_3 all different,

$$\begin{vmatrix} x^2 & x & y & 1 \\ x_1^2 & x_1 & y_1 & 1 \\ x_2^2 & x_2 & y_2 & 1 \\ x_3^2 & x_3 & y_3 & 1 \end{vmatrix} = 0$$

is a Cartesian equation for the parabola with vertical axis containing **R**, **S**, and **T**.

*** 36.** Determine the graph of the equation in Exercise 35 in case **R**, **S**, and **T** are collinear.

Applied Problems 4–3

1. A commemorative parabolic steel arch, 100 meters high, has its axis vertical and its feet 200 meters apart. Is the focus of the parabola above or below ground, and by how much?

2. A commemorative parabolic steel arch is planned, with its axis vertical and its feet 300 meters apart. If the focus of the parabola is to be 80 meters above ground, how high must the arch be?

3. A parabolic reflector is in the shape made by revolving an arc of a parabola, starting at the vertex, about the axis of the parabola. If the focus is 9 inches from the vertex, and the parabolic arc is 16 inches deep, how wide is the opening of the reflector?

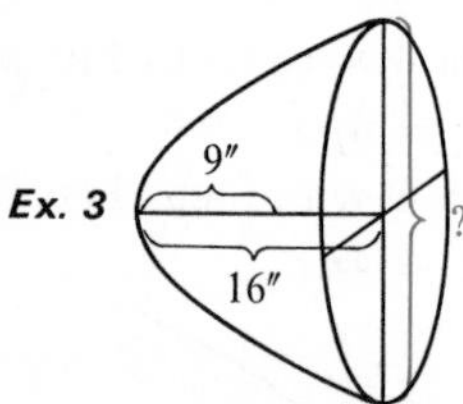

Ex. 3

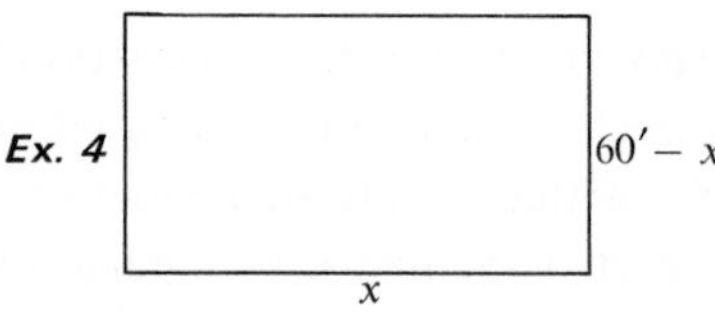

Ex. 4

4. A rectangular field is to be enclosed by 120 feet of fencing. Show that if y denotes the area when one side is of length x, then $y = 60x - x^2$. Plot the parabolic graph of this equation for $0 \leq x \leq 60$. For what value of x is the area greatest?

5. When a rock is thrown from a point **A**, it travels approximately along a parabolic arc with vertical axis. If the rock is thrown at an angle of 45° with the ground, then the focus of the parabola is on a horizontal line through **A**. Suppose a rock thrown at this angle attains a height of 40 feet above **A**. How far does the rock travel horizontally before returning to the same height as **A**?

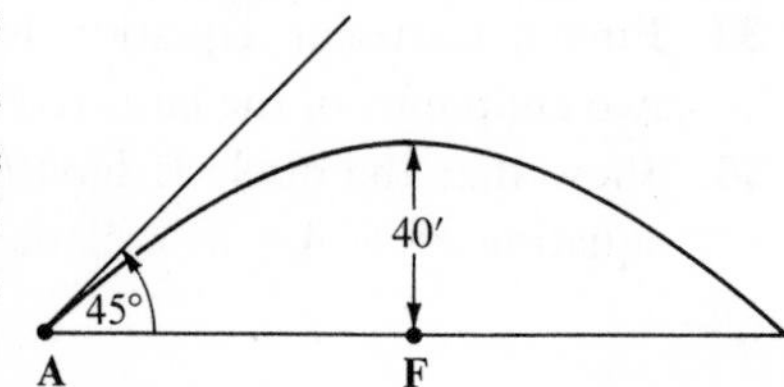

4–4 Ellipses

For two given points $\mathbf{F}_1$ and $\mathbf{F}_2$ in the plane, the set $\mathcal{E}$ of points $\mathbf{U}(x, y)$ in the plane such that the sum of the distances of $\mathbf{U}$ from $\mathbf{F}_1$ and $\mathbf{F}_2$ is a given constant, greater than $d(\mathbf{F}_1, \mathbf{F}_2)$, is called an **ellipse** (see Figure 4–9). The points

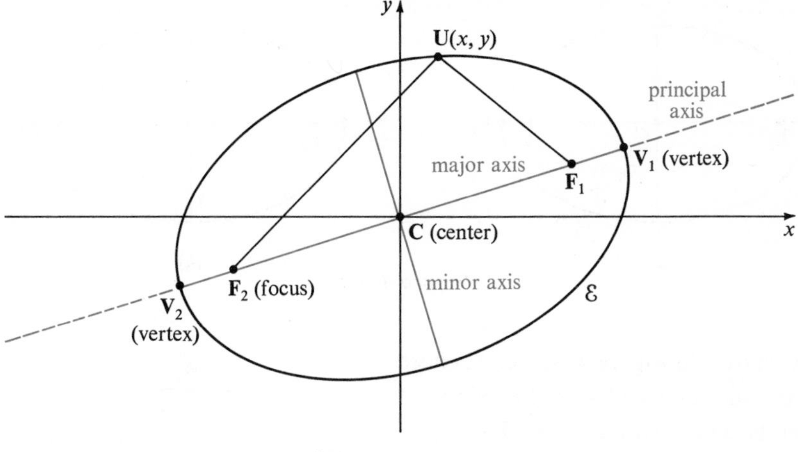

Figure 4–9

$\mathbf{F}_1$ and $\mathbf{F}_2$ are called the **foci** of the ellipse, and the line containing the foci is called the **principal axis** of the ellipse. The point $\mathbf{C}$ that bisects the segment $\overline{\mathbf{F}_1\mathbf{F}_2}$ is called the **center** of the ellipse. The points of intersection of the ellipse and its principal axis, $\mathbf{V}_1$ and $\mathbf{V}_2$ in Figure 4–9, are called the **vertices** of the ellipse, and the segment with the vertices as endpoints is called the **major axis** of the ellipse. The segment which is perpendicular to the major axis at the center and which has endpoints on the ellipse is called the **minor axis** of the ellipse.

You can imagine how you might construct an ellipse in laying out an oval flower bed. First, driving two wooden stakes into the ground, you would pass a

Figure 4–10

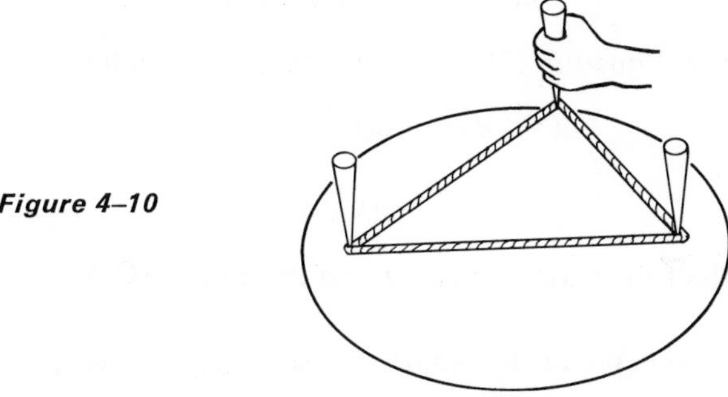

loop of rope, with some slack, about the two stakes. Then, holding the rope taut, you would mark the elliptical boundary of the bed, as indicated in Figure 4–10 above.

Some properties of an ellipse are apparent from its definition. Looking at Figure 4–11(a), for example, you can see that an ellipse must be symmetric (see page 130) with respect to its principal axis, and looking at Figure 4–11(b), you can see that it must also be symmetric with respect to the line containing its minor axis.

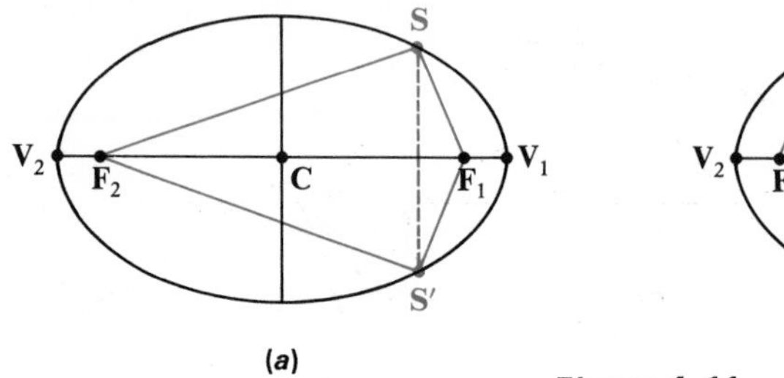

Figure 4–11

As shown in Figure 4–12, let c denote the distance of each focus of an ellipse $\mathcal{E}$ from its center, so that $d(\mathbf{F}_1, \mathbf{F}_2) = 2c$. Then let $2a$ denote the sum of the distances of each point $\mathbf{U}$ of $\mathcal{E}$ from $\mathbf{F}_1$ and $\mathbf{F}_2$:

$$d(\mathbf{F}_1, \mathbf{U}) + d(\mathbf{F}_2, \mathbf{U}) = 2a.$$

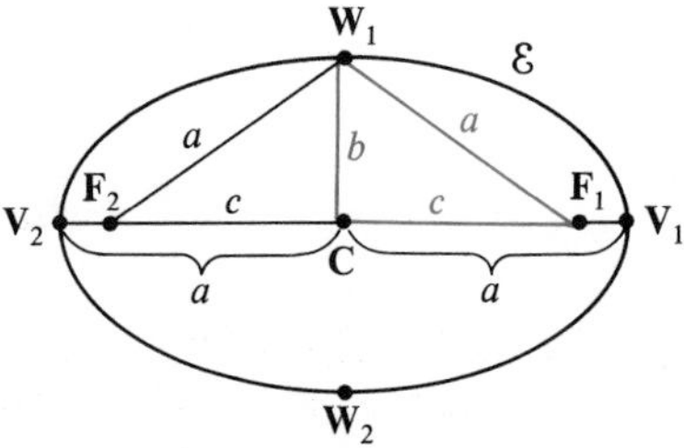

Figure 4–12

Since this sum must be greater than $d(\mathbf{F}_1, \mathbf{F}_2)$, you have $2a > 2c$, or $a > c$. Because the vertex $\mathbf{V}_1$, in particular, is a point of $\mathcal{E}$, you have

$$d(\mathbf{F}_1, \mathbf{V}_1) + d(\mathbf{F}_2, \mathbf{V}_1) = 2a. \qquad (1)$$

Notice, however, that by symmetry

$$d(\mathbf{F}_1, \mathbf{V}_1) = d(\mathbf{V}_2, \mathbf{F}_2). \qquad (2)$$

Substituting from Equation (2) into Equation (1), you obtain

$$d(\mathbf{V}_2, \mathbf{F}_2) + d(\mathbf{F}_2, \mathbf{V}_1) = 2a,$$

or

$$d(\mathbf{V}_2, \mathbf{V}_1) = 2a.$$

Thus the length of the major axis is $2a$, and the length $d(\mathbf{C}, \mathbf{V}_1)$ of a semi-major axis is a.

Since an endpoint $\mathbf{W}_1$ of the minor axis is on $\mathcal{E}$, you have

$$d(\mathbf{F}_1, \mathbf{W}_1) + d(\mathbf{F}_2, \mathbf{W}_1) = 2a,$$

or, by symmetry,

$$2d(\mathbf{F}_1, \mathbf{W}_1) = 2a,$$
$$d(\mathbf{F}_1, \mathbf{W}_1) = a.$$

Therefore, if b denotes the length $d(\mathbf{C}, \mathbf{W}_1)$ of a semi-minor axis of $\mathcal{E}$, then by the Pythagorean theorem you have

$$a^2 = b^2 + c^2. \tag{3}$$

You can find an equation for an ellipse with center at the origin and foci on the x-axis as follows: Suppose that the ellipse $\mathcal{E}$ (Figure 4–13) has foci $\mathbf{F}_1(c, 0)$ and $\mathbf{F}_2(-c, 0)$, and that for each point $\mathbf{U}(x, y)$ on $\mathcal{E}$ the sum of the distances of

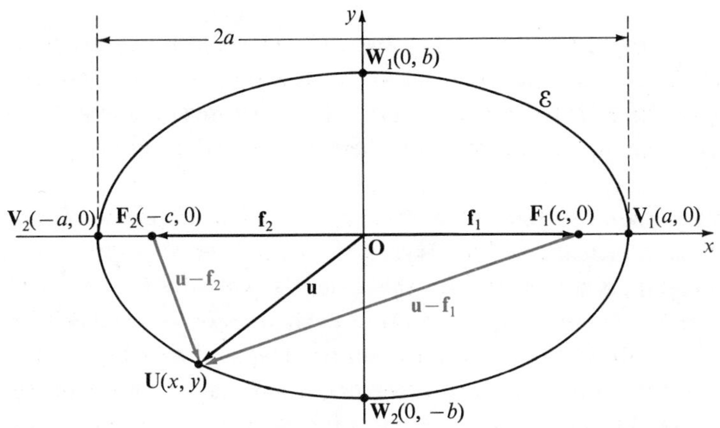

Figure 4–13

$\mathbf{U}$ from $\mathbf{F}_1$ and $\mathbf{F}_2$ is the constant $2a$, where $2a > 2c$, or $a > c$. Then a vector equation of $\mathcal{E}$ is

$$\|\mathbf{u} - \mathbf{f}_2\| + \|\mathbf{u} - \mathbf{f}_1\| = 2a.$$

Since $\mathbf{u} = (x, y)$, $\mathbf{f}_1 = (c, 0)$, and $\mathbf{f}_2 = (-c, 0)$, you have

$$\|(x, y) - (-c, 0)\| + \|(x, y) - (c, 0)\| = 2a,$$

$$\|(x + c, y)\| + \|(x - c, y)\| = 2a,$$

$$\sqrt{(x + c)^2 + y^2} + \sqrt{(x - c)^2 + y^2} = 2a.$$

If, now, you add $-\sqrt{(x - c)^2 + y^2}$ to both members of this equation and then square both members of the resulting equivalent equation, you obtain

$$(x + c)^2 + y^2 = 4a^2 - 4a\sqrt{(x - c)^2 + y^2} + (x - c)^2 + y^2.$$

Expanding binomials and simplifying, you have

$$a\sqrt{(x - c)^2 + y^2} = a^2 - cx. \tag{4}$$

If, next, both members of this equation are squared and equated, you find that

$$a^2(x^2 - 2cx + c^2 + y^2) = a^4 - 2a^2cx + c^2x^2.$$

Simplifying and factoring, you obtain

$$(a^2 - c^2)x^2 + a^2y^2 = a^2(a^2 - c^2).$$

Since $0 < c < a$, it follows that $c^2 < a^2$ and $a^2 - c^2 > 0$. If you set $b^2 = a^2 - c^2$, $b > 0$, then this last equation reduces to

$$b^2x^2 + a^2y^2 = a^2b^2,$$

or, on dividing by a^2b^2, to

$$\frac{x^2}{a^2} + \frac{y^2}{b^2} = 1. \tag{5}$$

The steps used in obtaining Equation (5) actually are reversible, although care must be exercised in discussing the square roots involved; the details will not be given here. It follows that a point $\mathbf{U}(x, y)$ is on the ellipse $\mathcal{E}$ if and only if its coordinates satisfy Equation (5); hence (5) is a (Cartesian) equation of the ellipse.

From Equation (5), you can see once again that an ellipse is symmetric with respect to its principal axis (in this case, the x-axis) and also with respect to the line containing its minor axis (in this case, the y-axis). That is, if the point $\mathbf{S}(r, t)$ is on the graph of Equation (5), then so are $\mathbf{S}'(r, -t)$ and $\mathbf{S}''(-r, t)$. Of course, the point $\mathbf{S}'''(-r, -t)$ is also on the graph. From Equation (5), you can also see that the endpoints of the major axis have coordinates $(a, 0)$ and $(-a, 0)$, and that the endpoints of the minor axis have coordinates $(0, b)$ and $(0, -b)$, as indicated in Figure 4–13.

Example 1. Find a Cartesian equation for the ellipse whose foci are $\mathbf{F}_1(3, 0)$ and $\mathbf{F}_2(-3, 0)$ and whose vertices are $\mathbf{V}_1(5, 0)$ and $\mathbf{V}_2(-5, 0)$. Sketch the graph of the equation.

Solution: From the given information, $a = 5$ and $c = 3$. Hence,

$$\begin{aligned} b^2 &= a^2 - c^2 \\ &= 25 - 9 = 16, \end{aligned}$$

and

$$b = 4.$$

Therefore, an equation for the ellipse is

$$\frac{x^2}{25} + \frac{y^2}{16} = 1.$$

A sketch of the ellipse is shown at the right.

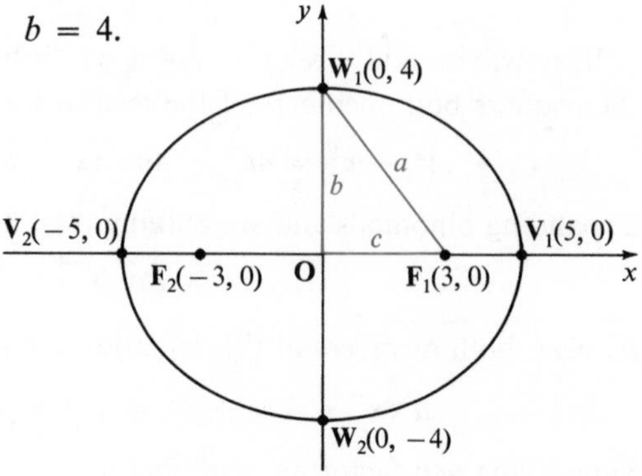

By the method used earlier, you can show that an equation for the ellipse with foci $\mathbf{F}_1(0, c)$ and $\mathbf{F}_2(0, -c)$ on the y-axis, and with sum of distances equal to $2a$, is of the form

$$\frac{x^2}{b^2} + \frac{y^2}{a^2} = 1, \qquad (6)$$

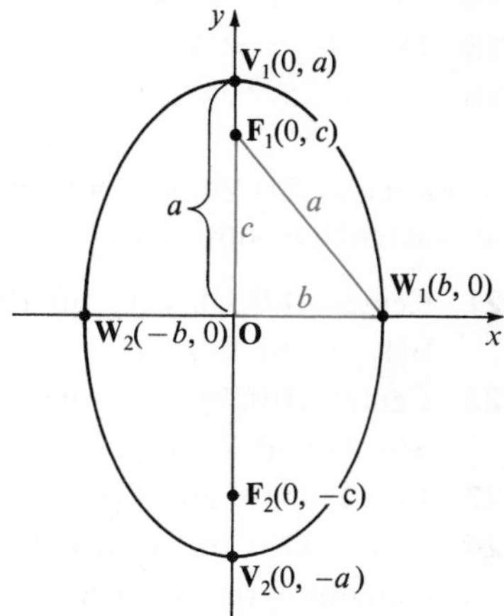

Figure 4–14

where $a > b > 0$ and $a^2 = b^2 + c^2$. As illustrated in Figure 4–14, the major axis of such an ellipse lies on the y-axis and the minor axis lies on the x-axis. The x- and y-intercepts of the ellipse are $\pm b$ and $\pm a$, respectively; hence the endpoints of the major axis have coordinates $(0, a)$ and $(0, -a)$ and the endpoints of the minor axis have coordinates $(b, 0)$ and $(-b, 0)$.

Exercises 4–4

In Exercises 1–8, find an equation of the form (5) or (6), given above, for the ellipse with the given foci $\mathbf{F}_1$ and $\mathbf{F}_2$ and vertices $\mathbf{V}_1$ and $\mathbf{V}_2$. Graph the equation. State the lengths of the major and minor axes.

1. $\mathbf{F}_1(4, 0), \mathbf{F}_2(-4, 0); \mathbf{V}_1(5, 0), \mathbf{V}_2(-5, 0)$
2. $\mathbf{F}_1(12, 0), \mathbf{F}_2(-12, 0); \mathbf{V}_1(13, 0), \mathbf{V}_2(-13, 0)$
3. $\mathbf{F}_1(\sqrt{5}, 0), \mathbf{F}_2(-\sqrt{5}, 0); \mathbf{V}_1(3, 0), \mathbf{V}_2(-3, 0)$
4. $\mathbf{F}_1(\sqrt{21}, 0), \mathbf{F}_2(-\sqrt{21}, 0); \mathbf{V}_1(5, 0), \mathbf{V}_2(-5, 0)$
5. $\mathbf{F}_1(0, 4), \mathbf{F}_2(0, -4); \mathbf{V}_1(0, 5), \mathbf{V}_2(0, -5)$
6. $\mathbf{F}_1(0, 3), \mathbf{F}_2(0, -3); \mathbf{V}_1(0, 5), \mathbf{V}_2(0, -5)$
7. $\mathbf{F}_1(0, \sqrt{3}), \mathbf{F}_2(0, -\sqrt{3}); \mathbf{V}_1(0, \sqrt{7}), \mathbf{V}_2(0, -\sqrt{7})$
8. $\mathbf{F}_1(0, 2), \mathbf{F}_2(0, -2); \mathbf{V}_1(0, \sqrt{13}), \mathbf{V}_2(0, -\sqrt{13})$

In Exercises 9–12, find an equation of the form (5) or (6), given above, for the ellipse having its center at the origin and having semi-major and semi-minor axes with lengths a and b as given. Also find the coordinates of the foci.

9. $a = 5, b = 3$; principal axis the x-axis
10. $a = 13, b = 12$; principal axis the x-axis
11. $a = \sqrt{13}, b = 3$; principal axis the y-axis
12. $a = \sqrt{7}, b = \sqrt{3}$; principal axis the y-axis

In Exercises 13–20, find the coordinates of the vertices $\mathbf{V}_1$ and $\mathbf{V}_2$ and foci $\mathbf{F}_1$ and $\mathbf{F}_2$ of the ellipse with the given equation.

13. $4x^2 + 9y^2 = 36$

14. $16x^2 + 9y^2 = 144$

15. $16x^2 + y^2 = 64$

16. $9x^2 + 25y^2 = 225$

17. $4x^2 + 36y^2 = 36$

18. $25x^2 + 4y^2 = 100$

19. $7x^2 + 3y^2 = 21$

20. $24x^2 + 5y^2 = 120$

In Exercises 21–24, find a Cartesian equation for the ellipse $\mathcal{E}$ with the given characteristics, and sketch the ellipse.

21. Center $\mathbf{O}(0, 0)$, axes on the coordinate axes, and containing the points $\mathbf{S}(1, 3)$ and $\mathbf{T}(4, 2)$.

22. Center $\mathbf{O}(0, 0)$, axes on the coordinate axes, and containing the points $\mathbf{S}(4, 3)$ and $\mathbf{T}(6, 2)$.

23. Foci $\mathbf{F}_1(4, 0)$ and $\mathbf{F}_2(-4, 0)$, and containing the point $\mathbf{S}(3, \frac{12}{5})$.

24. Center $\mathbf{O}(0, 0)$, principal axis the y-axis, distance between foci 24, and containing the point $\mathbf{S}(\frac{60}{13}, 5)$.

In Exercises 25–28, use the definition of an ellipse to find a Cartesian equation for the ellipse $\mathcal{E}$ with the given foci $\mathbf{F}_1$ and $\mathbf{F}_2$ and the given sum of the distances of each point on the ellipse from the foci.

* **25.** $\mathbf{F}_1(2, 4)$, $\mathbf{F}_2(-2, 4)$; sum of distances 6

* **26.** $\mathbf{F}_1(3, 2)$, $\mathbf{F}_2(-3, 2)$; sum of distances 8

* **27.** $\mathbf{F}_1(6, 1)$, $\mathbf{F}_2(6, -1)$; sum of distances 4

* **28.** $\mathbf{F}_1(-2, 4)$, $\mathbf{F}_2(-2, -4)$; sum of distances 12

The length of a segment which is perpendicular to the major axis of an ellipse, contains a focus of the ellipse, and has endpoints on the ellipse is

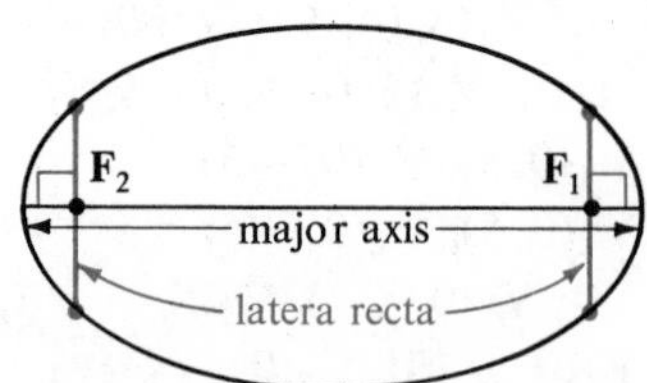

called the *focal width* of the ellipse. The segment itself is called a *latus rectum* (plural: *latera recta*) of the ellipse.

* **29.** Show that the focal width of the ellipse $\mathcal{E}$ with equation

$$\frac{x^2}{a^2} + \frac{y^2}{b^2} = 1$$

is $\dfrac{2b^2}{a}$.

* **30.** Show that the focal width of the ellipse $\mathcal{E}$ with equation

$$\frac{x^2}{a^2} + \frac{y^2}{b^2} = 1$$

is equal to $2b\sqrt{1 - e^2}$, where $e = \frac{c}{a}$.

* **31.** Let $\mathcal{E}$ be the set of all points **U** such that the ratio of the distance between **U** and the point $\mathbf{F}_1(ae, 0)$ [see Exercise 30] to the distance from **U** to the line $\mathfrak{D}_1$ with equation $x = \frac{a}{e}$ is equal to e. Show that $\mathcal{E}$ is the ellipse with equation

$$\frac{x^2}{a^2} + \frac{y^2}{b^2} = 1,$$

provided $e < 1$ and $b^2 = a^2(1 - e^2)$. (See diagram below.)

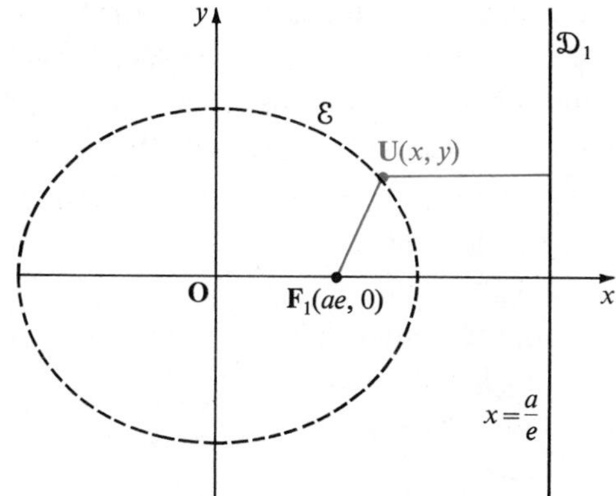

* **32.** Show that a Cartesian equation for the ellipse $\mathcal{E}$ with foci $\mathbf{F}_1(0, c)$ and $\mathbf{F}_2(0, -c)$ and vertices $\mathbf{V}_1(0, a)$ and $\mathbf{V}_2(0, -a)$ is

$$\frac{x^2}{b^2} + \frac{y^2}{a^2} = 1,$$

where $b^2 = a^2 - c^2 > 0$.

* **33.** Show that if $x_2^2 > x_1^2 > 0$ and $y_1^2 > y_2^2 > 0$, then

$$\begin{vmatrix} x^2 & y^2 & 1 \\ x_1^2 & y_1^2 & 1 \\ x_2^2 & y_2^2 & 1 \end{vmatrix} = 0$$

is a Cartesian equation of an ellipse or circle containing the points $\mathbf{S}(x_1, y_1)$ and $\mathbf{T}(x_2, y_2)$.

* **34.** Determine the graph of the equation in Exercise 33 in case $x_2^2 > x_1^2 > 0$ and $y_2^2 = y_1^2$.

Applied Problems 4–4

1. Suppose that a ladder which is 10 feet long rests against a vertical wall and that there is a mark on a rung of the ladder at a point 4 feet from the upper end of the ladder. Show that if the foot of the ladder slides away from the wall on a horizontal floor, causing the top of the ladder to slide down the wall, then the mark moves along an elliptical path.

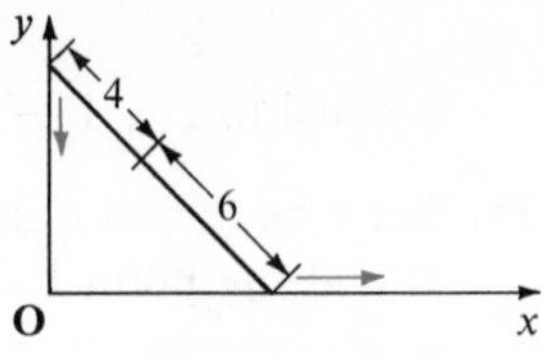

[*Hint:* Set up a coordinate system as indicated in the diagram above.]

2. The base of an auditorium is in the form of an ellipse 200 feet long and 160 feet wide. A pin dropped near one focus can clearly be heard near the other focus. How far apart are the foci?

3. The Earth moves in an elliptic orbit about the Sun, with the Sun at one of the foci. If the least distance of the Earth from the Sun is about 91,340,000 miles, and its greatest distance is about 94,450,000 miles, approximately how far is the Sun from the other focus of the ellipse? [*Hint:* The least and the greatest values occur when the Earth is on the major axis of the ellipse.]

4. An elliptical flower bed is to be made by driving two stakes in the ground 16 feet apart, passing a loop of rope of total length 36 feet about the stakes, and marking the boundary of the bed with a third stake while holding the rope taut. How long and how wide will the bed be?

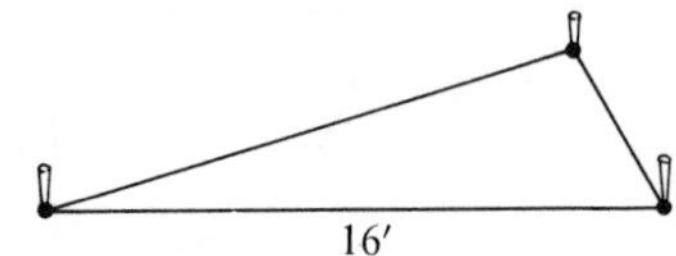

4–5 Hyperbolas

You learned in Section 4–4 that an ellipse is a locus of points determined by the *sum* of two distances. In this section you will study a locus determined by the *difference* of two distances: a *hyperbola*. For two given points $\mathbf{F}_1$ and $\mathbf{F}_2$ in the plane, the set $\mathcal{H}$ of points $\mathbf{U}(x, y)$ in the plane such that the absolute value of the difference of the distances of $\mathbf{U}$ from $\mathbf{F}_1$ and $\mathbf{F}_2$ is a given constant, less than $d(\mathbf{F}_1, \mathbf{F}_2)$, is a **hyperbola**. (See Figure 4–15.) Just as in the case of an ellipse, the terms **foci, principal axis, vertices,** and **center** for a hyperbola refer to the fixed points $\mathbf{F}_1$ and $\mathbf{F}_2$, the line containing these points, the points $\mathbf{V}_1$ and $\mathbf{V}_2$ where $\mathcal{H}$ intersects the principal axis, and the midpoint $\mathbf{C}$ of the segment $\overline{\mathbf{F}_1\mathbf{F}_2}$, respectively. The segment $\overline{\mathbf{V}_1\mathbf{V}_2}$ is called the **transverse axis** of the hyperbola. Notice from the diagram that a hyperbola consists of two separate branches.

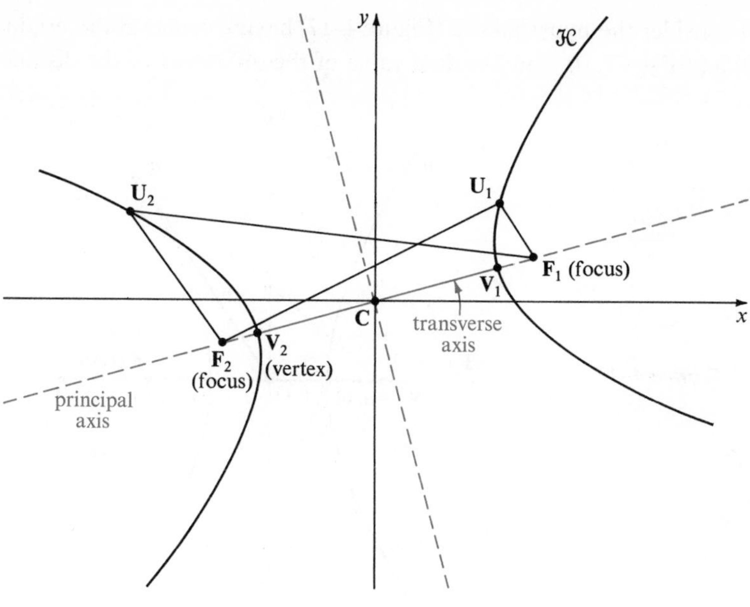

Figure 4–15

There are several methods for constructing a hyperbola. One method, which is similar to the method for constructing an ellipse discussed in Section 4–4, but somewhat more difficult to carry out, is the following: Tie a pencil near the middle of a piece of string, and pass the string around two thumbtacks inserted

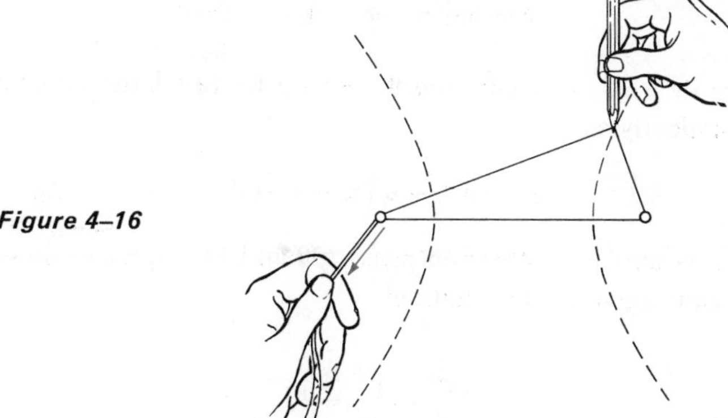

Figure 4–16

in a piece of cardboard. Using the pencil to keep the string taut, hold the two ends together and draw them toward you, as indicated in Figure 4–16. The pencil will mark an arc of one branch of the hyperbola. An arc of the other branch is obtained by interchanging the roles of the two thumbtacks.

Now consider the hyperbola $\mathcal{H}$ (Figure 4–17) having center at the origin, foci $\mathbf{F}_1(c, 0)$ and $\mathbf{F}_2(-c, 0)$, and absolute value of the difference of the distances of

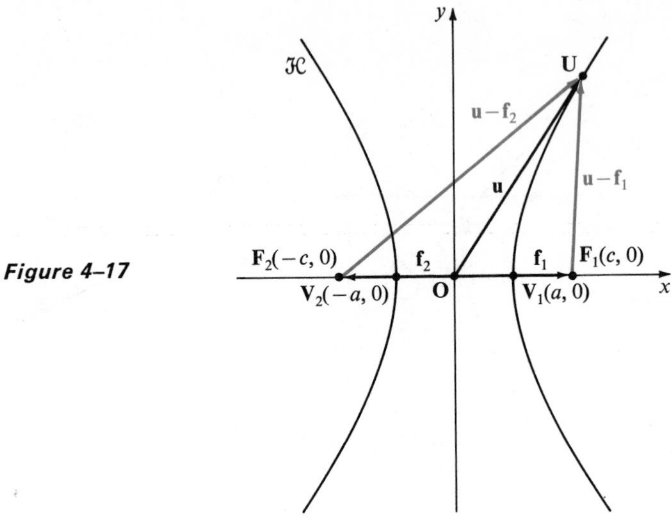

Figure 4–17

each point $\mathbf{U}(x, y)$ on $\mathcal{H}$ from $\mathbf{F}_1$ and $\mathbf{F}_2$ equal to the positive constant $2a$, $a < c$. Then, by the definition of *hyperbola*, a point $\mathbf{U}(x, y)$ lies on $\mathcal{H}$ if and only if

$$\left| \|\mathbf{u} - \mathbf{f}_1\| - \|\mathbf{u} - \mathbf{f}_2\| \right| = 2a,$$

or

$$\|\mathbf{u} - \mathbf{f}_1\| - \|\mathbf{u} - \mathbf{f}_2\| = \pm 2a.$$

Since $\mathbf{u} = (x, y)$, $\mathbf{f}_1 = (c, 0)$, and $\mathbf{f}_2 = (-c, 0)$, this latter equation can be written equivalently as

$$\sqrt{(x - c)^2 + y^2} - \sqrt{[x - (-c)]^2 + y^2} = \pm 2a.$$

By calculations similar to those on pages 139 and 140, you can show that this equation is equivalent to the equation

$$\frac{x^2}{a^2} - \frac{y^2}{b^2} = 1, \tag{1}$$

where

$$b^2 = c^2 - a^2 > 0 \text{ and } b > 0.$$

As with the corresponding equation of an ellipse (page 140), the steps are reversible, and accordingly a point $\mathbf{U}(x, y)$ is on $\mathcal{H}$ if and only if its coordinates satisfy Equation (1).

Similarly, a hyperbola with center at the origin, foci $\mathbf{F}_1(0, c)$ and $\mathbf{F}_2(0, -c)$, and absolute value of the difference of the distances equal to $2a$, $0 < a < c$, has Cartesian equation

$$\frac{y^2}{a^2} - \frac{x^2}{b^2} = 1. \tag{2}$$

The value of the constant a in Equations (1) and (2) might be greater than, equal to, or less than the value of b.

A hyperbola $\mathcal{H}$ with an equation of the form (1) or (2) is symmetric with respect to both coordinate axes. Thus if $\mathbf{S}(x, y)$ is a point of $\mathcal{H}$, then so are $\mathbf{S}'(x, -y)$, $\mathbf{S}''(-x, y)$, and $\mathbf{S}'''(-x, -y)$.

If you replace y with 0 in Equation (1), you obtain $\frac{x^2}{a^2} = 1$, or $x = \pm a$. Thus a and $-a$ are the x-intercepts of a hyperbola of the form (1). If you replace x with 0 in Equation (1), you obtain $\frac{y^2}{b^2} = -1$. Since $\frac{y^2}{b^2}$ is always nonnegative, the hyperbola has no y-intercepts.

In a similar manner, you can show that a hyperbola with an equation of the form (2) has y-intercepts a and $-a$ and no x-intercepts.

Now consider the equation

$$\frac{x^2}{a^2} - \frac{y^2}{b^2} = 0.$$

If you solve this equation for y^2 in terms of x, you obtain

$$y^2 = \frac{b^2}{a^2}x^2,$$

from which you have

$$y = \frac{b}{a}x \quad \text{or} \quad y = -\frac{b}{a}x.$$

The graphs of these equations are two lines which intersect at the origin and which have slope $\frac{b}{a}$ and $-\frac{b}{a}$, respectively.

If you solve Equation (1) on page 146 for y^2 in terms of x, you obtain

$$y^2 = \frac{b^2}{a^2}x^2 - b^2,$$

from which you have

$$y = \sqrt{\frac{b^2}{a^2}x^2 - b^2} \quad \text{or} \quad y = -\sqrt{\frac{b^2}{a^2}x^2 - b^2},$$

and this, in turn, is equivalent to

$$y = \frac{b}{a}\sqrt{x^2 - a^2} \quad \text{or} \quad y = -\frac{b}{a}\sqrt{x^2 - a^2}.$$

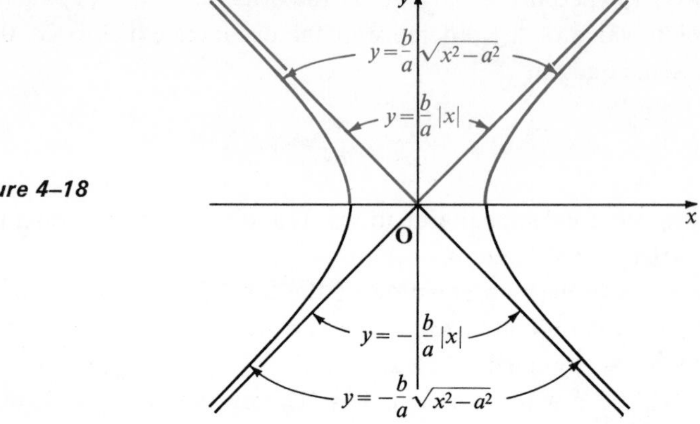

Figure 4–18

Now compare the corresponding values of y for the lines and for the hyperbola. As illustrated in Figure 4–18,

$$\frac{b}{a}\sqrt{x^2 - a^2} < \frac{b}{a}|x|$$

for all real numbers for which both members of the inequality are defined. If, however, $|x|$ is allowed to increase without bound, then the value of $\sqrt{x^2 - a^2}$ approaches closer and closer to that of $|x|$, and the value of $\frac{b}{a}\sqrt{x^2 - a^2}$ approaches closer and closer to that of $\frac{b}{a}|x|$. In fact, you have

$$\begin{aligned}\frac{b}{a}|x| - \frac{b}{a}\sqrt{x^2 - a^2} &= \frac{b}{a}(|x| - \sqrt{x^2 - a^2})\left(\frac{|x| + \sqrt{x^2 - a^2}}{|x| + \sqrt{x^2 - a^2}}\right)\\ &= \frac{b(x^2 - x^2 + a^2)}{a(|x| + \sqrt{x^2 - a^2})}\\ &= \frac{ab}{|x| + \sqrt{x^2 - a^2}},\end{aligned}$$

which can be made as small as you please by taking $|x|$ large enough.

As shown in Figure 4–18, you also have

$$-\frac{b}{a}\sqrt{x^2 - a^2} > -\frac{b}{a}|x|$$

for all real numbers for which both members of the inequality are defined; and if $|x|$ increases without bound, then the value of $-\frac{b}{a}\sqrt{x^2 - a^2}$ approaches closer and closer to that of $-\frac{b}{a}|x|$.

Therefore, you can conclude that as $|x|$ increases without bound, the value of $|y|$ for the hyperbola approaches closer and closer to that of $\frac{b}{a}|x|$, and the branches of the hyperbola become closer and closer to the lines with equations

$y = \frac{b}{a}x$ and $y = -\frac{b}{a}x$, respectively. These lines are called the **asymptotes** of the hyperbola.

More generally, whenever a curve approaches one or more lines in this manner, the lines are called **asymptotes** of the curve. We shall discuss asymptotes in more detail in Chapter 6.

The intercepts and asymptotes of a hyperbola, as well as symmetry considerations, are useful in sketching the hyperbola. As illustrated in Figure 4–19, the length of the transverse axis of a hyperbola with equation

$$\frac{x^2}{a^2} - \frac{y^2}{b^2} = 1 \qquad \text{or} \qquad \frac{y^2}{a^2} - \frac{x^2}{b^2} = 1$$

is $2a$.

The line segment through the center of a hyperbola, perpendicular to the principal axis, and with endpoints at a distance b, where $b^2 = c^2 - a^2$, from the center, is called the **conjugate axis** of the hyperbola. Thus, the length of the conjugate axis ($\overline{\mathbf{W}_1\mathbf{W}_2}$ in Figure 4–19) is $2b$.

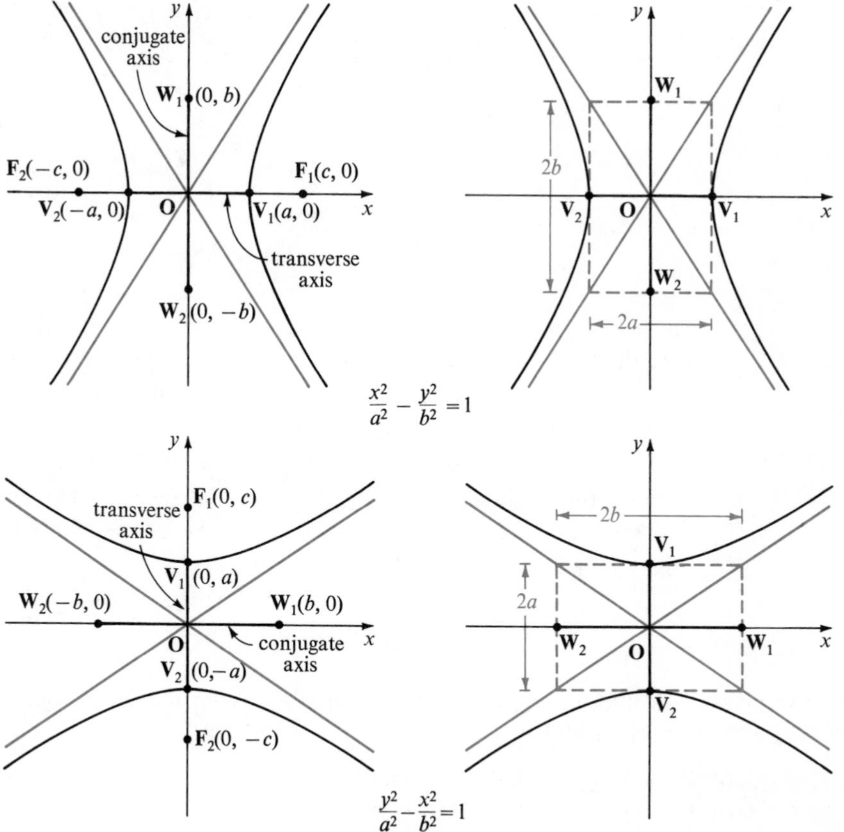

Figure 4–19

A rapid means of sketching a hyperbola of one of the types shown in Figure 4–19 is to construct a rectangle with center at the origin and with sides of length $2a$ and $2b$, where the sides of length $2a$ are parallel to the principal axis of the hyperbola. The diagonals of the rectangle are then segments of the asymptotes of the hyperbola, and the midpoints of the sides of length $2b$ are the vertices of the hyperbola.

Example 1. Sketch the graph of $\frac{x^2}{4} - \frac{y^2}{9} = 1$ using asymptotes, intercepts, and symmetry.

Solution: Notice first that the graph is a hyperbola with the x-axis as

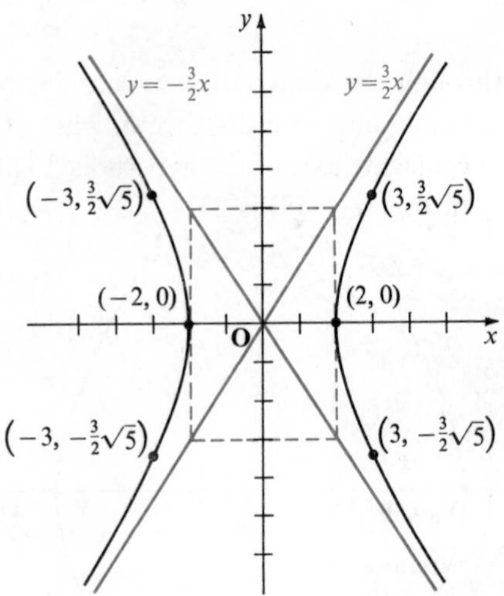

principal axis, and that $a = 2$ and $b = 3$. Then the asymptotes are the lines with equations $y = \frac{3}{2}x$ and $y = -\frac{3}{2}x$.

If you replace y with 0 in the given equation, you find that the x-intercepts are -2 and 2. There are no y-intercepts.

If you replace x with 3 or -3, you find that $y = \pm\frac{3}{2}\sqrt{5}$. Thus, the points $(3, \frac{3}{2}\sqrt{5})$, $(3, -\frac{3}{2}\sqrt{5})$, $(-3, \frac{3}{2}\sqrt{5})$, and $(-3, -\frac{3}{2}\sqrt{5})$ are on the graph. Using all these facts, you have the graph shown.

Exercises 4–5

In Exercises 1–8, find an equation for the hyperbola with the given foci $\mathbf{F}_1$ and $\mathbf{F}_2$ and vertices $\mathbf{V}_1$ and $\mathbf{V}_2$. Then graph the equation.

1. $\mathbf{F}_1(5, 0)$, $\mathbf{F}_2(-5, 0)$; $\mathbf{V}_1(3, 0)$, $\mathbf{V}_2(-3, 0)$
2. $\mathbf{F}_1(5, 0)$, $\mathbf{F}_2(-5, 0)$; $\mathbf{V}_1(4, 0)$, $\mathbf{V}_2(-4, 0)$
3. $\mathbf{F}_1(\sqrt{13}, 0)$, $\mathbf{F}_2(-\sqrt{13}, 0)$; $\mathbf{V}_1(3, 0)$, $\mathbf{V}_2(-3, 0)$
4. $\mathbf{F}_1(\sqrt{5}, 0)$, $\mathbf{F}_2(-\sqrt{5}, 0)$; $\mathbf{V}_1(\sqrt{3}, 0)$, $\mathbf{V}_2(-\sqrt{3}, 0)$

5. $\mathbf{F}_1(0, 5), \mathbf{F}_2(0, -5); \mathbf{V}_1(0, 4), \mathbf{V}_2(0, -4)$
6. $\mathbf{F}_1(0, 5), \mathbf{F}_2(0, -5); \mathbf{V}_1(0, 3), \mathbf{V}_2(0, -3)$
7. $\mathbf{F}_1(0, \sqrt{7}), \mathbf{F}_2(0, -\sqrt{7}); \mathbf{V}_1(0, \sqrt{3}), \mathbf{V}_2(0, -\sqrt{3})$
8. $\mathbf{F}_1(0, \sqrt{21}), \mathbf{F}_2(0, -\sqrt{21}); \mathbf{V}_1(0, \sqrt{5}), \mathbf{V}_2(0, -\sqrt{5})$

In Exercises 9–12, find an equation for the hyperbola with center at the origin, lengths of transverse and conjugate axes as given, and principal axis as specified.

Example. $2a = 8, 2b = 10$; x-axis

Solution: You have

$$2a = 8, a = 4;$$

and

$$2b = 10, b = 5.$$

Therefore, an equation of the hyperbola is

$$\frac{x^2}{16} - \frac{y^2}{25} = 1.$$

9. $2a = 10, 2b = 8$; x-axis
10. $2a = 14, 2b = 6$; x-axis
11. $2a = 26, 2b = 10$; y-axis
12. $2a = 2\sqrt{7}, 2b = 4$; y-axis

In Exercises 13–20, find the coordinates of the vertices and the foci of the hyperbola with the given equation. Then find equations for the asymptotes and sketch the hyperbola.

13. $\frac{x^2}{25} - \frac{y^2}{16} = 1$
14. $\frac{x^2}{25} - \frac{y^2}{9} = 1$
15. $\frac{x^2}{8} - \frac{y^2}{4} = 1$
16. $\frac{x^2}{144} - \frac{y^2}{25} = 1$
17. $\frac{y^2}{16} - \frac{x^2}{25} = 1$
18. $\frac{y^2}{13} - \frac{x^2}{4} = 1$
19. $\frac{y^2}{16} - \frac{x^2}{9} = 1$
20. $\frac{y^2}{12} - \frac{x^2}{8} = 1$

In Exercises 21–26, find a Cartesian equation for the hyperbola which has center at the origin and which satisfies the given conditions.

21. Foci $\mathbf{F}_1(4, 0)$ and $\mathbf{F}_2(-4, 0)$; containing $\mathbf{S}(14, 24)$.
22. Foci $\mathbf{F}_1(0, 4)$ and $\mathbf{F}_2(0, -4)$; one asymptote with equation $3y = x$.

* 23. Principal axis the x-axis; containing the points $\mathbf{S}(2, 1)$ and $\mathbf{T}(4, 3)$.

* 24. Principal axis the x-axis; containing the points $\mathbf{S}(3, 1)$ and $\mathbf{T}(9, 5)$.

* 25. Principal axis the y-axis; one end of the conjugate axis at $\mathbf{W}_1(3, 0)$; a vertex at the midpoint of the segment between the center and a focus.

* 26. Principal axis the y-axis; one focus $\mathbf{F}_1(0, 4)$; $c = 3a$.

The length of a segment which is perpendicular to the principal axis of a hyperbola, contains a focus of the hyperbola, and has endpoints on the hyperbola is called the *focal width* of the hyperbola. The segment itself is called a *latus rectum* of the hyperbola.

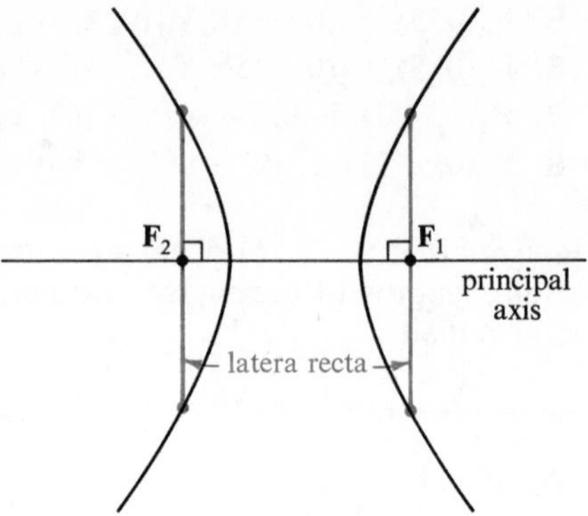

*** 27.** Find a Cartesian equation for the hyperbola $\mathcal{H}$ with center at the origin, focal width 36, and one focus $\mathbf{F}_2(-12, 0)$.

*** 28.** Show that the hyperbola $\mathcal{H}$ and the ellipse $\mathcal{E}$ with respective equations $3x^2 - y^2 = 12$ and $9x^2 + 25y^2 = 225$ have the same foci.

*** 29.** Show that the focal width of the hyperbola $\mathcal{H}$ with equation

$$\frac{x^2}{a^2} - \frac{y^2}{b^2} = 1$$

is $\dfrac{2b^2}{a}$.

*** 30.** Show that the focal width of the hyperbola $\mathcal{H}$ with equation

$$\frac{x^2}{a^2} - \frac{y^2}{b^2} = 1$$

is $2b\sqrt{e^2 - 1}$, where $e = \dfrac{c}{a}$.

*** 31.** Let $\mathcal{H}$ be the set of all points $\mathbf{U}$ such that the ratio of the distance between $\mathbf{U}$ and the point $\mathbf{F}_1(ae, 0)$ [see Exercise 30] to the distance from $\mathbf{U}$ to the line $\mathcal{D}_1$ with equation $x = \dfrac{a}{e}$ is equal to e. Show that $\mathcal{H}$ is the hyperbola with equation

$$\frac{x^2}{a^2} - \frac{y^2}{b^2} = 1,$$

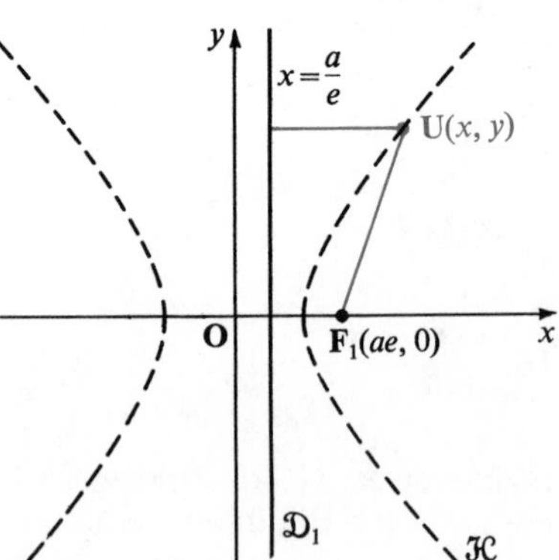

provided $e > 1$ and $b^2 = a^2(e^2 - 1)$.

*** 32.** Show that a Cartesian equation for the hyperbola with foci $\mathbf{F}_1(0, c)$ and $\mathbf{F}_2(0, -c)$, and difference of distances $2a$, $0 < a < c$, is

$$\frac{y^2}{a^2} - \frac{x^2}{b^2} = 1,$$

where $b^2 = c^2 - a^2 > 0$.

* **33.** Show that if $x_2^2 > x_1^2 > 0$ and $y_2^2 > y_1^2 > 0$, then

$$\begin{vmatrix} x^2 & y^2 & 1 \\ x_1^2 & y_1^2 & 1 \\ x_2^2 & y_2^2 & 1 \end{vmatrix} = 0$$

is an equation of a hyperbola containing the points $\mathbf{S}(x_1, y_1)$ and $\mathbf{T}(x_2, y_2)$.

* **34.** Determine the graph of the equation in Exercise 33 in case $(x_1, y_1) = (0, 0)$, but $(x_2, y_2) \neq (0, 0)$.

In Exercises 35–38, use the definition of a hyperbola to find a Cartesian equation for the hyperbola $\mathcal{H}$ with the given foci $\mathbf{F}_1$ and $\mathbf{F}_2$ and the given absolute value $2a$ of the difference of the distances of each point on the hyperbola from the foci.

* **35.** $\mathbf{F}_1(3, 1)$, $\mathbf{F}_2(-3, 1)$; $2a = 2$

* **36.** $\mathbf{F}_1(4, 4)$, $\mathbf{F}_2(4, -4)$; $2a = 6$

* **37.** $\mathbf{F}_1(4, 1)$, $\mathbf{F}_2(-4, 1)$; $2a = 4$

* **38.** $\mathbf{F}_1(2, 0)$, $\mathbf{F}_2(2, 10)$; $2a = 8$

An Alternative Definition of Conic Sections

4–6 The Focus-Directrix Property of Conic Sections

Earlier in this chapter, a parabola was defined in terms of the distances of each point on the curve from a fixed point (the focus) and a fixed line (the directrix). On the other hand, an ellipse and a hyperbola were defined in terms of the sum and the difference, respectively, of the distances of each point on the curve from two fixed points (the foci). It is also possible to define an ellipse and a hyperbola in terms of distances from a fixed point and a fixed line, just as was done for the parabola.

Consider the ellipse with center at the origin, a focus $\mathbf{F}_1(c, 0)$, and a vertex $\mathbf{V}_1(a, 0)$, $0 < c < a$, as shown in Figure 4–20. The distance $d(\mathbf{U}, \mathbf{F}_1)$ of each point $\mathbf{U}(x, y)$ on the ellipse from the focus $\mathbf{F}_1$ is given by

$$d(\mathbf{U}, \mathbf{F}_1) = \sqrt{(x - c)^2 + y^2}. \tag{1}$$

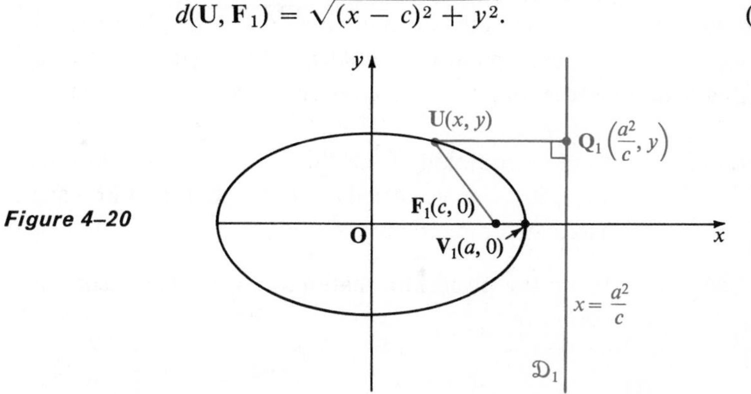

Figure 4–20

By Equation (4), page 139, the coordinates of each point **U** on the ellipse also satisfy

$$a\sqrt{(x-c)^2+y^2} = a^2 - cx,$$

or

$$\sqrt{(x-c)^2+y^2} = a - \frac{c}{a}x. \tag{2}$$

Substituting from (1) into (2), you have

$$d(\mathbf{U}, \mathbf{F}_1) = a - \frac{c}{a}x,$$

from which you obtain

$$d(\mathbf{U}, \mathbf{F}_1) = \frac{c}{a}\left(\frac{a^2}{c} - x\right). \tag{3}$$

Now observe that $\frac{a^2}{c} > a \geq x$, since $a > c > 0$ and $x \leq a$ for each point $\mathbf{U}(x, y)$ on the ellipse. Hence $\frac{a^2}{c} - x > 0$, and accordingly the factor $\frac{a^2}{c} - x$ in the right-hand member of Equation (3) is just the distance $d(\mathbf{U}, \mathfrak{D}_1)$ from a point $\mathbf{U}(x, y)$ on the ellipse to the line $\mathfrak{D}_1$ with equation $x = \frac{a^2}{c}$. That is, Equation (3) can be written as

$$d(\mathbf{U}, \mathbf{F}_1) = \frac{c}{a}d(\mathbf{U}, \mathfrak{D}_1). \tag{4}$$

The ratio $\frac{c}{a}$ is called the **eccentricity** of the ellipse and is customarily denoted by the letter e. Notice that

$$0 < e < 1$$

since $0 < c < a$. Substituting e for $\frac{c}{a}$ in Equation (4), you have

$$\frac{d(\mathbf{U}, \mathbf{F}_1)}{d(\mathbf{U}, \mathfrak{D}_1)} = e. \tag{5}$$

Since the argument just concluded is reversible, step by step, it follows that you can define an ellipse as the locus of all points in the plane such that the *ratio* of the distances of each point on the locus from a fixed point (the focus) and from a fixed line (the directrix) is a constant, e, with $0 < e < 1$.

Example 1. Find an equation of the ellipse $\mathcal{E}$ with center $\mathbf{O}(0, 0)$, one focus $\mathbf{F}_1(3, 0)$, and eccentricity $e = \frac{2}{3}$. Also find an equation of the associated directrix $\mathfrak{D}_1$.

Solution: From the given information, $c = 3$. Then, since $e = \frac{c}{a} = \frac{2}{3}$, you have

$$\frac{3}{a} = \frac{2}{3}, \quad \text{or} \quad a = \frac{9}{2}.$$

Next, using the fact that for an ellipse the values a, b, and c are related by

$$a^2 = b^2 + c^2,$$

you have

$$b^2 = a^2 - c^2 = \tfrac{81}{4} - 9 = \tfrac{45}{4}.$$

Therefore, since $a^2 = (\tfrac{9}{2})^2 = \tfrac{81}{4}$, an equation for $\mathcal{E}$ is

$$\frac{x^2}{\frac{81}{4}} + \frac{y^2}{\frac{45}{4}} = 1,$$

or

$$\frac{4x^2}{81} + \frac{4y^2}{45} = 1.$$

Substituting the values of a^2 and c in the equation

$$x = \frac{a^2}{c},$$

you obtain

$$x = \frac{\frac{81}{4}}{3},$$

or

$$x = \tfrac{27}{4},$$

as an equation of the associated directrix $\mathcal{D}_1$.

An ellipse with center $\mathbf{O}(0, 0)$, a focus $\mathbf{F}_1(c, 0)$, and a vertex $\mathbf{V}_1(a, 0)$ also has the focus $\mathbf{F}_2(-c, 0)$. Using this focus, as suggested by Figure 4–21, you can

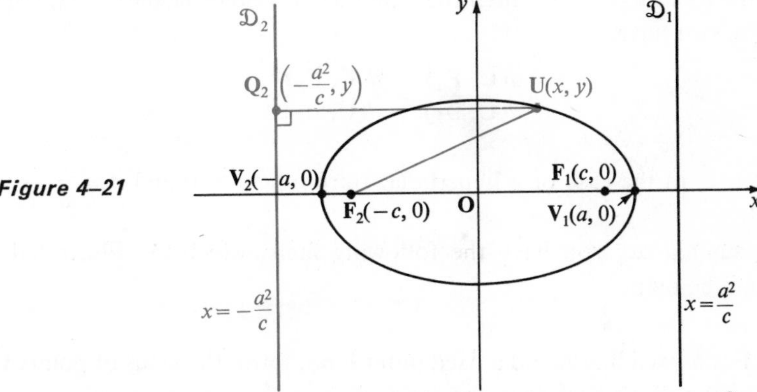

Figure 4–21

show (Exercise 32, page 160) that the line $\mathcal{D}_2$ with equation $x = -\dfrac{a^2}{c}$ is also a directrix for the ellipse. That is, you can show that for each point $\mathbf{U}(x, y)$ on the ellipse you have

$$\frac{d(\mathbf{U}, \mathbf{F}_2)}{d(\mathbf{U}, \mathcal{D}_2)} = e.$$

Thus each ellipse has two foci and two directrices, one directrix associated with each focus.

Using Figure 4–22, you can show (Exercise 33, page 160) by a method similar to that used above for the ellipse, that for a hyperbola with center at the origin,

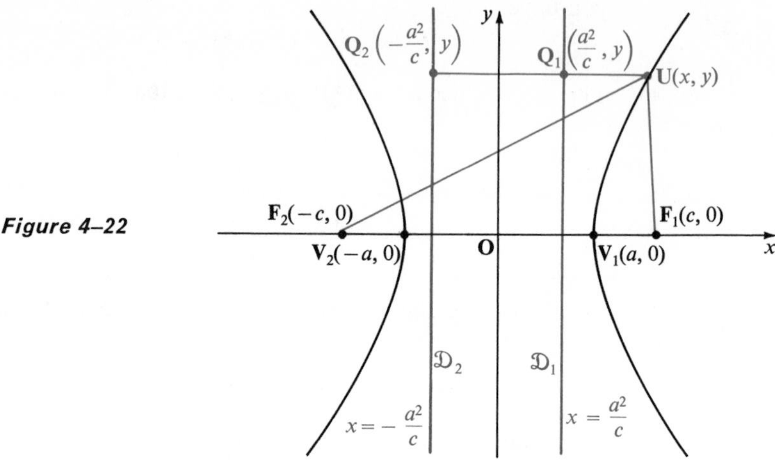

Figure 4–22

a focus $\mathbf{F}_1(c, 0)$, and a vertex $\mathbf{V}_1(a, 0)$, $0 < a < c$, the ratio of the distances of each point $\mathbf{U}(x, y)$ on the hyperbola from the focus $\mathbf{F}_1(c, 0)$ and from the line $\mathfrak{D}_1$ with equation $x = \frac{a^2}{c}$ is again given by Equation (5), page 154. If you replace the line $\mathfrak{D}_1$ with the line $\mathfrak{D}_2$ having equation $x = -\frac{a^2}{c}$, and the focus $\mathbf{F}_1(c, 0)$ with the focus $\mathbf{F}_2(-c, 0)$, then the equation still holds (Exercise 34, page 160). That is, you have

$$\frac{d(\mathbf{U}, \mathbf{F}_1)}{d(\mathbf{U}, \mathfrak{D}_1)} = \frac{d(\mathbf{U}, \mathbf{F}_2)}{d(\mathbf{U}, \mathfrak{D}_2)} = e.$$

Of course, in the case of a hyperbola, since $c > a > 0$ and $e = \frac{c}{a}$, you have $e > 1$.

To summarize, you have the following facts, which are illustrated in the diagram below:

■ For a fixed line $\mathfrak{D}$ and a fixed point $\mathbf{F}$ not on $\mathfrak{D}$, the locus of points $\mathbf{U}$ in the plane such that the ratio of the distances of $\mathbf{U}$ from $\mathbf{F}$ and $\mathfrak{D}$ is a constant, e, is

(a) an ellipse if $0 < e < 1$,

(b) a parabola if $e = 1$, and

(c) a hyperbola if $e > 1$.

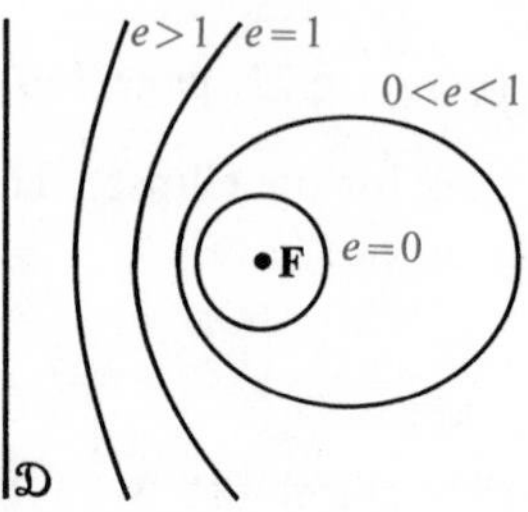

In each case, the constant e is called the *eccentricity* of the conic.

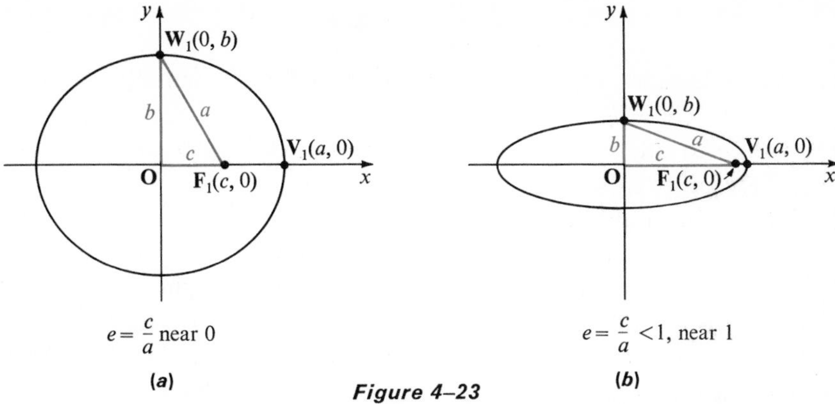

Figure 4–23

As suggested by Figure 4–23, the shape of an ellipse is related to its eccentricity. If the eccentricity, $e = \frac{c}{a}$, is near 0, then (relative to a) c, or $d(\mathbf{O}, \mathbf{F}_1)$, is very small and b is very nearly equal to a. Thus, the ellipse is almost circular [Figure 4–23(a)]. In fact, a circle is sometimes said to have eccentricity 0. If the eccentricity of the ellipse is near 1, then c is very nearly equal to a, and b is very small. In this case, the focal width (page 142) is small and the ellipse is elongated [Figure 4–23(b)]. All ellipses with the same eccentricity are similar figures (Exercise 28, page 159); that is, they have the same shape, and each is just an enlargement, or a reduction, of any other one.

The shape of a hyperbola is also related to its eccentricity. As suggested by Figure 4–24(a), if the eccentricity, $e = \frac{c}{a}$, is only slightly greater than 1, then c is nearly equal to a, and b is very small. Hence the focal width (page 152) is small and the hyperbola is "thin." If the eccentricity is large [Figure 4–24(b)], then (relative to c) a is small, and b is nearly equal to c. Hence the focal width is large and the hyperbola is "thick." All hyperbolas with the same eccentricity are similar figures (Exercise 29, page 159).

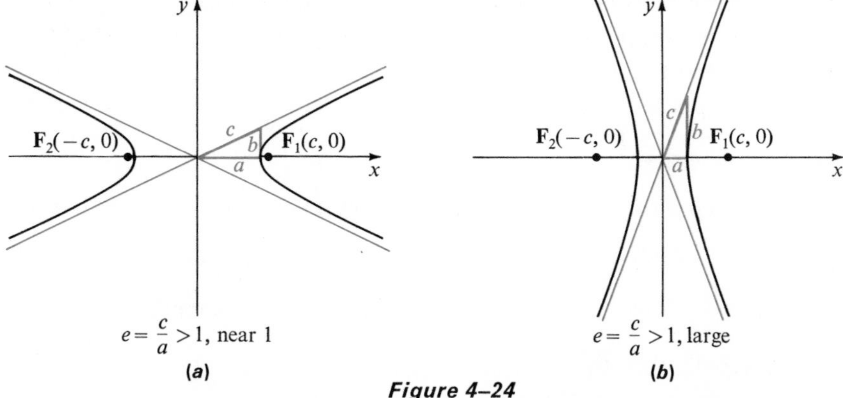

Figure 4–24

Exercises 4–6

1. Find an equation for the ellipse $\mathcal{E}$ with foci $\mathbf{F}_1(4, 0)$ and $\mathbf{F}_2(-4, 0)$, and eccentricity $e = \frac{4}{5}$.

2. Find an equation for the ellipse $\mathcal{E}$ with vertices $\mathbf{V}_1(5, 0)$ and $\mathbf{V}_2(-5, 0)$, and eccentricity $e = \frac{3}{5}$.

3. Find an equation for the ellipse $\mathcal{E}$ with center at the origin, one focus $\mathbf{F}_1(3, 0)$, and associated directrix the line $\mathfrak{D}_1$ with equation $x = \frac{16}{3}$.

4. Find an equation for the ellipse $\mathcal{E}$ with center at the origin, one end of the major axis at $\mathbf{V}_1(0, 13)$, and one directrix the line $\mathfrak{D}_1$ with equation $y = \frac{169}{12}$.

5. Find an equation for the ellipse $\mathcal{E}$ with foci $\mathbf{F}_1(5, 0)$ and $\mathbf{F}_2(-5, 0)$, and eccentricity $e = \frac{1}{3}$.

6. Find an equation for the ellipse $\mathcal{E}$ with vertices $\mathbf{V}_1(4, 0)$ and $\mathbf{V}_2(-4, 0)$, and eccentricity $e = \frac{1}{3}$.

7. Find an equation for the hyperbola $\mathcal{H}$ with foci $\mathbf{F}_1(3, 0)$ and $\mathbf{F}_2(-3, 0)$, and eccentricity $e = \frac{3}{2}$.

8. Find an equation for the hyperbola $\mathcal{H}$ with foci $\mathbf{F}_1(0, 2)$ and $\mathbf{F}_2(0, -2)$, and eccentricity $e = 3$.

9. Find an equation for the hyperbola $\mathcal{H}$ with center at the origin, one focus $\mathbf{F}_1(4, 0)$, and associated directrix the line $\mathfrak{D}_1$ with equation $x = \frac{3}{4}$.

10. Find an equation for the hyperbola $\mathcal{H}$ with center at the origin, one vertex $\mathbf{V}_1(4, 0)$, and associated directrix the line $\mathfrak{D}_1$ with equation $x = \frac{16}{7}$.

11. Find an equation for the hyperbola $\mathcal{H}$ having as one directrix the line with equation $x = \frac{1}{2}$ and as asymptotes the lines with equations $y = \pm 3x$.

12. Find an equation for the hyperbola $\mathcal{H}$ having as one directrix the line with equation $y = 2$ and as asymptotes the lines with equations $y = \pm x$.

13. If $a = b$ in the equation for a hyperbola with center at the origin and foci on a coordinate axis, then the curve is called an **equilateral hyperbola**. Find the eccentricity of an equilateral hyperbola.

14. Show that the asymptotes of an equilateral hyperbola (see Exercise 13) are perpendicular.

15. Find the eccentricity of an ellipse if the line containing one end of the minor axis and a focus is perpendicular to the line containing the same end of the minor axis and the other focus.

16. Show that if p is the distance from a focus of an ellipse to the corresponding directrix, then $p = \dfrac{b^2}{c}$.

* **17.** Show that if p is the distance from a focus to the corresponding directrix of an ellipse, then the length of the major axis is $\dfrac{2ep}{(1 - e^2)}$.

* **18.** Show that if p is the distance from a focus to the corresponding directrix of a hyperbola, then the length of the transverse axis is $\dfrac{2ep}{(e^2 - 1)}$.

* **19.** The asymptotes of a hyperbola with horizontal transverse axis have slopes $\pm m$, respectively, with $m > 0$. Express the eccentricity of the hyperbola in terms of m.

* **20.** Show that for the ellipse $\mathcal{E}$ with equation

$$\frac{x^2}{a^2} + \frac{y^2}{b^2} = 1 \qquad (a > b)$$

the lengths of the segments from the foci to a point $\mathbf{U}(x, y)$ on $\mathcal{E}$ are $a \pm ex$.

* **21.** Show that for the hyperbola $\mathcal{H}$ with equation

$$\frac{x^2}{a^2} - \frac{y^2}{b^2} = 1,$$

the lengths of the segments from the foci to a point $\mathbf{U}(x, y)$ on $\mathcal{H}$ are $|ex \pm a|$.

* **22.** Find an equation for the hyperbola $\mathcal{H}$ with foci $\mathbf{F}_1(6, 1)$ and $\mathbf{F}_2(-4, 1)$, and eccentricity $e = \frac{3}{2}$.

* **23.** Find an equation for the hyperbola $\mathcal{H}$ with vertices $\mathbf{V}_1(6, 1)$ and $\mathbf{V}_2(-4, 1)$, and eccentricity $e = \frac{3}{2}$.

* **24.** Find an equation for the hyperbola $\mathcal{H}$ with center $\mathbf{C}(2, 3)$, one focus $\mathbf{F}_1(6, 3)$, and associated directrix the line $\mathfrak{D}_1$ with equation $x = 3$.

* **25.** Find an equation for the ellipse with center $\mathbf{C}(2, 3)$, one focus $\mathbf{F}_1(6, 3)$, and associated directrix the line $\mathfrak{D}_1$ with equation $x = \frac{33}{4}$.

* **26.** Find equations for the directrices of the hyperbola $\mathcal{H}$ with equation

$$x^2 - y^2 + 4x + 4y - 48 = 0.$$

* **27.** Find equations for the directrices of the ellipse $\mathcal{E}$ with equation

$$x^2 + 5y^2 + 6x - 40y - 167 = 0.$$

* **28.** Show that all ellipses with the same eccentricity are similar figures. [*Hint:* Show that if the ellipses $\dfrac{x^2}{a_1^2} + \dfrac{y^2}{b_1^2} = 1$ and $\dfrac{x^2}{a_2^2} + \dfrac{y^2}{b_2^2} = 1$ have the same eccentricity and $a_2 = ka_1$, then $b_2 = kb_1$. Next show that if $\mathbf{U}_1(x, y)$ is on the first ellipse then $\mathbf{U}_2(kx, ky)$ is on the second ellipse.]

* **29.** Show that all hyperbolas with the same eccentricity are similar figures. [*Hint:* See Exercise 28.]

* **30.** Show that all parabolas are similar figures. [*Hint:* For parabolas $y^2 = 4p_1x$ and $y^2 = 4p_2x$, show that if $\mathbf{U}_1(x, y)$ is on the first parabola then $\mathbf{U}_2\left(\dfrac{p_2}{p_1}x, \dfrac{p_2}{p_1}y\right)$ is on the second.]

* **31.** Show that all circles are similar figures. [*Hint:* See Exercise 30.]

* **32.** Show that for all points $\mathbf{U}(x, y)$ on the ellipse with foci $\mathbf{F}_1(c, 0)$ and $\mathbf{F}_2(-c, 0)$, and vertices $\mathbf{V}_1(a, 0)$ and $\mathbf{V}_2(-a, 0)$, you have

$$\frac{d(\mathbf{U}, \mathbf{F}_2)}{d(\mathbf{U}, \mathfrak{D}_2)} = e,$$

where $e = \dfrac{c}{a}$ and $\mathfrak{D}_2$ is the line with equation $x = -\dfrac{a^2}{c}$.

[*Hint:* See Exercise 20.]

* **33.** Show that for all points $\mathbf{U}(x, y)$ on the hyperbola with foci $\mathbf{F}_1(c, 0)$ and $\mathbf{F}_2(-c, 0)$, and vertices $\mathbf{V}_1(a, 0)$ and $\mathbf{V}_2(-a, 0)$, you have

$$\frac{d(\mathbf{U}, \mathbf{F}_1)}{d(\mathbf{U}, \mathfrak{D}_1)} = e,$$

where $e = \dfrac{c}{a}$ and $\mathfrak{D}_1$ is the line with equation $x = \dfrac{a^2}{c}$.

[*Hint:* See Exercise 21.]

* **34.** Show that for all points $\mathbf{U}(x, y)$ on the hyperbola with foci $\mathbf{F}_1(c, 0)$ and $\mathbf{F}_2(-c, 0)$, and vertices $\mathbf{V}_1(a, 0)$ and $\mathbf{V}_2(-a, 0)$, you have

$$\frac{d(\mathbf{U}, \mathbf{F}_2)}{d(\mathbf{U}, \mathfrak{D}_2)} = e,$$

where $e = \dfrac{c}{a}$ and $\mathfrak{D}_2$ is the line with equation $x = -\dfrac{a^2}{c}$.

Chapter Summary

1. A **locus** is a set of points. The set is ordinarily specified by giving geometric conditions satisfied by the points of the set.
2. An **equation of a locus** is an equation that is satisfied by the coordinates of every point on the locus but is not satisfied by the coordinates of any other point.
3. A **circle** is the set of all points in the plane at a given distance (the radius) from a given point (the center). An equation for a circle $\mathcal{C}$ can be written in the form

$$(x - h)^2 + (y - k)^2 = r^2,$$

where the point $\mathbf{C}(h, k)$ is the center, and r is the radius, of $\mathcal{C}$. If the binomials are expanded, this equation can be written in the form

$$x^2 + y^2 + Dx + Ey + F = 0,$$

where D, E, and F are constants.
4. A **parabola** is the set of all points in the plane such that each point in the set is at the same distance from a given point $\mathbf{F}$ (the **focus**) and a given line $\mathfrak{D}$ not containing $\mathbf{F}$ (the **directrix**). The line that contains the focus $\mathbf{F}$ and is perpendicular to the directrix $\mathfrak{D}$ is called the **axis** of the parabola. A parabola is symmetric with respect to its axis.

5. The standard form of the equation for a parabola which has the origin as vertex and the x- or the y-axis as axis is

$$y^2 = 4px \quad \text{or} \quad x^2 = 4py,$$

respectively, where p is the directed distance from the vertex to the focus.

6. An **ellipse** is the set of all points in the plane such that the sum of the distances of any point in the set from two given points $\mathbf{F}_1$ and $\mathbf{F}_2$ (the **foci**) is a given constant, greater than $d(\mathbf{F}_1, \mathbf{F}_2)$. The line containing the foci is the **principal axis** of the ellipse, and the midpoint of the segment connecting the foci is the **center** of the ellipse. The points of intersection of the ellipse and the principal axis are the **vertices** of the ellipse. The line segment joining the vertices is the **major axis** of the ellipse, and the line segment which is perpendicular to the major axis at its midpoint, and has endpoints on the ellipse, is the **minor axis** of the ellipse.

7. The standard form of an equation for an ellipse with center at the origin and foci on the x-axis or the y-axis is

$$\frac{x^2}{a^2} + \frac{y^2}{b^2} = 1 \quad \text{or} \quad \frac{x^2}{b^2} + \frac{y^2}{a^2} = 1,$$

respectively, where a is the length of a semi-major axis and $b = \sqrt{a^2 - c^2}$ (where c is the distance from the center to a focus) is the length of a semi-minor axis.

8. A **hyperbola** is the set of all points in the plane such that the absolute value of the difference of the distances of any point in the set from two given points $\mathbf{F}_1$ and $\mathbf{F}_2$ (the foci) is a given constant, less than $d(\mathbf{F}_1, \mathbf{F}_2)$. The definitions of the terms **principal axis, center**, and **vertices** for a hyperbola are similar to those for an ellipse (see page 144). The segment joining the vertices is the **transverse axis** of the hyperbola, and the segment of length $2b$ (see item 9) which is perpendicular to, and bisected by, the transverse axis at its midpoint is the **conjugate axis** of the hyperbola.

9. The standard form of the equation for a hyperbola with center at the origin and foci on the x-axis or the y-axis is

$$\frac{x^2}{a^2} - \frac{y^2}{b^2} = 1 \quad \text{or} \quad \frac{y^2}{a^2} - \frac{x^2}{b^2} = 1,$$

respectively, where a is the length of a semi-transverse axis and $b = \sqrt{c^2 - a^2}$ (where c is the distance from the center to a focus) is the length of a semi-conjugate axis. The lines with equations

$$y = \pm\frac{b}{a}x \quad \text{or} \quad y = \pm\frac{a}{b}x,$$

respectively, are the asymptotes of the hyperbola.

10. Ellipses, parabolas, and hyperbolas can be described as loci of points such that the ratio of the distances of each point on the locus from a fixed point (a focus) and a fixed line (the associated directrix) is a constant, namely $\frac{c}{a}$, or e. The constant e is called the **eccentricity** of the conic. The locus is an ellipse, a parabola, or a hyperbola according as $0 < e < 1$, $e = 1$, or $e > 1$. The eccentricity of a circle can be considered to be 0. Ellipses and hyperbolas each have two foci and two directrices; for an ellipse or a hyperbola with center at the origin and foci on the x-axis, the equations for the directrices are $x = \frac{a^2}{c}$ and $x = -\frac{a^2}{c}$. A parabola has just one focus and one directrix.

11. If the eccentricity e of an ellipse is near 0, then the ellipse is nearly circular; if e is near 1 (but less than 1), then the ellipse is relatively long and thin. If the eccentricity e of a hyperbola is near 1 (but greater than 1), then the focal width of the hyperbola is relatively small; if e is large, then the focal width is large. All parabolas are similar figures, and all circles are similar figures. All ellipses with the same eccentricity are similar figures and all hyperbolas with the same eccentricity are similar figures.

Chapter Review Exercises

1. Find an equation for the locus of all points in the plane that are equidistant from **S**(2, 5) and **T**(−2, −3).

2. Find an equation for the locus of all points in the plane such that the line segment joining any point of the locus with **S**(1, 5) is perpendicular to the line segment joining the point with **T**(7, −3).

3. Find an equation for the locus of all points in the plane such that the slope of the line passing through any point of the locus and **S**(4, 5) is one-half of the slope of the line passing through the point and **T**(−2, −3).

4. Find an equation for the circle with radius 7 and center **C**(−1, −3).

5. Find the coordinates of the center and the radius of the circle with equation

$$x^2 + y^2 - 6x + 8y + 12 = 0.$$

6. Find an equation for the circle passing through the points **Q**(0, −2), **S**(7, −3), and **T**(8, 4).

7. Find an equation for the circle with center **S**(4, −3) that is tangent to the line with equation $x - y = 0$.

8. Find an equation for the parabola with vertex at the origin and focus **F**(−2, 0).

9. Find an equation for the parabola with focus $\mathbf{F}(0, -5)$ and directrix the line $\mathfrak{D}$ with equation $y = 3$.

10. Find the coordinates of the focus and an equation for the directrix of the parabola which has vertex at the origin and which passes through the points $\mathbf{S}(-3, -4)$ and $\mathbf{T}(3, -4)$.

11. Find an equation for the parabola with focus $\mathbf{F}(2, 3)$ and directrix the line $\mathfrak{D}$ with equation $x = -6$.

12. Find an equation of the form $y = ax^2 + bx + c$ for the parabola containing the points $\mathbf{Q}(1, 0)$, $\mathbf{S}(0, -7)$, and $\mathbf{T}(-4, 5)$.

13. Find an equation for the ellipse with center at the origin, one focus $\mathbf{F}_1(5, 0)$, and one vertex $\mathbf{V}_1(13, 0)$.

14. Find an equation for the ellipse with center at the origin, principal axis on the y-axis, $a = \sqrt{5}$, and $b = 2$. Also find the coordinates of the foci, and determine the eccentricity.

15. Find the coordinates of the foci and vertices of the ellipse with equation

$$25x^2 + 9y^2 = 225.$$

16. Find an equation for the hyperbola with center at the origin, one focus $\mathbf{F}_1(13, 0)$, and one vertex $\mathbf{V}_1(5, 0)$.

17. Find an equation for the hyperbola with center at the origin, principal axis on the y-axis, transverse axis of length 6, and conjugate axis of length 8.

18. Find the coordinates of the vertices and foci of the hyperbola with equation

$$\frac{y^2}{8} - \frac{x^2}{17} = 1.$$

Also find equations of the asymptotes.

19. Find an equation for the ellipse with vertices $\mathbf{V}_1(5, 0)$ and $\mathbf{V}_2(-5, 0)$, and eccentricity $e = \frac{3}{5}$.

20. Find equations for the directrices of the ellipse with equation

$$\frac{x^2}{16} + \frac{y^2}{9} = 1.$$

21. Find an equation for the ellipse with foci $\mathbf{F}_1(7, 0)$ and $\mathbf{F}_2(-3, 0)$, and eccentricity $e = \frac{3}{4}$.

22. Find an equation for the hyperbola with vertices $\mathbf{V}_1(0, 6)$ and $\mathbf{V}_2(0, -6)$, and eccentricity $e = \frac{5}{3}$.

23. Find an equation for the hyperbola with center at the origin, one focus $\mathbf{F}_1(0, 10)$, and corresponding directrix the line $\mathfrak{D}_1$ with equation $y = \frac{8}{5}$.

24. Find an equation for the hyperbola having the line with equation $x = 1$ as a directrix and the lines with equations $y = \pm\frac{2}{3}x$ as asymptotes.

Dandelin Spheres

It is a rather surprising fact, since the ancient Greeks were quite familiar with many properties of conic sections, that the following elegant geometric configuration (see Figure 4, page 166) was unknown until 1822. It was discovered by two Belgian mathematicians, Lambert Adolphe Jacques Quetelet (1796–1874) and Germinal Pierre Dandelin (1794–1847).

Notice first, as preliminary observations, that for a point O (Figure 1) outside a sphere $\mathcal{S}$ the lines that pass through O and are tangent to $\mathcal{S}$ are the

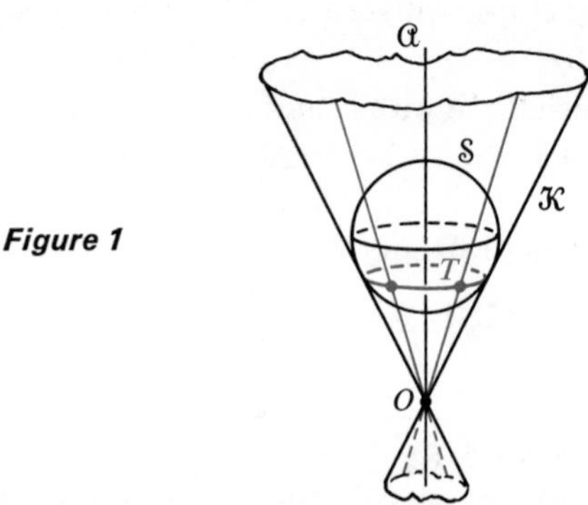

Figure 1

generators of a right circular cone $\mathcal{K}$ with vertex O, that the points T of tangency constitute a circle $\mathcal{C}$ in a plane perpendicular to the axis $\mathcal{A}$ of the cone, and that the length $d(O, T)$ is the same for all points T on $\mathcal{C}$.

For two such spheres $\mathcal{S}_1$ and $\mathcal{S}_2$ (Figure 2) tangent to $\mathcal{K}$ in circles $\mathcal{C}_1$ and $\mathcal{C}_2$, a generator of $\mathcal{K}$ meets $\mathcal{C}_1$ and $\mathcal{C}_2$ in points T_1 and T_2, respectively. The part of the cone between two parallel planes cutting it is called a *frustum*.

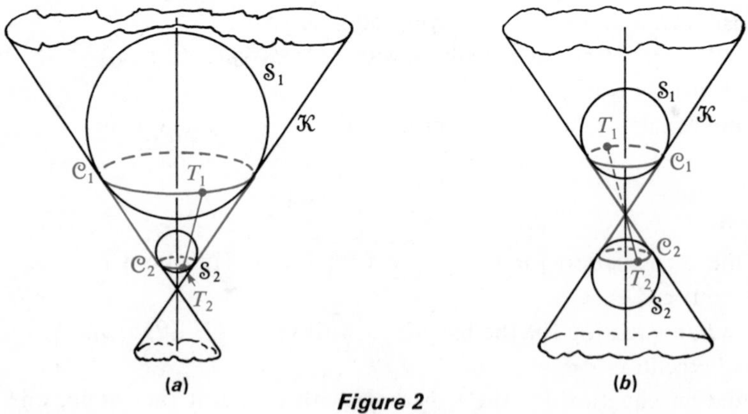

Figure 2

The distance $d(T_1, T_2)$ is the same for all generators of $\mathcal{K}$, whether the spheres $\mathcal{S}_1$ and $\mathcal{S}_2$ are in the same nappe of $\mathcal{K}$ [Figure 2(a)] or not [Figure 2(b)]. We can denote this constant distance by $2a$. Then $2a$ is the slant height of the frustum of the cone $\mathcal{K}$, with bases bounded by $\mathcal{C}_1$ and $\mathcal{C}_2$.

Now let $\mathcal{K}$ (Figure 3), be a right circular cone, with vertex V, whose generators make an angle α,

$$0 < m^{\mathrm{R}}(\alpha) < \frac{\pi}{2},$$

with the axis $\mathcal{A}$ of $\mathcal{K}$. Let a plane $\mathcal{P}$ be constructed so that it crosses the axis and cuts only one nappe $\mathcal{N}_1$ of the cone, making a closed curve as indicated below. Let the curve $\mathcal{E}$ be the intersection of $\mathcal{P}$ and $\mathcal{K}$. We are going to show by direct construction that $\mathcal{E}$ is an ellipse.

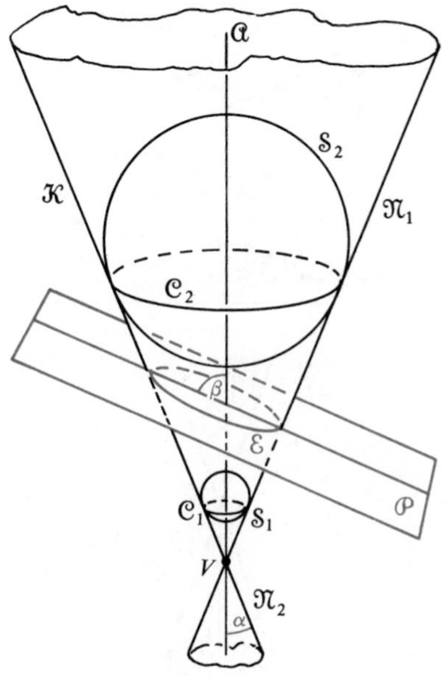

Figure 3

Let $\mathcal{S}_1$ (Figure 3) be a very small sphere having a circle of contact $\mathcal{C}_1$ with $\mathcal{N}_1$ on the same side of $\mathcal{P}$ as the vertex of $\mathcal{K}$, and let $\mathcal{S}_2$ be a huge sphere with circle of contact $\mathcal{C}_2$ on the opposite side of $\mathcal{P}$. You can imagine $\mathcal{S}_1$ as expanding and $\mathcal{S}_2$ as contracting, until they just touch $\mathcal{P}$ in points F_1 and F_2, respectively (Figure 4 on page 166). In these final positions, the spheres $\mathcal{S}_1$ and $\mathcal{S}_2$ are called *Dandelin spheres*.

Suppose $\mathcal{S}_1$ and $\mathcal{S}_2$ are Dandelin spheres and (as shown in Figure 4) let $\mathcal{P}_1$ and $\mathcal{P}_2$ be the planes determined by $\mathcal{C}_1$ and $\mathcal{C}_2$. Let $\mathcal{P}_1$ and $\mathcal{P}_2$ cut $\mathcal{P}$ in lines $\mathcal{D}_1$ and $\mathcal{D}_2$, respectively.

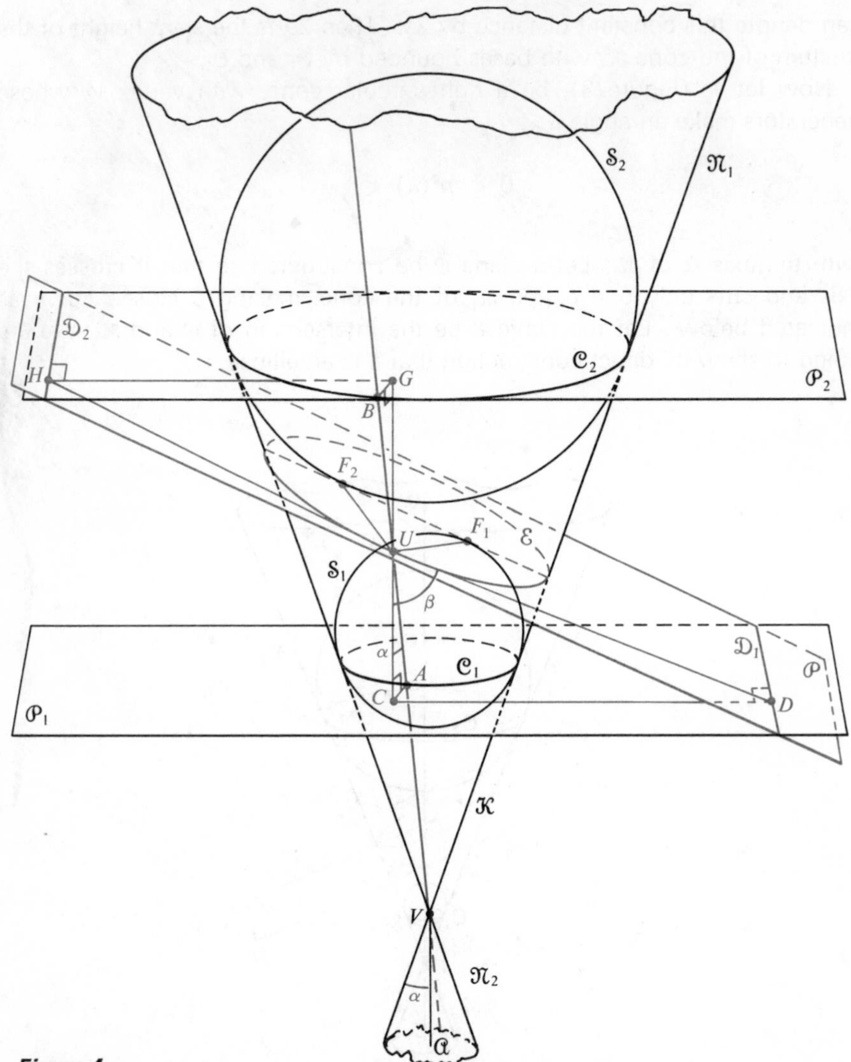

Figure 4

For any point U on $\mathcal{E}$, let the generator of $\mathcal{K}$ through U cut $\mathcal{C}_1$ and $\mathcal{C}_2$ in points A and B, respectively. Since the lines UF_1 and UA are tangent to $\mathcal{S}_1$, and the lines UF_2 and UB are tangent to $\mathcal{S}_2$, you have

$$d(U, F_1) = d(U, A)$$

and

$$d(U, F_2) = d(U, B).$$

Adding, you obtain

$$d(U, F_1) + d(U, F_2) = d(U, A) + d(U, B) = d(A, B).$$

This can be written as

$$d(U, F_1) + d(U, F_2) = 2a, \tag{1}$$

where $2a = d(A, B)$ is the slant height of the frustum of the cone $\mathcal{K}$, with bases bounded by $\mathcal{C}_1$ and $\mathcal{C}_2$. Further, for any point R on $\mathcal{P}$ inside $\mathcal{E}$ you have

$$d(R, F_1) + d(R, F_2) < 2a,$$

and for any point S on $\mathcal{P}$ outside $\mathcal{E}$ you have

$$d(S, F_1) + d(S, F_2) > 2a,$$

as suggested by Figure 5(a) and Figure 5(b). Therefore, $\mathcal{E}$ is the locus of

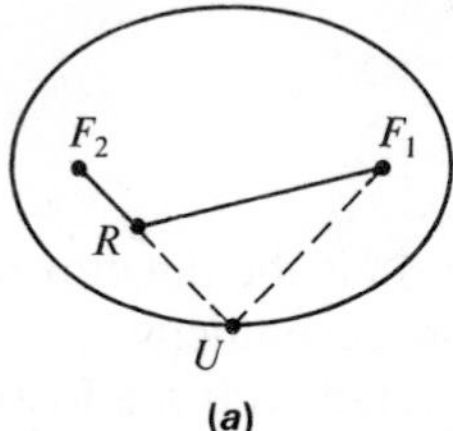

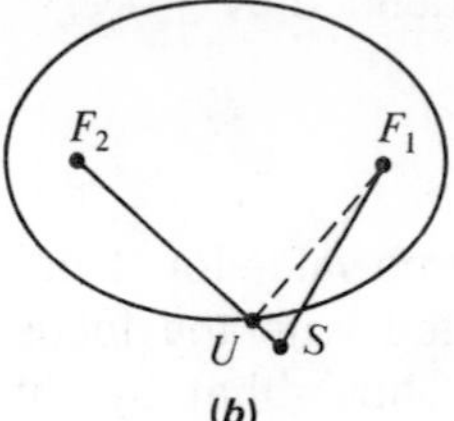

(a) ***Figure 5*** **(b)**

all points on $\mathcal{P}$ satisfying Equation (1); accordingly, by definition (page 137), $\mathcal{E}$ is an ellipse with foci F_1 and F_2. The relationship in Equation (1) gives a geometric interpretation of the constant $2a$ in the definition of an ellipse; namely, $2a$ is the slant height of the frustum.

Let the point C in Figure 4 be the foot of the perpendicular from U to the plane $\mathcal{P}_1$, and let D be the foot of the perpendicular from C to the line $\mathcal{D}_1$. Since UC is a perpendicular to the plane $\mathcal{P}_1$, it is parallel to the axis. Thus the angle between UC and UA is equal to α, the angle between the axis and the cone. Thus from right triangle AUC you have

$$\cos \alpha = \frac{d(U, C)}{d(U, A)},$$

or, since $d(U, A) = d(U, F_1)$,

$$\cos \alpha = \frac{d(U, C)}{d(U, F_1)}.$$

Let β be the angle that $\mathcal{P}$ makes with the axis. From right triangle UDC, you have

$$\cos \beta = \frac{d(U, C)}{d(U, D)},$$

or, since $d(U, D) = d(U, \mathcal{D}_1)$,

$$\cos \beta = \frac{d(U, C)}{d(U, \mathcal{D}_1)}.$$

Therefore,

$$\frac{d(U, F_1)}{d(U, \mathfrak{D}_1)} = \frac{d(U, F_1)/d(U, C)}{d(U, \mathfrak{D}_1)/d(U, C)}$$

$$= \frac{1/\cos\alpha}{1/\cos\beta}$$

$$= \frac{\cos\beta}{\cos\alpha}.$$

Accordingly you have

$$d(U, F_1) = e[d(U, \mathfrak{D}_1)],$$

where e, defined by

$$e = \frac{\cos\beta}{\cos\alpha},$$

has the same value for all points U on $\mathcal{E}$. Thus $\mathfrak{D}_1$ is the directrix of the ellipse $\mathcal{E}$ associated with the focus F_1, and e is the eccentricity of $\mathcal{E}$. A similar argument shows that $\mathfrak{D}_2$ is the directrix associated with the focus F_2. Since

$$0 < m^{\mathbf{R}}(\alpha) < m^{\mathbf{R}}(\beta) < \frac{\pi}{2}$$

you have

$$1 > \cos\alpha > \cos\beta > 0,$$

and therefore

$$0 < \frac{\cos\beta}{\cos\alpha} < 1,$$

that is,

$$0 < e < 1.$$

In Figure 4 on page 166, imagine that the cone $\mathcal{K}$ remains fixed but that the plane $\mathcal{P}$ rotates toward a horizontal position, so that $m^{\mathbf{R}}(\beta)$ approaches $\frac{\pi}{2}$ (with axis of rotation the horizontal line in $\mathcal{P}$ that intersects the axis $\mathcal{A}$ of the cone). Do you see that as the spheres $\mathcal{S}_1$ and $\mathcal{S}_2$ and the planes $\mathcal{P}_1$ and $\mathcal{P}_2$ adjust accordingly, the directrices $\mathfrak{D}_1$ and $\mathfrak{D}_2$ recede indefinitely and the foci F_1 and F_2 approach a common point? In the limiting position, $\mathcal{P}$ is parallel to $\mathcal{P}_1$ and $\mathcal{P}_2$, and $\mathcal{E}$ becomes a circle. The circle has no directrices. Since now $m^{\mathbf{R}}(\beta) = \frac{\pi}{2}$, and $\cos\frac{\pi}{2} = 0$, for the circle you have

$$e = \frac{0}{\cos\alpha} = 0.$$

In the other direction, imagine that $m^{\mathbf{R}}(\beta)$ approaches $m^{\mathbf{R}}(\alpha)$, using the same rotational axis. Then $\mathcal{S}_2$ moves farther and farther away. In the limiting position, $\mathcal{S}_2$ is removed altogether, and F_2 and $\mathcal{D}_2$ disappear with it. Now $\mathcal{P}$ is parallel to a generator of $\mathcal{K}$, and the intersection of $\mathcal{P}$ and the cone is a parabola. Since now $m^{R}(\beta) = m^{R}(\alpha)$, and $\cos \beta = \cos \alpha$, for the parabola you have

$$e = \frac{\cos \beta}{\cos \alpha} = 1.$$

As $\mathcal{P}$ rotates still farther, with $m^{R}(\alpha) > m^{R}(\beta) > 0$, $\mathcal{P}$ intersects the other nappe, $\mathcal{N}_2$, of $\mathcal{K}$, and an argument similar to the one given above for the ellipse shows that the intersection of $\mathcal{P}$ with $\mathcal{K}$ is a hyperbola. Since now $\frac{\pi}{2} > m^{R}(\alpha) > m^{R}(\beta) > 0$, and $0 < \cos \alpha < \cos \beta < 1$, for the hyperbola you have

$$e = \frac{\cos \beta}{\cos \alpha} > 1.$$

As an exercise, you might draw and discuss figures, analogous to Figure 4, for the circle, parabola, and hyperbola. You might also consider figures for the various degenerate conic sections; in particular, for the case of two parallel lines, you might hold $\mathcal{S}_1$ fixed and let the vertex V of $\mathcal{K}$ recede indefinitely.

Chapter 5

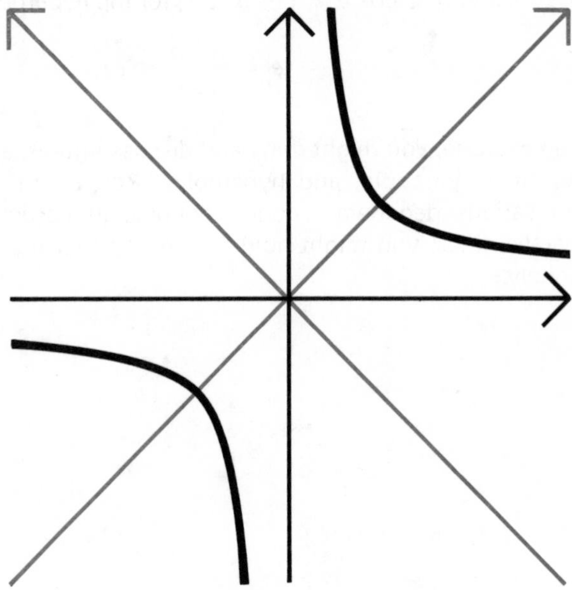

Translations and rotations of axes can be used to help identify and sketch conic sections that are not in standard position. These topics, as well as a discussion of tangents to conic sections, are included in this chapter.

Transformation of Coordinates

Translations and Rotations

5–1 Translation of Axes

In a number of different kinds of mathematical problems, the situation can be clarified, or the calculations simplified, by changing the location of the coordinate axes. Such a change results in a *transformation of coordinates*. In this chapter, you will learn about two basic types of transformations of coordinates, *translations* and *rotations*.

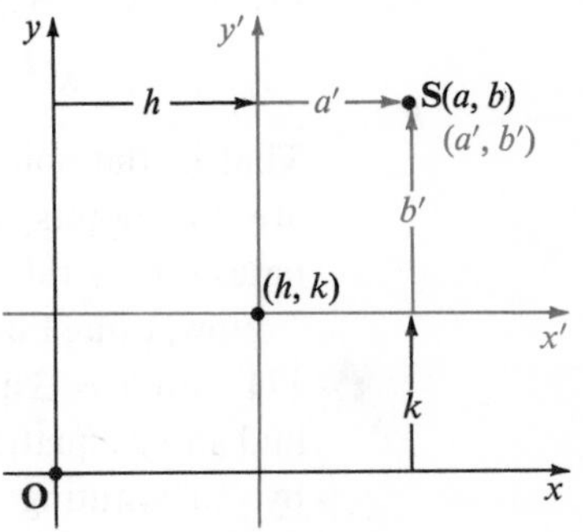

Figure 5–1

Figure 5–1 shows two sets of Cartesian axes in the plane, with the respective axes parallel to each other and in the same direction. The x'- and y'-axes can be viewed as the result of "sliding" the x- and y-axes across the plane, while keeping them parallel to their original orientations, until the origin coincides with the point whose xy-coordinates are (h, k). Such a "sliding," or **translation**, of axes assigns to each point $\mathbf{S}(a, b)$ in the plane a new set of coordinates, (a', b'). As the figure suggests, you have $a = a' + h$ and $b = b' + k$. Thus the xy-coordinates are related to the $x'y'$-coordinates by the equations

$$\begin{aligned} x &= x' + h, \\ y &= y' + k, \end{aligned} \qquad (1)$$

or, equivalently, by the equations

$$\begin{aligned} x' &= x - h, \\ y' &= y - k. \end{aligned} \qquad (2)$$

Equations (1) and (2) can be used very effectively to study circles whose centers are not at the origin, and parabolas, ellipses, and hyperbolas whose respective axes are parallel to the coordinate axes, but which are located in positions other than the standard positions discussed in Chapter 4.

Example 1. A parabola has vertex $\mathbf{V}(3, 4)$ and focus $\mathbf{F}(3, 6)$. Find a Cartesian equation for the parabola.

Solution: First make a sketch showing the coordinate axes and the given points. Since the vertex and focus of a parabola are located on its axis, the given parabola has the line with equation $x = 3$ as its axis.

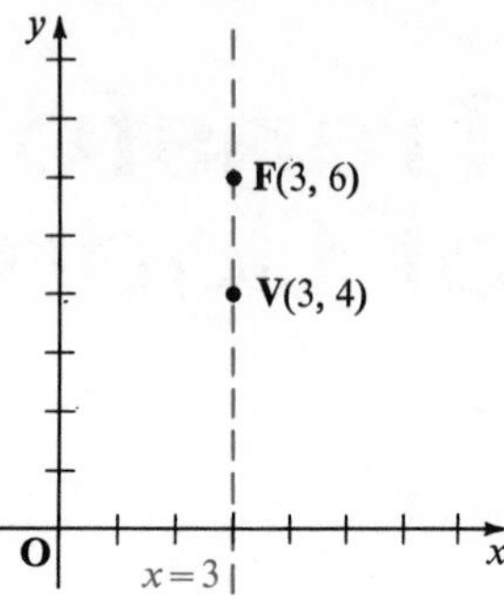

Next, sketch a set of $x'y'$-axes with origin at the vertex $\mathbf{V}(3, 4)$ of the parabola. An $x'y'$-equation of the parabola is then

$$x'^2 = 4py',$$

or, since $6 - 4 = 2 = p$,

$$x'^2 = 8y'. \qquad (3)$$

That is, the points on the parabola are the points whose $x'y'$-coordinates satisfy this equation.

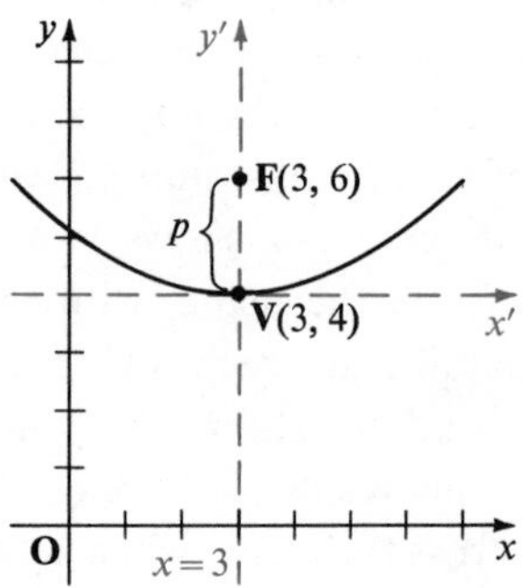

Now, using Equations (2) on page 171 with $h = 3$ and $k = 4$, you can find an xy-equation for the parabola by substituting the values $x' = x - 3$ and $y' = y - 4$ in Equation (3). Thus, the required equation is

$$(x - 3)^2 = 8(y - 4),$$

or

$$x^2 - 6x - 8y + 41 = 0.$$

Any equation of the form

$$Ax^2 + Cy^2 + Dx + Ey + F = 0 \qquad (A \text{ and } C \text{ not both } 0) \qquad (4)$$

ordinarily has a graph that either is a circle or is an ellipse, parabola, or hyperbola with (principal) axis parallel to a coordinate axis. In certain exceptional cases, its graph may be a point, a line, two parallel lines, two intersecting lines, or the empty set.

By completing the squares in x and y if both A and C are different from 0, or otherwise by completing one square and combining the remaining linear and constant terms, and then using Equations (2) on page 171, you can readily identify and sketch the curve represented by an equation of the form (4).

Example 2. Find an $x'y'$-equation for the graph of

$$9x^2 - 4y^2 + 36x - 24y - 36 = 0$$

so that the center of the graph is at the origin in the $x'y'$-system. Sketch the curve.

Solution: Complete the squares in x and y and simplify:

$$\begin{aligned} 9x^2 + 36x - 4y^2 - 24y &= 36, \\ 9(x^2 + 4x) - 4(y^2 + 6y) &= 36, \\ 9(x^2 + 4x + 4) - 4(y^2 + 6y + 9) &= 36 + 36 - 36, \\ 9(x + 2)^2 - 4(y + 3)^2 &= 36. \end{aligned}$$

Then divide both members by 36. You obtain

$$\frac{(x + 2)^2}{4} - \frac{(y + 3)^2}{9} = 1,$$

or

$$\frac{[x - (-2)]^2}{4} - \frac{[y - (-3)]^2}{9} = 1.$$

Now, using Equations (2) on page 171 with $h = -2$ and $k = -3$, you can write this last equation in terms of $x'y'$-coordinates as

$$\frac{x'^2}{4} - \frac{y'^2}{9} = 1.$$

This is an equation of a hyperbola with center at the origin in the $x'y'$-system. Its graph is shown at the right.

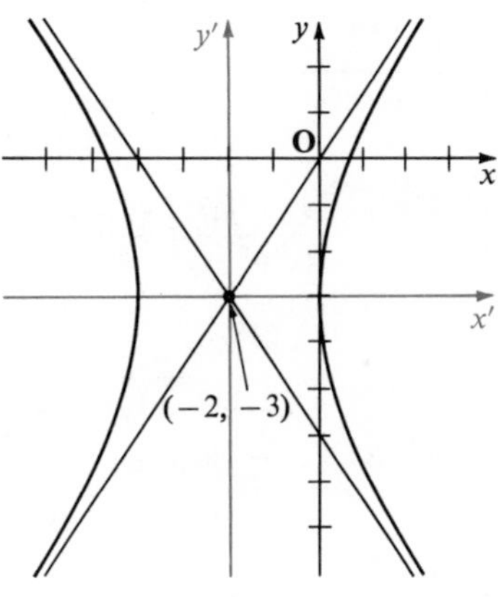

Example 3. Determine the coordinates of the vertex **V** and the focus **F** of the parabola with equation

$$y^2 + 4x + 6y + 1 = 0.$$

Solution: First complete the square in y and simplify:

$$\begin{aligned} y^2 + 6y &= -4x - 1, \\ y^2 + 6y + 9 &= -4x - 1 + 9, \\ (y + 3)^2 &= -4x + 8, \\ (y + 3)^2 &= -4(x - 2), \end{aligned}$$

or

$$[y - (-3)]^2 = -4(x - 2).$$

(*Solution continued*)

Next, use Equations (2) on page 171 with $h = 2$ and $k = -3$ to find an $x'y'$-equation in standard form for the parabola:

$$y'^2 = -4x'.$$

In the $x'y'$-system, this is an equation of the parabola whose vertex **V** has coordinates $(x', y') = (0, 0)$ and whose focus **F** has coordinates $(x', y') = (-1, 0)$.

Finally, use Equations (1) on page 171 with $h = 2$ and $k = -3$ to find the xy-coordinates of the vertex and focus: For the vertex **V**, you have

$$(x, y) = \big(0 + 2, 0 + (-3)\big) = (2, -3),$$

and for the focus **F**, you have

$$(x, y) = \big(-1 + 2, 0 + (-3)\big) = (1, -3).$$

The process of completing the square(s) and simplifying the resulting equation that was used in Examples 2 and 3 can be applied to any equation of the form (4). You can always work toward an equation of one of the forms shown in the table below. (This table should be understood but not memorized.)

Type	*Equation*	*Graph*	*Center*	*Vertices*	*Foci*
(1)	$(x-h)^2 + (y-k)^2 = r^2$	Circle	(h, k)	—	—
$r > 0$ $a > b > 0$	$\frac{(x-h)^2}{a^2} + \frac{(y-k)^2}{b^2} = 1$	Ellipse	(h, k)	$(h \pm a, k)$	$(h \pm \sqrt{a^2 - b^2}, k)$
	$\frac{(y-k)^2}{a^2} + \frac{(x-h)^2}{b^2} = 1$	Ellipse	(h, k)	$(h, k \pm a)$	$(h, k \pm \sqrt{a^2 - b^2})$
(2)	$(x-h)^2 = 4p(y-k)$	Parabola	—	(h, k)	$(h, k + p)$
$p \neq 0$	$(y-k)^2 = 4p(x-h)$	Parabola	—	(h, k)	$(h + p, k)$
(3) $a > 0$	$\frac{(x-h)^2}{a^2} - \frac{(y-k)^2}{b^2} = 1$	Hyperbola	(h, k)	$(h \pm a, k)$	$(h \pm \sqrt{a^2 + b^2}, k)$
$b > 0$	$\frac{(y-k)^2}{a^2} - \frac{(x-h)^2}{b^2} = 1$	Hyperbola	(h, k)	$(h, k \pm a)$	$(h, k \pm \sqrt{a^2 + b^2})$

In exceptional cases, however, when you have reduced the second-degree terms to a form appearing in the left-hand member of an equation in the table, you will not be able to express the right-hand member in the form shown.

Thus, for equations of Type 1, in place of r^2 (with $r > 0$) or 1 in the right-hand member, you might have a *nonpositive* constant q. If $q = 0$, then the graph is just the point (h, k); and if $q < 0$, then the graph is the empty set, $\emptyset$.

For equations of Type 2, in place of $4p(x - h)$ or $4p(y - k)$, with $p \neq 0$, in the right-hand member, you might again have a constant q. If $q > 0$, then the graph is two (parallel) horizontal lines or two (parallel) vertical lines; if $q = 0$, then the graph is one horizontal line or one vertical line; and if $q < 0$, then the graph is $\emptyset$.

For equations of Type 3, in place of 1 in the right-hand member, you might have 0. The graph is then a pair of intersecting lines.

From the preceding discussion, the following facts emerge:

■ The graph of $Ax^2 + Cy^2 + Dx + Ey + F = 0$ (where A and C are not both 0) is a(n):

1. *Circle* if $A = C$. In exceptional cases, the graph may be a point or $\emptyset$.
2. *Ellipse* if $A \neq C$ and A and C have the same sign ($AC > 0$). In exceptional cases, the graph may be a point or $\emptyset$.
3. *Parabola* if either $A = 0$ or $C = 0$. In exceptional cases, the graph may be two parallel lines, a line, or $\emptyset$.
4. *Hyperbola* if A and C are opposite in sign ($AC < 0$). In exceptional cases, the graph may be two intersecting straight lines.

The types of loci listed above are called **conic sections**, or **conics**, because they result from the intersection of a plane and a right circular cone of two nappes (Figures 5–2 and 5–3).

The Nondegenerate Conic Sections

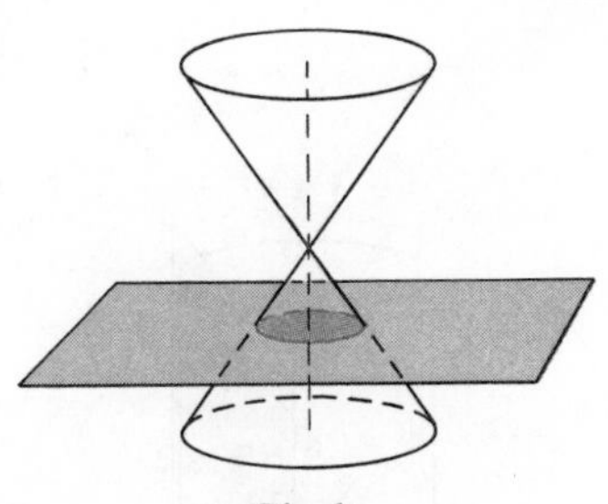

Circle

Plane perpendicular to the axis, cutting one nappe.

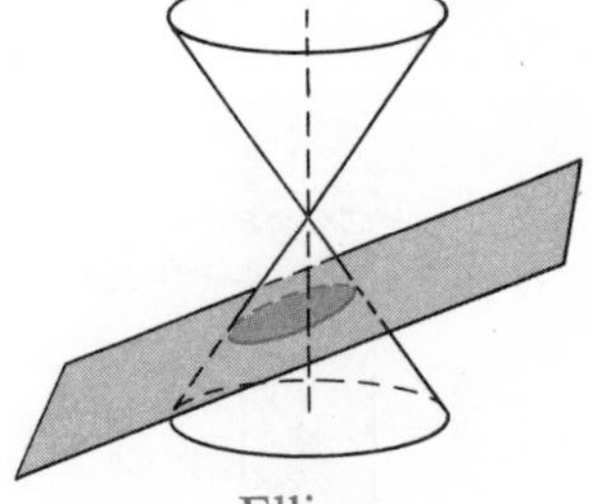

Ellipse

Plane oblique to the axis, cutting one nappe.

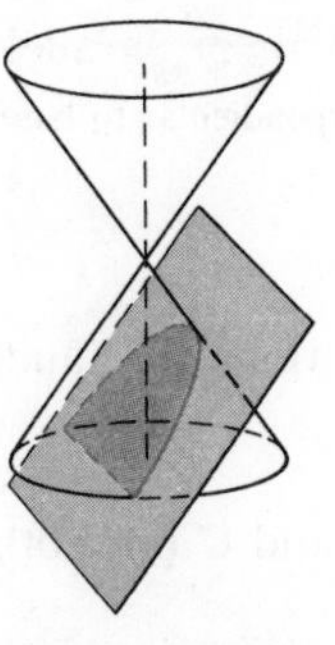

Parabola

Plane parallel to an element.

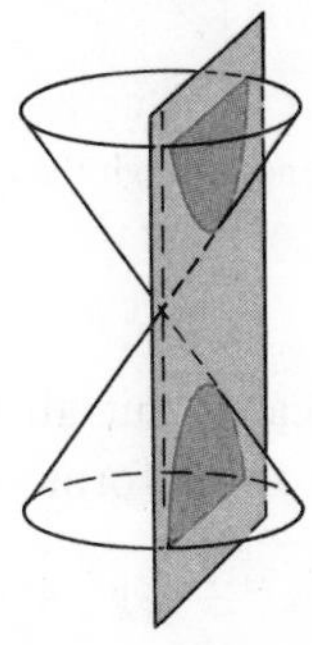

Hyperbola

Plane cutting both nappes.

Figure 5–2

The nonempty exceptional cases result when the plane passes through the vertex of the cone (Figure 5–3). (For the case of two parallel lines, the cone must be replaced with a right circular cylinder, which can be thought of as a right circular cone whose vertex is infinitely far off.) The exceptional loci are called **degenerate conic sections**.

The Degenerate Conic Sections

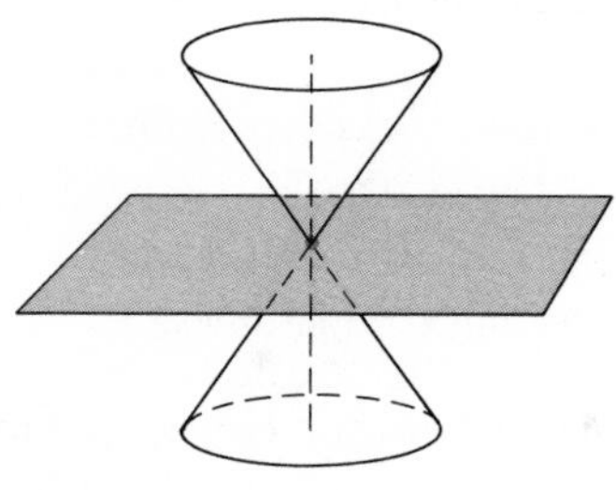

Point
Plane cutting only at the vertex.

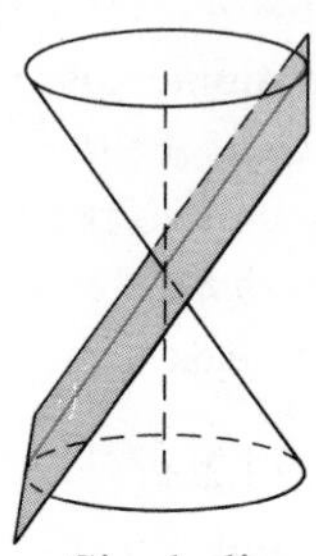

Single line
Plane tangent along an element.

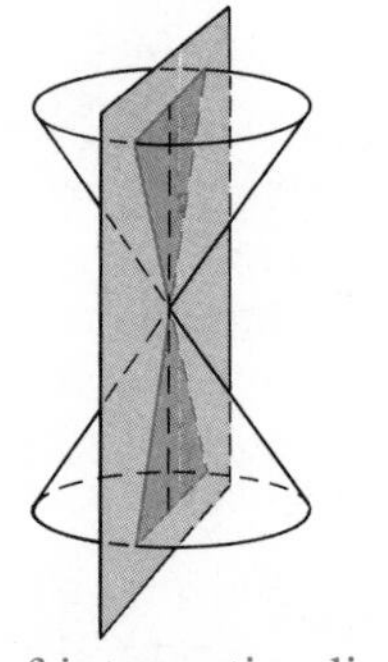

Pair of intersecting lines
Plane through the axis.

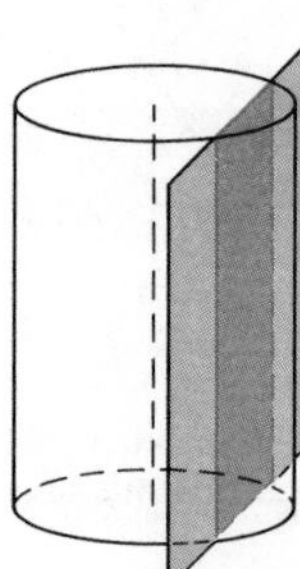

Pair of parallel lines
Plane perpendicular to base of cylinder.

Figure 5–3

If you carry out all the simplifications or reductions to standard form of equations of the form

$$Ax^2 + Cy^2 + Dx + Ey + F = 0 \qquad (A \text{ and } C \text{ not both } 0),$$

you will find (Exercises 39–45, page 179) the results summarized in the following table, which is intended for reference and again is not to be memorized.

Type	Condition			Graph
Elliptic type $AC > 0$	$A(4ACF - CD^2 - AE^2) < 0$	$A = C$		a circle
		$A \neq C$		an ellipse
	$A(4ACF - CD^2 - AE^2) > 0$			the empty set
	$4ACF - CD^2 - AE^2 = 0$			a point
Parabolic type $AC = 0$	$A \neq 0$, $C = 0$	$E \neq 0$		a parabola
		$E = 0$	$4AF - D^2 < 0$	two parallel lines
			$4AF - D^2 = 0$	one line
			$4AF - D^2 > 0$	the empty set
	$A = 0$, $C \neq 0$	$D \neq 0$		a parabola
		$D = 0$	$4CF - E^2 < 0$	two parallel lines
			$4CF - E^2 = 0$	one line
			$4CF - E^2 > 0$	the empty set
Hyperbolic type $AC < 0$	$4ACF - CD^2 - AE^2 \neq 0$			a hyperbola
	$4ACF - CD^2 - AE^2 = 0$			two intersecting lines

Exercises 5–1

In Exercises 1–6, an *xy*-equation and the *xy*-coordinates of a point **S** are given. Find an *x′y′*-equation for the graph of the given equation if the origin of the *x′y′*-system is at the point **S**.

Example. $y^2 - 6y + 5 = 4x$; $\mathbf{S}(-1, 3)$

Solution: If the origin of the $x'y'$-system is at the point $\mathbf{S}(-1, 3) = \mathbf{S}(h, k)$, then the xy- and the $x'y'$-coordinates are related by the equations

$$x = x' + h, \qquad y = y' + k,$$

or

$$x = x' - 1, \qquad y = y' + 3.$$

Substituting $x' - 1$ for x and $y' + 3$ for y in the given equation, and then simplifying the resulting equation, you obtain

$$\begin{aligned} (y' + 3)^2 - 6(y' + 3) + 5 &= 4(x' - 1), \\ y'^2 + 6y' + 9 - 6y' - 18 + 5 &= 4x' - 4, \\ y'^2 &= 4x'. \end{aligned}$$

1. $x - 4y^2 + 16y - 7 = 0$; $\mathbf{S}(-3, 4)$

2. $x^2 - y^2 - 6x + 10y - 20 = 0$; $\mathbf{S}(3, 5)$

3. $x^2 + 4y^2 - 6x - 16y - 11 = 0$; $\mathbf{S}(3, 2)$

4. $x^2 + y^2 - 8x + 10y - 4 = 0$; $\mathbf{S}(4, -5)$
5. $x^2 + 2y^2 - 2x + 16y + 33 = 0$; $\mathbf{S}(1, -4)$
6. $4x^2 - y^2 - 12x - 6y + 24 = 0$; $\mathbf{S}(\frac{3}{2}, -3)$

In Exercises 7–18, find a Cartesian equation for the conic section satisfying the given conditions. Sketch the curve.

7. Parabola with vertex $\mathbf{V}(3, -2)$ and focus $\mathbf{F}(3, 4)$.
8. Parabola with vertex $\mathbf{V}(-\frac{3}{2}, 2)$ and focus $\mathbf{F}(1, 2)$.
9. Parabola with focus $\mathbf{F}(2, 4)$ and directrix the line with equation $x = 8$.
10. Parabola with focus $\mathbf{F}(-1, 2)$ and directrix the line with equation $y = -4$.
11. Ellipse with foci $\mathbf{F}_1(1, 2)$ and $\mathbf{F}_2(9, 2)$, and major axis of length 10.
12. Ellipse with foci $\mathbf{F}_1(-3, 2)$ and $\mathbf{F}_2(-3, 6)$, and major axis of length 12.
13. Ellipse with vertices $\mathbf{V}_1(8, 2)$ and $\mathbf{V}_2(-4, 2)$, and one focus $\mathbf{F}_1(6, 2)$.
14. Ellipse with center $\mathbf{C}(-2, 4)$, one vertex $\mathbf{V}_1(3, 4)$, and associated focus $\mathbf{F}_1(2, 4)$.
15. Hyperbola with foci $\mathbf{F}_1(3, 2)$ and $\mathbf{F}_2(3, -6)$, and transverse axis of length 4.
16. Hyperbola with foci $\mathbf{F}_1(-6, -3)$ and $\mathbf{F}_2(4, -3)$, and one vertex $\mathbf{V}_2(3, -3)$.
17. Hyperbola with vertices $\mathbf{V}_1(6, 2)$ and $\mathbf{V}_2(-2, 2)$, and eccentricity $\frac{7}{5}$.
18. Hyperbola with center $\mathbf{C}(3, 2)$, one focus $\mathbf{F}_1(3, 7)$, and eccentricity $\frac{5}{3}$.

In Exercises 19–30, find an $x'y'$-equation for the graph of the given xy-equation so that the center or, in the case of a parabola, the vertex, of the graph is at the origin in the $x'y'$-system. Sketch the curve, showing both sets of coordinate axes, and identify the type of curve.

19. $x^2 + y^2 + 4x - 10y - 36 = 0$
20. $x^2 - 4y^2 + 4x + 32y - 64 = 0$
21. $2x^2 + y^2 + 8x - 8y - 48 = 0$
22. $x^2 - 4x - 4y + 16 = 0$
23. $4x + y^2 + 4y - 4 = 0$
24. $4x^2 + 9y^2 + 8x + 36y + 4 = 0$
25. $4x^2 + 4y^2 - 12x + 8y - 3 = 0$
26. $8x - y^2 - 8y = 0$
27. $9x^2 + 4y^2 - 18x + 24y + 45 = 0$
28. $x^2 - y^2 + 6x + 4y + 5 = 0$
29. $4x^2 - 9y^2 - 16x + 18y - 7 = 0$
30. $4y^2 - 3x^2 + 8y - 12x - 16 = 0$

In Exercises 31–36, sketch the graph of the given equation.

31. $x^2 + y^2 - 4x + 6y + 13 = 0$
32. $4x^2 - 9y^2 + 16x + 18y + 7 = 0$
33. $x^2 + 2y^2 - 4y + 3 = 0$
34. $x^2 - 8x + 15 = 0$
35. $y^2 + 6y + 9 = 0$
36. $x^2 + 10x + 30 = 0$

37. Use a translation of axes to remove first-degree terms from the equation $xy + 4x - 8y + 6 = 0$; that is, find an $x'y'$-equation for the graph of the given xy-equation such that the first-degree terms in x' or y' have coefficient 0. (*Hint:* Replace x and y with $x' + h$ and $y' + k$, respectively, and find suitable values for h and k.)

38. Use the method suggested in Exercise 37 to remove first-degree terms from the equation

$$xy + ax + by + c = 0.$$

Exercises 39–45 refer to the equation $Ax^2 + Cy^2 + Dx + Ey + F = 0$, where A and C are not both 0.

* **39.** Show that if $A = 0$, $C \neq 0$, and $D \neq 0$, then the equation is equivalent to one of the form

$$(y - k)^2 = 4p(x - h).$$

* **40.** Show that if $A \neq 0$, $C = 0$, and $E \neq 0$, then the equation is equivalent to one of the form

$$(x - h)^2 = 4p(y - k).$$

* **41.** Show that if $A = C \neq 0$, then the equation is equivalent to one of the form

$$(x - h)^2 + (y - k)^2 = q.$$

What must be true of q if the equation is to have a nonempty graph in $\mathcal{R}^2$?

* **42.** Show that if $A \neq C$, but $AC > 0$, then the equation is equivalent to an equation of one of the forms

$$\frac{(x - h)^2}{a^2} + \frac{(y - k)^2}{b^2} = 1 \quad \text{or} \quad \frac{(y - k)^2}{a^2} + \frac{(x - h)^2}{b^2} = 1,$$

$a > b > 0$, provided $A(4ACF - CD^2 - AE^2) < 0$.

* **43.** Show that if $AC < 0$, then the equation is equivalent to an equation of one of the forms

$$\frac{(x - h)^2}{a^2} - \frac{(y - k)^2}{b^2} = 1 \quad \text{or} \quad \frac{(y - k)^2}{a^2} - \frac{(x - h)^2}{b^2} = 1,$$

$a > 0$, $b > 0$, provided $4ACF - CD^2 - AE^2 \neq 0$.

* **44.** In Exercise 42, why must $A(4ACF - CD^2 - AE^2)$ be negative? What can be said about the graph if $A(4ACF - CD^2 - AE^2)$ is 0? What if it is positive?

* **45.** In Exercise 43, why must $4ACF - CD^2 - AE^2$ not equal zero? What can be said about the graph if $4ACF - CD^2 - AE^2 = 0$?

* **46.** By a translation of axes, points $\mathbf{S}(x_1, y_1)$ and $\mathbf{T}(x_2, y_2)$ are given new coordinates (x_1', y_1') and (x_2', y_2'), respectively, in accordance with Equations (2) on page 171. Verify algebraically that

$$\sqrt{(x_2' - x_1')^2 + (y_2' - y_1')^2} = \sqrt{(x_2 - x_1)^2 + (y_2 - y_1)^2},$$

and interpret this fact geometrically.

* **47.** By a translation of axes, (distinct) points $\mathbf{S}(x_1, y_1)$, $\mathbf{T}(x_2, y_2)$, and $\mathbf{U}(x_3, y_3)$ are given new coordinates $\mathbf{S}'(x_1', y_1')$, $\mathbf{T}'(x_2', y_2')$, and $\mathbf{U}'(x_3', y_3')$, respectively. If $\mathbf{s}$, $\mathbf{t}$, $\mathbf{u}$, $\mathbf{s}'$, $\mathbf{t}'$, and $\mathbf{u}'$ are the vectors corresponding to $\mathbf{S}$, $\mathbf{T}$, $\mathbf{U}$, $\mathbf{S}'$, $\mathbf{T}'$, and $\mathbf{U}'$, respectively, verify algebraically that

$$\frac{(\mathbf{t}' - \mathbf{s}') \cdot (\mathbf{u}' - \mathbf{s}')}{\|\mathbf{t}' - \mathbf{s}'\| \, \|\mathbf{u}' - \mathbf{s}'\|} = \frac{(\mathbf{t} - \mathbf{s}) \cdot (\mathbf{u} - \mathbf{s})}{\|\mathbf{t} - \mathbf{s}\| \, \|\mathbf{u} - \mathbf{s}\|},$$

and interpret this fact geometrically.

5–2 Rotation of Axes

In Section 5–1, you saw that you can use a translation of axes to transform a given equation into a simpler (equivalent) equation whose graph is easy to recognize. This is also true of a *rotation of axes.*

If the coordinate axes are rotated about the origin while all points in the plane are considered fixed, then every point (or vector) except the origin will have a new set of coordinates (or components). These new coordinates can be determined by using facts from trigonometry, as developed below.

Figure 5–4 shows two sets of Cartesian axes in the plane. As illustrated, the x'- and y'-axes have been obtained by rotating the x- and y-axes about the origin through an angle ϕ. Suppose that $\mathbf{S}$ is any point in the plane, and that $\mathbf{S}$ has coordinates (x, y) with respect to the x- and y-axes and (x', y') with respect to the x'- and y'-axes. If the direction angle of the vector $\mathbf{s}$ with respect to the x-axis is θ, then, as shown, the direction angle of $\mathbf{s}$ with respect to the x'-axis is $\theta - \phi$. Thus, if the magnitude of $\mathbf{s}$ is r, you have

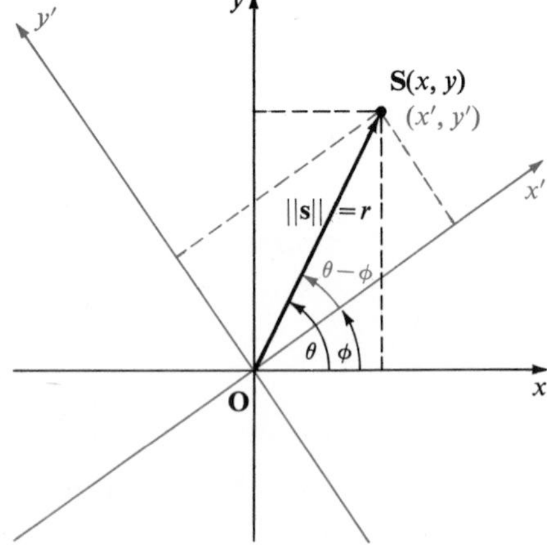

Figure 5–4

$$x = r\cos\theta, \qquad y = r\sin\theta$$

and

$$x' = r\cos(\theta - \phi), \qquad y' = r\sin(\theta - \phi). \tag{*}$$

Now, recall from trigonometry that

$$\cos(\theta - \phi) = \cos\theta\cos\phi + \sin\theta\sin\phi,$$

and

$$\sin(\theta - \phi) = \sin\theta\cos\phi - \cos\theta\sin\phi.$$

Hence Equations (*) can be written as

$$x' = r\cos\theta\cos\phi + r\sin\theta\sin\phi,$$
$$y' = r\sin\theta\cos\phi - r\cos\theta\sin\phi,$$

or, since $r\cos\theta = x$ and $r\sin\theta = y$, as

$$x' = x\cos\phi + y\sin\phi, \qquad y' = -x\sin\phi + y\cos\phi. \tag{1}$$

Equations (1), in turn, can be written equivalently as

$$x = x'\cos\phi - y'\sin\phi, \qquad y = x'\sin\phi + y'\cos\phi. \tag{2}$$

(See Exercise 23, page 186.) Thus, under a rotation of axes, the xy-coordinates of a point are related to the $x'y'$-coordinates by Equations (1) or, equivalently, by Equations (2).

Example 1. Consider a rotation of the coordinate axes for which $m^\circ(\phi) = 30$. If the xy-coordinates of the points **S** and **T** are $(4, -2)$ and $(3, 1)$, respectively, find the $x'y'$-coordinates of **S** and **T**.

Solution: Use Equations (1) above with $\sin\phi = \sin 30^\circ = \frac{1}{2}$ and $\cos\phi = \cos 30^\circ = \frac{\sqrt{3}}{2}$ to find the $x'y'$-coordinates.

For **S**, you have:

$$x' = (4)\left(\frac{\sqrt{3}}{2}\right) + (-2)(\tfrac{1}{2}), \qquad y' = (-4)(\tfrac{1}{2}) + (-2)\left(\frac{\sqrt{3}}{2}\right);$$

$$x' = 2\sqrt{3} - 1, \qquad y' = -2 - \sqrt{3}.$$

For **T**, you have:

$$x' = (3)\left(\frac{\sqrt{3}}{2}\right) + (1)(\tfrac{1}{2}), \qquad y' = (-3)(\tfrac{1}{2}) + (1)\left(\frac{\sqrt{3}}{2}\right);$$

$$x' = \frac{3\sqrt{3} + 1}{2}, \qquad y' = \frac{-3 + \sqrt{3}}{2}.$$

Thus the $x'y'$-coordinates of **S** and **T** are,

$$(2\sqrt{3} - 1, -2 - \sqrt{3}) \quad \text{and} \quad \left(\frac{3\sqrt{3} + 1}{2}, \frac{-3 + \sqrt{3}}{2}\right),$$

respectively.

If you solved Exercise 46, page 180, you algebraically verified the geometrically obvious fact that a translation of axes does not alter distances between points. Of course, this is also true for a rotation of axes (Exercise 24, page 186), as the following example suggests.

Example 2. For the points **S** and **T** in Example 1, page 181, verify that you get the same result if you calculate $d(\mathbf{S}, \mathbf{T})$ in either coordinate system.

Solution: If (x_1, y_1) and (x_2, y_2) represent the xy-coordinates, and (x_1', y_1') and (x_2', y_2') represent the $x'y'$-coordinates, of **S** and **T**, respectively, then you have:

$$(x_1, y_1) = (4, -2)$$

$$(x_2, y_2) = (3, 1)$$

$$d(\mathbf{S}, \mathbf{T}) = \sqrt{(3-4)^2 + [1-(-2)]^2}$$

$$= \sqrt{(-1)^2 + 3^2}$$

$$= \sqrt{10}$$

$$(x_1', y_1') = (2\sqrt{3} - 1, -2 - \sqrt{3})$$

$$(x_2', y_2') = \left(\frac{3\sqrt{3}+1}{2}, \frac{-3+\sqrt{3}}{2}\right)$$

$$d(\mathbf{S}, \mathbf{T}) = \sqrt{\left(\frac{3-\sqrt{3}}{2}\right)^2 + \left(\frac{1+3\sqrt{3}}{2}\right)^2}$$

$$= \sqrt{\frac{9-6\sqrt{3}+3}{4} + \frac{1+6\sqrt{3}+27}{4}}$$

$$= \sqrt{10}$$

Thus $d(\mathbf{S}, \mathbf{T})$ is the same in either coordinate system.

Similarly, if you solved Exercise 47, page 180, you algebraically verified the geometric fact that a translation of axes does not alter angles. This is also true, of course, for rotations of axes (Exercise 25, page 186) since these transformations merely assign new coordinates to each point and do not alter the size or shape of geometric figures.

The following example illustrates how a rotation of axes can be used to transform an equation from an unfamiliar form to a form in which you can recognize its graph by inspection.

Example 3. Determine an $x'y'$-equation for the graph of $xy = 4$ under a rotation of axes about the origin with $m°(\phi) = 45$. Use the result to identify and sketch the graph of the original equation.

Solution: You have $\cos 45° = \dfrac{1}{\sqrt{2}}$ and $\sin 45° = \dfrac{1}{\sqrt{2}}$. If you use these values in Equations (2), page 181, you obtain

$$x = \frac{x' - y'}{\sqrt{2}} \qquad \text{and} \qquad y = \frac{x' + y'}{\sqrt{2}}.$$

Substituting these expressions for x and y in the equation $xy = 4$, you have

$$\left(\frac{x' - y'}{\sqrt{2}}\right)\left(\frac{x' + y'}{\sqrt{2}}\right) = 4,$$

$$\frac{x'^2 - y'^2}{2} = 4,$$

$$x'^2 - y'^2 = 8,$$

or

$$\frac{x'^2}{8} - \frac{y'^2}{8} = 1.$$

You should recognize that the graph of this last equation is a hyperbola with the x'-axis as its principal axis. Therefore the graph of the original equation is a hyperbola whose principal axis makes an angle of 45° with the positive ray of the x-axis. Its graph is shown below.

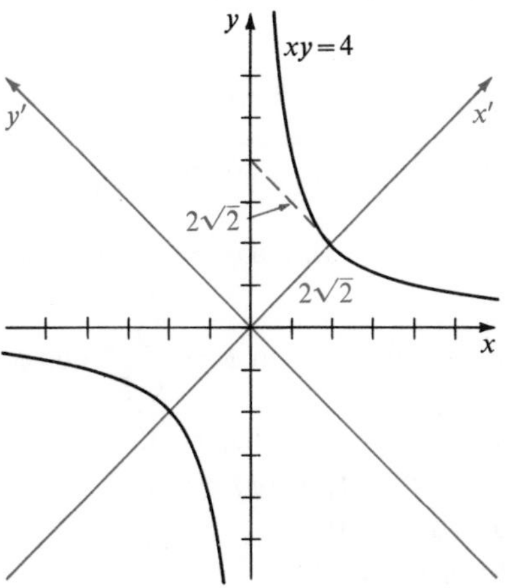

Exercises 5–2

In Exercises 1–6, find the $x'y'$-coordinates of the point whose xy-coordinates are given, if the coordinate axes are rotated about the origin through the angle ϕ whose measure is given.

Example. $(2, -3)$; $120°$

Solution: Since $\cos 120° = -\frac{1}{2}$ and $\sin 120° = \frac{\sqrt{3}}{2}$, Equations (1), page 181, take the form

$$x' = -\tfrac{1}{2}x + \frac{\sqrt{3}}{2}y, \qquad y' = -\frac{\sqrt{3}}{2}x - \tfrac{1}{2}y.$$

Thus, for $(2, -3)$ you have

$$x' = -\tfrac{1}{2}(2) + \frac{\sqrt{3}}{2}(-3) = -1 - \frac{3\sqrt{3}}{2},$$

$$y' = -\frac{\sqrt{3}}{2}(2) - \tfrac{1}{2}(-3) = -\sqrt{3} + \tfrac{3}{2}.$$

Hence, the required coordinates are

$$\left(\frac{-2 - 3\sqrt{3}}{2}, \frac{3 - 2\sqrt{3}}{2}\right).$$

1. $(1, 3)$; $30°$

2. $(2, 3)$; $45°$

3. $(-2, 4)$; $240°$

4. $(-3, 7)$; $135°$

5. $(-6, -3)$; $315°$

6. $(0, -4)$; $225°$

In Exercises 7–16, use Equations (2), page 181, to determine an $x'y'$-equation for the graph of the given xy-equation under a rotation of axes through an angle ϕ as specified.

7. $3x - 4y + 10 = 0$; $\sin\phi = \frac{3}{5}$, $\cos\phi = \frac{4}{5}$

8. $x - 2y - 3 = 0$; $\sin\phi = \frac{-5}{13}$, $\cos\phi = \frac{12}{13}$

9. $8x^2 - 4xy + 5y^2 - 36 = 0$; $\sin\phi = \frac{2}{\sqrt{5}}$, $\cos\phi = \frac{1}{\sqrt{5}}$

10. $5x^2 + 4xy + 8y^2 - 36 = 0$; $\sin\phi = \frac{2}{\sqrt{5}}$, $\cos\phi = \frac{1}{\sqrt{5}}$

11. $2x^2 + 3xy + 2y^2 - 7 = 0$; $m°(\phi) = 45$

12. $x^2 - 2xy + y^2 - 3x = 0$; $m°(\phi) = 45$

13. $x^2 + y^2 = 25$; $m°(\phi) = 60$

14. $4x^2 + 9y^2 = 36$; $m°(\phi) = 90$

15. $11x^2 - 24xy + 4y^2 + 30x + 40y - 45 = 0$; $\sin\phi = \frac{4}{5}$, $\cos\phi = \frac{3}{5}$
16. $9x^2 + 24xy + 16y^2 - 20x + 15y - 100 = 0$; $\sin\phi = \frac{4}{5}$, $\cos\phi = \frac{3}{5}$

In Exercises 17–22, rotate the x- and y-axes so that the line with the given xy-equation has the given slope m relative to the x'- and y'-axes, and find an $x'y'$-equation for the line. (Give two answers.)

Example. $2x - y = 3$; $m = 0$

Solution: Using Equations (2), page 181, you can write the given equation as

$$2(x'\cos\phi - y'\sin\phi) - (x'\sin\phi + y'\cos\phi) = 3.$$

Grouping x'- and y'-terms, you have

$$2x'\cos\phi - x'\sin\phi - 2y'\sin\phi - y'\cos\phi = 3,$$
$$x'(2\cos\phi - \sin\phi) - y'(2\sin\phi + \cos\phi) = 3. \qquad (*)$$

The slope of the graph of this equation is

$$m = \frac{2\cos\phi - \sin\phi}{2\sin\phi + \cos\phi}.$$

Since you wish this slope to be 0, it follows that you must have

$$2\cos\phi - \sin\phi = 0,$$

or

$$\frac{\sin\phi}{\cos\phi} = \tan\phi = 2.$$

Because $\tan\phi$ is positive, you can choose ϕ to be in either Quadrant I or Quadrant III. The two situations are illustrated in the diagrams at the top of the following page.

To identify values for $\cos\phi$ and $\sin\phi$ in Quadrant I, you can sketch a right triangle with angle ϕ such that $\tan\phi = \frac{2}{1}$. It follows that the hypotenuse is of length $\sqrt{1^2 + 2^2} = \sqrt{5}$, and hence $\sin\phi = \dfrac{2}{\sqrt{5}}$ and $\cos\phi = \dfrac{1}{\sqrt{5}}$. Substituting these values in Equation (*), you have

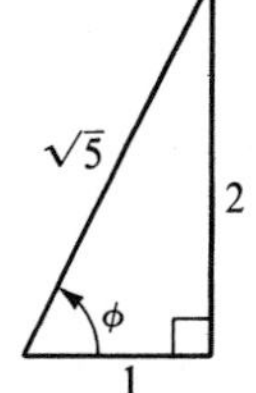

$$x'\left[2\left(\frac{1}{\sqrt{5}}\right) - \frac{2}{\sqrt{5}}\right] - y'\left[\frac{4}{\sqrt{5}} + \frac{1}{\sqrt{5}}\right] = 3,$$

or

$$y' = -\tfrac{3}{5}\sqrt{5}.$$

(Solution continued)

For ϕ in Quadrant III, you have $\sin \phi = -\frac{2}{\sqrt{5}}$ and $\cos \phi = -\frac{1}{\sqrt{5}}$, and Equation (*) becomes $y' = \frac{3}{5}\sqrt{5}$. Therefore, the line can be represented with slope 0 by either

$$y' = -\tfrac{3}{5}\sqrt{5} \quad \text{or} \quad y' = \tfrac{3}{5}\sqrt{5}.$$

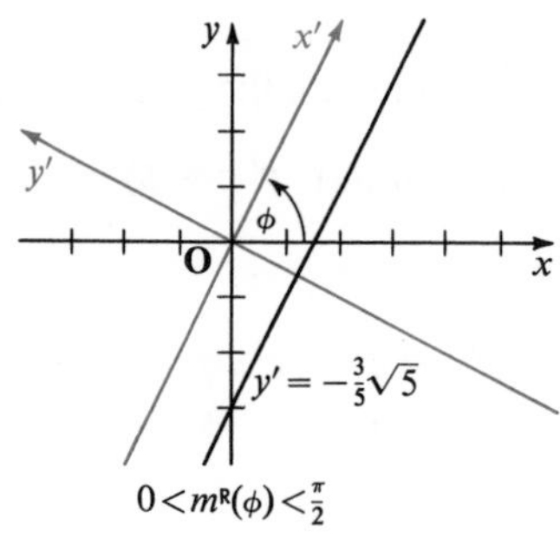

$0 < m^R(\phi) < \frac{\pi}{2}$

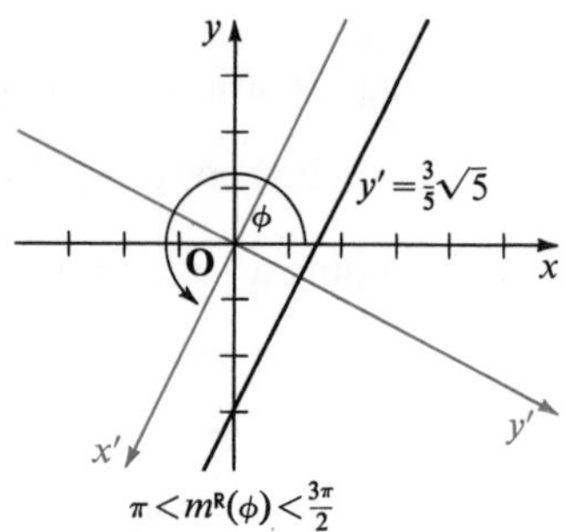

$\pi < m^R(\phi) < \frac{3\pi}{2}$

17. $2y - x = 4$; $m = 0$

18. $x + y = -2$; $m = 0$

19. $3x + y = 6$; $m = 3$

20. $x + y = 4$; $m = \frac{1}{2}$

21. $3x + y = 2$; no slope (line vertical)

22. $x - y = 6$; no slope (line vertical)

* **23.** Show that if $x' = x\cos\phi + y\sin\phi$ and $y' = -x\sin\phi + y\cos\phi$, then $x = x'\cos\phi - y'\sin\phi$ and $y = x'\sin\phi + y'\cos\phi$, and conversely. (*Hint:* Solve the first two equations simultaneously for x and y by using the fact that $\sin^2\phi + \cos^2\phi = 1$.)

* **24.** By a rotation of axes, points $\mathbf{S}(x_1, y_1)$ and $\mathbf{T}(x_2, y_2)$ are given new coordinates, (x_1', y_1') and (x_2', y_2'), respectively, in accordance with Equations (1) on page 181. Verify algebraically that

$$\sqrt{(x_2' - x_1')^2 + (y_2' - y_1')^2} = \sqrt{(x_2 - x_1)^2 + (y_2 - y_1)^2},$$

and interpret this fact geometrically.

* **25.** By a rotation of axes, points $\mathbf{S}(x_1, y_1)$, $\mathbf{T}(x_2, y_2)$, and $\mathbf{U}(x_3, y_3)$ are given new coordinates, $\mathbf{S}'(x_1', y_1')$, $\mathbf{T}'(x_2', y_2')$, and $\mathbf{U}'(x_3', y_3')$, respectively. If $\mathbf{s}, \mathbf{t}, \mathbf{u}, \mathbf{s}', \mathbf{t}'$, and $\mathbf{u}'$ are the vectors corresponding to $\mathbf{S}, \mathbf{T}, \mathbf{U}, \mathbf{S}', \mathbf{T}'$, and $\mathbf{U}'$, respectively, verify algebraically that

$$\frac{(\mathbf{t}' - \mathbf{s}') \cdot (\mathbf{u}' - \mathbf{s}')}{\|\mathbf{t}' - \mathbf{s}'\| \, \|\mathbf{u}' - \mathbf{s}'\|} = \frac{(\mathbf{t} - \mathbf{s}) \cdot (\mathbf{u} - \mathbf{s})}{\|\mathbf{t} - \mathbf{s}\| \, \|\mathbf{u} - \mathbf{s}\|},$$

and interpret this fact geometrically.

Applications of Transformations

5–3 The General Second-Degree Equation

An equation of the form

$$Ax^2 + Bxy + Cy^2 + Dx + Ey + F = 0, \tag{1}$$

where A, B, C, D, E, and F are constants, with A, B, and C not all equal to 0, is called a **second-degree**, or **quadratic, equation** in two variables. If $B = 0$, then the equation takes the form

$$Ax^2 + Cy^2 + Dx + Ey + F = 0;$$

and, as you saw in Section 5–1, you can identify its graph by completing the squares in x and y, and then using an appropriate translation of axes.

If $B \neq 0$ in Equation (1) above, then you cannot identify the graph directly by this method. You can, however, use a rotation of axes to obtain an equation for the graph of (1) that does not contain an $x'y'$-term, and then proceed as before.

If the axes are rotated through an angle of radian measure ϕ, then the old and new coordinates are related by

$$\begin{aligned} x &= x' \cos \phi - y' \sin \phi \\ \text{and} \qquad y &= x' \sin \phi + y' \cos \phi. \end{aligned} \tag{2}$$

If these values are substituted in Equation (1), then the resulting equation is of the form

$$A'x'^2 + B'x'y' + C'y'^2 + D'x' + E'y' + F' = 0, \tag{3}$$

where (Exercise 19, page 191)

$$A' = A \cos^2 \phi + B \cos \phi \sin \phi + C \sin^2 \phi, \tag{4}$$
$$B' = 2(C - A) \sin \phi \cos \phi + B(\cos^2 \phi - \sin^2 \phi), \tag{5}$$
$$C' = A \sin^2 \phi - B \cos \phi \sin \phi + C \cos^2 \phi, \tag{6}$$
$$D' = D \cos \phi + E \sin \phi, \tag{7}$$
$$E' = -D \sin \phi + E \cos \phi, \tag{8}$$
$$F' = F. \tag{9}$$

To eliminate the $x'y'$-term in (3), that is, to have $B' = 0$, you set the right-hand member of Equation (5) equal to 0 and solve the resulting equation,

$$2(C - A) \sin \phi \cos \phi + B(\cos^2 \phi - \sin^2 \phi) = 0, \tag{10}$$

for ϕ. Recall from trigonometry that for all real values of ϕ,

$$2 \sin \phi \cos \phi = \sin 2\phi \qquad \text{and} \qquad \cos^2 \phi - \sin^2 \phi = \cos 2\phi.$$

Thus, Equation (10) is equivalent to

$$(C - A) \sin 2\phi + B \cos 2\phi = 0. \tag{11}$$

Now, there are two cases to consider: (1) $A = C$ and (2) $A \neq C$.

Case (1): If $A = C$, then Equation (11) becomes

$$B \cos 2\phi = 0,$$

from which you have

$$\cos 2\phi = 0.$$

Values of 2ϕ satisfying this equation are of the form

$$\frac{\pi}{2} + k\pi, \qquad k \text{ an integer},$$

and therefore values of ϕ are of the form

$$\frac{\pi}{4} + k\left(\frac{\pi}{2}\right), \qquad k \text{ an integer}.$$

The least positive value of this form for ϕ is $\frac{\pi}{4}$, and this is the value ordinarily used.

Case (2): If $A \neq C$, then Equation (11) is equivalent to

$$(A - C) \sin 2\phi = B \cos 2\phi,$$

$$\frac{\sin 2\phi}{\cos 2\phi} = \frac{B}{A - C},$$

or

$$\tan 2\phi = \frac{B}{A - C}.$$

In this case, you again have a choice of values for 2ϕ, and you can restrict 2ϕ to any interval of length π you wish. Ordinarily, you select a value in the interval $0 < 2\phi < \pi$, or $0 < \phi < \frac{\pi}{2}$. Thus the angle of rotation can be restricted to being a positive acute angle.

Example 1. Use a rotation of axes to identify the graph of

$$x^2 - 3xy + 5y^2 - 4 = 0,$$

and sketch the graph.

Solution: Since $A = 1$, $B = -3$, and $C = 5$, you have

$$\tan 2\phi = \frac{-3}{1 - 5} = \frac{3}{4}.$$

Choose 2ϕ so that $0 < 2\phi < \pi$. Then since $\tan 2\phi > 0$, you know that $0 < 2\phi < \frac{\pi}{2}$, or $0 < \phi < \frac{\pi}{4}$.

Sketching an appropriate right triangle, as shown at the right, you can see by inspection that $\cos 2\phi = \frac{4}{5}$, $\sin 2\phi = \frac{3}{5}$. To find values for $\cos \phi$ and $\sin \phi$, recall from trigonometry that

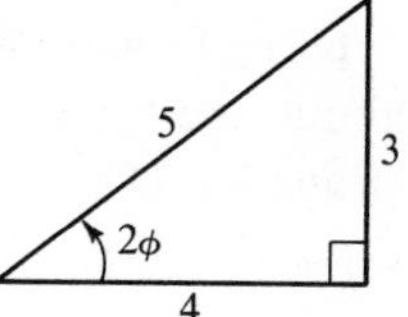

$$\cos \phi = \pm\sqrt{\frac{1 + \cos 2\phi}{2}},$$

$$\sin \phi = \pm\sqrt{\frac{1 - \cos 2\phi}{2}}.$$

Since $0 < \phi < \frac{\pi}{4}$, $\cos \phi$ and $\sin \phi$ are positive; thus you must choose the $+$ sign in each case. You have

$$\cos \phi = \sqrt{\frac{1 + \cos 2\phi}{2}} = \sqrt{\frac{1 + \frac{4}{5}}{2}} = \frac{3}{\sqrt{10}},$$

$$\sin \phi = \sqrt{\frac{1 - \cos 2\phi}{2}} = \sqrt{\frac{1 - \frac{4}{5}}{2}} = \frac{1}{\sqrt{10}}.$$

Using these values in Equations (2), page 181, you find the rotation equations to be

$$x = \frac{3x' - y'}{\sqrt{10}} \quad \text{and} \quad y = \frac{x' + 3y'}{\sqrt{10}}.$$

Substituting these expressions for x and y in the equation $x^2 - 3xy + 5y^2 - 4 = 0$, you obtain

$$5x'^2 + 55y'^2 = 40,$$

or

$$\frac{x'^2}{8} + \frac{y'^2}{\frac{8}{11}} = 1.$$

This is an equation of an ellipse; its graph is shown below.

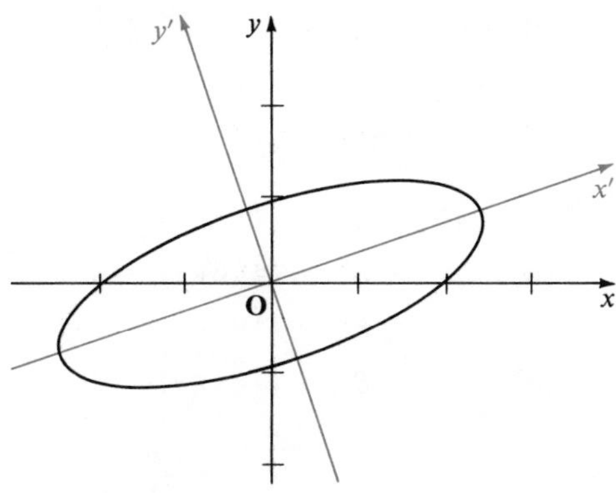

Using Equations (4)–(6) on page 187, you can obtain some direct information about the graph of the general second-degree equation in two variables (Equation (1), page 187). Note first that by adding corresponding members of Equations (4) and (6) you obtain

$$A' + C' = A(\cos^2 \phi + \sin^2 \phi) + C(\cos^2 \phi + \sin^2 \phi) = A + C.$$

Next, you can verify (Exercise 20, page 191) using (4), (5), and (6) that

$$4A'C' - B'^2 = 4AC - B^2.$$

The number $4AC - B^2$ is called the **characteristic** of Equation (1), and $4A'C' - B'^2$ is, thus, the characteristic of the transformed equation under a rotation. The fact that $4A'C' - B'^2 = 4AC - B^2$ is expressed by saying that the characteristic is **invariant** under a rotation of axes. Of course, since $A' + C' = A + C$, the number $A + C$ is also invariant under a rotation of axes.

One purpose of a rotation of axes is to produce a transformed equation in which $B' = 0$. By the invariance of the characteristic, if $B' = 0$, you have

$$4A'C' = 4AC - B^2. \tag{12}$$

You have already learned (page 177) that the graph of

$$A'x^2 + C'y^2 + D'x + E'y + F' = 0,$$

where A' and C' are not both 0, is of elliptic type if $A'C' > 0$, of parabolic type if $A'C' = 0$, and of hyperbolic type if $A'C' < 0$. Thus, by Equation (12):

■ The graph of $Ax^2 + Bxy + Cy^2 + Dx + Ey + F = 0$ is
(1) of elliptic type if $4AC - B^2 > 0$,
(2) of parabolic type if $4AC - B^2 = 0$, and
(3) of hyperbolic type if $4AC - B^2 < 0$.

Example 2. Name the type of graph of

(a) $3x^2 - 14xy + 9x - 7y + 13 = 0$
(b) $6x^2 - 2xy + y^2 + 2x + 3y - 41 = 0$
(c) $3x^2 - 12xy + 3y^2 + 6x - 4y + 8 = 0$
(d) $3x^2 - 6xy + 3y^2 - 14x + 22y - 7 = 0$

Solution:
(a) $4AC - B^2 = (4)(3)(0) - 196 = -196 < 0$; hyperbolic
(b) $4AC - B^2 = (4)(6)(1) - 4 = 20 > 0$; elliptic
(c) $4AC - B^2 = (4)(3)(3) - 144 = -108 < 0$; hyperbolic
(d) $4AC - B^2 = (4)(3)(3) - 36 = 0$; parabolic

Exercises 5–3

In Exercises 1–6, find values for $\sin\phi$ and $\cos\phi$ such that ϕ defines a rotation of axes which eliminates the $x'y'$-term.

1. $x^2 + 2xy + y^2 + 2x - 4y + 5 = 0$
2. $2x^2 - 5xy + 2y^2 - 7x + 8y - 32 = 0$
3. $2x^2 + \sqrt{3}xy + 5y^2 + x - 3y + 8 = 0$
4. $3xy - \sqrt{3}y^2 + 7x - 4y + 10 = 0$
5. $2x^2 + 3xy - 2y^2 + 5x - 4y - 13 = 0$
6. $x^2 + 4xy + 5y^2 - 8x + 3y + 12 = 0$

In Exercises 7–12, use the characteristic to identify the type of the graph of the given equation. Then use a rotation of axes to eliminate the $x'y'$-term. Sketch the graph and show both the x- and y-axes and the x'- and y'-axes. [*Hint:* Use Equations (4)–(9) on page 187.]

7. $3x^2 + 2xy + 3y^2 = 16$
8. $x^2 - 3xy + y^2 = 5$
9. $3x^2 + 4\sqrt{3}xy - y^2 = 15$
10. $2x^2 + 3xy - 2y^2 = 25$
11. $7x^2 - 4xy + 4y^2 = 240$
12. $5x^2 - 12xy = 10$

In Exercises 13–18, use a rotation of axes to eliminate the $x'y'$-term and then use a translation to eliminate the first-degree terms of the resulting equation. [*Hint:* Use Equations (4)–(9) on page 187.]

13. $3x^2 + 10xy + 3y^2 - 2x - 14y - 5 = 0$
14. $4x^2 + 4xy + y^2 - 24x + 38y - 139 = 0$
15. $x^2 - \sqrt{3}xy + 2\sqrt{3}x - 3y - 3 = 0$
16. $3xy - 4y^2 + x - 2y + 1 = 0$
17. $x^2 + 2xy + y^2 + 2x - 4y + 5 = 0$
18. $2x^2 - 5xy + 2y^2 - 7x + 8y - 32 = 0$

Exercises 19–23 refer to the general quadratic equation

$$Ax^2 + Bxy + Cy^2 + Dx + Ey + F = 0, \qquad A^2 + B^2 + C^2 \neq 0.$$

* **19.** Verify Equations (4)–(9) on page 187.

* **20.** Show that $4AC - B^2$ is invariant under a rotation of axes.

* **21.** Show that $A + C$ and $4AC - B^2$ are invariant under a translation of axes.

The expression $\Delta = 4ACF - B^2F - AE^2 - CD^2 + BDE$ referred to in Exercises 22 and 23 is called the **discriminant** of the general quadratic equation. The discriminant Δ is invariant under translation and rotation of axes.

* **22.** Show that if $D = 0$, $E = 0$, and $4AC - B^2 \neq 0$, then

$$F = \frac{\Delta}{4AC - B^2}.$$

Explain why F is invariant under rotation of axes if $D = 0$, $E = 0$, and $4AC - B^2 \neq 0$.

* **23.** Show that the discriminant $\delta = b^2 - 4ac$ of the quadratic equation $ax^2 + bx + c = 0$, $a \neq 0$, can be written in the form

$$\delta = -\begin{vmatrix} 2a & b \\ b & 2c \end{vmatrix},$$

and that the discriminant Δ (see page 191) can be written in the form

$$\Delta = \frac{1}{2}\begin{vmatrix} 2A & B & D \\ B & 2C & E \\ D & E & 2F \end{vmatrix}.$$

5–4 Tangents to Conic Sections

Consider a portion of a given curve $\mathcal{C}$ (Figure 5–5) containing the points $\mathbf{S}_1$ and $\mathbf{S}_2$, and the line $\mathcal{L}_2$ passing through $\mathbf{S}_1$ and $\mathbf{S}_2$. If you visualize $\mathbf{S}_2$ as a particle sliding along the curve toward $\mathbf{S}_1$, then you can imagine that the distance between $\mathbf{S}_1$ and $\mathbf{S}_2$ grows smaller and smaller, and thus decreases to an arbitrarily small value. Then you can say that $\mathbf{S}_2$ **approaches** $\mathbf{S}_1$ along $\mathcal{C}$ and that $\mathbf{S}_1$ is the **limit** of $\mathbf{S}_2$ under these circumstances.

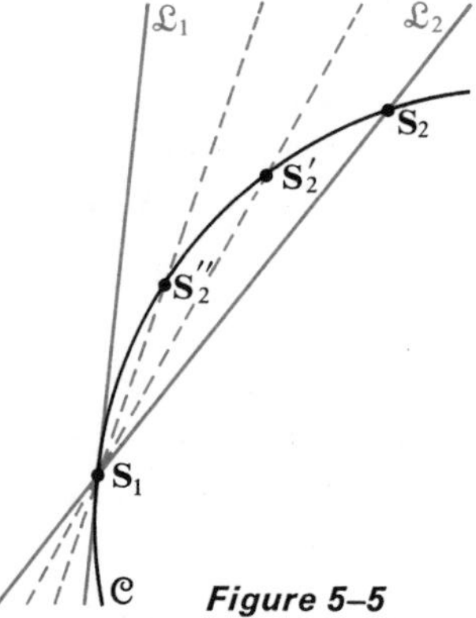

Figure 5–5

From Figure 5–5, you can see that as $\mathbf{S}_2$ approaches $\mathbf{S}_1$ along $\mathcal{C}$ the position of the line $\mathcal{L}_2$ might approach that of a line $\mathcal{L}_1$. In such a case, the line $\mathcal{L}_1$ is said to be **tangent,** or to be **the tangent,** to the curve $\mathcal{C}$ at $\mathbf{S}_1$.

In this section, you will be concerned with the problem of finding an equation of the line tangent to a conic section $\mathcal{C}$ at a specified point on $\mathcal{C}$. As an approach to this problem, consider the special case of a conic section $\mathcal{C}$ that passes through the origin and the process of determining an equation of a line tangent to $\mathcal{C}$ at the origin.

First observe that the graph $\mathcal{C}$ of any second-degree equation contains the origin if and only if its constant term is 0. That is, the graph $\mathcal{C}$ of

$$Ax^2 + Bxy + Cy^2 + Dx + Ey + F = 0 \tag{1}$$

contains the origin if and only if $F = 0$.

Next observe that for values of k such that $0 < |k| < 1$, not only is $k^2 < |k|$, but as $|k|$ decreases, k^2 decreases even faster. For example, as k takes on the values $\frac{1}{10}$, $\frac{1}{100}$, and $\frac{1}{1000}$, k^2 takes on the values $\frac{1}{100}$, $\frac{1}{10,000}$, and $\frac{1}{1,000,000}$, respectively. Similarly, for $0 < |h| < 1$ and $0 < |k| < 1$, if h and k both approach 0, then hk approaches 0 much faster than the greater of $|h|$ and $|k|$.

Now let us examine the behavior of the various terms of Equation (1), with $F = 0$, when x and y both approach 0, that is, when the point $\mathbf{S}(x, y)$ on the curve $\mathcal{C}$ approaches the origin $\mathbf{O}$ along $\mathcal{C}$. You should observe that the terms Ax^2, Bxy, and Cy^2 decrease in absolute value much more rapidly than does the greater of $|x|$ and $|y|$. Thus, the coordinates of points $\mathbf{S}(x, y)$ that lie on $\mathcal{C}$ and are *near the origin* not only satisfy Equation (1) but *very nearly* satisfy the equation $Dx + Ey = 0$.

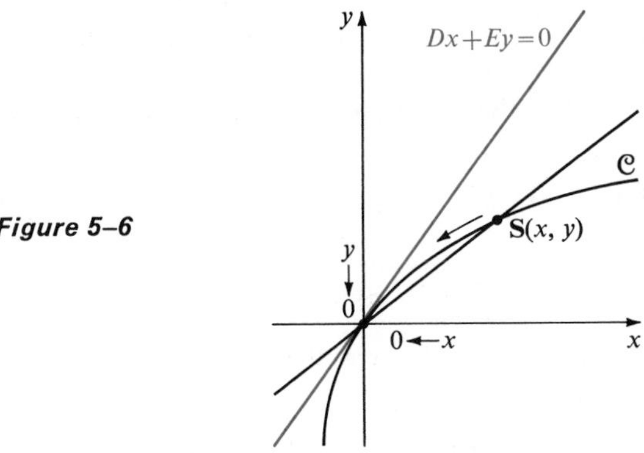

Figure 5–6

Suppose that D and E are not both zero. Then, as illustrated in Figure 5–6, it seems plausible to conclude that as a point $\mathbf{S}(x, y)$ on $\mathcal{C}$ approaches $\mathbf{O}$ along $\mathcal{C}$, the line through $\mathbf{S}$ and $\mathbf{O}$ approaches the line with equation $Dx + Ey = 0$ as a limiting position. This reasoning leads to the following result, which will not be proved here:

■ An equation for the line tangent at the origin to the graph of $Ax^2 + Bxy + Cy^2 + Dx + Ey = 0$, where at least one of D and E is different from 0, is $Dx + Ey = 0$.

Example 1. Find an equation for the tangent to the graph of

$$3x^2 + 4y^2 + 2x - 3y = 0$$

at the origin.

Solution: By inspection, $D = 2$ and $E = -3$. Therefore, the required equation is $2x - 3y = 0$.

The method discussed above actually applies to graphs of polynomial equations in general. For example, the graph $\mathcal{G}$ of

$$x^6 - 7x^3y^2 + y^4 + 2x - 3y = 0$$

contains the origin $\mathbf{O}$ since the constant term is 0. An equation of the line tangent to $\mathcal{G}$ at $\mathbf{O}$ is $2x - 3y = 0$.

Frequently, of course, it is desired to find an equation for the tangent to a conic section at some point other than the origin. Although there are techniques (of differential calculus) for doing this directly, it is simpler for you now to translate the axes so that the required point is the new origin. You can then find an equation for the tangent by the principle stated above. A second translation back to the original axes gives the required equation of the tangent.

Example 2. Find an equation for the line $\mathcal{L}$ that is tangent at the point $\mathbf{S}(2, -1)$ to the ellipse with equation $x^2 + 4y^2 = 8$.

Solution: First observe that the point $\mathbf{S}(2, -1)$ does, indeed, lie on the ellipse, since $(2)^2 + 4(-1)^2 = 8$.

Then, if you take $(2, -1)$ as the new origin, the translation equations are

$$x = x' + 2 \qquad \text{and} \qquad y = y' - 1.$$

Substituting these values for x and y in the xy-equation of the ellipse, you obtain

$$(x' + 2)^2 + 4(y' - 1)^2 = 8,$$
$$x'^2 + 4x' + 4 + 4y'^2 - 8y' + 4 = 8,$$
$$x'^2 + 4y'^2 + 4x' - 8y' = 0.$$

Now observe that this equation has 0 as its constant term, and hence that its graph contains the origin of the $x'y'$-system. Accordingly, by inspection,

$$4x' - 8y' = 0,$$

or

$$x' - 2y' = 0,$$

is an equation of the line $\mathcal{L}$ that is tangent to the ellipse at the origin of the $x'y'$-system. If you next translate back to the xy-system using the equations

$$x' = x - 2 \qquad \text{and} \qquad y' = y + 1,$$

you obtain

$$(x - 2) - 2(y + 1) = 0,$$

or

$$x - 2y - 4 = 0,$$

as an xy-equation for $\mathcal{L}$.

For a comparison of techniques, you might use the present general method to solve Example 4 on page 127.

Exercises 5–4

In Exercises 1–8, find a Cartesian equation for the tangent to the given curve at the origin.

1. $x^2 - 3x + 4y = 0$

2. $y^2 + 7x - 5y = 0$

3. $x^2 + y^2 - 4x + 8y = 0$

4. $x^2 + y^2 + 3x + 5y = 0$

5. $3x^2 + 2y^2 + 6x - 9y = 0$

6. $4x^2 + 9y^2 - 4x + 6y = 0$

7. $x^2 - y^2 + 5x + 2y = 0$

8. $9y^2 - 6x^2 + 2x - y = 0$

In Exercises 9–28, use the technique of Example 2, page 194, to find an xy-equation for the tangent to the curve with the given equation at the point with the given coordinates.

9. $x^2 + 2x - 3y + 1 = 0;\ (2, 3)$

10. $y^2 - 5x + 4y + 3 = 0;\ (3, 2)$

11. $x^2 + y^2 = 5;\ (-2, -1)$

12. $x^2 + y^2 = 25;\ (4, -3)$

13. $4x^2 + 9y^2 = 72;\ (-3, -2)$

14. $6x^2 + 3y^2 = 54;\ (1, -4)$

15. $x^2 - y^2 = -5;\ (2, -3)$

16. $5y^2 - 2x^2 = 18;\ (-1, -2)$

17. $x^2 + y^2 + 2x - 2y - 6 = 0;\ (1, 3)$

18. $x^2 + y^2 - 3x + 7y + 2 = 0;\ (2, 0)$

19. $\dfrac{(x - 2)^2}{4} + \dfrac{(y + 3)^2}{9} = 1;\ (4, -3)$

20. $3x^2 + y^2 + 4x - 3y + 1 = 0;\ (-1, 3)$

21. $4x^2 - 3y^2 + 6x - 2y + 12 = 0;\ (2, -4)$

22. $y^2 - 2x^2 + 3x - 4y + 2 = 0;\ (-2, -2)$

23. $x^2 + 4xy - 3y^2 - 1 = 0;\ (2, 3)$

24. $10x^2 - 24xy + 3y^2 + 2x - 3y + 12 = 0;\ (1, 1)$

25. $3x^2 - 14xy + 9x - 7y + 30 = 0;\ (1, 2)$

26. $6x^2 - 2xy + y^2 + 2x + 3y - 60 = 0;\ (3, 0)$

27. $3x^2 - 12xy + 3y^2 + 6x - 4y - 55 = 0;\ (2, -1)$

28. $3x^2 - 6xy + 3y^2 - 14x + 22y = 0;\ (3, -3)$

* **29.** Show that the tangent to the curve with equation

$$Ax^2 + Bxy + Cy^2 + Dx + Ey + F = 0$$

at any point $\mathbf{S}(x_1, y_1)$ on the curve has an equation of the form

$$Ax_1x + B\left(\frac{x_1y + y_1x}{2}\right) + Cy_1y + D\left(\frac{x + x_1}{2}\right) + E\left(\frac{y + y_1}{2}\right) + F = 0.$$

[*Hint:* Translate axes, with $(h, k) = (x_1, y_1)$.]

* **30.** Show that if the line $\mathcal{L}$ is tangent to the parabola $\mathcal{P}$ with equation $y^2 = 4px$ at the point $\mathbf{A}(x_1, y_1)$ on $\mathcal{P}$, then $\mathcal{L}$ intersects the x-axis at the point $\mathbf{B}(-x_1, 0)$. [*Hint:* Use Exercise 29.]

* **31.** Use the result of Exercise 29 to show that the slope m of the tangent is given by

$$m = -\frac{2Ax_1 + By_1 + D}{Bx_1 + 2Cy_1 + E}.$$

An Alternative Method

5–5 Tangents by the Method of Discriminants (Optional)

In Section 5–4, you saw that the tangent line $\mathcal{L}_1$ to a curve $\mathcal{C}$ at the point $\mathbf{S}_1$ on $\mathcal{C}$ can be thought of as the limiting position (if there is such a limiting position) of the line $\mathcal{L}_2$ through the points $\mathbf{S}_1$ and $\mathbf{S}_2$, as $\mathbf{S}_2$ approaches $\mathbf{S}_1$ along $\mathcal{C}$. In this section, we shall look at this situation from a slightly different viewpoint. Referring to Figure 5–5, page 192, you can see that the line $\mathcal{L}_2$ intersects the curve $\mathcal{C}$ in the two points $\mathbf{S}_1$ and $\mathbf{S}_2$. As $\mathbf{S}_2$ approaches $\mathbf{S}_1$, and $\mathcal{L}_2$ approaches $\mathcal{L}_1$, the two points of intersection of $\mathcal{L}_2$ with $\mathcal{C}$ become closer and closer together until, intuitively speaking, they "become one point" when $\mathcal{L}_2$ reaches the limiting position of $\mathcal{L}_1$. Thus, you can think intuitively of the point of tangency as a "double point of intersection" of a line with the curve. This concept is analogous to that of a "double root" of a quadratic equation.

Recall from algebra that the quadratic formula

$$x = \frac{-b \pm \sqrt{b^2 - 4ac}}{2a}$$

specifies the roots of the equation $ax^2 + bx + c = 0$, $a \neq 0$, in terms of the real-valued coefficients a, b, and c. The value of the discriminant, $b^2 - 4ac$, determines the nature of the roots. Thus, for $b^2 - 4ac > 0$, there are two unequal real roots, while if $b^2 - 4ac = 0$, there is one real root, which can be viewed intuitively as a "double root" if you think of the discriminant as a variable that approaches 0. The graphical situation is shown in Figure 5–7 below.

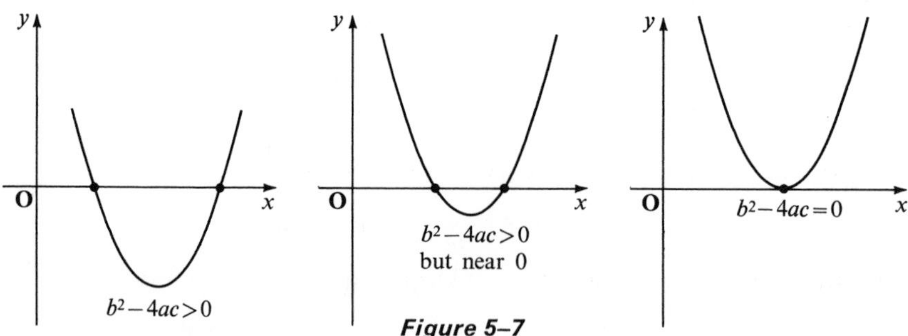

Figure 5–7

You can combine the notions of a "double point of intersection" and a "double root" of a quadratic equation to construct a computational device for determining tangents to conic sections, as illustrated in the following examples.

A line and a nondegenerate conic section have either 0, 1, or 2 points in common. They have exactly one point in common if and only if either the line is tangent to the conic section, or the conic section is a parabola and the line is parallel to its axis, or the conic section is a hyperbola and the line is parallel to (but not coincident with) one of its asymptotes.

Example 1. Two lines through the point $\mathbf{R}(-3, 4)$ are tangent to the parabola $\mathcal{P}$ with equation $y^2 = 16x$. Find equations of these lines.

Solution: Notice first that the vertical line through the point $\mathbf{R}(-3, 4)$ has no point in common with the parabola, since for each point on the parabola you have $x = \frac{y^2}{16} \geq 0$ and, consequently, $x \neq -3$. Any nonvertical line through $\mathbf{R}(-3, 4)$ has an equation of the form

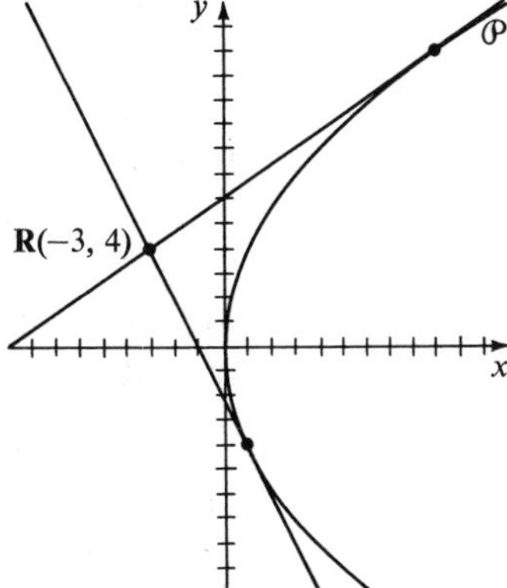

$$y - 4 = m(x + 3),$$

or

$$y = mx + 3m + 4. \quad (1)$$

If this equation is solved simultaneously with the equation $y^2 = 16x$, the resulting solution(s) will be the coordinates of the point(s) of intersection of the line and the parabola.

If $m = 0$ in Equation (1), the line is parallel to the axis of the parabola; there is just one point of intersection in this case, but it is not a point of tangency.

If you solve Equation (1) for x in terms of y and m, $m \neq 0$, you obtain

$$x = \frac{y - 3m - 4}{m}.$$

Substituting for x in the equation $y^2 = 16x$, you have

$$y^2 = 16\left(\frac{y - 3m - 4}{m}\right),$$

$$my^2 = 16y - 48m - 64,$$

or

$$my^2 - 16y + 48m + 64 = 0. \quad (2)$$

Comparing this equation with the equation $ay^2 + by + c = 0$, $a \neq 0$, you see that $a = m$, $b = -16$, and $c = 48m + 64$. The discriminant of (2) is then

$$\begin{aligned}(-16)^2 - 4(m)(48m + 64) &= 256 - 192m^2 - 256m \\ &= -192m^2 - 256m + 256.\end{aligned}$$

Now, the solutions of Equation (2) are the y-coordinates of the points of intersection of the line with the parabola. For tangency, you want these points to be the same point; that is, you want Equation (2), with $m \neq 0$, to have a single solution. For this to

(*Solution continued*)

be true, the discriminant of (2) must be 0:

$$-192m^2 - 256m + 256 = 0,$$

or

$$-64(3m^2 + 4m - 4) = 0.$$

This equation is equivalent to

$$(3m - 2)(m + 2) = 0,$$

from which you have

$$m = \tfrac{2}{3} \quad \text{or} \quad m = -2.$$

Equations for the lines through $\mathbf{R}(-3, 4)$ with slope $\frac{2}{3}$ and -2, respectively, are

$$y - 4 = \tfrac{2}{3}(x + 3) \quad \text{and} \quad y - 4 = -2(x + 3),$$

or

$$2x - 3y + 18 = 0 \quad \text{and} \quad 2x + y + 2 = 0;$$

and therefore these are equations of the two required tangent lines.

In Example 1, the equation for the lines, $y - 4 = m(x + 3)$, represents a **family** of lines through the point $\mathbf{R}(-3, 4)$. (See Figure 5–8.) The slope m is called the *parameter* of the family since each value of m determines a member of the family, just as the parameter r in a parametric vector equation for a line (page 49) determines a point on the line. The vertical line with equation $x = -3$ is also considered to be a member of the family.

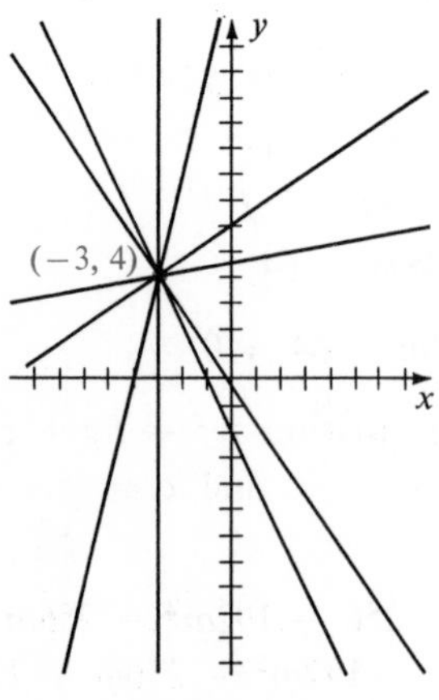

Figure 5–8

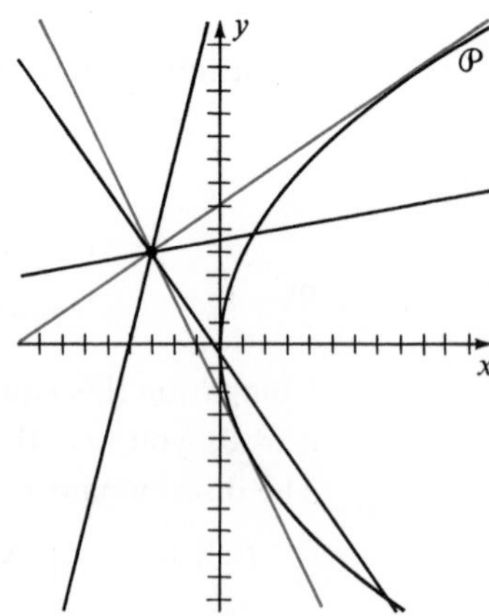

Figure 5–9

The problem posed in Example 1 can be viewed as selecting from this family of lines those two which are tangent to the parabola (colored lines in Figure 5–9).

The next example involves another family of lines, a family each of whose members has the same slope.

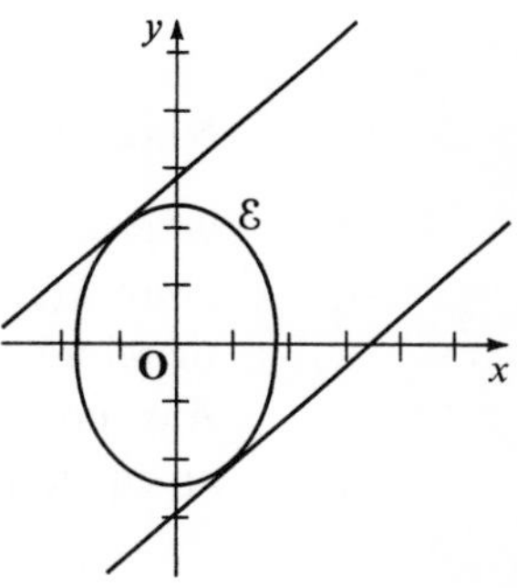

Example 2. Find equations for the lines with slope $\frac{5}{6}$ that are tangent to the ellipse $\mathcal{E}$ with equation

$$5x^2 + 3y^2 - 17 = 0.$$

Solution: Each member of the family of lines with slope $\frac{5}{6}$ has an equation of the form $y = \frac{5}{6}x + b$. Replacing y with $\frac{5}{6}x + b$ in the equation for the ellipse gives you

$$5x^2 + 3(\tfrac{5}{6}x + b)^2 - 17 = 0,$$

$$5x^2 + 3\left(\frac{5x + 6b}{6}\right)^2 - 17 = 0,$$

$$5x^2 + \tfrac{3}{36}(25x^2 + 60bx + 36b^2) - 17 = 0,$$

$$60x^2 + 25x^2 + 60bx + 36b^2 - 204 = 0,$$

or

$$85x^2 + 60bx + 36b^2 - 204 = 0.$$

The discriminant of this equation is

$$(60b)^2 - 4(85)(36b^2 - 204),$$

or

$$240(-36b^2 + 289).$$

Setting the second factor equal to 0, you have

$$-36b^2 + 289 = 0,$$

or

$$b^2 = \tfrac{289}{36}.$$

Then

$$b = \tfrac{17}{6} \quad \text{or} \quad b = -\tfrac{17}{6}.$$

Replacing b in the equation $y = \frac{5}{6}x + b$ with $\frac{17}{6}$ and $-\frac{17}{6}$, in turn, produces equations for the required tangents,

$$5x - 6y + 17 = 0 \quad \text{and} \quad 5x - 6y - 17 = 0.$$

The method of discriminants can be used to find an equation of the tangent to a conic section at a given point of the curve, as illustrated in the following example.

Example 3. Determine an equation of the tangent at the point **S**(3, 4) to the hyperbola $\mathcal{H}$ with equation

$$x^2 - \frac{y^2}{2} = 1.$$

Solution: Notice first that **S** is on $\mathcal{H}$, since $3^2 - \dfrac{4^2}{2} = 9 - 8 = 1$. The vertical line through **S** also has the point **T**(3, −4) in common with $\mathcal{H}$; therefore, this line is not tangent to $\mathcal{H}$. Any nonvertical line through **S** has an equation of the form

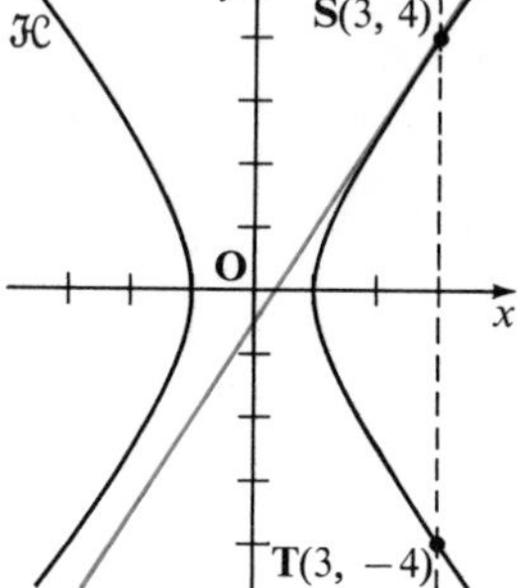

$$y - 4 = m(x - 3),$$

or

$$y = 4 + m(x - 3).$$

Substituting for y from this equation into the given equation for $\mathcal{H}$, you obtain

$$x^2 - \frac{[4 + m(x - 3)]^2}{2} = 1, \quad \text{or}$$

$$(2 - m^2)x^2 + (-8m + 6m^2)x + (-9m^2 + 24m - 18) = 0. \tag{3}$$

If $2 - m^2 = 0$, then the line is parallel to an asymptote of the hyperbola since the slopes of the asymptotes are $\pm\sqrt{2}$; there is just one point of intersection for $m = \pm\sqrt{2}$, but it is not a point of tangency.

If $2 - m^2 \neq 0$, then Equation (3) has two solutions for x unless its discriminant is zero:

$$(-8m + 6m^2)^2 - 4(2 - m^2)(-9m^2 + 24m - 18) = 0.$$

This equation simplifies to

$$16(2m - 3)^2 = 0, \tag{4}$$

which is equivalent to

$$2m - 3 = 0, \quad \text{or} \quad m = \tfrac{3}{2}.$$

The line through **S** with slope $\frac{3}{2}$ has equation

$$y = 4 + \tfrac{3}{2}(x - 3),$$

or

$$3x - 2y - 1 = 0,$$

and therefore this is an equation of the required tangent line.

In Example 3, it actually was not necessary to check first that the given point **S** is on the given conic section $\mathcal{H}$. The fact that there is only one value of m in the solution set of Equation (4) implies that **S** is on $\mathcal{H}$. If there are two values $m \in \mathcal{R}$ in the solution set, as in Example 1 on page 197, then the given point is "outside" the given conic section; that is, any connected portion of the conic section would have a convex appearance when viewed from the point. If there is no $m \in \mathcal{R}$ in the solution set (for example, if the equation for m turns out to be $m^2 + 1 = 0$), then the point is "inside" the conic section.

For a comparison of techniques, you might use the present more general method to solve Example 4 on page 127.

Exercises 5–5

In Exercises 1–8, find equations of all lines through the given point **S** that are also tangent to the curve with the given equation.

1. $\mathbf{S}(0, 0)$; $y = x^2 + 4$
2. $\mathbf{S}(0, 0)$; $y = x^2 + 1$
3. $\mathbf{S}(0, 0)$; $x = y^2 + 9$
4. $\mathbf{S}(0, 0)$; $x = -y^2 - 1$
5. $\mathbf{S}(-3, 3)$; $y^2 + 3x = 0$
6. $\mathbf{S}(-1, 0)$; $x^2 + y + x = 0$
7. $\mathbf{S}(0, -3)$; $x^2 + y + 1 = 0$
8. $\mathbf{S}(3, 0)$; $y^2 - 4x + 8 = 0$

In Exercises 9–14, find equations of all lines with the given slope that are also tangent to the curve with the given equation.

9. $m = 1$; $x^2 + y^2 = 8$
10. $m = \frac{3}{2}$; $x^2 + y^2 = 13$
11. $m = 1$; $y^2 = 4x - 8$
12. $m = 1$; $y^2 = 4x$
13. $m = 2$; $3x^2 + 2y = 0$
14. $m = -\frac{1}{2}$; $y^2 - 6x = 0$

In Exercises 15–20, find equations for all lines having the specified characteristics.

15. Tangent to the curve with equation $x^2 + 4y^2 - 4x + 6y = 0$, and a member of the family with equations of the form $y = \frac{2}{3}x + b$.

16. Tangent to the curve with equation $y^2 - 3x + 2y + 1 = 0$, and a member of the family with equations of the form $y = \frac{2}{3}x + b$.

17. Tangent to the curve with equation $y = 4 - x^2$, and containing the point **R**(0, 5).

18. Tangent to the curve with equation $x^2 + y^2 + 4y = 16$, and containing the point **R**(2, 4).

19. Tangent to the curve with equation $x^2 + 4y^2 = 4$, and parallel to the graph of $x - 2y = 0$.

20. Tangent to the curve with equation $x^2 + y^2 = 40$, and parallel to the graph of $3x - y = 4$.

* **21.** Show that if the line with equation $y = mx + k$ is tangent to the parabola with equation $y^2 = 4px$, then $k = \dfrac{p}{m}$.

*** 22.** Show that if the line with equation $y = mx + k$ is tangent to the ellipse with equation

$$\frac{x^2}{a^2} + \frac{y^2}{b^2} = 1,$$

then

$$k = \pm\sqrt{a^2m^2 + b^2}.$$

*** 23.** Show that if the line with equation $y = mx + k$ is tangent to the hyperbola with equation

$$\frac{x^2}{a^2} - \frac{y^2}{b^2} = 1,$$

then

$$k = \pm\sqrt{a^2m^2 - b^2}.$$

Chapter Summary

1. If the x- and y-axes are translated so that the origin of a new coordinate system, an $x'y'$-system, is at the point whose xy-coordinates are (h, k), then the old coordinates (x, y) and the new coordinates (x', y') of a point are related by the equations

$$x = x' + h$$
$$y = y' + k$$

or, equivalently, by the equations

$$x' = x - h$$
$$y' = y - k.$$

2. **Translations** and **rotations** of axes can often be used to transform an equation from an unfamiliar form to a form in which the nature and properties of its graph can be determined by inspection.

3. By completing the squares in both x and y, you can write an equation of the form

$$Ax^2 + Cy^2 + Dx + Ey + F = 0, \qquad A \neq 0 \text{ and } C \neq 0,$$

equivalently in the form

$$A(x - h)^2 + C(y - k)^2 = q, \qquad q \text{ a constant.}$$

If $A \neq 0$ and $C = 0$, then you can write the equation equivalently in the form $A(x - h)^2 = q(y - k)$ or $A(x - h)^2 = q$. If $A = 0$ and $C \neq 0$, the forms are $C(y - k)^2 = q(x - h)$ and $C(y - k)^2 = q$. Then, using a translation of axes, with $x' = x - h$ and $y' = y - k$, you can transform these equations into standard forms whose graphs are easily recognized.

4. The intersection of a plane and a right circular cone of two nappes is called a **conic section**. If the plane does not pass through the vertex of the cone, then the intersection is either a circle, an ellipse, a parabola, or a hyperbola. If the plane does pass through the vertex of the cone, then the intersection is a point, a line, or a pair of intersecting lines; these intersections are called *degenerate conic sections*. Other degenerate conic sections are a pair of distinct parallel lines and $\emptyset$.

5. Each of the loci described in Item 4 has an equation of the form

$$Ax^2 + Bxy + Cy^2 + Dx + Ey + F = 0,$$

where A, B, and C are not all 0. Conversely, each equation of this form has one of these loci as its graph.

6. If the x- and y-axes are rotated about the origin through an angle ϕ, then the new coordinates (x', y') of a point are related to the old coordinates (x, y) by the equations

$$\begin{aligned} x' &= x\cos\phi + y\sin\phi \\ y' &= -x\sin\phi + y\cos\phi \end{aligned}$$

or, equivalently, by the equations

$$\begin{aligned} x &= x'\cos\phi - y'\sin\phi \\ y &= x'\sin\phi + y'\cos\phi. \end{aligned}$$

7. For an equation of the form

$$Ax^2 + Bxy + Cy^2 + Dx + Ey + F = 0, \qquad A, B, C \text{ not all } 0,$$

you can obtain an equivalent $x'y'$-equation in which the coefficient of $x'y'$ is 0 by rotating the axes through an angle ϕ. If $A \neq C$, then ϕ is determined by the equation

$$\tan 2\phi = \frac{B}{A - C}.$$

If $A = C$, then $\tan 2\phi$ is undefined, and you use $\phi = \dfrac{\pi}{4}$.

8. The **characteristic** $4AC - B^2$ of a second-degree equation in two variables is **invariant** under a rotation of axes; that is,

$$4A'C' - B'^2 = 4AC - B^2.$$

The expression $A + C$ is also invariant under rotation.

9. The graph of a second-degree equation in two variables is
a circle or an ellipse if $4AC - B^2 > 0$ (a point, $\emptyset$),
a parabola if $4AC - B^2 = 0$ (a line, two parallel lines, $\emptyset$),
a hyperbola if $4AC - B^2 < 0$ (two intersecting lines).
The graphs listed parenthetically are the degenerate forms.

10. If a line and a nondegenerate conic section have exactly one point in common, then, with the following two exceptions, the line is a **tangent** to the conic section at the common point.
Exceptions: A line parallel to the axis of a parabola, and a line parallel to, but not coincident with, an asymptote of a hyperbola.

11. An equation for the line tangent at the origin to the graph of

$$Ax^2 + Bxy + Cy^2 + Dx + Ey = 0 \qquad (F = 0),$$

where at least one of D and E is different from 0, is

$$Dx + Ey = 0.$$

12. To find an equation for the tangent to a conic section at a point $\mathbf{S}(h, k)$ other than the origin:

1. Translate the axes so that the origin of the $x'y'$-system is at the point $\mathbf{S}(h, k)$, and find an $x'y'$-equation for the conic section under this translation.
2. Use the result stated in Item 11 to find an $x'y'$-equation for the tangent.
3. Transform this equation into the required equation by translating the axes back to their original position.

13. (Optional) For a given point **S**, equations for all tangents through **S** to a given conic section $\mathcal{C}$ can be found by the method of discriminants. There are two such tangents if **S** is "outside" $\mathcal{C}$, one if **S** is on $\mathcal{C}$, and none if **S** is "inside" $\mathcal{C}$.

Chapter Review Exercises

In Exercises 1–4, find an $x'y'$-equation for the graph of the given xy-equation if the origin of the $x'y'$-system is at the point whose xy-coordinates are given.

1. $4x^2 - 9y^2 + 24x - 36y - 60 = 0$; $(-3, -2)$
2. $x^2 + y^2 - 8x - 12y + 27 = 0$; $(4, 6)$
3. $x^2 - 4x + 3y + 2 = 0$; $(2, -1)$
4. $y^2 - 8y - 4x + 12 = 0$; $(-1, 4)$

In Exercises 5–8, find a Cartesian equation for the conic section satisfying the given conditions. Sketch the curve.

5. Circle with center $\mathbf{S}(3, -4)$ and radius 5.
6. Ellipse with foci $\mathbf{F}_1(-1, 8)$ and $\mathbf{F}_2(-1, 2)$, and minor axis of length 8.
7. Parabola with vertex $\mathbf{V}(2, 7)$ and focus $\mathbf{F}(2, 3)$.
8. Hyperbola with center $\mathbf{S}(2, 4)$, vertical transverse axis of length 10, and eccentricity $\frac{13}{5}$.

9. Find the $x'y'$-coordinates of the point whose xy-coordinates are $(5, -3)$ if the axes are rotated through an angle of $30°$.

10. Find an $x'y'$-equation for the graph of $2x^2 + 3xy - 2y^2 = 25$ if the axes are rotated through an angle of $45°$.

11. Find an $x'y'$-equation for the graph of $8x^2 + 12xy + 13y^2 = 884$ if the axes are rotated through an angle ϕ such that

$$\cos \phi = \frac{2}{\sqrt{13}} \quad \text{and} \quad \sin \phi = \frac{3}{\sqrt{13}}.$$

12. Rotate the x- and y-axes so that the line $\mathcal{L}$ with xy-equation $3x + 4y = 7$ has slope 0 relative to the x'- and y'-axes, and find an $x'y'$-equation for the line.

13. Use the characteristic $4AC - B^2$ to name the type of graph of each of the following:
(**a**) $3x^2 - 12xy - 3y^2 + 2x - 4y + 5 = 0$;
(**b**) $x^2 - 7xy + 12y^2 - 3x + 8y - 21 = 0$;
(**c**) $2x^2 + 5xy - 4y^2 + x - 2y + 17 = 0$.

14. Use a rotation of axes to obtain an $x'y'$-equation with no $x'y'$-term that is equivalent to the equation

$$4x^2 + 4xy + y^2 - 24x + 38y - 139 = 0.$$

Then use a translation of axes to obtain an equivalent $x''y''$-equation with no first-degree terms.

15. Find an equation of the tangent at the origin to the curve with equation

$$8x^2 - 6xy + 9y^2 + 13x - 24y = 0.$$

16. Find an equation of the line that is tangent at the point $\mathbf{S}(2, -2)$ to the curve with equation $3x^2 + 8xy - 4y^2 + 5x - 7y + 12 = 0$.

17. Find equations for the tangents through the origin to the curve with equation $x = y^2 - y + 1$.

18. Find equations for all lines with slope -1 that are tangent to the curve with equation $x^2 + y^2 = 25$.

Reflection Properties of Conic Sections

The giant radio telescopes in Parkes, Australia, and in Jodrell Bank, England, like the one in Andover, Maine, and all other radio telescopes, are *reflecting* telescopes. This means that *by reflection* they concentrate faint incoming parallel radio waves into one focal point, as illustrated below at the left. These

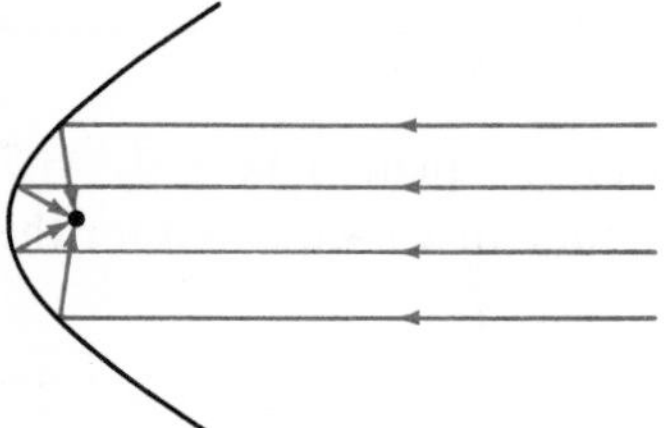

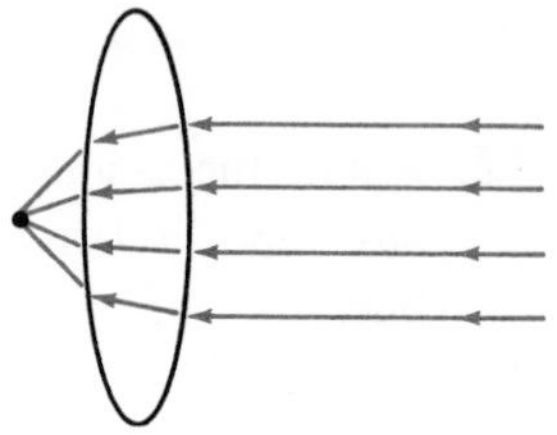

reflecting telescopes are used in studying distant galactic systems, quasars, and so on, and in observing and helping to direct the flights of space craft.

Most large light-gathering astronomical telescopes also use this same reflection principle, rather than the refraction principle illustrated on the right, above.

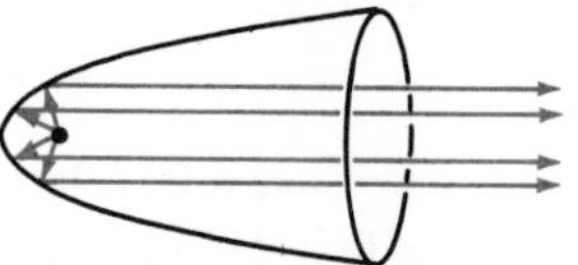

The reflection principle is used in reverse in automobile headlights, which throw out a beam of nearly parallel rays from a small electric light bulb.

Ideally, the reflecting surface of a reflecting telescope is in the form of a connected portion of a paraboloid of revolution, that is, of a surface generated by revolving a parabola about its axis (see Chapter 10). Some of the largest reflecting radio telescopes, however, are spread out over such an extensive region on the earth's surface that they must be made up of a large number of small reflecting panes at discrete locations on a paraboloid of revolution. The axis and focus of the parabola are the axis and focus, respectively, of the paraboloid, and incoming waves parallel to the axis are reflected in such a way as to pass through the focus.

"Focus," incidentally, is the Latin word for "hearth" or "fireplace." It was introduced into science by Johann Kepler (1571–1630) in 1604.

Let us see now how it is that the parabola (and consequently the paraboloid of revolution) has the reflection property indicated above. From physics, you need to know only that when light, or a radio wave, is reflected from a smooth surface, the angle $90° - \beta$ of reflection is of the same measure as the angle $90° - \alpha$ of incidence.

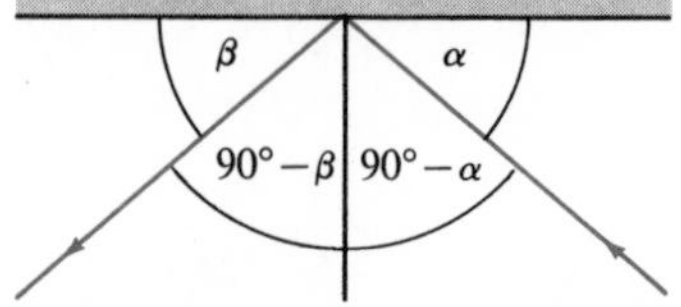

The parabola $\mathcal{P}$ with equation $y^2 = 4px$, $p > 0$, has the x-axis as its axis. Let the line $\mathcal{L}$, parallel to that axis, intersect $\mathcal{P}$ at the point $A(x_1, y_1)$, as shown in the figure below, and let the line $\mathfrak{J}$ be tangent to $\mathcal{P}$ at A. Let the angle between $\mathfrak{L}$ and $\mathfrak{J}$ be denoted by α, the angle between $\mathfrak{J}$ and the line joining A and the

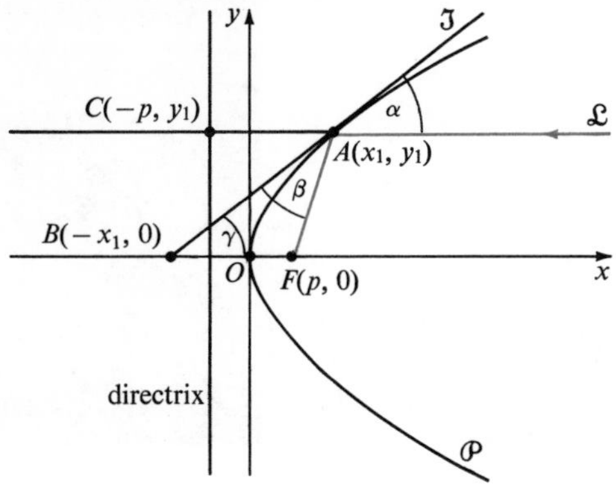

focus $F(p, 0)$ of $\mathcal{P}$ by β, and the angle between $\mathfrak{J}$ and the axis of $\mathcal{P}$ by γ, as shown. Then, of course

$$m^\circ(\gamma) = m^\circ(\alpha), \tag{1}$$

since $\mathfrak{L}$ is parallel to the x-axis.

Let $\mathfrak{J}$ intersect the axis of the parabola at the point B. You saw in Exercise 30, page 195, that the coordinates of B are $(-x_1, 0)$.

Let $\mathfrak{L}$ intersect the directrix of the parabola at the point C. Then the coordinates of C are $(-p, y_1)$.

Notice from the figure that $d(A, C) = x_1 + p$ and also that $d(B, F) = x_1 + p$. Therefore,

$$d(B, F) = d(A, C). \tag{2}$$

Notice also that, by the definition of a parabola,

$$d(A, F) = d(A, C). \tag{3}$$

From Equations (2) and (3) it follows that

$$d(A, F) = d(B, F),$$

and accordingly that ABF is an isosceles triangle with vertex at F. Therefore,

$$m^\circ(\beta) = m^\circ(\gamma). \tag{4}$$

From Equations (1) and (4) it follows that

$$m^\circ(\beta) = m^\circ(\alpha),$$

and hence

$$m^\circ(90^\circ - \beta) = m^\circ(90^\circ - \alpha).$$

Therefore, by the reflection principle of physics, incoming waves parallel to the axis are reflected in such a way as to pass through the focus.

This giant radio telescope, located in Parkes, Australia, has a diameter of 210 feet and can receive radio waves from the solar system, the galaxy, and extragalactic nebulae.

The ellipse and hyperbola also have interesting reflection properties. These can be established analytically by the same method as was used for the parabola, but a more intuitive argument will be given here.

If a man wanted to run as quickly as possible from point A to point B, touching wall MN on the way, as pictured in the diagram below, what path should

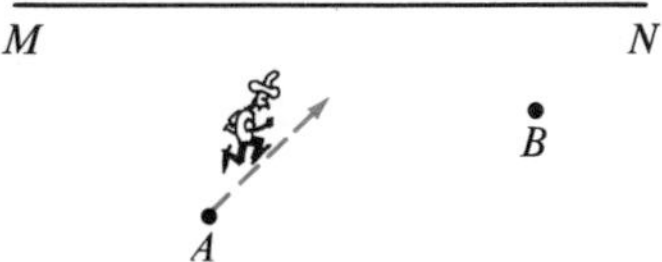

he follow? Assuming that the terrain is uniform, and neglecting turn-around time, he would naturally run along linear segments, from A to a point C on MN, and then from C to B. But how should he choose point C? The answer is simple when you think of the "reflection" B' of B in MN, that is, the point on the line through B perpendicular to MN, at the same distance as B from MN, but on the other side of MN. The man should run from A directly toward B' until he reaches a point C on MN, and then should run from C directly to B.

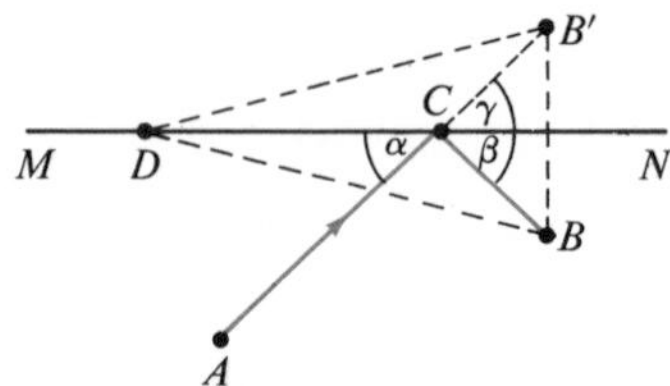

In the diagram above you can see that

$$\overline{AD} + \overline{DB'} > \overline{AC} + \overline{CB'}$$

where D is any point on MN other than C, and therefore that

$$\overline{AD} + \overline{DB} > \overline{AC} + \overline{CB}.$$

In the diagram, you can also see that

$$m^\circ(\beta) = m^\circ(\gamma),$$

and that

$$m^\circ(\gamma) = m^\circ(\alpha),$$

and therefore that

$$m^\circ(\beta) = m^\circ(\alpha) \quad \text{and} \quad m^\circ(90^\circ - \beta) = m^\circ(90^\circ - \alpha).$$

Thus the point C is also characterized by the fact that the angle $90^\circ - \beta$ of reflection is of the same measure as the angle $90^\circ - \alpha$ of incidence.

Think now of the points A and B as being foci of an ellipse $\mathcal{E}$ and of MN as being a line tangent to $\mathcal{E}$ at the point C. If the runner must touch MN at

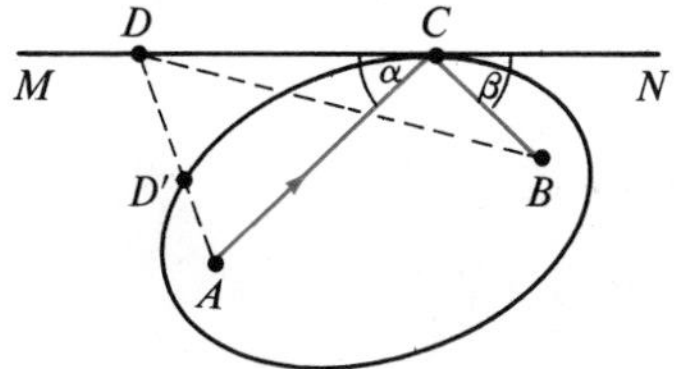

some point in going from A to B, what point of MN should he choose? In the figure, you can easily see that

$$\overline{AD} + \overline{DB} > \overline{AD'} + \overline{D'B}.$$

But by the definition of an ellipse, you also have

$$\overline{AD'} + \overline{D'B} = \overline{AC} + \overline{CB}.$$

Therefore

$$\overline{AD} + \overline{DB} > \overline{AC} + \overline{CB},$$

and accordingly the runner should choose the point of tangency in order to complete the course in the least possible time.

It follows, then, that *a line tangent to an ellipse makes angles of equal measure with the focal radii to the point of tangency.*

In the foregoing illustration, the runner might be replaced with a ray of light and the wall with a reflecting surface since, by Fermat's principle of physics, light also travels from A to B via the wall in such a way as to minimize total elapsed time.

Thus combustible material placed near one focus of an elliptical reflector can readily be ignited by placing a source of heat at the other focus.

The solar furnace shown under construction below, and in final form at the right, is located in Odeillo in the French Pyrenees. The reflecting surface of this 54 by 40 meter parabolic mirror is composed of many small mirrors.

The same principle applies to "whispering galleries." A pin dropped near a focal point in the elliptical Mormon Tabernacle in Salt Lake City can easily be heard at the other focal point a considerable distance away.

There is an old story that the walls of the Ratskeller in Bremen were rather elliptical in shape and the city fathers managed to keep remarkably well informed concerning local events by quietly drinking wine at a table near one focal point while the burghers enjoyed themselves at a table near the other focus.

A line MN tangent to a hyperbola also makes angles of equal measure with the focal radii to the point S of tangency; that is, in the figure below,

$$m^\circ(\alpha) = m^\circ(\gamma). \tag{5}$$

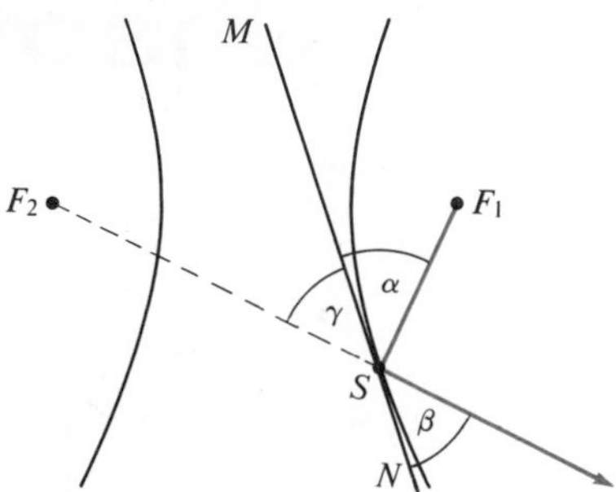

If β is the complement of the angle of reflection of a ray from F_1 to S in the hyperbolic mirror shown, then

$$m^\circ(90^\circ - \beta) = m^\circ(90^\circ - \alpha)$$

since the angle of reflection has the same measure as the angle of incidence. Thus,

$$m^\circ(\beta) = m^\circ(\alpha). \tag{6}$$

Therefore, by Equations 5 and 6,

$$m^\circ(\beta) = m^\circ(\gamma),$$

and accordingly the ray is reflected along an extension of the segment $\overline{F_2S}$. Thus the ray *appears* to issue from F_2.

One important application of the hyperbolic mirror is in the Cassegrain telescope, which employs both this mirror and a parabolic mirror. The hidden focus of the Cassegrain mirror coincides with that of the parabolic mirror and therefore light coming from the parabolic mirror which appears to be converging to one focus is reflected totally in the other focus.

Chapter 6

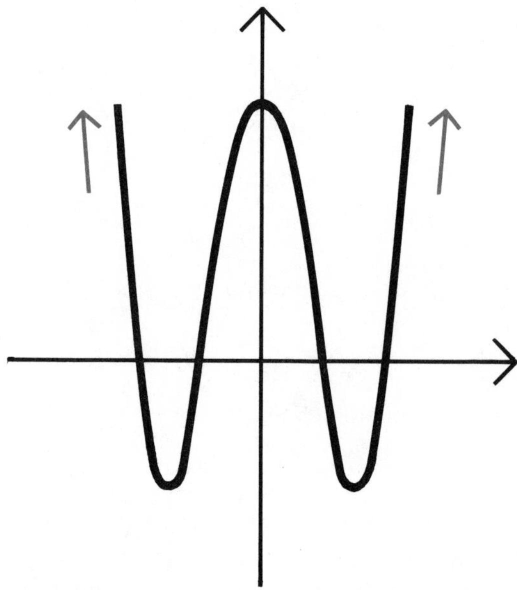

In this chapter, various techniques for graphing polynomial and rational functions are presented. Addition of ordinates and graphs of parametric equations are also discussed.

Curve Sketching

Graphing Polynomial and Rational Functions

6–1 Polynomial Functions

In your earlier work, you probably learned that a **function** f is a set of ordered pairs (a, b) in which no two second components correspond to the same first component. The set of first components is called the **domain** of the function, and the set of second components is called the **range** of the function. The pairing of elements of the domain with those of the range is often accomplished by means of a formula, such as $x^2 + 2x$, where the variable x denotes an element in the domain and the variable $y = x^2 + 2x$ denotes the associated element in the range. The particular element y associated with a given element x in the function f is denoted by $f(x)$ (read "f of x"). Other symbols such as F and g are also used to name functions.

In this book, it will be assumed, unless otherwise stated, that the domain of each function under discussion is the set of all real numbers for which the formula yields a real-number element in the range. Thus, the formulas $x^2 + 2x$ and $\dfrac{x}{(x-1)(x+2)}$ define the functions

$$f = \{(x, f(x)) \colon f(x) = x^2 + 2x\}$$

and

$$F = \left\{(x, F(x)) \colon F(x) = \frac{x}{(x-1)(x+2)}\right\},$$

where the domain of the function f is $\mathcal{R}$, while that of the function F is the set of all real numbers excluding 1 and -2.

In Chapter 2, you studied linear functions determined by equations of the form $y = mx + b$, which may be written as

$$l(x) = a_0x + a_1, \qquad a_0 \neq 0.$$

In Chapter 4, you studied quadratic functions having equations of the form

$$q(x) = a_0x^2 + a_1x + a_2, \qquad a_0 \neq 0.$$

These are special cases of a more general kind of function called a *polynomial function*. Any equation of the form

$$P(x) = a_0x^n + a_1x^{n-1} + a_2x^{n-2} + \cdots + a_{n-1}x + a_n, \tag{1}$$

where the a's are constants, $a_0 \neq 0$, and n is a nonnegative integer, defines a **polynomial function** P. The right-hand member of Equation (1) is called a **polynomial of degree** n. For example, the polynomial $x^3 - 4$ is of degree 3, and the polynomial 7 is of degree 0. (The constant 0 is called the **zero polynomial**; no degree is assigned to it.)

Since in Equation (1) the replacement set for x is $\mathcal{R}$ and the $a_i \in \mathcal{R}$, the function P has a graph in $\mathcal{R}^2$ and is called a **real polynomial function**. For convenience, we shall plot the graphs of functions in an xy-coordinate system, no matter what designation has been given the function. Thus, we may use "y-intercept" to refer to $P(0)$ or $f(0)$ or any other value corresponding to $x = 0$.

In graphing real polynomial functions of degree greater than 2, it is usually necessary to lean somewhat more heavily on the plotting of points than in the case of linear or quadratic functions. There are, however, several general properties of polynomial functions that can help you sketch their graphs.

1. *Dominance of leading term.* For values of x with $|x|$ sufficiently large, the leading term, a_0x^n, of the polynomial in the right-hand member of

$$P(x) = a_0x^n + a_1x^{n-1} + \cdots + a_n, \qquad a_0 \neq 0,$$

"dominates" the other terms in the polynomial, in the sense that it is greater (by an unlimitedly great amount) in absolute value than the sum of all the rest of the terms. As a result, you can tell the general shape of the graph of a real polynomial function in regions relatively remote from the origin by examining the degree of the polynomial and the leading coefficient. If n is *even* (but not zero) and a_0 is *negative*, then the graph behaves as suggested in Figure 6–1(a).

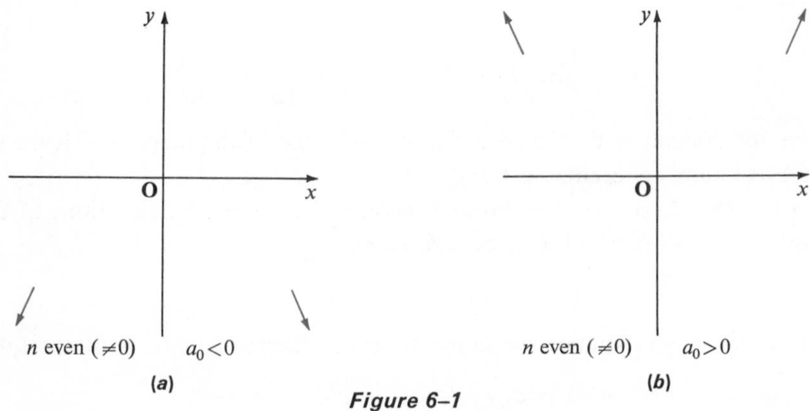

Figure 6–1

But if n is *even* (but not zero) and a_0 is *positive*, then the graph behaves as suggested in Figure 6–1(b).

Examples are given in Figure 6–2.

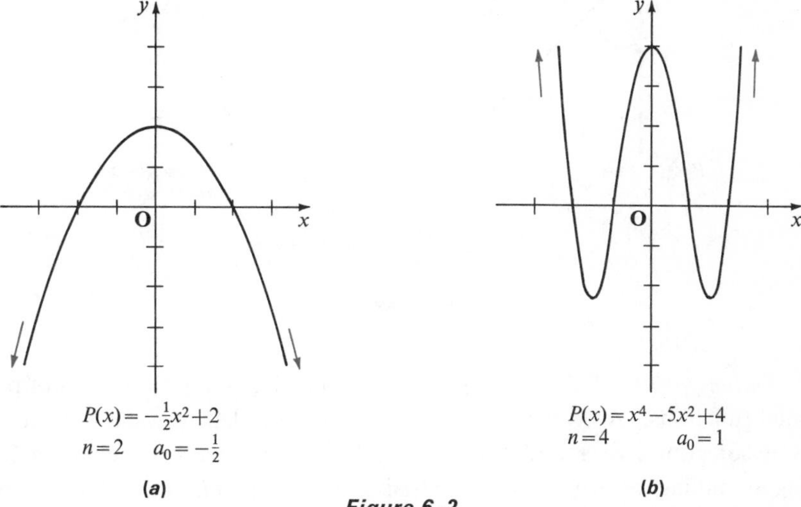

Figure 6–2

On the other hand, if the polynomial function is of *odd* degree, then, sufficiently far from the origin, it displays the characteristics suggested in Figure 6–3.

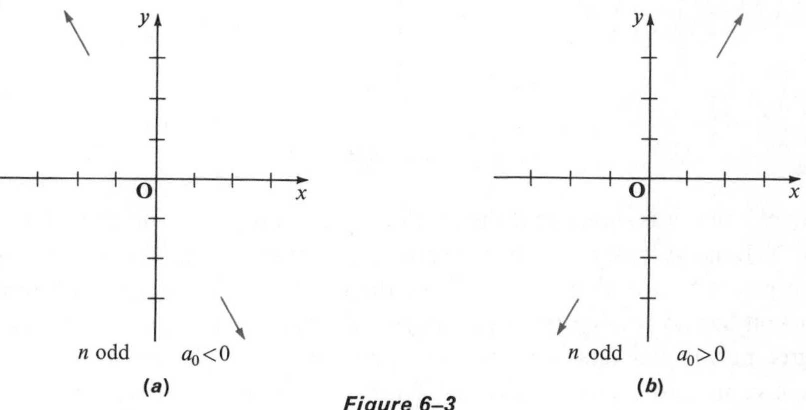

Figure 6–3

Examples are given in Figure 6–4 on the following page.

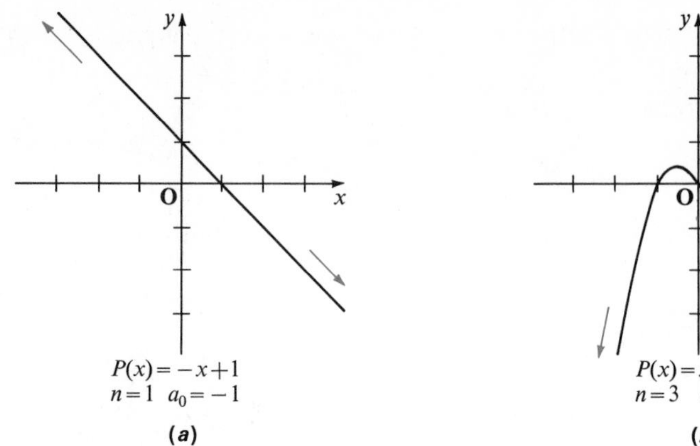

Figure 6–4

2. *Turning points.* Figure 6–5 shows some portions of typical graphs of polynomial functions. As you scan any such graph from left to right, you can see that it sometimes rises and sometimes falls. Points where it changes from rising to falling or from falling to rising are called **turning points** or **local maximum** (or **minimum**) **points.**

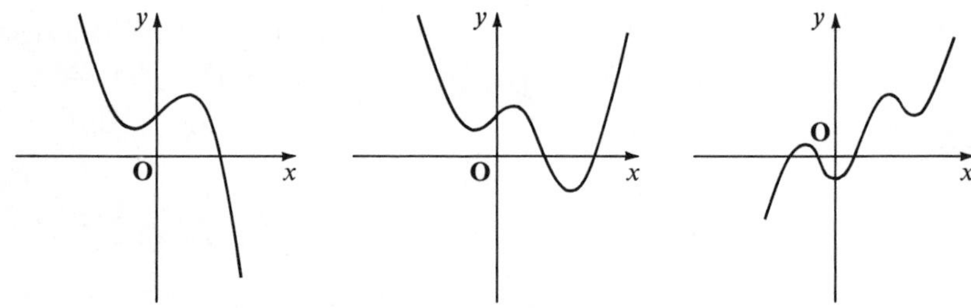

Figure 6–5

In calculus, it is shown that the graph of a real polynomial function of degree n, $n \geq 1$, has at most $n - 1$ turning points, and otherwise has fewer than $n - 1$ such points by a multiple of 2. Thus, the graph of a first-degree polynomial function has no turning point [see Figure 6–4(a)], and the graph of a second-degree polynomial function has one turning point [see Figure 6–2(a)]. The graph of a third-degree polynomial function may have 2 turning points [see Figure 6–4(b)] or no turning point (for example, the graph of $P(x) = x^3$). The graph of a fourth-degree polynomial function may have 3 turning points [see Figure 6–2(b)] or 1 turning point (for example, $P(x) = x^4$). Similarly, the graph of a fifth-degree polynomial function can have 4, 2, or 0 turning points.

3. *Intercepts.* The graph of

$$P(x) = a_0x^n + a_1x^{n-1} + \cdots + a_n, \qquad a_0 \neq 0,$$

clearly has a_n as y-intercept.

Values of x for which

$$a_0x^n + a_1x^{n-1} + \cdots + a_n = 0 \tag{2}$$

are the x-intercepts of the graph. Thus, they are the **roots** of the equation $P(x) = 0$. These values of x can sometimes be obtained by factoring. If they cannot actually be found, they can often be approximated. Observe that a polynomial function P is *continuous* (its graph has no breaks or gaps for $x \in \mathcal{R}$). Therefore, if for $a < b$, $P(a)$ and $P(b)$ are opposite in sign, then there must be at least one value c, $a < c < b$, for which $P(c) = 0$. Figure 6–6 suggests why this is so.

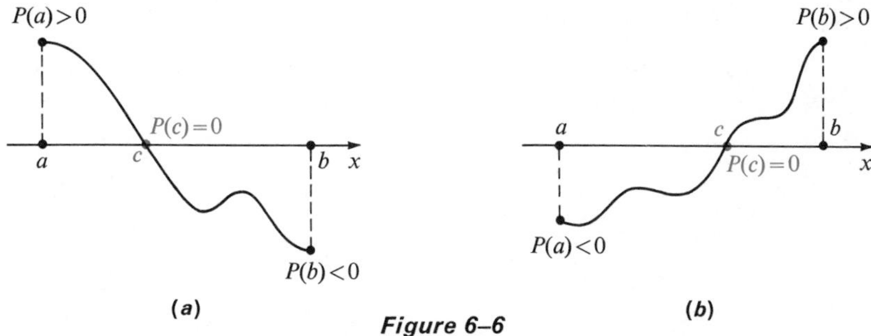

Figure 6–6

In algebra, you learned that a polynomial equation of the form (2) can have at most n real roots. It follows that the graph of a polynomial function of degree n can have *at most n* x-intercepts. As suggested by Figures 6–3 and 6–4, a polynomial function of odd degree must have at least one x-intercept.

The following example illustrates how you can use the foregoing facts to sketch the graph of a polynomial function.

Example. Sketch the graph of the function f defined by

$$f(x) = x^3 - 2x^2 - x + 2.$$

Solution: Observe first that the coefficient of the leading term of the polynomial in the right-hand member is positive and that the term is of degree 3. This means that the graph has the general configuration shown, with at most two turning points. By inspection of the defining equation, you see that the y-intercept is 2.

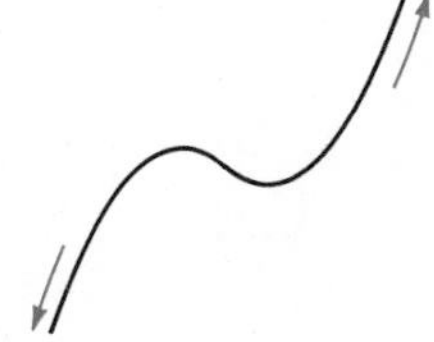

(*Solution continued*)

Upon factoring the polynomial in the right-hand member of the defining equation, you obtain

$$\begin{aligned} x^3 - 2x^2 - x + 2 &= x^2(x - 2) - (x - 2) \\ &= (x - 2)(x^2 - 1) \\ &= (x - 2)(x - 1)(x + 1). \end{aligned}$$

Since this expression equals 0 when x equals 2, 1, or -1, these numbers are x-intercepts. Thus, we have the following coordinates of points on the graph:

x	0	2	1	-1
y	2	0	0	0

To obtain a more accurate sketch, you can identify a few additional points. Thus:

$$\begin{aligned} f(-\tfrac{3}{2}) &= (-\tfrac{3}{2})^3 - 2(-\tfrac{3}{2})^2 - (-\tfrac{3}{2}) + 2 \\ &= -\tfrac{27}{8} - \tfrac{9}{2} + \tfrac{3}{2} + 2 = -4\tfrac{3}{8} \end{aligned}$$

$$\begin{aligned} f(-\tfrac{1}{2}) &= (-\tfrac{1}{2})^3 - 2(-\tfrac{1}{2})^2 - (-\tfrac{1}{2}) + 2 \\ &= -\tfrac{1}{8} - \tfrac{1}{2} + \tfrac{1}{2} + 2 = 1\tfrac{7}{8} \end{aligned}$$

$$f(\tfrac{3}{2}) = (\tfrac{3}{2})^3 - 2(\tfrac{3}{2})^2 - (\tfrac{3}{2}) + 2 = \tfrac{27}{8} - \tfrac{9}{2} - \tfrac{3}{2} + 2 = -\tfrac{5}{8}$$

$$f(3) = 3^3 - 2(3^2) - 3 + 2 = 27 - 18 - 3 + 2 = 8$$

When the points $(-\frac{3}{2}, -4\frac{3}{8})$, $(-\frac{1}{2}, 1\frac{7}{8})$, $(\frac{3}{2}, -\frac{5}{8})$, and $(3, 8)$, determined above, are added, you obtain the sketch shown.

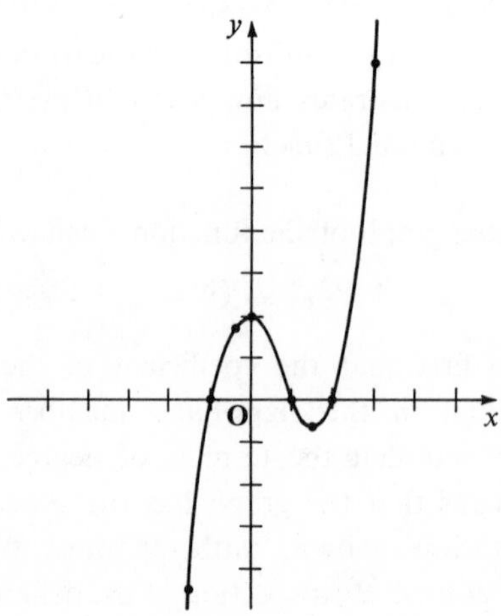

Exercises 6–1

In Exercises 1–10, the given equation defines a function whose graph has one of the following general shapes. In each case, identify the appropriate shape or shapes. (In some cases, there may be two possibilities.)

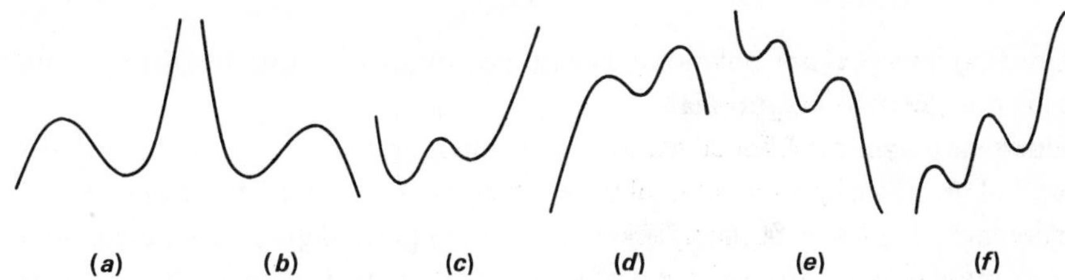

(*a*) (*b*) (*c*) (*d*) (*e*) (*f*)

Example. $F(x) = -3x^3 + 2x^2 - x + 1$

Solution: The graph of the equation can have at most two turning points, and the dominant term of the polynomial has a negative coefficient. The only possibility, therefore, is (b).

1. $F(x) = -2x^4 + x^2 - 3x + 2$
2. $G(x) = 5x^3 - 2x^2 + 3x - 1$
3. $H(x) = 2x^5 - 3x^2 + x - 5$
4. $F(x) = -4x^3 + 3x^2 - 2x + 1$
5. $G(x) = -x^5 + 2x^3 + x^2 - 3$
6. $H(x) = x^4 - 2x^3 + 3x^2 - 2x + 1$
7. $F(x) = \frac{1}{2}x^3 - 3x^2 + 2x + 1$
8. $G(x) = -\frac{2}{3}x^5 + x^4 - 3x^3 + 2$
9. $H(x) = 4x^4 + x^3 - 2x^2 + 3x$
10. $J(x) = 2x^5 - 3x^4 + 2x^2 - 5x + 1$

In Exercises 11–26, sketch the graph of the function defined by the given equation.

11. $f(x) = x^3$
12. $g(x) = -2x^3$
13. $f(x) = x^2 - x^3$
14. $g(x) = \frac{1}{4}(x^3 - x^2)$
15. $F(x) = x^3 - 4x^2 + 3x$
16. $G(x) = \frac{1}{2}(x^3 - x)$
17. $f(x) = x^3 - 2x^2 - 5x + 6$
18. $h(x) = x^3 - 3x^2 - x + 3$
19. $f(x) = x^4$
20. $F(x) = -\frac{1}{4}x^4$
21. $f(x) = x^4 - x^3 - 2x^2 + 3x - 3$
22. $g(x) = x - x^4$
23. $f(x) = (x - 1)^2(x - 2)$
24. $g(x) = (x - 1)(4 - x^2)$
25. $f(x) = x(1 - x)^3$
26. $g(x) = (x^2 - 1)(x - 2)^2$

* **27.** Show that the graph of $y = ax^4 + bx^2 + c$ has the property that if it contains the point (x_1, y_1), then it also contains the point $(-x_1, y_1)$. Any function whose graph has this property is called an **even function**. Notice that the graph of an even function is symmetric with respect to the y-axis.

* **28.** Show that the graph of $y = ax^5 + cx^3 + ex$ has the property that if it contains the point (x_1, y_1), then it also contains $(-x_1, -y_1)$. Any function whose graph has this property is called an **odd function**. An odd function is sometimes said to be **symmetric with respect to the origin**.

6–2 Rational Functions: Vertical Asymptotes

A **rational function** is a function defined by an equation of the form

$$F(x) = \frac{f(x)}{g(x)},$$

where $f(x)$ and $g(x)$ are polynomials with no common factors involving x, and $g(x)$ is not the zero polynomial.

Unlike a polynomial function, a rational function may have a domain that is not all of $\mathcal{R}$. Clearly, any value of x for which $g(x) = 0$ must be excluded from the domain. If $g(c) = 0$, then $f(c) \neq 0$ because $f(x)$ and $g(x)$ have no common factors involving x. The existence of such a real number c imparts a characteristic feature to the graph of a rational function, as can be seen as follows. Because a polynomial function is continuous, if $g(c) = 0$ and $f(c) \neq 0$, then, in the neighborhood of c, $|g(x)|$ must be very small and $\left|\frac{f(x)}{g(x)}\right|$ must be very large. If, for a particular value of x near c, $f(x)$ and $g(x)$ are of the same sign, then $F(x)$ is large and positive; but if $f(x)$ and $g(x)$ are opposite in sign, then $F(x)$ is large in absolute value but negative.

For example, the absolute value of the function f defined by

$$f(x) = \frac{x}{(x - 1)(x + 2)}$$

becomes larger and larger for values of x closer and closer to 1 and -2, since when x is in the neighborhood of 1 or -2, the denominator of the fraction is very small (i.e., near 0), while the numerator is near 1 or -2. Thus the graph of f approaches the graphs of $x = 1$ and $x = -2$ as suggested by the arrows in Figure 6–7. Of course, the curve cannot behave in all of these ways, since it is the graph of a function. You can identify the directions for this particular graph by examining the signs (+ or −) associated with the numerator and

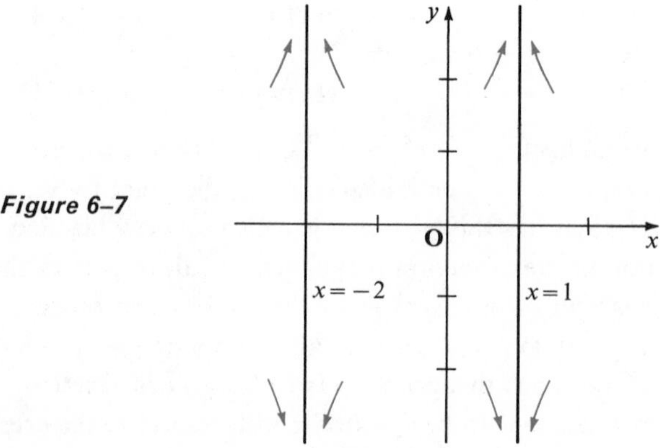

Figure 6–7

denominator of the fraction in the defining equation and thereby determining the sign of the ordinate y over various intervals for x. A convenient device for this purpose is the *sign graph.*

Figure 6–8 shows a sign graph for the fraction $\frac{x}{(x-1)(x+2)}$. The sign graph is constructed by showing, along the top line, the interval of real numbers

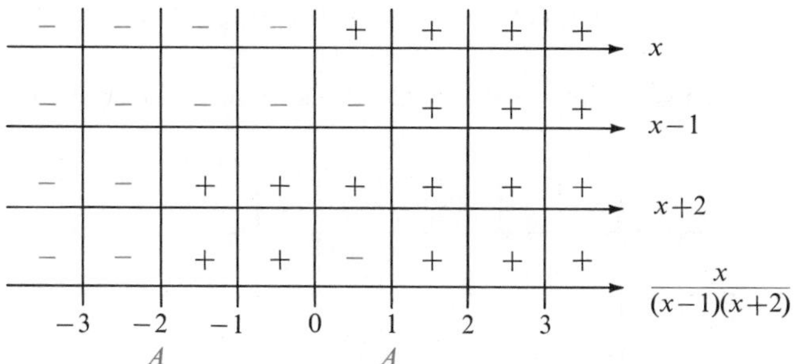

Figure 6–8

where x is positive ($x > 0$) and where x is negative ($x < 0$). The second and third lines similarly show where $x - 1$ is positive ($x > 1$) and negative ($x < 1$), and where $x + 2$ is positive ($x > -2$) and negative ($x < -2$), respectively. The signs on the bottom line are obtained by then observing those intervals where there are even and odd numbers of negative factors, respectively. The bottom line identifies those intervals of the x-axis where $\frac{x}{(x-1)(x+2)}$ is positive and those intervals where it is negative. The letter A indicates the x-values excluded from the domain of the function.

From the information in Figure 6–8, you can identify the regions which contain no point of the graph. These are the shaded regions shown in Figure 6–9. Combining this information with that given in Figure 6–7, you have the rough sketch of the graph shown in Figure 6–9. You will learn the behavior of the graph to the left of $x = -2$ and to the right of $x = 1$ in the next section.

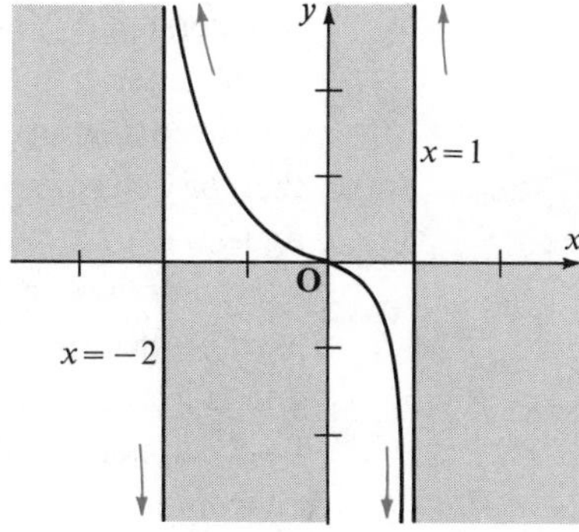

Figure 6–9

Lines such as the graphs of $x = -2$ and $x = 1$ in Figure 6–9 are called *vertical asymptotes.* In general, an **asymptote** is a line which the graph approaches unlimitedly closely. (Recall Section 4–5.)

Relations other than rational functions can have graphs with vertical asymptotes.

Example. Identify the equations of the vertical asymptotes for the graph of the function defined by $f(x) = \dfrac{x - 1}{\sqrt{9 - x^2}}$. Make a sketch showing restricted regions for the graph of the relation.

Solution: By inspection, you see that the denominator of the defining

−	−	−	−	−	−	+	+	+	+	$x-1$
n	n	+	+	+	+	+	+	n	n	$\sqrt{9-x^2}$
n	n	−	−	−	−	+	+	n	n	$\frac{x-1}{\sqrt{9-x^2}}$

$-4 \quad -3 \quad -2 \quad -1 \quad 0 \quad 1 \quad 2 \quad 3 \quad 4$

(A at -3 and at 3)

fraction is zero when $x = 3$ and when $x = -3$. Since the numerator is not 0 at either of these values of x, equations of the vertical asymptotes are $x = 3$ and $x = -3$. For

$$\frac{x - 1}{\sqrt{9 - x^2}},$$

the sign graph requires an additional consideration, since for $|x| > 3$ the radicand $9 - x^2$ is negative and the denominator is not defined. You might use the small letter n to show this. The restricted regions of the plane can then be determined as shown in the figure at the right. The small curved arrows in color suggest the asymptotic behavior of the graph, and a rough sketch of the graph has been made, using the following coordinates:

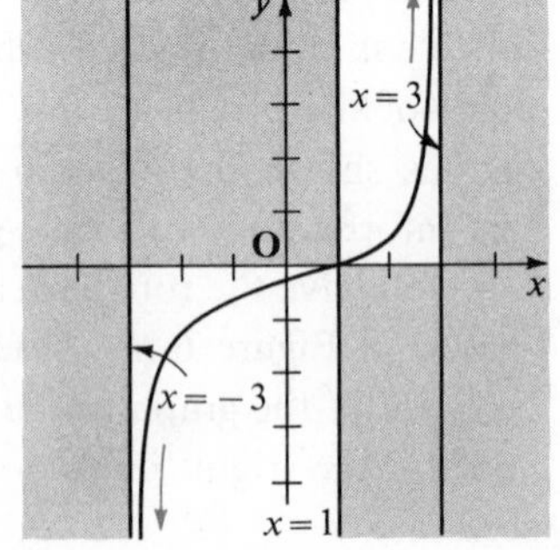

x	-2	-1	0	1	2
y	-1.3	-0.7	$-\frac{1}{3}$	0	0.4

Exercises 6–2

In Exercises 1–28, (a) determine the equations of any vertical asymptotes and (b) use a sign graph to help you sketch the restricted regions of the plane for the relation whose defining equation is given. Use small arrows to suggest asymptotic behavior of the graph.

1. $f(x) = \dfrac{3}{x}$

2. $g(x) = \dfrac{-2}{x}$

3. $h(x) = \dfrac{-4}{x-2}$

4. $j(x) = \dfrac{3}{x+3}$

5. $f(x) = \dfrac{2}{x(x+1)}$

6. $g(x) = \dfrac{3}{x(x-3)}$

7. $h(x) = \dfrac{4}{x^2-4}$

8. $f(x) = \dfrac{2}{9-x^2}$

9. $g(x) = \dfrac{2}{(x-1)(x+3)}$

10. $r(x) = \dfrac{1}{(x+2)(x-3)}$

11. $s(x) = \dfrac{x}{x+2}$

12. $t(x) = \dfrac{x}{x-3}$

13. $a(x) = \dfrac{-x}{x-2}$

14. $b(x) = \dfrac{-x}{x+1}$

15. $c(x) = \dfrac{x}{(x+1)(x-2)}$

16. $d(x) = \dfrac{x}{(x-3)(x+1)}$

17. $F(x) = \dfrac{2x}{(2x+1)(x-2)}$

18. $G(x) = \dfrac{3x}{(2x+3)(x-1)}$

19. $H(x) = \dfrac{x+4}{x^2-9}$

20. $F(x) = \dfrac{x-3}{x^2-4}$

21. $G(x) = \dfrac{2x+1}{x^2-3x-10}$

22. $H(x) = \dfrac{3x-2}{x^2+4x-12}$

* **23.** $M(x) = \dfrac{x}{\sqrt{4-x^2}}$

* **24.** $N(x) = \dfrac{x+1}{\sqrt{x^2-16}}$

* **25.** $Q(x) = \dfrac{x-1}{x\sqrt{x^2-4}}$

* **26.** $R(x) = \dfrac{\sqrt{x^2-4}}{x(x^2-1)}$

* **27.** $y^2 = \dfrac{x}{(x-3)(x+1)}$

* **28.** $y^2 = \dfrac{x}{\sqrt{x^2-2x-8}}$

6–3 Rational Functions: Horizontal and Oblique Asymptotes

The graph of a rational function may also have horizontal asymptotes. Such asymptotes can sometimes be identified for a rational function such as that defined by

$$y = f(x) = \frac{1}{x - 3}$$

by solving the defining equation for x in terms of y. Thus, you would obtain

$$x = \frac{3y + 1}{y},$$

and it is evident that the graph of $y = 0$ (the x-axis) is an asymptote for the curve, because the numerator is close to 1 whenever the denominator is close to 0.

There is, however, a more convenient way to identify horizontal asymptotes, namely, by examining the behavior of $f(x)$ as x increases indefinitely in absolute value. It can be shown that the graph of the function defined by

$$y = \frac{a_0x^n + a_1x^{n-1} + \cdots + a_n}{b_0x^m + b_1x^{m-1} + \cdots + b_m}, \tag{1}$$

where $a_0, b_0 \neq 0$ and m and n are positive integers, has a horizontal asymptote at:

a. $y = 0$, if $n < m$;

b. $y = \dfrac{a_0}{b_0}$, if $n = m$.

c. If $n > m$, the graph has no horizontal asymptote.

You can see that the foregoing facts are plausible if you divide the numerator and denominator of the fraction in Equation (1) by x^m to obtain

$$y = \frac{\dfrac{a_0}{x^{m-n}} + \dfrac{a_1}{x^{m-n+1}} + \cdots + \dfrac{a_n}{x^m}}{b_0 + \dfrac{b_1}{x} + \cdots + \dfrac{b_m}{x^m}}.$$

If, now, $n < m$, then x^{m-n} is a positive integral power of x, and as $|x|$ grows large, each term containing a power of x in its denominator approaches 0; hence, y approaches 0, and the graph of $y = 0$ (the x-axis) is an asymptote. A similar consideration for the case $m = n$ leads to the conclusion that the graph of $y = \dfrac{a_0}{b_0}$ is an asymptote. If $n > m$, then as $|x|$ grows large, so does $|y|$.

In addition to identifying vertical and horizontal asymptotes and restricted regions, other aids to graphing rational functions are:

1. The x- and y-intercepts of the graph.
2. The sign of y-values for x-values near asymptotes.
3. A few selected points on the graph.

Example 1. Complete the sketch of the graph of

$$y = \frac{x}{(x-1)(x+2)}$$

begun in Section 6–2.

Solution: Since $y = \dfrac{x}{x^2 + x - 2}$, you see that the degree of the denominator is greater than the degree of the numerator and so the graph has the x-axis as a horizontal asymptote. You can also find coordinates of a few points such as the following:

x	-3	-1	$\frac{1}{2}$	2
y	$-\frac{3}{4}$	$\frac{1}{2}$	$-\frac{2}{5}$	$\frac{1}{2}$

Thus, you have the graph as shown.

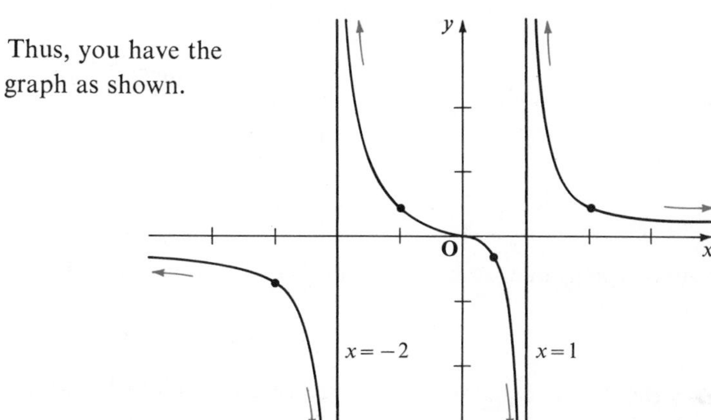

Example 2. Graph the rational function f defined by $y = f(x) = \dfrac{2x - 4}{x + 2}$.

Solution: By inspection, you see that there is a vertical asymptote at $x = -2$. Also, from statement b, page 224, there is a horizontal asymptote at $y = \frac{2}{1}$, or 2. A sign graph, or a careful mental consideration of the signs of the expressions $2x - 4$ and $x + 2$, leads to the figure at the right showing asymptotes and restricted regions shaded in gray. If, by dividing $2x - 4$ by $x + 2$, you write the defining equation in the form

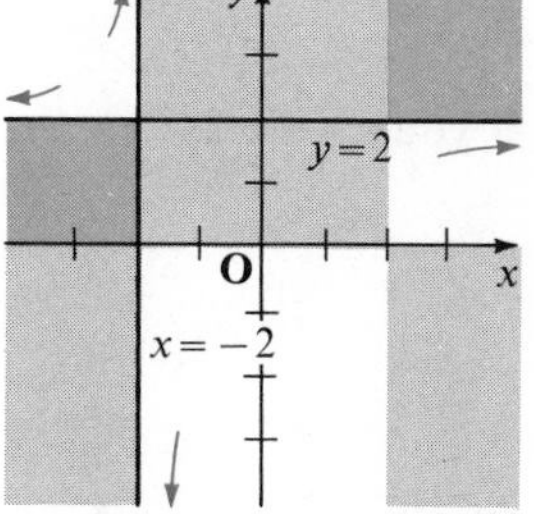

$$y = 2 - \frac{8}{x + 2},$$

you can see that if $x > -2$, then $y < 2$, and if $x < -2$, then $y > 2$. Hence the regions shaded in color are also restricted,

(Solution continued)

and the graph must approach the asymptotes as indicated by the arrows in the figure.

If $x = 0$, then $f(x) = -2$, and the only y-intercept is -2. If $f(x) = 0$, then $x = 2$, and 2 is the only x-intercept.

Finally, by plotting a few points such as $(-4, 6)$ and $(-5, \frac{14}{3})$ to locate the position of the graph in the second quadrant, you can complete the sketch of the graph as shown below.

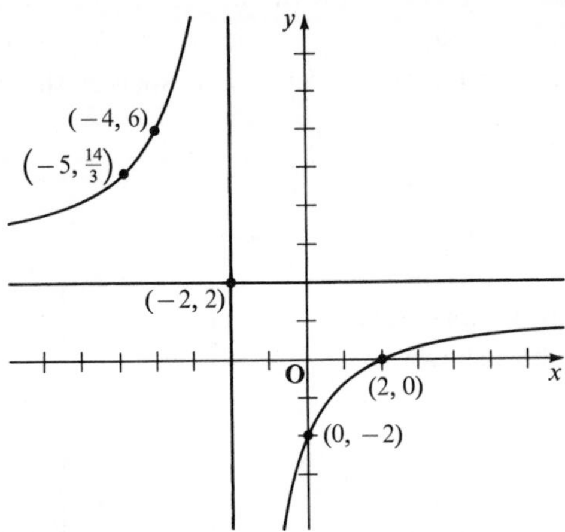

It is worthwhile observing that when the equation in Example 2 is written equivalently as

$$xy + 2y - 2x + 4 = 0, \qquad x \neq -2,$$

it is clearly the equation of a hyperbola, and, under a suitable translation of axes, the graph would have an equation of the form $x'y' = k$. You can see that the center of the hyperbola is at $(-2, 2)$.

In case the degree of the numerator is greater by one than the degree of the denominator, you have a special situation with respect to asymptotes. Consider the rational function defined by

$$y = r(x) = \frac{2x^2 + 3x + 5}{x + 3}. \tag{2}$$

If you divide $2x^2 + 3x + 5$ by $x + 3$, you have

$$\begin{array}{r} 2x - 3 \\ x + 3\overline{)2x^2 + 3x + 5} \\ \underline{2x^2 + 6x } \\ -3x + 5 \\ \underline{-3x - 9} \\ 14 \end{array}$$

so that

$$r(x) = 2x - 3 + \frac{14}{x + 3}.$$

If now $|x|$ grows large, $\left|\frac{14}{x+3}\right|$ becomes small, and it is evident that $r(x)$ grows close to $2x - 3$, from above if x is positive, and from below if x is negative. Thus the graph of $a(x) = 2x - 3$ is an asymptote. Notice that although the asymptote is a line, it is not parallel to either coordinate axis; it is an *oblique asymptote* of the graph of Equation (2). Notice also that when you write the equation equivalently in the form $2x^2 - xy + 3x - 3y + 5 = 0$, $x \neq -3$, you can recognize it as an equation of a hyperbola.

Example 3. Graph the rational function defined by

$$y = f(x) = \frac{x^2 - 2x - 3}{x - 1}.$$

Solution: By inspection, you see that there is a vertical asymptote at $x = 1$, and no horizontal asymptotes. If $x = 0$, then $f(x) = 3$, and so 3 is the only y-intercept. If $f(x) = 0$, then $x^2 - 2x - 3 = 0$, $(x - 3)(x + 1) = 0$, and $x = 3$ or $x = -1$. Hence, 3 and -1 are x-intercepts. Finally, since

$$\frac{x^2 - 2x - 3}{x - 1} = x - 1 - \frac{4}{x - 1},$$

the graph of $a(x) = x - 1$ is an oblique asymptote, and the graph appears as shown below.

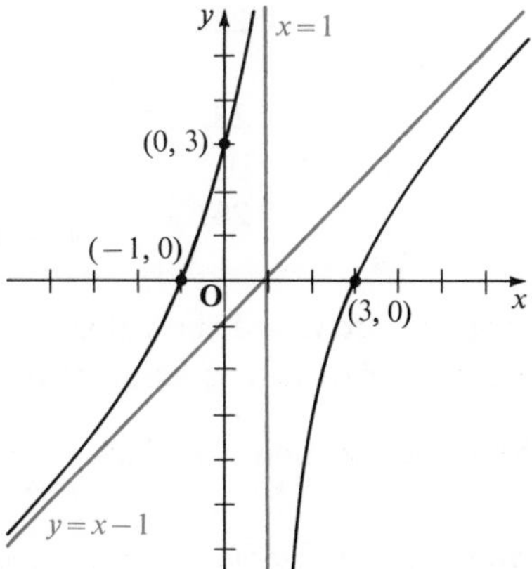

If $n > m + 1$, in Equation (1), page 224, then the quotient polynomial $Q(x)$ obtained by dividing the numerator by the denominator will be of second or higher degree. In such a case, the graph of the rational function approaches the graph of the polynomial function Q unlimitedly closely.

Exercises 6–3

In Exercises 1–18, graph the rational function defined by the given equation.

1. $f(x) = \dfrac{4}{x-2}$

2. $g(x) = \dfrac{8}{x+2}$

3. $f(x) = \dfrac{x}{x-1}$

4. $g(x) = \dfrac{x}{x+3}$

5. $F(x) = \dfrac{2x}{x-4}$

6. $G(x) = \dfrac{3x}{2x+4}$

7. $g(x) = \dfrac{x+1}{x-1}$

8. $h(x) = \dfrac{x-2}{x+2}$

9. $F(x) = \dfrac{x}{(x+1)(x-1)}$

10. $G(x) = \dfrac{x}{(x+2)(x-2)}$

11. $f(x) = \dfrac{x+1}{x^2+2x-3}$

12. $g(x) = \dfrac{x-1}{x^2+x-6}$

13. $f(x) = \dfrac{2x^2}{x^2-x-6}$

14. $g(x) = \dfrac{3x^2}{x^2-3x-10}$

15. $f(x) = \dfrac{x^2+x+1}{x-1}$

16. $h(x) = \dfrac{x^2-2x+2}{x+2}$

17. $g(x) = \dfrac{4x^3-16x^2-x+4}{(x-3)^2}$

*18. $f(x) = \dfrac{1+2x-x^2-2x^3}{(x-2)^2}$

In Exercises 19–24, the equation determines a relation that is not a function. Graph the relation.

*19. $y^2 = \dfrac{x}{4-x}$

*20. $y^2 = \dfrac{x^3-8}{2x}$

*21. $y^2 = \dfrac{4x^2}{x^2-4}$

*22. $y^2 = \dfrac{3x^2-x^3}{x+1}$

*23. $y^2 = \dfrac{(x-1)^2}{9-x^2}$

*24. $y^2 = \dfrac{x^2}{4-x^2}$

*25. Translate the axes appropriately, and show that the graph of the equation in Example 2, page 225, then has an equation of the form $x'y' = k$.

*26. Repeat Exercise 25 for the equation $y = f(x) = \dfrac{x-a}{x-b}$, $a \neq b$.

*27. Show that the function f defined by $f(x) = \dfrac{1}{1+x^2}$ is an even function and sketch its graph. [*Hint:* See Exercise 27, page 219.]

*28. Show that the function f defined by $f(x) = \dfrac{x}{1+x^2}$ is an odd function and sketch its graph. [*Hint:* See Exercise 28, page 219.]

Graphing Other Functions

6–4 Addition of Ordinates

In addition to the techniques for graphing functions that we have discussed thus far, there is another method which may help in some situations. It is **addition,** or **composition, of ordinates.** This method can be employed advantageously if the function f, which you want to graph, can be written as the sum of functions f_1 and f_2 whose graphs are easy to obtain. The graph of f can then be constructed by drawing the graphs of f_1 and f_2 in the same coordinate plane and graphically adding corresponding ordinates, as illustrated in Example 1 below. (This method can also be used, although not as easily, if the function f is the sum of more than two functions.)

Example 1. Construct the graph of $f(x) = x^3 + x^2$ by addition of ordinates.

Solution: Let $f_1(x) = x^3$ and $f_2(x) = x^2$. Then for any real number a, you have $f(a) = f_1(a) + f_2(a)$; that is, the ordinate of f at a is the sum of the ordinates of f_1 and f_2 at a. Draw the graphs of $f_1(x) = x^3$ and $f_2(x) = x^2$ in the same coordinate plane, as illustrated below, and then graphically add the directed distances $f_1(x)$ and $f_2(x)$ for a sufficient number of values of x to determine the shape of the graph. (Different scales have been used here for the x-axis and the y-axis to provide greater clarity.)

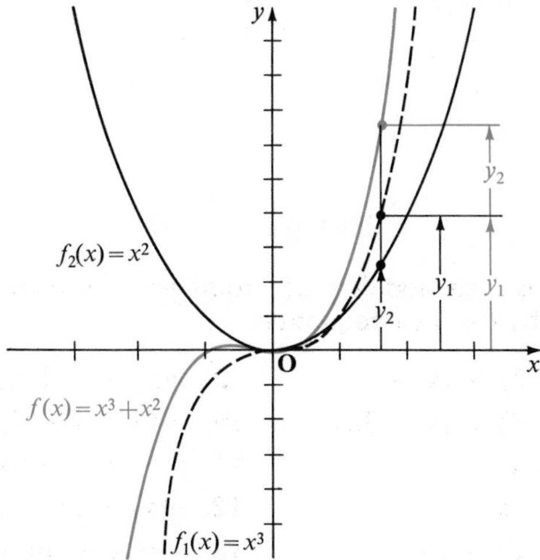

Of course, in general you would not use addition of ordinates for constructing graphs of polynomial functions. If, however, a defining equation contains one or more terms which are not monomials, you will often find this procedure convenient.

Example 2. Graph the function defined by $y = x + \sin x$.

Solution: Notice that x must be a real number. Thus, if you think of $\sin x$ in terms of measures of an angle, you should recall from your study of trigonometry that the measure should be presumed to be in radians, since there is no degree sign shown. You can construct the graph by first graphing $y = \sin x$ and then adding the corresponding value x to each ordinate, as shown below.

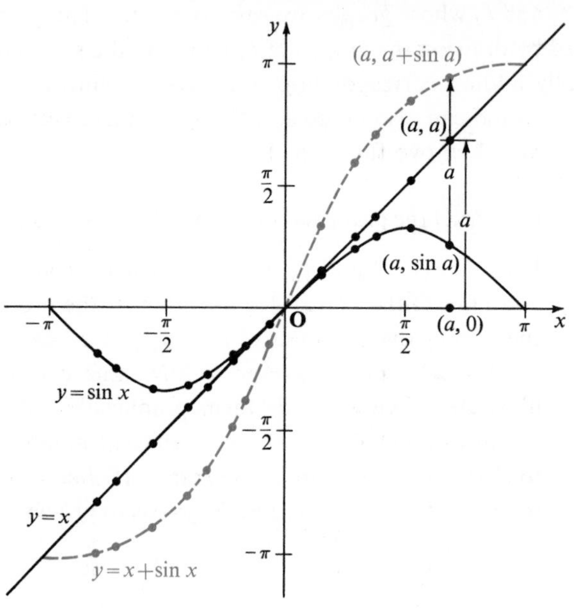

Exercises 6–4

In Exercises 1–16, use addition of ordinates to sketch the graph of the function defined by the given equation.

1. $f(x) = (x + 2) + (x - 3)$

2. $g(x) = (x - 5) + (4 - 3x)$

3. $h(x) = x^2 + x$

4. $f(x) = 2x^2 - x$

5. $g(x) = x^3 + (x - 1)$

6. $h(x) = (x + 3) - x^3$

7. $f(x) = x^2 + \dfrac{1}{x}$

8. $g(x) = x - \dfrac{1}{x^2}$

9. $h(x) = x + 2^x$

10. $j(x) = x - 2^{-x}$

11. $y = x + \cos x$

12. $g(x) = 2x - \sin x$

13. $y = x + \log_{10} x$

14. $f(x) = x - \log_{10} x$

***15.** $y = \frac{1}{3}x^2 + \cos 2x$

***16.** $h(x) = \frac{1}{3}x^2 + \sin 2x$

6–5 Parametric Equations

You learned in Section 2–1 that the equations

$$x = x_1 + r(x_2 - x_1) \qquad \text{and} \qquad y = y_1 + r(y_2 - y_1)$$

are called *parametric equations* of the line through the points $\mathbf{S}(x_1, y_1)$ and $\mathbf{T}(x_2, y_2)$ and that the variable r is called a *parameter*.

More generally, consider the pair of parametric equations

$$x = f(r) \qquad \text{and} \qquad y = g(r),$$

where r is a member of a given set S. For each value of r in S, these equations determine a value for x and a corresponding value for y. These paired values for x and y are then the Cartesian coordinates of a point, and the set of points determined in this way constitutes the graph of the parametric equations. (As mentioned on page 213, the domain S of r will generally be taken to be the set of all real numbers for which $x = f(r)$ and $y = g(r)$ are real numbers.)

Example 1. Sketch the graph of the parametric equations

$$x = r^2 - 3r + 1 \qquad \text{and} \qquad y = r + 1.$$

Solution: Since you need to find ordered pairs (x, y) with which to determine the graph, you can begin by making a table of values using suitable replacements for r:

r	-4	-3	-2	-1	0	1	2	3	4	5
x	29	19	11	5	1	-1	-1	1	5	11
y	-3	-2	-1	0	1	2	3	4	5	6

Then you can locate the points whose xy-coordinates are given by the table and sketch the graph.

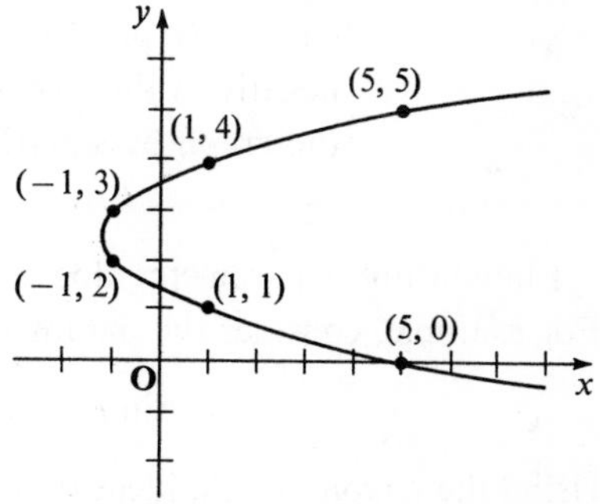

In certain cases, parametric equations can be combined into a single Cartesian equation by algebraically eliminating the parameter. For example, consider the equations for Example 1 above:

$$x = r^2 - 3r + 1 \qquad \text{and} \qquad y = r + 1.$$

If you solve the second equation for r in terms of y, and use the resulting expression as a replacement for r in the first equation, then you have

$$x = (y - 1)^2 - 3(y - 1) + 1,$$
$$x = y^2 - 2y + 1 - 3y + 3 + 1,$$
$$x = y^2 - 5y + 5.$$

You should recognize this equation as one whose graph is a parabola.

Example 2. Find a Cartesian equation for the graph of

$$x = 2\cos r, \qquad y = 5\sin r,$$

by eliminating the parameter.

Solution: If you solve the parametric equations for $\cos r$ and $\sin r$, respectively, you have

$$\frac{x}{2} = \cos r \quad \text{and} \quad \frac{y}{5} = \sin r.$$

Then, if you square both members of each of these equations and add the corresponding members, you obtain

$$\frac{x^2}{4} + \frac{y^2}{25} = \cos^2 r + \sin^2 r.$$

Since $\cos^2 r + \sin^2 r = 1$ for all r, you then have, as a Cartesian equation of the graph,

$$\frac{x^2}{4} + \frac{y^2}{25} = 1.$$

Notice that since, as r takes on values from 0 to 2π, the original equations produce all possible combinations of positive and negative values of x and y, no new points have been added to the graph by squaring.

Eliminating a parameter does not always yield an equivalent solution set. For example, consider the parametric equations

$$x = \sin r \quad \text{and} \quad y = 1 - \cos 2r.$$

Using the trigonometric identity

$$\cos 2r = 1 - 2\sin^2 r,$$

you find that

$$y = 1 - (1 - 2\sin^2 r),$$
$$y = 2\sin^2 r,$$

or

$$y = 2x^2.$$

The graph of this equation is a parabola; it contains all points (x, y) such that $x \in \mathbb{R}$, $y \geq 0$, and $y = 2x^2$. However, for $x = \sin r$, $|x|$ must be less than or equal to 1, and since $y = 1 - \cos 2r$, $0 \leq y \leq 2$. Thus, any point on the graph of the parametric equations is a point on the graph of $y = 2x^2$, but the converse is not true. Therefore, without further conditions being placed on the domain and range, the Cartesian equation does not have the same graph as the parametric equations. The graphs are related as shown in Figure 6–10.

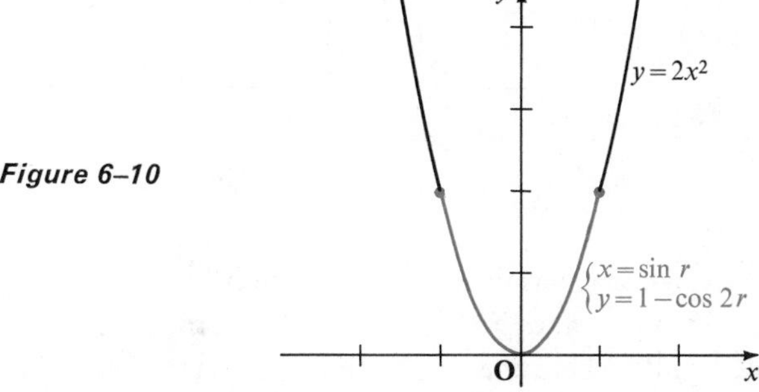

Figure 6–10

There exist an unlimited number of parametric equations for any given curve in the plane, each depending on a specific parameter. You can obtain one set of such equations by using the simple relationship $y = rx$, where r is a parameter, in conjunction with a Cartesian equation for the curve.

Example 3. Find a set of parametric equations for the circle with equation $x^2 + y^2 - 2x = 0$.

Solution: Let $y = rx$, and substitute in the given equation. You obtain

$$\begin{aligned} x^2 + (rx)^2 - 2x &= 0, \\ x^2 + r^2x^2 - 2x &= 0, \\ x^2(1 + r^2) - 2x &= 0, \\ x[x(1 + r^2) - 2] &= 0. \end{aligned}$$

It follows that

$$x = 0 \quad \text{or} \quad x(1 + r^2) - 2 = 0.$$

If $x = 0$, then $y = rx = 0$. The second of these equations can be solved to give values of x in terms of r, yielding

$$x = \frac{2}{1 + r^2}.$$

Since $y = rx$, you finally have the set of parametric equations

$$x = \frac{2}{1 + r^2}, \qquad y = \frac{2r}{1 + r^2}.$$

Notice that the parametric equations obtained in Example 3 fail to provide for the fact that the origin is on the curve, since there is no value of r for which $x = 0$. To see why this occurs, look at Figure 6–11. In using the

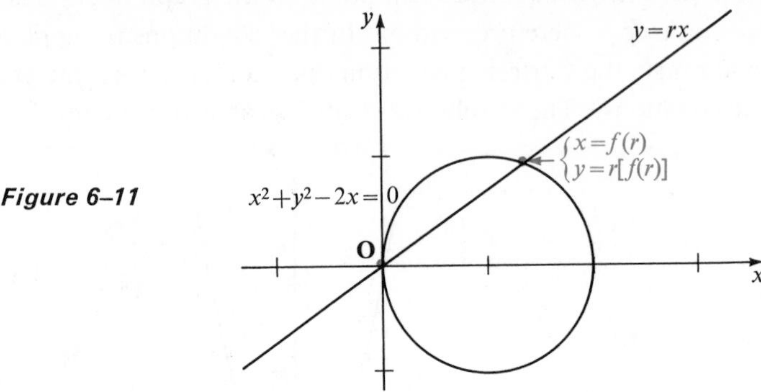

Figure 6–11

relationship $y = rx$ in conjunction with the equation $x^2 + y^2 - 2x = 0$, you are, in effect, expressing x and y in terms of the parameter r (the slope of $y = rx$) for those points where the two graphs intersect. The values of x for which such intersections occur are determined by

$$x[x(1 + r^2) - 2] = 0.$$

Now, for every slope r the line intersects the circle at two points, one of which is the origin. The x-coordinate of the origin is given by $x = 0$, and the x-coordinate of the other point is given by the equation $x(1 + r^2) - 2 = 0$. The only exception to the line's intersecting the curve at two points occurs in the case in which the line is tangent to the circle at the origin. In this case, since a vertical line has no slope, the single point of intersection (the origin) cannot be represented by coordinates of the form $(f(r), rf(r))$, and to this extent the solution given for Example 3 is deficient.

In this example we were able to identify the geometric meaning of the parameter r as the slope of the secant line through the origin. In many important problems, the parameter has such a geometric (or physical) significance.

Exercises 6–5

In Exercises 1–10, sketch the graph of the parametric equations, using the method of Example 1, page 231.

1. $x = 2t + 1,\ y = 2 - t$

2. $x = 3t - 2,\ y = 3 - 2t$

3. $x = 2 + \dfrac{1}{t},\ y = t + 1$

4. $x = \dfrac{2 + t}{t},\ y = \dfrac{3 - t}{t}$

5. $x = t + 1,\ y = t^2 + 4$

6. $x = t^2 - 3,\ y = t + 1$

7. $x = t^2 + 1,\ y = t^2 - 1$

8. $x = t^2,\ y = 8 - t^2$

9. $x = t^2,\ y = t^3$

10. $x = |t|,\ y = t^2$

In Exercises 11–18, sketch the graph of the parametric equations over the interval $0 \leq m^\circ(r) < 360$ or $0 \leq m^R(r) < 2\pi$.

11. $x = \sin r,\ y = \cos r$

12. $x = \sin r,\ y = \sin 2r$

13. $x = 1 - \cos r,\ y = 1 + \sin r$

14. $x = \sec r,\ y = \tan r$

15. $x = 2 + \cos r,\ y = \cos 2r$

16. $x = \sin r,\ y = \cos 2r - 1$

17. $x = \cos \frac{r}{2},\ y = 1 + \cos r$

18. $x = \sin r,\ y = \cos 2r$

In Exercises 19–26, eliminate the parameter r in the pair of equations to obtain a Cartesian equation.

19. $x = r - 3,\ y = 3 + r$

20. $x = 2r + 1,\ y = 3 - r$

21. $x = r + 2,\ y = r^2 + 6r + 11$

22. $x = r^2 - r + 2,\ y = 2 - r$

23. $x = \frac{1}{r},\ y = r + \frac{1}{r}$

24. $x = r + 2,\ y = \frac{2}{r(r + 4)}$

25. $x = r^2 + r,\ y = r^2 - r$

26. $x = r^2 + 2r,\ y = r^2 - 2$

In Exercises 27–34, eliminate the parameter r to obtain a Cartesian equation. Is the graph of the Cartesian equation the same as the graph of the parametric equations? Explain your answer.

27. $x = 3 \sin r,\ y = 3 \cos r$

28. $x = 4 \cos r,\ y = 3 \sin r$

29. $x = \sin r,\ y = \cos 2r$

30. $x = \cos r,\ y = \cos 2r$

31. $x = \cos 2r,\ y = 2 \sin r \cos r$

32. $x = 1 - \cos 2r,\ y = \cos r$

33. $x = 1 - \cos r,\ y = \cos 2r$

34. $x = \sin r + \cos r,\ y = \sin r - \cos r$

In Exercises 35–42, use the relationship $y = rx$ to obtain parametric equations. Is the graph of the parametric equations the same as the graph of the Cartesian equation? Explain your answer.

35. $2x + y = 6$

36. $3x - 4y = 12$

37. $x^2 + y^2 = 9$

38. $x^2 - y^2 = 16$

39. $x^2 + y^2 - 2x - 2y = 0$

40. $2x^2 + 3y^2 = 36$

41. $x^3 + y^3 - 3xy = 0$

42. $x^3 + xy^2 + y^2 - 3x^2 = 0$

* **43.** If the relationship $y = x + r$ is used to obtain parametric equations for the graph of $y = f(x)$, what is the geometric significance of the parameter r?

* **44.** If the relationship $y = r(x + 1)$ is used to obtain parametric equations for the graph of $y = f(x)$, what is the geometric significance of the additive constant 1?

* **45.** Use the relationship $y = x + r$ to obtain parametric equations for the circle with Cartesian equation $x^2 + y^2 - 2x = 0$. Explain why the graph of the parametric equations is or is not the same as the graph of the Cartesian equation.

* **46.** Use the relationship $y = r(x + 1)$ to obtain parametric equations for the circle with Cartesian equation $x^2 + y^2 - 2x = 0$. Explain why the graph of the parametric equations is or is not the same as the graph of the Cartesian equation.

Chapter Summary

1. Any equation of the form

$$P(x) = a_0x^n + a_1x^{n-1} + a_2x^{n-2} + \cdots + a_{n-1}x + a_n,$$

where the a's are constants, $a_0 \neq 0$, and n is a nonnegative integer, defines a **polynomial function.** As an aid to graphing a polynomial function, you should consider the **dominant term**, the possible numbers of **turning points**, and the ***x*- and *y*-intercepts.**

2. Any equation of the form

$$F(x) = \frac{f(x)}{g(x)},$$

where $f(x)$ and $g(x)$ are polynomials with no common factors involving x, and $g(x)$ is not the zero polynomial, defines a **rational function.** As an aid to graphing a rational function, you should consider the ***x*- and *y*-intercepts, asymptotes, restricted regions** of the plane, and a few selected points on the graph.

3. Some functions have graphs most easily constructed by adding corresponding ordinates of graphs of other functions. This technique of graphing is called **addition**, or **composition, of ordinates.**

4. You can graph parametric equations by assigning values to the parameter in order to determine ordered pairs (x, y) and then plotting the associated points. Alternatively, it is sometimes more convenient to **eliminate the parameter** and simply graph the Cartesian equation you obtain. In the latter case, you must be careful to state any restrictions on x and y so that the Cartesian equation will have the same graph as the parametric equations.

Chapter Review Exercises

1. What are the possible numbers of turning points on the graph of

$$y = 2x^7 + 3x^6 - 2x + 5?$$

2. Graph the function f defined by $f(x) = 2x^3 - 9x^2 + 4x + 15$.

3. Determine equations of any vertical asymptotes of the graph of the rational function f defined by

$$f(x) = \frac{2x^2}{(x-3)(x+1)},$$

and make a sketch showing all restricted regions of the plane.

4. Determine equations of any **(a)** horizontal or **(b)** oblique asymptotes of the graph of the function specified in Exercise 3, and sketch the graph.

5. Use addition of ordinates to graph the function g defined by

$$g(x) = x^2 + (2x - 3).$$

6. Graph the parametric equations

$$x = r - 1, \qquad y = r^2.$$

7. Eliminate the parameter r from the equations

$$x = r + 2, \qquad y = r^2 - 1,$$

to obtain an equivalent Cartesian equation.

8. Use the relationship $y = rx$ to obtain parametric equations for the graph of $2x^2 - 4y^2 = 1$.

Maps of the Plane onto Itself

In studying the graph of an equation such as

$$x^2 + y^2 - 8x + 6y + 21 = 0,$$

you have learned to write the equation equivalently as

$$(x - 4)^2 + (y + 3)^2 = 4. \tag{1}$$

Then by means of a substitution,

$$\begin{aligned} x' &= x - 4, \\ y' &= y + 3, \end{aligned} \tag{2}$$

you again write the equation equivalently, in terms of new coordinates, as

$$x'^2 + y'^2 = 4. \tag{3}$$

In doing this, you effect a *change of coordinate axes*, as indicated in the figure at the left below.

Equations (1) and (3) designate the same set of points, namely the circle with center at the point S having coordinates $(x, y) = (4, -3)$, or $(x', y') = (0, 0)$, and of radius 2. In the discussion above, then, Equations (2) have been used to obtain Equation (3) as an "alias" (another name) for the set of points formerly named by Equation (1).

There is another, equally important, interpretation of Equations (2): These equations define a *mapping* of the entire plane onto itself. Only one set of axes is considered in this mapping, and each point $U(x, y)$ is mapped onto a

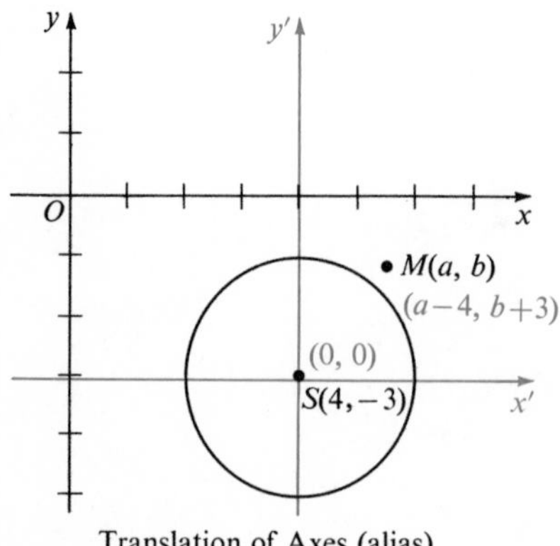

Translation of Axes (alias)

Translation of the Plane (alibi)

corresponding point $U'(x', y')$, where x' and y' are given by Equations (2). Thus the point $M(a, b)$ is mapped onto the point $N(a - 4, b + 3)$. Since, for each point U in the plane, the x-coordinate of U' is 4 less than the x-coordinate of U, and the y-coordinate of U' is 3 more than the y-coordinate of U, you can see in particular that the map of the circle defined by Equation (1) is the circle with center at the origin and of radius 2, as shown in the figure on page 238.

In a mapping such as that indicated above, it is convenient to think of points as being physically transported to corresponding points in the map. You accordingly speak of the mapping as a *transformation of the plane*. With this interpretation, Equations (2) can be said to have been used to obtain an "alibi" (another place) for the set of points named by Equation (1).

In general, the equations

$$\begin{aligned} x' &= x + h \\ y' &= y + k \end{aligned}$$

can be considered in either of two ways. They are equations for the *translation of axes* by the vector amount $(-h, -k)$, in which the x'- and y'-axes have the same directions as the x- and y-axes, respectively, and the $x'y'$-origin is at the point having xy-coordinates $(-h, -k)$. They are also equations for the *translation of the plane* by the vector amount (h, k), in which each point $M(a, b)$ is mapped onto a corresponding point $N(a + h, b + k)$.

In the same way, the equations

$$\begin{aligned} x' &= x \cos \theta - y \sin \theta \\ y' &= x \sin \theta + y \cos \theta \end{aligned} \tag{4}$$

can be interpreted as equations either for a *rotation of axes* (alias), as illustrated on the left below, or for a *rotation of the plane* (alibi), in which every point in the plane is mapped onto a corresponding point, as illustrated on the right below.

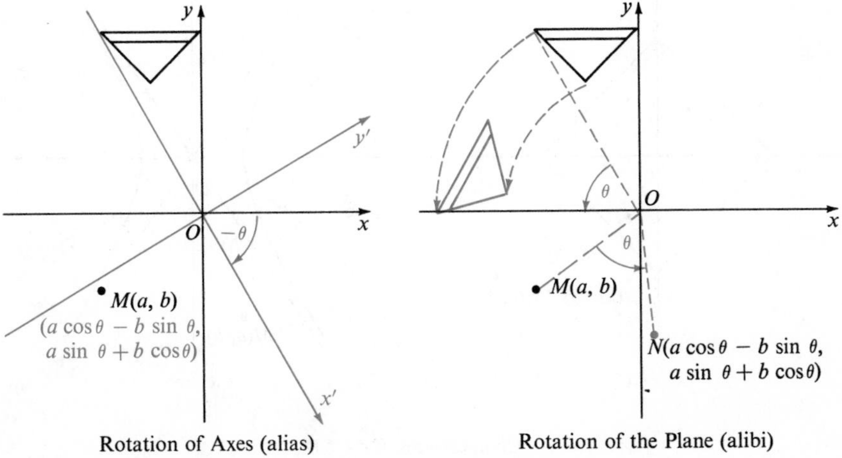

Rotation of Axes (alias)

Rotation of the Plane (alibi)

Equations and illustrations of some other basic transformations of the plane are the following.

Reflections in the axes:

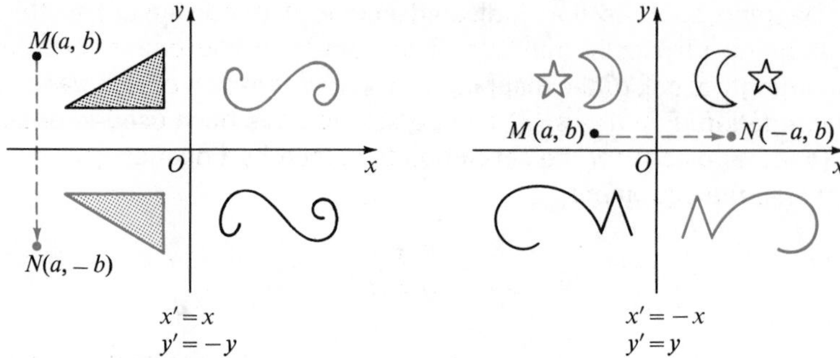

One-dimensional strains:

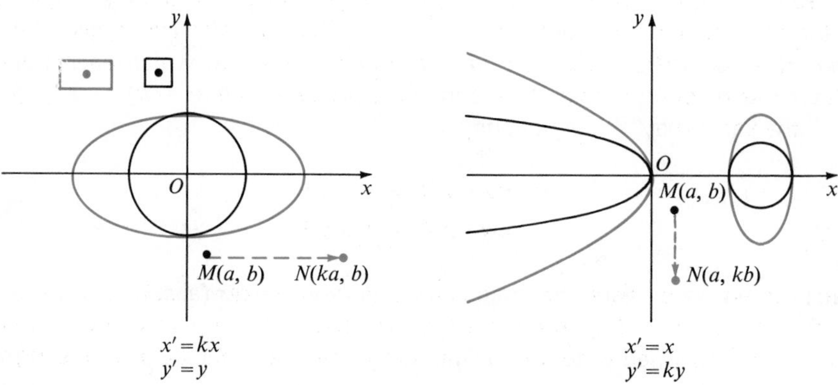

Elongations, $k > 1$

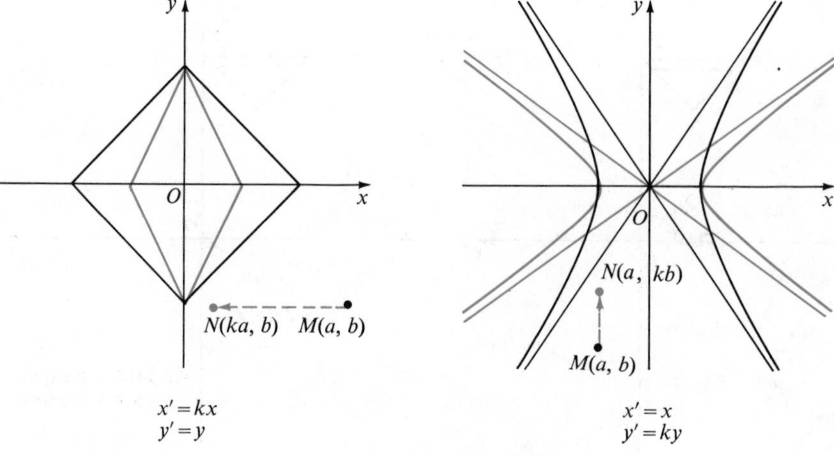

Compressions, $0 < k < 1$

Simple shears:

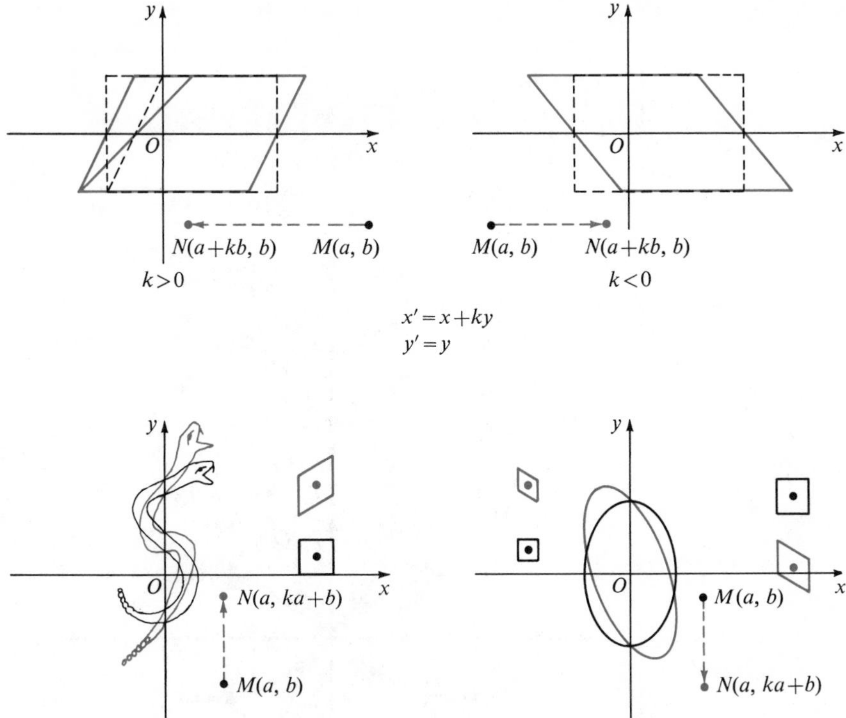

$$x' = x$$
$$y' = kx + y$$

Transformations obtained by a succession of these basic transformations of the plane will be considered in the essay at the end of Chapter 7.

Chapter 7

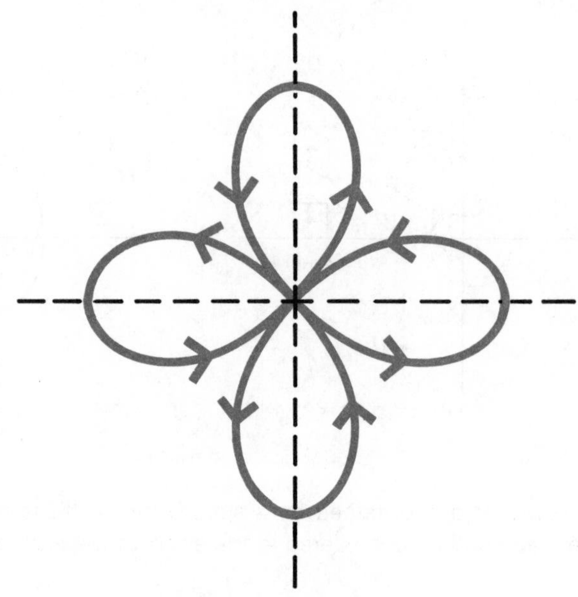

In some situations in mathematics and its applications, polar coordinates are more useful than Cartesian coordinates. In this chapter, polar coordinates are introduced and polar equations and their graphs studied.

Polar Coordinates

Polar Equations and Their Graphs

7–1 Polar Coordinate System

In the preceding chapters, you have been concerned with a Cartesian coordinate system in the plane. In this chapter, we shall discuss another useful coordinate system in the plane: a *polar coordinate system.*

To see how a polar coordinate system can be established, consider a ray $\mathcal{A}$ with initial point **O** (see Figure 7–1). For any point **S** in the plane, let $\mathcal{B}$ be a ray through **S** with initial point **O**, and let θ be an angle with $\mathcal{A}$ as initial side and $\mathcal{B}$ as terminal side. Let us agree that the counterclockwise sense of rotation is positive. Then, as illustrated in Figure 7–1, the point **S** is precisely determined

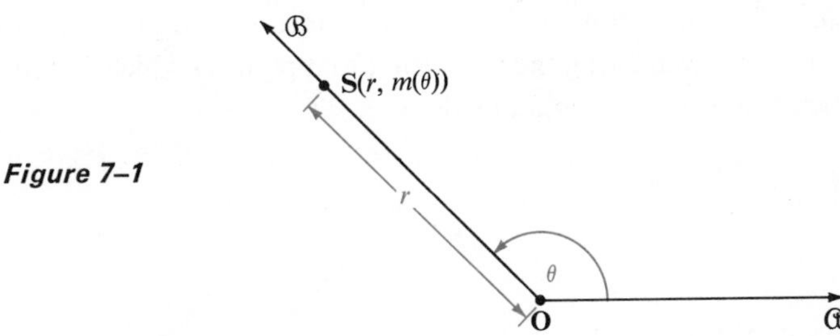

Figure 7–1

by the coordinates $(r, m(\theta))$, where r is the distance from **O** to **S**, and $m(\theta)$ is the measure of θ in some specified units, ordinarily degrees or radians. (Note that θ is not restricted to being an angle whose measure satisfies $0 \leq m^\circ(\theta) < 360$, as in the case of vectors.) The coordinates $(r, m(\theta))$, or, more briefly, (r, θ), are called a pair of **polar coordinates** of **S**. The ray $\mathcal{A}$ is called the **polar axis**, and the point **O** is called the **pole**. Rays with the pole as initial point are called **polar rays**.

The direction angle θ in Figure 7–1 is coterminal with infinitely many other angles having the polar axis as initial side. Since any of these angles can be used to specify the position of the ray $\mathcal{B}$, the point **S** can be specified by an infinite number of different ordered pairs of polar coordinates. For example, if **S** has coordinates

$$(3, 120°),$$

then it also has coordinates

$$(3, 120° - 360°), \text{ or } (3, -240°),$$
$$(3, 120° + 360°), \text{ or } (3, 480°),$$

and, in general,

$$(3, 120° + k(360°)), \text{ where } k \text{ is an integer.}$$

Example 1. Graph the point **S** having $(2, -85°)$ as polar coordinates.

Solution: First draw an angle θ with measure $-85°$. Then, on the terminal side of θ, measure 2 units from the pole, and label the point **S** with its coordinates.

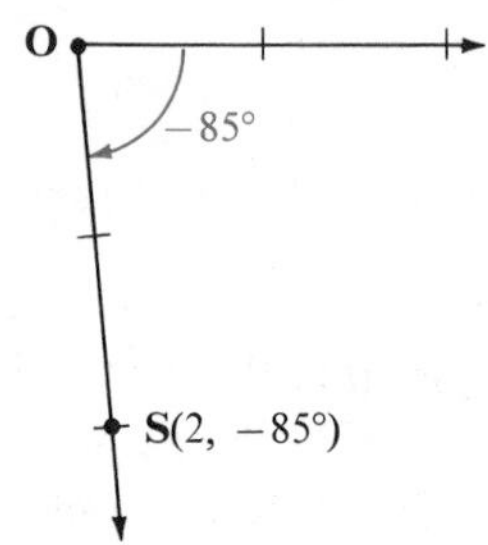

In the discussion on page 243, we considered the distance r to be an undirected distance, that is, a nonnegative number. In some situations, however, it is useful to think of r as a *directed* distance, which may, in some cases, be negative; and from now on, we shall do this. As shown in Figure 7–2(a), it is customary to interpret a positive distance r as a distance measured from **O** along the terminal side of the direction angle θ, and to interpret a negative distance r as a distance of $|r|$ measured along the ray with initial point **O** which has the *opposite* direction from that of the terminal side of θ.

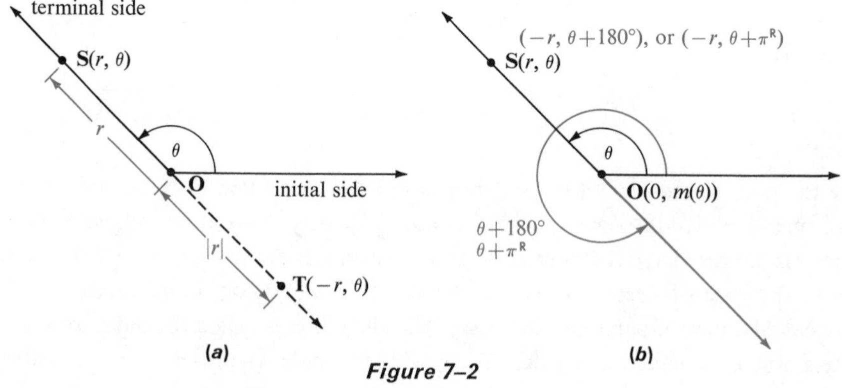

Figure 7–2

Thus, as illustrated in Figure 7–2(b), if a point **S** can be specified by the polar coordinates (r, θ), then it can also be specified by the coordinates $(-r, \theta + 180°)$ or $(-r, \theta + \pi^R)$. Note that the pole can be specified by *any* ordered pair of the form $(0, m(\theta))$, where the direction angle θ is arbitrary.

Example 2. Graph the point **S** with polar coordinates $(-1, 30°)$, and give three additional pairs of coordinates for **S**.

Solution: First draw an angle θ with measure 30°. Then extend the terminal side of θ through the pole, and measure 1 unit from the pole

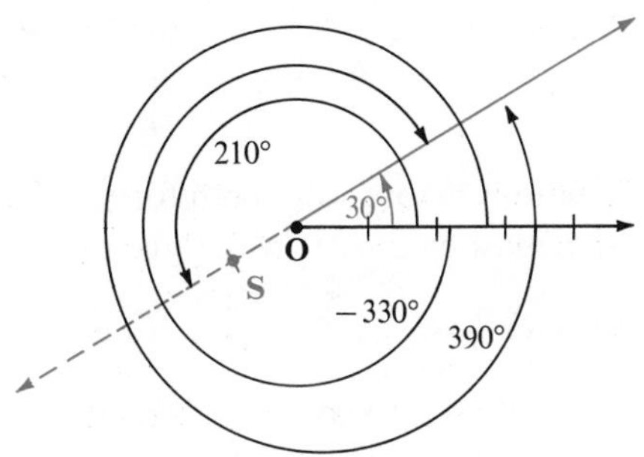

along this extension. The point reached is **S**.

Three additional pairs of coordinates for **S** are

$$(-1, 30° + 360°), \text{ or } (-1, 390°);$$

$$(-1, 30° - 360°), \text{ or } (-1, -330°);$$

and

$$(-(-1), 30° + 180°), \text{ or } (1, 210°).$$

If, as shown in Figure 7–3, the pole of a polar coordinate system is at the origin

Figure 7–3

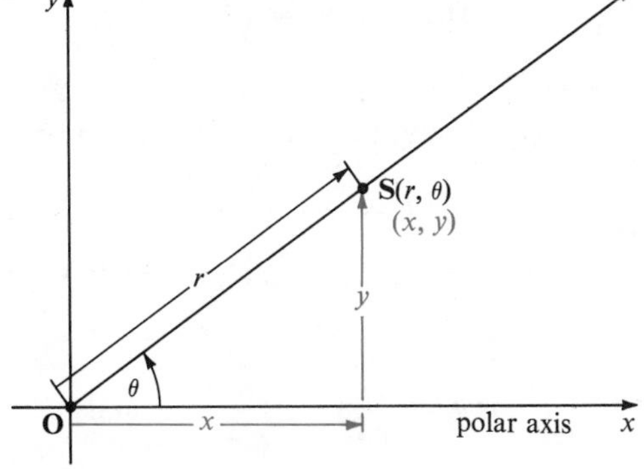

of a Cartesian coordinate system, and the polar axis coincides with the non-negative x-axis, then the Cartesian coordinates (x, y) and the polar coordinates

(r, θ) of a point **S** in the plane are related by the equations

$$x = r\cos\theta \quad \text{and} \quad y = r\sin\theta, \tag{1}$$

and also, for $x^2 + y^2 \neq 0$, by the equations

$$r = \pm\sqrt{x^2+y^2},\ \cos\theta = \frac{x}{\pm\sqrt{x^2+y^2}},\ \sin\theta = \frac{y}{\pm\sqrt{x^2+y^2}}. \tag{2}$$

You can check that Equations (1) are valid both for $r \geq 0$ and for $r < 0$. In Equations (2), the same sign must be chosen in all three equations. For example, if r is negative, then you have $r = -\sqrt{x^2+y^2}$, $\cos\theta = \dfrac{x}{-\sqrt{x^2+y^2}}$, and $\sin\theta = \dfrac{y}{-\sqrt{x^2+y^2}}$. The first equation in (2) holds also, of course, for $x^2 + y^2 = 0$.

Example 3. **(a)** Find the Cartesian coordinates of the point with polar coordinates $(3, 120°)$.

(b) Find two pairs of polar coordinates, one pair with $r > 0$ and the other pair with $r < 0$, for the point with Cartesian coordinates $(\sqrt{2}, -\sqrt{2})$.

Solution: **(a)** Using Equations (1), you have

$$x = r\cos\theta = 3\cos 120° = 3(-\tfrac{1}{2}) = -\tfrac{3}{2},$$

$$y = r\sin\theta = 3\sin 120° = 3\left(\frac{\sqrt{3}}{2}\right) = \frac{3\sqrt{3}}{2}.$$

Thus, $(x, y) = \left(-\dfrac{3}{2}, \dfrac{3\sqrt{3}}{2}\right)$.

(b) Using Equations (2), you have

$$\begin{aligned} r = \pm\sqrt{x^2+y^2} &= \pm\sqrt{(\sqrt{2})^2 + (-\sqrt{2})^2} \\ &= \pm\sqrt{2+2} \\ &= \pm\sqrt{4} = \pm 2; \end{aligned}$$

$$\cos\theta = \frac{\sqrt{2}}{\pm 2} = \pm\frac{\sqrt{2}}{2} \quad \text{and} \quad \sin\theta = \frac{-\sqrt{2}}{\pm 2} = \mp\frac{\sqrt{2}}{2}.$$

Thus, for $r > 0$, you have

$$r = 2, \quad \cos\theta = \frac{\sqrt{2}}{2}, \quad \text{and} \quad \sin\theta = \frac{-\sqrt{2}}{2}.$$

Since $\cos 315° = \frac{\sqrt{2}}{2}$ and $\sin 315° = \frac{-\sqrt{2}}{2}$, a pair of coordinates with $r > 0$ is $(2, 315°)$.

Similarly, for $r < 0$, you have

$$r = -2, \qquad \cos\theta = \frac{-\sqrt{2}}{2}, \qquad \text{and} \qquad \sin\theta = \frac{\sqrt{2}}{2};$$

and hence a pair of coordinates with $r < 0$ is $(-2, 135°)$.

You are familiar with the use of rectangular-coordinate graph paper from your earlier courses. There is also a special polar-coordinate graph paper for graphing points whose polar coordinates are given. This graph paper consists of (1) a set of rays radiating from the pole and (2) a set of concentric circles with the pole as center (see Figure 7–4). The measures of angles determined by

Figure 7–4

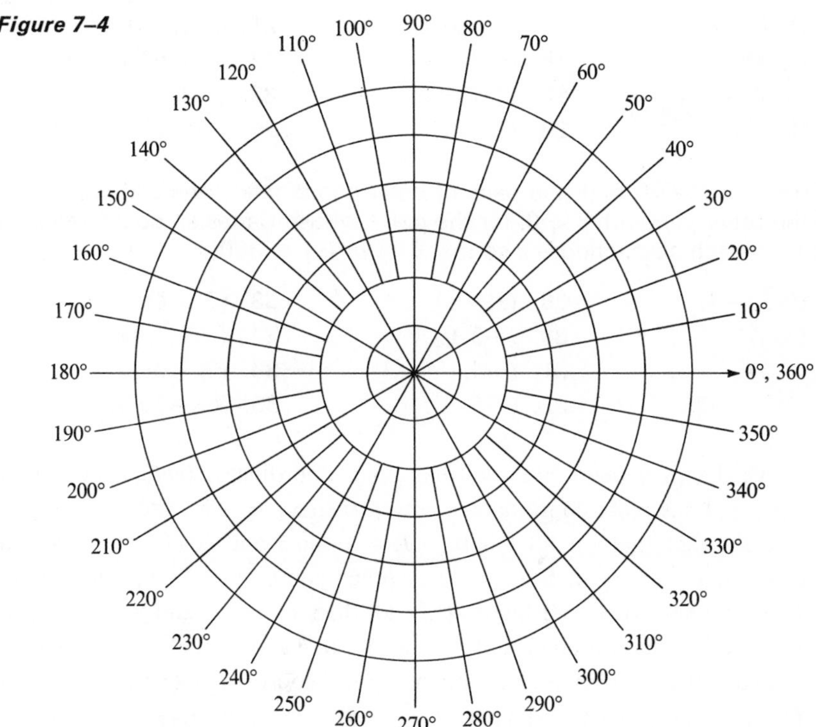

the rays are frequently shown in multiples of 10° or, alternatively, $\frac{\pi^{R}}{18}$, and the radii of the circles are multiples of some convenient unit of linear measure. To locate a point with coordinates (r, θ), then, you simply count r units out from the pole along the ray labeled $m°(\theta)$.

Exercises 7–1

In Exercises 1–12, graph the point with the given polar coordinates, and specify three additional pairs of polar coordinates for the point.

1. $(3, 25°)$

2. $(4, 45°)$

3. $(2, 120°)$

4. $(3, 150°)$

5. $\left(2, \frac{\pi^R}{2}\right)$

6. $\left(3, \frac{3\pi^R}{4}\right)$

7. $\left(1, \frac{7\pi^R}{6}\right)$

8. $\left(4, \frac{7\pi^R}{4}\right)$

9. $(-1, 40°)$

10. $(-3, 100°)$

11. $\left(-2, \frac{5\pi^R}{6}\right)$

12. $\left(-4, \frac{11\pi^R}{6}\right)$

In Exercises 13–24, find the Cartesian coordinates of the point with the given polar coordinates.

13. $(2, 45°)$
14. $(3, 30°)$
15. $(6, 90°)$
16. $(5, 270°)$
17. $(3, 180°)$
18. $(4, 360°)$
19. $(2, 135°)$
20. $(3, 240°)$
21. $(-2, 330°)$
22. $(-1, 60°)$
23. $(-3, 150°)$
24. $(-5, 225°)$

In Exercises 25–36, find two pairs of polar coordinates, one pair with $r > 0$ and the other pair with $r < 0$, for the point whose Cartesian coordinates are given. In each case, choose θ so that $0 \leq m°(\theta) < 360$.

25. $(\sqrt{3}, -3)$
26. $(4, 4)$
27. $(0, 5)$
28. $(2\sqrt{3}, -2)$
29. $(3, -3)$
30. $(-3, 0)$
31. $(-4, -4\sqrt{3})$
32. $(-5, -5)$
33. $(0, -4)$
34. $(-\sqrt{3}, 1)$
35. $(2, -2)$
36. $(6, -2\sqrt{3})$

* **37.** Prove that if a point $S(x_1, y_1)$ has polar coordinates (r, θ), and a point $T(x_2, y_2)$ has polar coordinates $(-r, \theta)$, then $x_1 = -x_2$ and $y_1 = -y_2$.

* **38.** Use the fact that for all θ, $\sin(-\theta) = -\sin\theta$ and $\cos(-\theta) = \cos\theta$ to prove that if a point $S(x_1, y_1)$ has polar coordinates (r, θ) and a point $T(x_2, y_2)$ has polar coordinates $(r, -\theta)$, then $x_1 = x_2$, and $y_1 = -y_2$.

* **39.** Use the fact that for all θ, $\sin(\pi^R - \theta) = \sin\theta$ and $\cos(\pi^R - \theta) = -\cos\theta$ to prove that is a point $S(x_1, y_1)$ has polar coordinates (r, θ) and a point $T(x_2, y_2)$ has polar coordinates $(r, \pi^R - \theta)$ then $x_1 = -x_2$ and $y_1 = y_2$.

* **40.** Use the fact that for all θ, $\sin\left(\frac{\pi^R}{2} - \theta\right) = \cos\theta$ and $\cos\left(\frac{\pi^R}{2} - \theta\right) = \sin\theta$ to prove that if a point $S(x_1, y_1)$ has polar coordinates (r, θ) and a point $T(x_2, y_2)$ has polar coordinates $\left(r, \frac{\pi^R}{2} - \theta\right)$, then $x_1 = y_2$ and $y_1 = x_2$.

7–2 Polar Equations

On page 246, you saw that polar and Cartesian coordinates of a point are related by the equations

$$x = r\cos\theta, \quad y = r\sin\theta, \tag{1}$$

and, for $x^2 + y^2 \neq 0$, by

$$r = \pm\sqrt{x^2 + y^2}, \ \cos\theta = \frac{x}{\pm\sqrt{x^2 + y^2}}, \ \sin\theta = \frac{y}{\pm\sqrt{x^2 + y^2}}. \tag{2}$$

These equations can be used to transform the equation of a locus from Cartesian to polar form and from polar to Cartesian form, respectively.

Example 1. Transform the equation $y = x + 3$ into a polar equation of the form $r = f(\theta)$.

Solution: Substituting from Equations (1) above into the equation $y = x + 3$, you have

$$r\sin\theta = r\cos\theta + 3,$$

from which you obtain

$$r\sin\theta - r\cos\theta = 3,$$

$$r(\sin\theta - \cos\theta) = 3,$$

or

$$r = \frac{3}{\sin\theta - \cos\theta}.$$

Notice in Example 1 that values of θ for which $\cos\theta = \sin\theta$, that is, values for which $m°(\theta) = 45° + k(180°)$, k an integer, are not in the domain of θ.

Example 2. Transform the polar equation

$$r = \frac{1}{1 - \cos\theta}$$

into a Cartesian equation, and sketch the graph of the equation.

Solution: Excluding values of θ for which $1 - \cos\theta = 0$, you can rewrite the given equation equivalently as

$$r(1 - \cos\theta) = 1,$$

or

$$r - r\cos\theta = 1.$$

(*Solution continued*)

Then, using Equations (1) and (2), page 249, you can substitute $\pm\sqrt{x^2 + y^2}$ for r, and x for $r \cos \theta$, in this equation to obtain

$$\pm\sqrt{x^2 + y^2} - x = 1,$$

or

$$\pm\sqrt{x^2 + y^2} = 1 + x.$$

Squaring both members, you have

$$x^2 + y^2 = 1 + 2x + x^2, \quad \text{or}$$

$$y^2 = 2x + 1.$$

You will recognize this last equation as the equation of a parabola. Its graph is shown at the right.

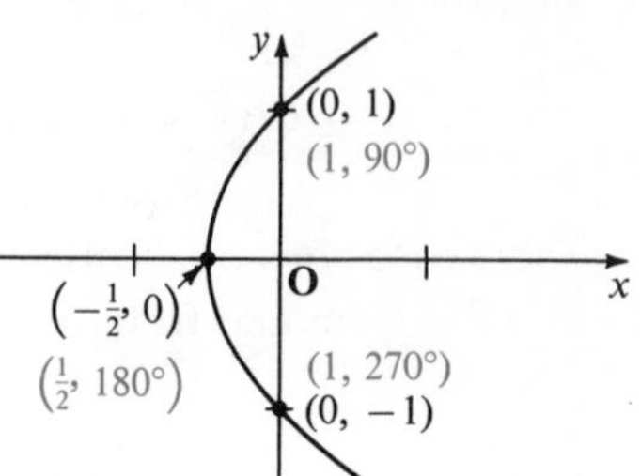

Exercises 7–2

In Exercises 1–12, transform the given Cartesian equation into a polar equation of the form $r = f(\theta)$.

1. $x^2 + y^2 = 25$
2. $x^2 + y^2 - 16 = 0$
3. $x^2 + y = 0$
4. $3x - y^2 = 0$
5. $y = x - 4$
6. $x + 3y = 2$
7. $3x^2 + 4y^2 = 5$
8. $4x^2 + 9y^2 = 36$
9. $x^2 - y^2 = 16$
10. $9x^2 - 4y^2 = 36$
11. $xy = 4$
12. $xy + 9 = 0$

In Exercises 13–24, transform the given polar equation into a Cartesian equation, and sketch the graph.

13. $r = 4$
14. $r = 3$
15. $r = 4 \sin \theta$
16. $r = 8 \cos \theta$
17. $r = \dfrac{5}{\sin \theta - \cos \theta}$
18. $r = \dfrac{4}{\sin \theta + \cos \theta}$
19. $r = \dfrac{3}{3 - \cos \theta}$
20. $r = \dfrac{4}{4 + \sin \theta}$
21. $r = \dfrac{3}{2 - \sin \theta}$
22. $r = \dfrac{5}{3 + \cos \theta}$
23. $r = \dfrac{2}{5 - \cos \theta}$
24. $r = \dfrac{3}{6 - \sin \theta}$

* **25.** Use the fact that for all θ_1 and θ_2,

$$\cos \theta_2 \cos \theta_1 + \sin \theta_2 \sin \theta_1 = \cos (\theta_2 - \theta_1)$$

to show that the distance between the points with coordinates (r_1, θ_1) and (r_2, θ_2) is given by $d = \sqrt{r_2^2 + r_1^2 - 2r_2r_1 \cos (\theta_2 - \theta_1)}$.

* **26.** Use the result of Exercise 25 to show that a polar equation for the circle with radius a and center $\mathbf{C}(r_1, \theta_1)$ is

$$r^2 - 2rr_1 \cos (\theta - \theta_1) + r_1^2 - a^2 = 0.$$

7–3 Graphs of Polar Equations

The **graph of a polar equation** in a polar-coordinate plane is the set of all points in the plane having *at least one* pair of polar coordinates (r, θ) that satisfies the equation. Notice that a point on the graph may have some pairs of polar coordinates that do not satisfy the equation. For example, the polar coordinates $(-3, 180°)$ do not satisfy the polar equation $r = 3$. Nevertheless, the point **R** having these coordinates is on the graph of the equation, because **R** also has coordinates $(3, 0°)$.

One way to sketch the graph of a polar equation is simply to plot points. Since many polar equations involve trigonometric functions, a table of function values such as that on page 387 can sometimes be useful. In many cases, however, the values shown in the abbreviated table below will suffice to determine the graph. For corresponding angles in the second, third, and fourth quadrants, simple sign adjustments will produce the necessary values.

$m°(\theta)$	$\sin\theta$	$\cos\theta$	$\tan\theta$
0	0	1	0
30	$\frac{1}{2} = 0.500$	$\frac{\sqrt{3}}{2} \doteq 0.866$	$\frac{\sqrt{3}}{3} \doteq 0.577$
45	$\frac{\sqrt{2}}{2} \doteq 0.707$	$\frac{\sqrt{2}}{2} \doteq 0.707$	1
60	$\frac{\sqrt{3}}{2} \doteq 0.866$	$\frac{1}{2} = 0.500$	$\sqrt{3} \doteq 1.732$
90	1	0	Not defined

Example 1. Sketch the graphs of

(a) $r = 3$, **(b)** $\theta = 45°$, **(c)** $r = 3\cos\theta$, **(d)** $r = \sin 2\theta$.

Solution: **(a)** Since r is constant, it is the same for all values of θ. Any point whose coordinates are of the form $(3, \theta)$ is on the graph, and conversely. Thus, the graph is the circle with center **O** and radius 3, as shown below.

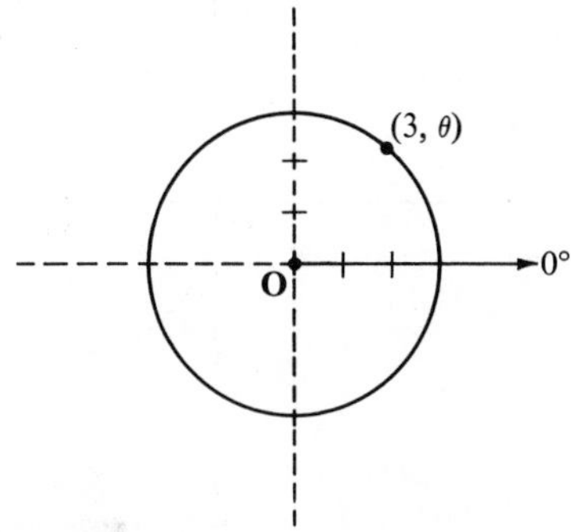

(Solution continued)

(b) Since θ is constant, it is the same for all values of r. Any point whose coordinates are of the form $(r, 45°)$ is on the graph, and conversely. Thus, the graph is a line passing through the pole and making an angle of 45° with the polar axis, as illustrated below. Points for which $r > 0$ lie in the first quadrant, and points for which $r < 0$ lie in the third quadrant.

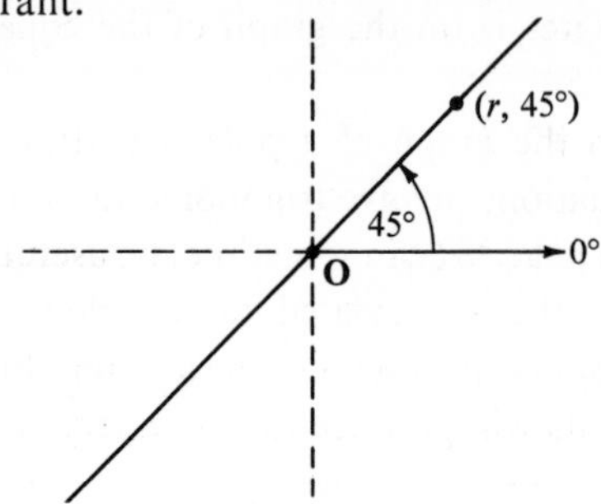

(c) To graph the equation $r = 3 \cos \theta$, first make a table of values, as shown below. As you locate the points whose

θ	0^R	$\frac{\pi}{6}^R$	$\frac{\pi}{3}^R$	$\frac{\pi}{2}^R$	$\frac{2\pi}{3}^R$	$\frac{5\pi}{6}^R$	π^R	$\frac{7\pi}{6}^R$	$\frac{4\pi}{3}^R$	$\frac{3\pi}{2}^R$	$\frac{5\pi}{3}^R$	$\frac{11\pi}{6}^R$	$2\pi^R$
$\cos \theta$	1	$\frac{\sqrt{3}}{2}$	$\frac{1}{2}$	0	$-\frac{1}{2}$	$-\frac{\sqrt{3}}{2}$	-1	$-\frac{\sqrt{3}}{2}$	$-\frac{1}{2}$	0	$\frac{1}{2}$	$\frac{\sqrt{3}}{2}$	1
$r = 3 \cos \theta$	3	$\frac{3\sqrt{3}}{2}$	$\frac{3}{2}$	0	$-\frac{3}{2}$	$-\frac{3\sqrt{3}}{2}$	-3	$-\frac{3\sqrt{3}}{2}$	$-\frac{3}{2}$	0	$\frac{3}{2}$	$\frac{3\sqrt{3}}{2}$	3

coordinates are given in this table, you will note that you have two pairs of coordinates determining each point; for example, $(-3, \pi^R)$ determines the same point as $(3, 0^R)$. The graph is a circle which has center $\mathbf{S}(\frac{3}{2}, 0^R)$ and which passes through **O**, as shown.

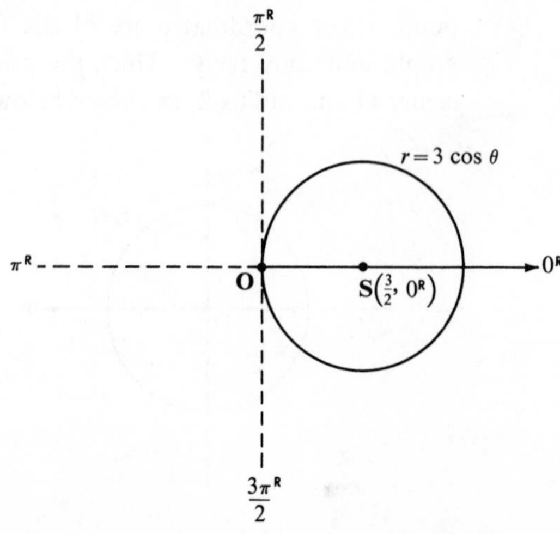

(d) Prepare a table of values, as shown below.

① ②

θ	0°	30°	45°	60°	90°	120°	135°	150°	180°
2θ	0°	60°	90°	120°	180°	240°	270°	300°	360°
$r = \sin 2\theta$	0	$\frac{\sqrt{3}}{2}$	1	$\frac{\sqrt{3}}{2}$	0	$-\frac{\sqrt{3}}{2}$	-1	$-\frac{\sqrt{3}}{2}$	0

③ ④

θ	180°	210°	225°	240°	270°	300°	315°	330°	360°
2θ	360°	420°	450°	480°	540°	600°	630°	660°	720°
$r = \sin 2\theta$	0	$\frac{\sqrt{3}}{2}$	1	$\frac{\sqrt{3}}{2}$	0	$-\frac{\sqrt{3}}{2}$	-1	$-\frac{\sqrt{3}}{2}$	0

The graph of this equation is a so-called "four-leafed rose." The circled numerals indicate which leaf contains the points whose coordinates are given in the corresponding portion of the table shown above. The arrows on the graph indicate the smooth path of a tracing point as θ goes from 0° to 360°.

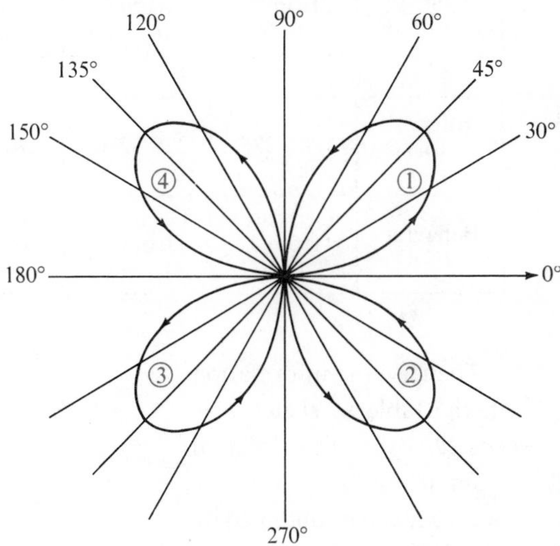

An alternative, and generally easier, way to graph this type of polar equation is to sketch the trigonometric function involved in a Cartesian plane and then to use the information about the behavior of y obtained from this *auxiliary graph* to deduce the behavior of r over successive intervals of angle measure.

Example 2. Sketch the graph of $r = \cos 2\theta$.

Solution: First sketch two cycles of $y = \cos 2x$ in a Cartesian plane, as

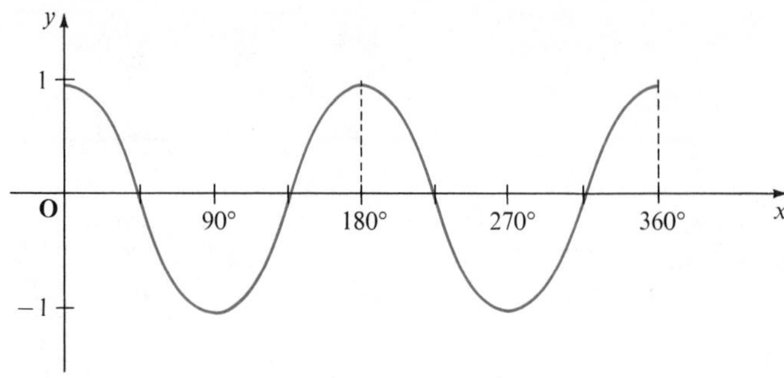

shown. By inspection of this graph, you can verify the information in the chart below.

Interval for $m°(\theta)$	0°–45°	45°–90°	90°–135°	135°–180°
Behavior of r	Decreases from 1 to 0	Decreases from 0 to −1	Increases from −1 to 0	Increases from 0 to 1

Interval for $m°(\theta)$	180°–225°	225°–270°	270°–315°	315°–360°
Behavior of r	Decreases from 1 to 0	Decreases from 0 to −1	Increases from −1 to 0	Increases from 0 to 1

Then use the information in the table to sketch $r = \cos 2\theta$ over the interval from 0° to 45°. Greater accuracy can be obtained by plotting a few points, as shown at the right.

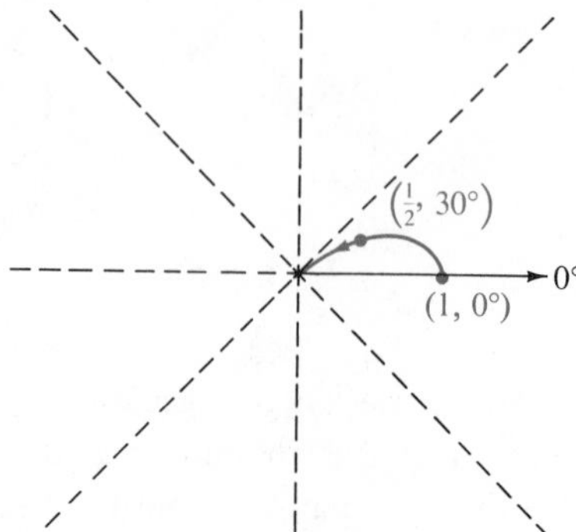

Next, continue the sketch for the interval from 45° to 90°, taking advantage of the symmetry of the graph of $y = \cos 2x$.

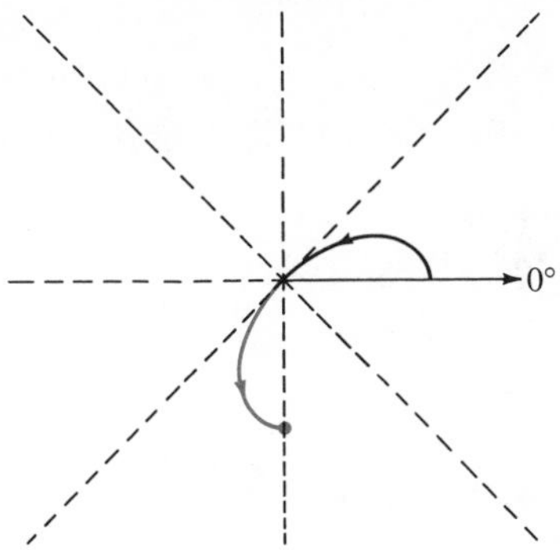

By continuing this procedure, you finally obtain the complete graph as shown. This curve is another example of a "four-leafed rose."

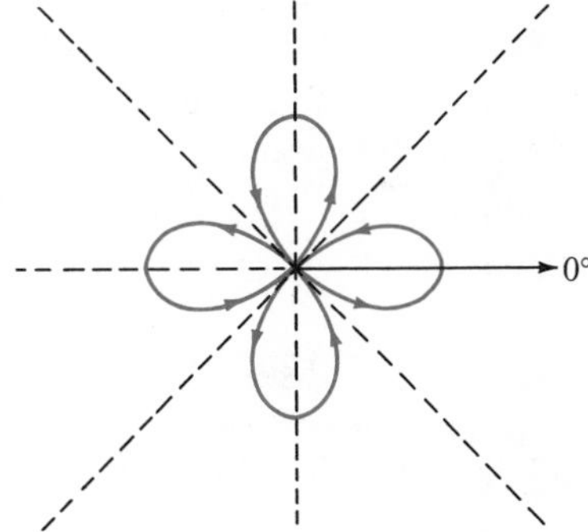

In sketching polar graphs, the following facts concerning symmetry are sometimes useful (see Exercises 37–40, page 248):

1. If replacing θ with $-\theta$ in a polar equation does not change the equation, then the graph of the equation is symmetric with respect to the line containing the polar axis.
2. If replacing r with $-r$ in a polar equation does not change the equation, then the graph of the equation is symmetric with respect to the pole.
3. If replacing θ with $\pi^{R} - \theta$ in a polar equation does not change the equation, then the graph of the equation is symmetric with respect to the line having equation $\theta = \frac{\pi^{R}}{2}$.
4. If replacing θ with $\frac{\pi^{R}}{2} - \theta$ in a polar equation does not change the equation, then the graph of the equation is symmetric with respect to the line having equation $\theta = \frac{\pi^{R}}{4}$.

Exercises 7–3

In Exercises 1–30, graph the given polar equation.

1. $r = 4$

2. $r = 5$

3. $\theta = 60°$

4. $\theta = 135°$

5. $r = 2 \sin \theta$

6. $r = 4 \cos \theta$

7. $r = \sin 3\theta$
(three-leafed rose)

8. $r = \cos 3\theta$
(three-leafed rose)

9. $r = 2(1 + \sin \theta)$
(cardioid)

10. $r = 3(1 - \cos \theta)$
(cardioid)

11. $r = 4 - 2 \sin \theta$
(limaçon)

12. $r = 2 - 4 \sin \theta$
(limaçon)

13. $r = 2 \cos (\theta + 45°)$

14. $r = 3 \sin (\theta - 60°)$

15. $r = 2 \sec \theta$

16. $r = 3 \csc \theta$

17. $r^2 = 2 \cos \theta$
(lemniscate)

18. $r^2 = \cos 2\theta$
(lemniscate)

19. $r = \tan \theta$

20. $r = \cot \theta$

21. $r = \dfrac{4}{2 - \cos \theta}$

22. $r = \dfrac{2}{1 - \cos \theta}$

23. $r = \dfrac{4}{1 + 2 \cos \theta}$

24. $r = \dfrac{2}{1 + 2 \cos \theta}$

25. $r = \dfrac{2}{1 + 2 \sin \theta}$

26. $r = \dfrac{3}{1 - 2 \sin \theta}$

* **27.** $(r^2 - 2r \cos \theta + 1)(r^2 + 2r \cos \theta + 1) = 4$ (oval of Cassini)

* **28.** $r = \tan \theta \sin \theta$ (cissoid)

* **29.** $r = m^{\mathrm{R}}(\theta)$ (spiral of Archimedes)

* **30.** $r = 2^{\frac{1}{2} m^{\mathrm{R}}(\theta)}$ (logarithmic spiral)

7–4 Intersections of Graphs of Polar Equations

If you can solve a system of Cartesian equations analytically, you can find the coordinates of all points of intersection of the graphs of the equations. Because the polar coordinates of a point are not unique, this is not necessarily true in the case of a system of polar equations. The graphs of a system of polar equations may intersect in a point, or points, whose coordinates are not in the solution set of the system, as the following example shows. Thus, in considering systems of polar equations, you must distinguish between those ordered pairs constituting solutions of the system and those identifying points of intersection of the graphs.

Example. For the equations

$$r = 3 + 6 \cos \theta \quad \text{and} \quad r = 3,$$

(a) determine all (r, θ) which satisfy the system of equations;
(b) determine the points of intersection of the graphs of the two equations, over the interval $0 \leq m^\circ(\theta) \leq 360$.

Solution: **(a)** If you substitute 3 for r in the equation $r = 3 + 6 \cos \theta$, you have

$$3 = 3 + 6 \cos \theta,$$

from which you obtain

$$\cos \theta = 0,$$

and

$$m^\circ(\theta) = 90 \text{ or } 270.$$

Therefore, $(3, 90^\circ)$ and $(3, 270^\circ)$ are the pairs (r, θ) which satisfy both equations.

(b) Sketch the graphs of both equations in the same polar coordinate system.

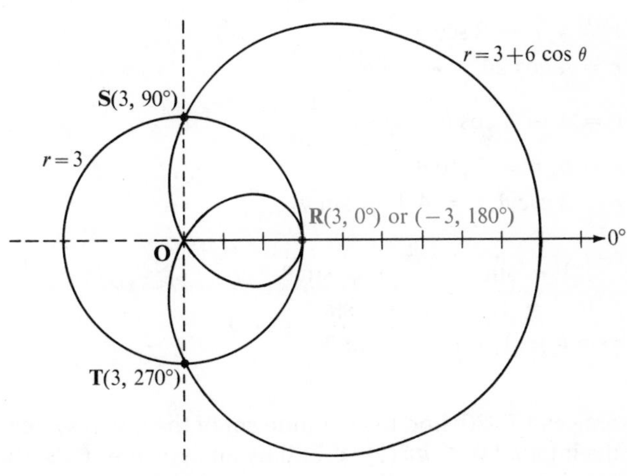

By inspection, you can see that the points **S**$(3, 90^\circ)$ and **T**$(3, 270^\circ)$ are points of intersection, as you found in part (a). You can also see, however, that there is a third point, **R**, of intersection. One pair of coordinates for this point is

(*Solution continued*)

$(3, 0°)$, which satisfies the equation $r = 3$ but does not satisfy the equation $r = 3 + 6 \cos \theta$. Another pair of coordinates for the point **R** is $(-3, 180°)$, which satisfies the equation $r = 3 + 6 \cos \theta$ but does not satisfy the equation $r = 3$. Since both of these ordered pairs determine **R**, either could be used to name the point of intersection. Therefore, the points of intersection over the specified interval are **R**$(3, 0°)$, **S**$(3, 90°)$, and **T**$(3, 270°)$.

Exercises 7–4

In Exercises 1–8, find all (r, θ), with $0 \leq m°(\theta) < 360$, that satisfy the given system of polar equations.

1. $r = \sin \theta, r = \cos \theta$
2. $r = 1 + \sin \theta, r = 1 + \cos \theta$
3. $r = \cos \theta, r = 1 - \cos \theta$
4. $r = \sin \theta, r = 1 - \sin \theta$
5. $r = 1, r = \sin 2\theta$
6. $r = \sqrt{3}, r = 2 \cos 2\theta$
7. $r = \sin \theta, r = \csc \theta$
8. $r = \cos \theta, r = \sec \theta$

In Exercises 9–16, find coordinates of all points of intersection of the graphs of the given system of polar equations over the interval $0 \leq m°(\theta) < 360$.

9. $r = 4, r = 2 \sec \theta$
10. $r = 4 \cos 2\theta, r = \sin \theta$
11. $r = 4 - 4 \cos \theta, r = \dfrac{4}{1 - \cos \theta}$
12. $r = 6, r = 3 \csc \theta$
13. $r = 3 \csc \theta, r = 4(1 + \sin \theta)$
14. $r = \dfrac{2}{1 - \sin \theta}, r = \dfrac{6}{1 + \sin \theta}$
15. $r \sin \theta = 1, r = 3 - 2 \sin \theta$
16. $r \sin \theta = 1, r = 2 - \sin \theta$

In Exercises 17–20, find the solution set of the given system of polar equations over the interval $0 \leq m°(\theta) < 360$ by analytic methods; then find coordinates of the points of intersection of the graphs of the equation over this interval by graphic methods. Explain any discrepancies.

17. $r = \cos \theta, r = 1 + 2 \cos \theta$
18. $r = \sin \theta, r = 1 + 2 \sin \theta$
19. $r^2 = \sin 2\theta, r^2 = \cos 2\theta$
20. $r^2 = \sin 2\theta, r = \sqrt{2} \sin \theta$

Polar Equations for Loci

7–5 Polar Equations for Conic Sections

In Section 4–6, you saw that parabolas, ellipses, and hyperbolas can be defined in terms of the ratio of the distances of each point on the curve from a focus and a corresponding directrix. In this section, polar equations of these curves will be discussed.

Consider first a conic section having a focus at the pole and a line $\mathfrak{D}$ with Cartesian equation of the form $x = -p$ $(p > 0)$ as corresponding directrix, as shown in Figure 7–5.

Figure 7–5

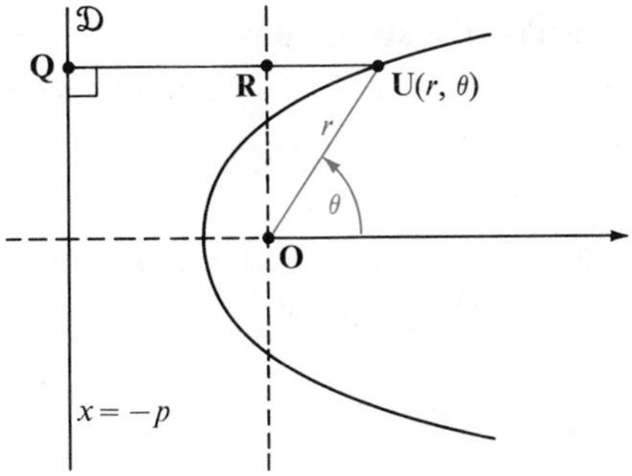

Recall from Section 4–6 that if **U** is any point on the conic section, then the eccentricity e $(e > 0)$ of the conic section is given by

$$e = \frac{d(\mathbf{O}, \mathbf{U})}{d(\mathfrak{D}, \mathbf{U})}.$$

From this equation, you have

$$d(\mathbf{O}, \mathbf{U}) = e[d(\mathfrak{D}, \mathbf{U})]. \qquad (1)$$

In terms of polar coordinates, however,

$$d(\mathbf{O}, \mathbf{U}) = |r|;$$

and, if $l(\mathbf{Q}, \mathbf{R})$ and $l(\mathbf{R}, \mathbf{U})$ represent the directed distances from **Q** to **R**, and from **R** to **U**, respectively, then

$$\begin{aligned} d(\mathfrak{D}, \mathbf{U}) &= |l(\mathbf{Q}, \mathbf{R}) + l(\mathbf{R}, \mathbf{U})| \\ &= |p + r\cos\theta|. \end{aligned}$$

Hence, Equation (1) can be written equivalently as

$$|r| = e|p + r\cos\theta|,$$

or

$$r = \pm(ep + er\cos\theta).$$

It can be shown, however, that the equations

$$r = ep + er\cos\theta \qquad \text{and} \qquad r = -(ep + er\cos\theta)$$

actually represent the same set of points (see Exercise 20, page 262). Thus the locus can be represented by the single equation

$$r = ep + er\cos\theta,$$

or

$$r = \frac{ep}{1 - e\cos\theta}. \tag{2}$$

Equation (2) is called the **standard polar form of the equation of a conic section** with a focus at the origin and corresponding directrix the line with equation $x = -p$.

Example. Name the curve, find a Cartesian equation for a directrix, and sketch the graph of

$$r = \frac{4}{2 - \cos\theta}.$$

Solution: Since you wish the denominator of the right-hand member to be in the form $1 - e\cos\theta$, you can multiply both the numerator and the denominator of the right-hand member by $\frac{1}{2}$ to obtain

$$r = \frac{2}{1 - \frac{1}{2}\cos\theta}. \tag{3}$$

Then, by inspection, you can see that $e = \frac{1}{2}$. Thus, since $\frac{1}{2} < 1$, the curve is an ellipse.

Next, comparing Equations (2) and (3), you can see that

$$ep = 2,$$

or, since $e = \frac{1}{2}$,

$$\tfrac{1}{2}p = 2.$$

Thus,

$$p = 4,$$

and a Cartesian equation for a directrix is

$$x = -4.$$

To sketch the graph, first find coordinates for a few points on the curve. Substituting 0, 90, 180, and 270 for $m°(\theta)$ in the given equation, you obtain $r = 4, 2, \frac{4}{3}$, and 2, respectively. Hence the points $(4, 0°)$, $(2, 90°)$, $(\frac{4}{3}, 180°)$, and $(2, 270°)$ are on the ellipse. Knowing these points, you can sketch the curve with reasonable accuracy, as shown below.

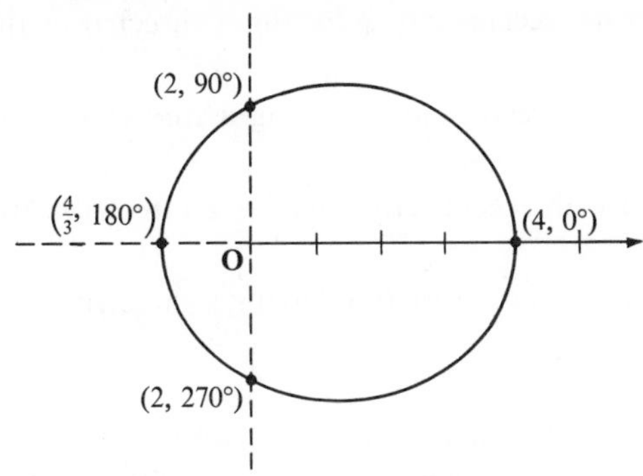

Next, for a conic section with focus at the pole and with corresponding directrix the line with equation $x = p$ $(p > 0)$, the polar form can similarly be shown (Exercise 17, page 262) to be

$$r = \frac{ep}{1 + e\cos\theta}.$$

For a conic section with focus at the pole and corresponding directrix parallel to the polar axis, the form is (Exercise 18, page 262)

$$r = \frac{ep}{1 \pm e\sin\theta},$$

where p is positive, and the sign $+$ is selected if the directrix is above the focus and the sign $-$ is selected if the directrix is below the focus.

Exercises 7–5

In Exercises 1–10, name the curve, find a Cartesian equation for a directrix, and sketch the graph of the given equation.

1. $r = \dfrac{2}{1 - \cos\theta}$

2. $r = \dfrac{4}{1 - \cos\theta}$

3. $r = \dfrac{6}{2 + \cos\theta}$

4. $r = \dfrac{8}{1 + 3\cos\theta}$

5. $r = \dfrac{10}{1 + \sin\theta}$

6. $r = \dfrac{4}{3 + 2\sin\theta}$

7. $r = \dfrac{16}{4 - 5\sin\theta}$

8. $r = \dfrac{12}{1 - 4\sin\theta}$

9. $r = \dfrac{8}{5 + 4\sin\theta}$

10. $r = \dfrac{10}{5 - 2\cos\theta}$

In Exercises 11–16, write a polar equation for the specified conic section having the pole as the focus corresponding to the given directrix.

11. A parabola having a directrix with Cartesian equation $x = -4$.
12. A parabola having a directrix with Cartesian equation $y = 2$.
13. An ellipse with eccentricity $\frac{3}{4}$ having a directrix with Cartesian equation $y = -6$.
14. An ellipse with eccentricity $\frac{2}{3}$ having a directrix with Cartesian equation $x = 4$.
15. A hyperbola with eccentricity 2 having a directrix with Cartesian equation $x = -4$.
16. A hyperbola with eccentricity $\frac{3}{2}$ having a directrix with Cartesian equation $y = 6$.

*** 17.** Use an argument similar to that on pages 259–260 to show that a polar equation for a conic section with eccentricity e, a focus at the pole, and corresponding directrix having Cartesian equation $x = p$ $(p > 0)$ is

$$r = \frac{ep}{1 + e\cos\theta}.$$

*** 18.** Show that a polar equation for a conic section with eccentricity e, with a focus at the pole, and with corresponding directrix having Cartesian equation $y = -p$ $(p > 0)$ is $r = \dfrac{ep}{1 - e\sin\theta}$.

*** 19.** Determine the locus of the equation

$$r = \frac{ep}{1 + e\cos(\theta - \alpha)},$$

where α is a constant angle.

*** 20.** Show that the equations

$$r = \frac{ep}{1 - e\cos\theta} \quad \text{and} \quad r = \frac{-ep}{1 + e\cos\theta}$$

have the same graph. [*Hint:* A point with polar coordinates (r, θ) also has polar coordinates $(-r, \theta + \pi)$, and $\cos(\theta + \pi) = -\cos\theta$ for all θ.]

7–6 Locus Problems in Polar Coordinates

Equations in polar form for various loci can be derived in a manner similar to that used to derive Cartesian equations for loci (Section 4–1). In fact, the derivation of Equation (2), page 260, is an example of this. In many cases, a polar equation for a curve may be simpler in form than a corresponding Cartesian equation.

Example 1. Let **O** be a fixed point on a circle $\mathcal{C}$ of diameter a, and for each *directed* line $\mathcal{L}$ through **O** let **S** be the second point in which $\mathcal{L}$ intersects $\mathcal{C}$. (If $\mathcal{L}$ is tangent to $\mathcal{C}$, let $\mathbf{S} = \mathbf{O}$.) Then locate point **T** on $\mathcal{L}$ so that $l(\mathbf{S}, \mathbf{T})$, the *directed* distance from **S** to **T**, is equal to a. (Note that $l(\mathbf{S}, \mathbf{T}) = -l(\mathbf{T}, \mathbf{S})$.) Determine a polar equation for the locus of all such points **T**.

Solution: Establish a polar coordinate system with the pole at **O** and the polar axis lying along the diameter containing **O**, as shown in the diagrams below. Let **M** be the second point in which the circle intersects the polar axis.

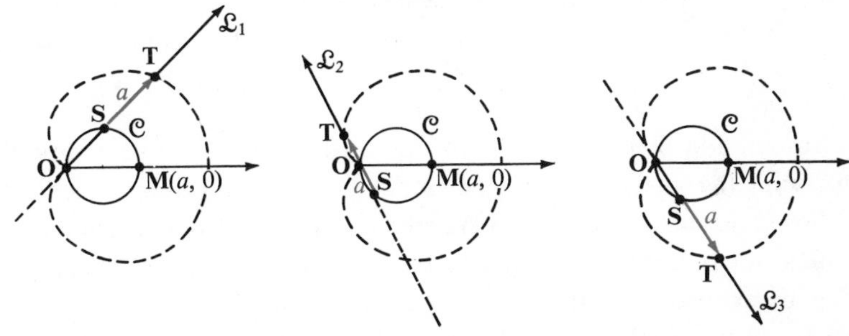

By definition, a point $\mathbf{T}(r, \theta)$ of the directed line $\mathcal{L}$ is on the locus if and only if

$$l(\mathbf{O}, \mathbf{T}) = l(\mathbf{O}, \mathbf{S}) + a.$$

Three positions ($\mathcal{L}_1$, $\mathcal{L}_2$, and $\mathcal{L}_3$) for the line $\mathcal{L}$, and corresponding points **S** and **T**, are shown in the diagrams above.

From the diagrams below, you can see that $l(\mathbf{O}, \mathbf{T}) = r$ and

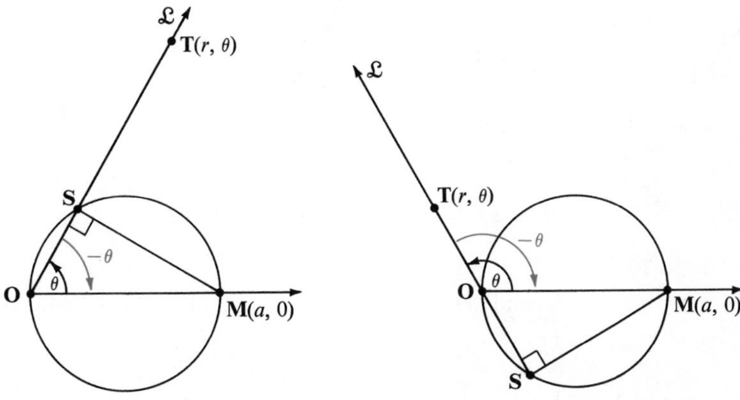

$l(\mathbf{O}, \mathbf{S}) = l(\mathbf{O}, \mathbf{M}) \cos(-\theta) = a \cos \theta$. Thus, substituting r for

(*Solution continued*)

$l(\mathbf{O}, \mathbf{T})$ and $a \cos \theta$ for $l(\mathbf{O}, \mathbf{S})$ in the equation

$$l(\mathbf{O}, \mathbf{T}) = l(\mathbf{O}, \mathbf{S}) + a,$$

you have

$$r = a \cos \theta + a = a(1 + \cos \theta).$$

Hence

$$r = a(1 + \cos \theta)$$

is the desired equation of the locus.

This heart-shaped curve (see the first set of diagrams on page 263) is called a **cardioid.** A cardioid is a special case of a **limaçon**, a curve defined by $l(\mathbf{S}, \mathbf{T}) = k$, where k is a nonzero constant, not necessarily equal to the diameter a of the circle.

In deriving polar equations for loci, it is useful to have a formula in polar form for the distance between two points. You can obtain such a formula as follows: Suppose that $\mathbf{S}(x_1, y_1)$ and $\mathbf{T}(x_2, y_2)$ are any two points in the plane. Then, by Equations (1), page 246, the Cartesian coordinates (x_1, y_1) and (x_2, y_2) can be expressed in the form $(r_1 \cos \theta_1,\ r_1 \sin \theta_1)$ and $(r_2 \cos \theta_2,\ r_2 \sin \theta_2)$, respectively. Replacing x_1, x_2, y_1, and y_2 in the Cartesian distance formula

$$d(\mathbf{S}, \mathbf{T}) = \sqrt{(x_2 - x_1)^2 + (y_2 - y_1)^2}$$

with the corresponding expressions in terms of r and θ, you have

$$d(\mathbf{S}, \mathbf{T}) = \sqrt{(r_2 \cos \theta_2 - r_1 \cos \theta_1)^2 + (r_2 \sin \theta_2 - r_1 \sin \theta_1)^2},$$

which (see Exercise 25, page 250) simplifies to

$$d(\mathbf{S}, \mathbf{T}) = \sqrt{r_2^2 + r_1^2 - 2r_2r_1(\cos \theta_2 \cos \theta_1 + \sin \theta_2 \sin \theta_1)}.$$

Since

$$\cos \theta_2 \cos \theta_1 + \sin \theta_2 \sin \theta_1 = \cos (\theta_2 - \theta_1),$$

you have the required formula

■ $$d(\mathbf{S}, \mathbf{T}) = \sqrt{r_2^2 + r_1^2 - 2r_2r_1 \cos (\theta_2 - \theta_1)}.$$

Notice that this formula expresses the familiar Law of Cosines in trigonometry.

Example 2. Find a polar equation for the locus of all points $\mathbf{U}$ in the plane located 2 units from the point $\mathbf{S}(4, 30°)$.

Solution: Make a sketch. Using the distance formula with $(r_2, \theta_2) = (r, \theta)$ and $(r_1, \theta_1) = (4, 30°)$, you have

$$d(\mathbf{S}, \mathbf{U}) = \sqrt{r^2 + 4^2 - 2(r)(4)\cos(\theta - 30°)}.$$

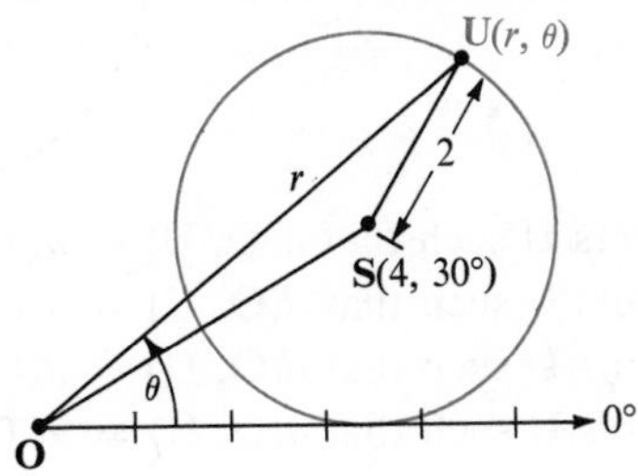

Then, since $d(\mathbf{S}, \mathbf{U}) = 2$, you have

$$\sqrt{r^2 + 16 - 8r\cos(\theta - 30°)} = 2,$$
$$r^2 + 16 - 8r\cos(\theta - 30°) = 4,$$

or

$$r^2 - 8r\cos(\theta - 30°) + 12 = 0.$$

Exercises 7–6

In Exercises 1–8, find a polar equation for the given locus of points $\mathbf{U}(r, \theta)$.

1. The locus of the midpoints **U** of all chords drawn from the pole to points on the circle with equation $r = 2\cos\theta$.

2. The locus of all points **U** equidistant from the pole and the line with Cartesian equation $x = a$.

3. The locus of all points **U** half as far from the pole as they are from the line with Cartesian equation $y = b$.

4. The locus of all points **U** the sum of whose distances from the pole and $\mathbf{S}(4, 0°)$ is 6.

5. The locus of all points **U** the difference of whose distances from the pole and $\mathbf{S}(4, 90°)$ is 2.

6. The locus of the midpoints **U** of all segments of length 3 with one endpoint on the polar axis and the other on the line with equation $\theta = 90°$.

7. The locus of all points **U** such that the directed distance $l(\mathbf{S}, \mathbf{T})$ described in Example 1, page 263, is a constant k, where $k > a$.

8. The locus of all points **U** such that the directed distance $l(\mathbf{S}, \mathbf{T})$ described in Example 1, page 263, is a constant k, where $k < a$.

Exercises 9–12 refer to the diagram shown at the right. Find the specified loci.

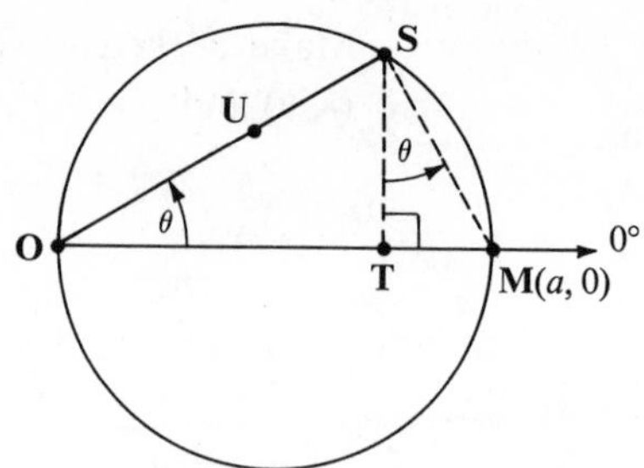

9. The locus of all points **U** such that $d(\mathbf{O}, \mathbf{U}) = d(\mathbf{T}, \mathbf{M})$.
10. The locus of all points **U** such that $d(\mathbf{O}, \mathbf{U}) = d(\mathbf{O}, \mathbf{T}) - d(\mathbf{T}, \mathbf{M})$.
11. The locus of all points **U** such that $d(\mathbf{O}, \mathbf{U}) = d(\mathbf{O}, \mathbf{S}) - d(\mathbf{O}, \mathbf{T})$.
12. The locus of all points **U** such that $d(\mathbf{O}, \mathbf{U}) = d(\mathbf{T}, \mathbf{S})$.

Chapter Summary

1. Points in the plane can be specified by **polar coordinates** as well as by Cartesian coordinates. The polar and Cartesian coordinates of a point are related by the equations

$$x = r\cos\theta \qquad \text{and} \qquad y = r\sin\theta,$$

and also, for $x^2 + y^2 \neq 0$, by the equations

$$r = \pm\sqrt{x^2 + y^2}, \ \cos\theta = \frac{x}{\pm\sqrt{x^2 + y^2}}, \ \sin\theta = \frac{y}{\pm\sqrt{x^2 + y^2}},$$

where the same sign must be chosen in all three equations.

2. The **graph of a polar equation** is the set of all points having at least one pair of polar coordinates that satisfies the equation. You can graph polar equations by plotting points or by using an **auxiliary graph** to deduce the behavior of r over successive intervals of angle measure.

3. Graphs of polar equations may intersect even though the polar equations have no common solution.

4. Parabolas, ellipses, and hyperbolas having a focus at the pole and a line with Cartesian equation of the form $x = \pm p$ or $y = \pm p$ as corresponding directrix have polar equations of the form

$$r = \frac{ep}{1 \pm e\cos\theta} \qquad \text{or} \qquad r = \frac{ep}{1 \pm e\sin\theta},$$

respectively.

5. Loci can be specified by polar equations.

Chapter Review Exercises

1. Find the Cartesian coordinates of the point with polar coordinates $(3, 60°)$.

2. Find a pair of polar coordinates with $r > 0$ and $0 \leq m°(\theta) < 360$ for the point with Cartesian coordinates $\left(\frac{1}{\sqrt{2}}, -\frac{1}{\sqrt{2}}\right)$.

3. Transform the Cartesian equation $y^2 = 4x$ into a polar equation.

4. Transform the polar equation $r = 3 \cos \theta$ into a Cartesian equation.

5. Graph the equation $r = 2 + 2 \cos \theta$.

6. Solve the system

$$r = \cos \theta$$
$$r = \cos^2 \theta - 2$$

over the interval $0 \leq m°(\theta) < 360$ by analytic methods.

7. Find coordinates of all points of intersection of the graphs of the equations $r = \sin \theta$ and $r = 1 + \cos^2 \theta$ over the interval $0 \leq m°(\theta) < 360$.

8. Write a polar equation for the ellipse with eccentricity $\frac{1}{2}$ which has the pole as focus and the line with Cartesian equation $y = -2$ as corresponding directrix.

9. Find a Cartesian equation for a directrix of the hyperbola with equation

$$r = \frac{12}{1 - 2 \sin \theta}.$$

10. Find a polar equation for the locus of all points equidistant from the pole and the line with Cartesian equation $x = -4$.

Affine Transformations

In the essay at the end of Chapter 6, you considered certain simple transformations of the plane. Still other transformations of the plane can be generated by successively applying these simple, or basic, transformations.

Thus if the simple strain T_1 defined by

$$\begin{aligned} x' &= kx \\ y' &= y \end{aligned} \tag{1}$$

is followed by the simple strain T_2 defined by

$$\begin{aligned} x'' &= x' \\ y'' &= ky' \end{aligned} \tag{2}$$

—that is, if points $U(x, y)$ are mapped onto points $U'(x', y')$, where x' and y' are given by Equations (1), and then the points $U'(x', y')$ are mapped onto points $U''(x'', y'')$, where x'' and y'' are given by Equations (2)—then

$$\begin{aligned} x'' &= kx \\ y'' &= ky. \end{aligned}$$

This transformation is called the *product* T of T_1 and T_2 and is designated by $T = T_2T_1$. The notation T_2T_1 means first apply transformation T_1, and then apply transformation T_2 to the result of transformation T_1.

The transformation T_2T_1 above is a *transformation of similitude:* It is a *stretching* if $|k| > 1$, as illustrated at the right, and is a *shrinking* if $0 < |k| < 1$.

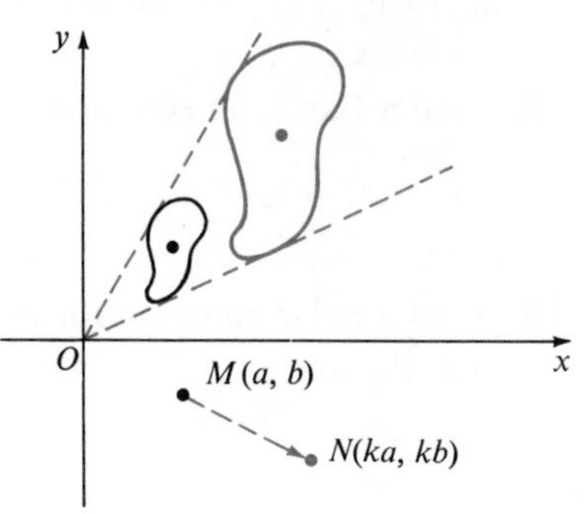

The statement that two transformations of the plane are *equal* means that each point of the plane is mapped onto the same point by both transformations. Thus for the strains T_1 and T_2 defined by Equations (1) and (2) above, you can readily see that $T_2T_1 = T_1T_2$, even though the two products are not formed in the same order.

If a transformation T is obtained by a succession of any number n of the basic transformations of the plane discussed in the essay for Chapter 6 (see pages 238–241), then you say that T has been factored into the following product,

$$T = T_nT_{n-1} \cdots T_2T_1.$$

Every transformation of the plane that can be factored into a product of basic transformations in this way actually can be represented by equations of the form

$$\begin{aligned} x' &= ax + by + h, \\ y' &= cx + dy + k, \end{aligned} \tag{3}$$

where a, b, c, d, h, and k are constants and the *determinant of coefficients*, $\begin{vmatrix} a & b \\ c & d \end{vmatrix}$ is different from 0. A transformation of the form (3), with nonzero determinant of coefficients, is called an *affine transformation.* In particular, each of the basic transformations of the plane is an affine transformation.

Conversely, every affine transformation can be expressed as a product of basic transformations of the sorts previously discussed. (In fact, every rotation can be expressed as a product of basic transformations of the other simple kinds, and accordingly rotations might be omitted from the list of basic transformations.)

It is interesting to note that the commutative law does not, in general, hold for the product of affine transformations. That is, you do not always have $T_1T_2 = T_2T_1$ for affine transformations T_1 and T_2. To see that this is so, let the transformation T_1 be the reflection of the plane in the x-axis, and let T_2 be the rotation of the plane through an angle of measure 90°. You can readily check that T_2T_1 maps the point $S(1, 0)$ onto the point $T(0, 1)$, while T_1T_2 maps S onto the point $R(0, -1)$, and $R \neq T$.

The associative law, $T_3(T_2T_1) = (T_3T_2)T_1$, however, does hold for products of affine transformations, as you can easily see from a geometrical consideration of the transformations.

Every affine transformation T has an *inverse affine transformation* T^{-1} that "undoes" what T "does." It is easy to see that this statement is true for the basic transformations: A translation by the vector amount (h, k) has the translation by the vector amount $(-h, -k)$ as its inverse, and a rotation through an angle θ has the rotation through the angle $-\theta$ as its inverse. A strain having factor k is undone by a strain having factor $\frac{1}{k}$, and a simple shear having factor k is undone by a simple shear having factor $-k$. (See pages 238–241.)

The product transformation $T^{-1}T$ therefore maps each point of the plane onto itself, and accordingly it is equal to the *identity transformation* I. For each affine transformation T, you have the algebraic identities

$$T^{-1}T = I = TT^{-1}$$

and

$$IT = T = TI.$$

To show that the product T_2T_1 of any two basic transformations has an inverse, notice that

$$(T_1^{-1}T_2^{-1})(T_2T_1) = T_1^{-1}(T_2^{-1}T_2)T_1 = T_1^{-1}IT_1 = T_1^{-1}T_1 = I.$$

Thus the inverse of T_2T_1 is $T_1^{-1}T_2^{-1}$. Similarly, the inverse of $T_3T_2T_1$ is $T_1^{-1}T_2^{-1}T_3^{-1}$. Continuing in this way, you can show by mathematical induction that every product of basic transformations has an inverse, and furthermore that every affine transformation has an affine inverse.

You can also show *directly* that every affine transformation has an affine inverse by solving Equations (3) for x and y in terms of x' and y'. Since the determinant of coefficients is different from 0, you obtain:

$$\begin{aligned} x &= \frac{d}{\begin{vmatrix} a & b \\ c & d \end{vmatrix}} x' - \frac{b}{\begin{vmatrix} a & b \\ c & d \end{vmatrix}} y' - \frac{dh - bk}{\begin{vmatrix} a & b \\ c & d \end{vmatrix}} \\ y &= \frac{-c}{\begin{vmatrix} a & b \\ c & d \end{vmatrix}} x' + \frac{a}{\begin{vmatrix} a & b \\ c & d \end{vmatrix}} y' + \frac{ch - ak}{\begin{vmatrix} a & b \\ c & d \end{vmatrix}} \end{aligned} \tag{4}$$

Since the transformation (3) maps points $U(x, y)$ onto points $U'(x', y')$, and the transformation (4) maps the points U' back onto the original points U, the transformation (4) is the inverse of transformation (3).

The notion of a *group of transformations* is important in many branches of mathematics and also in applications to atomic physics, crystallography, and so on. Any nonempty set $\mathcal{G}$ of transformations of the plane is called a *group of transformations* if it has the following properties.

1. *Closure.* If $T_1 \in \mathcal{G}$ and $T_2 \in \mathcal{G}$, then $T_2T_1 \in \mathcal{G}$.
2. *Associativity.* If $T_1 \in \mathcal{G}$, $T_2 \in \mathcal{G}$, and $T_3 \in \mathcal{G}$, then $T_3(T_2T_1) = (T_3T_2)T_1$.
3. *Inverse.* If $T \in \mathcal{G}$, then T has an inverse $T^{-1} \in \mathcal{G}$.

Notice that since $T^{-1}T \in \mathcal{G}$ and $T^{-1}T = I$, every group $\mathcal{G}$ of transformations of the plane contains the identity transformation I.

The set of all affine transformations of the plane, for example, is a group of transformations. If the identity transformation is included, then the set of all translations, the set of all rotations, the set of all one-dimensional strains in the x-direction, and so on, are groups.

A subset of a given group might also form a group. For example, the set consisting of the identity transformation I together with the rotations $R_{\frac{\pi}{2}}$, R_π, and $R_{\frac{3\pi}{2}}$ of the plane through angles of radian measure $\frac{\pi}{2}$, π, and $\frac{3\pi}{2}$, respectively, is a group when multiplication is defined as shown in the table at the right. Notice that this group is commutative; that is, for all T_1 and all T_2 in this group, you have $T_1T_2 = T_2T_1$.

	I	$R_{\frac{\pi}{2}}$	R_π	$R_{\frac{3\pi}{2}}$
I	I	$R_{\frac{\pi}{2}}$	R_π	$R_{\frac{3\pi}{2}}$
$R_{\frac{\pi}{2}}$	$R_{\frac{\pi}{2}}$	R_π	$R_{\frac{3\pi}{2}}$	I
R_π	R_π	$R_{\frac{3\pi}{2}}$	I	$R_{\frac{\pi}{2}}$
$R_{\frac{3\pi}{2}}$	$R_{\frac{3\pi}{2}}$	I	$R_{\frac{\pi}{2}}$	R_π

One of the most important concepts in mathematics is that of an *invariant*. Geometry, in fact, is sometimes called the study of properties that are invariant under given classes of transformations.

Translations, rotations, and reflections are *rigid-motion transformations;* the size and the shape of plane figures are invariant under these transformations. Transformations of similitude leave shape but not size invariant. Like the familiar Mercator map, all of these mappings leave the measure of angles invariant.

Area is invariant under shear transformations, but neither length nor measure of angles is invariant either under shear transformations or under strain transformations. Are there, then, *any* properties of plane figures that are invariant under *all* affine transformations?

If in the linear equation

$$Ax + By + C = 0, \tag{5}$$

where A and B are not both 0, you substitute the expressions in the right-hand members of Equations (4) for x and y, respectively, you can see that a linear equation in x' and y' results. (You should check that the coefficients of x' and y' cannot both be 0.) Conversely, if in the linear equation in x' and y' you substitute the expressions in the right-hand members of Equations (3), you regain Equation (5). It follows that *the property of being a line is invariant under affine transformations!* You can show similarly that the property of being a circle or an ellipse, the property of being a parabola, and the property of being a hyperbola are all invariant under affine transformations.

Notice that the invariants indicated above are *nonmetric properties*, unlike length, area, and angle measure, which are *metric properties*. *Affine geometry* is the study of properties that are invariant under affine transformations. *Metric geometry*, on the other hand, is the study of properties that are invariant under rigid-motion transformations. *Euclidean geometry* includes both metric geometry and the study of properties that are invariant under similarity transformations.

Chapter 8

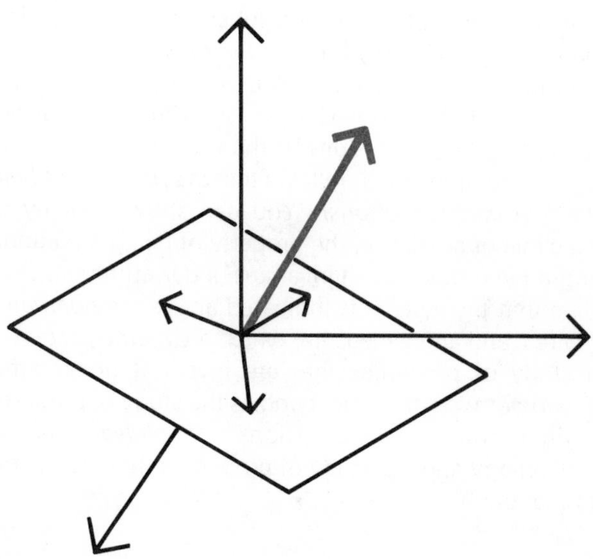

In this chapter, the algebra and geometry of three-dimensional vectors are considered. Direction cosines are introduced, and the cross product of vectors is defined and discussed.

Vectors in Space

Geometry of Ordered Triples

8–1 Cartesian Coordinates; Distance between Points

You are familiar with the definition of the Cartesian product $A \times B$ of sets A and B, namely,

$$A \times B = \{(x, y): x \in A \text{ and } y \in B\}.$$

A similar definition applies to the Cartesian product $A \times B \times C$ of sets A, B, and C:

$$A \times B \times C = \{(x, y, z): x \in A, y \in B, \text{ and } z \in C\}.$$

The symbol (x, y, z) represents an **ordered triple**.

In this chapter, you will be concerned with the Cartesian product

$$\mathcal{R} \times \mathcal{R} \times \mathcal{R} = \mathcal{R}^3,$$

that is,

$$\{(x, y, z): x \in \mathcal{R}, y \in \mathcal{R}, \text{ and } z \in \mathcal{R}\}.$$

As you will see, each ordered triple (x, y, z) of real numbers can be uniquely assigned to a point in space, and each point in space can be uniquely assigned to an ordered triple of real numbers, by means of a rectangular Cartesian, or Cartesian, coordinate system in three dimensions. Thus, there is a one-to-one correspondence between the set of points in space and $\mathcal{R}^3$, and the geometric representation of $\mathcal{R}^3$ is the entire three-dimensional space.

To see one way in which a three-dimensional Cartesian coordinate system can be established, think of your right hand oriented in such a way that your thumb points to the right, your index finger points upward, and your second finger points toward you (Figure 8–1). If, now, you think of a coordinate system in space with a first number line, the x-axis, in the direction of your second

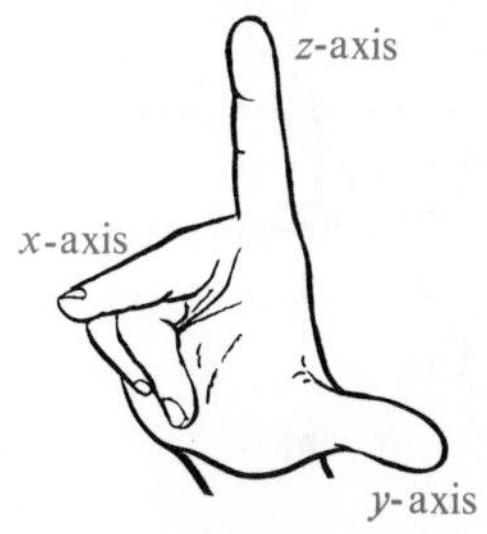

Figure 8–1

finger, a second number line, the y-axis, in the direction of your thumb, and a third number line, the z-axis, in the direction of your index finger, you will have a mental picture of a right-hand Cartesian coordinate system [Figure 8–2(a)]. For a left-hand system [Figure 8–2(b)], just replace your right hand with your left hand and repeat the process (with your thumb pointing to the left and your fingers labeled as before).

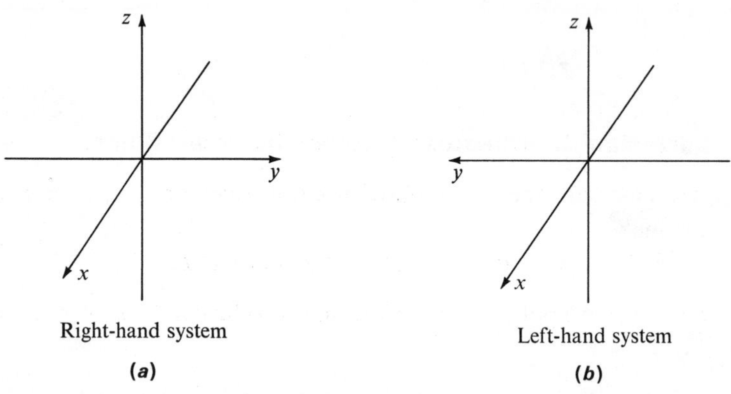

Figure 8–2

Notice that if you imagine an ordinary wood screw with axis along the z-axis, then a twist from the positive x-direction toward the positive y-direction will advance the screw in the positive z-direction in a right-hand system [Figure 8–3(a)], but not in a left-hand system. Again, if you curl the fingers of your right hand from the positive x-direction toward the positive y-direction, then in a right-hand system your thumb will point in the positive z-direction [Figure 8–3(b)]. In this book, only right-hand systems will be used from now on.

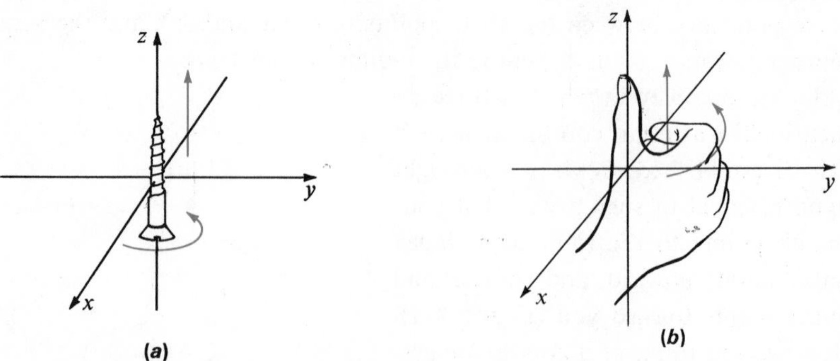

Figure 8–3

More formally, to establish such a three-dimensional Cartesian coordinate system, you select three mutually perpendicular number lines in space intersecting at a point **O**, using the same scale on each of these lines, with zero point at **O**. The point **O** is then called the **origin**, and the lines are called the **coordinate axes**. The first, second, and third coordinate axes are ordinarily associated with the variables x, y, and z, respectively, as described on pages 273 and 274.

The three planes determined by the coordinate axes are called the **coordinate planes**. In sketching the coordinate system, the plane containing the x- and y-axes (the xy-plane) is ordinarily thought of as being horizontal. The plane containing the x- and z-axes (the xz-plane) and the plane containing the y- and z-axes (the yz-plane) are then visualized as being in vertical orientation. This coordinate system is usually drawn as if you were looking toward the origin, with the positive x-axis pointed toward you but somewhat to your left, the positive y-axis pointed toward your right, and the positive z-axis pointed upward, so that the negative portions of the three axes are on the opposite sides of the coordinate planes from you. To indicate this, the negative portions of the axes are frequently shown as dashed lines (Figure 8–4).

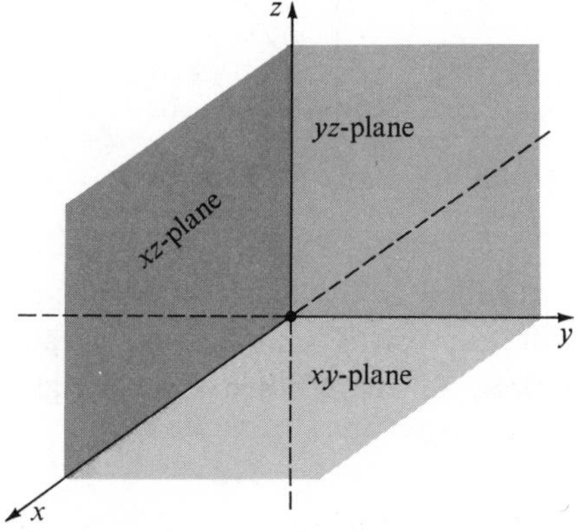

Figure 8–4

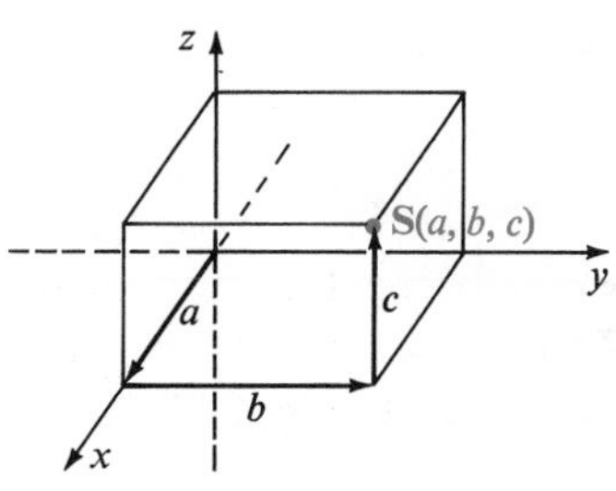

Figure 8–5

In a three-dimensional Cartesian coordinate system, a point is located by specifying the directed distances from the coordinate planes to the point. Thus, the point **S** in Figure 8–5 has coordinates (a, b, c), where a, b, and c are the directed distances to the point from the yz-, xz-, and xy-planes, respectively. You can see from Figure 8–5 that a point in space uniquely determines its coordinates, and, conversely, that an ordered triple of real numbers uniquely determines a point in space. Thus two ordered triples (a, b, c) and (x, y, z) correspond to the same point if and only if they are equal, that is, if and only if $a = x$, $b = y$, and $c = z$.

The three coordinate planes separate space into eight parts called **octants.** In general, only the octant to the front upper right is assigned a number (I). The other octants can be identified by the signs of the components of the ordered triples associated with them, such as $(+, -, +)$ or $(-, -, +)$.

The distance between two points in space can be found by applying the (plane) Pythagorean theorem twice. Namely, if $\mathbf{S}(x_1, y_1, z_1)$ and $\mathbf{T}(x_2, y_2, z_2)$ are two points in space (Figure 8–6), then the distance $d(\mathbf{S}, \mathbf{T})$ between these two points is

$$d(\mathbf{S}, \mathbf{T}) = \sqrt{(x_2 - x_1)^2 + (y_2 - y_1)^2 + (z_2 - z_1)^2}\,. \quad (1)$$

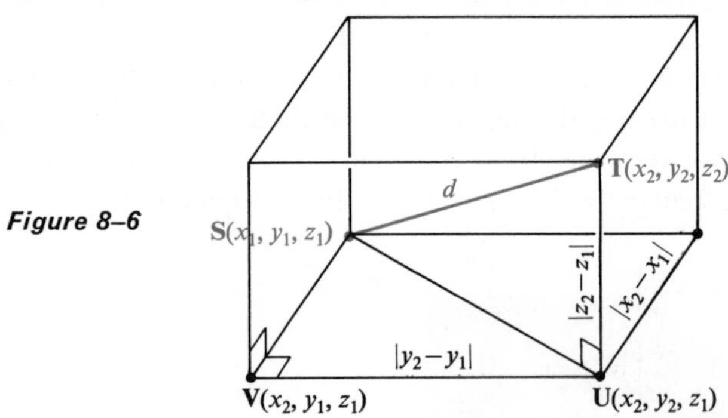

Figure 8–6

To see why this is so, you can proceed as follows. By constructing planes parallel to the coordinate planes through each of **S** and **T**, you locate $\mathbf{U}(x_2, y_2, z_1)$ and $\mathbf{V}(x_2, y_1, z_1)$. The points **S**, **T**, and **U** then form a right triangle. Using the Pythagorean theorem, you have

$$d(\mathbf{S}, \mathbf{T}) = \sqrt{[d(\mathbf{S}, \mathbf{U})]^2 + [d(\mathbf{U}, \mathbf{T})]^2}.$$

Next, using the distance formula, you have

$$\begin{aligned} d(\mathbf{S}, \mathbf{U}) &= \sqrt{[d(\mathbf{S}, \mathbf{V})]^2 + [d(\mathbf{V}, \mathbf{U})]^2} \\ &= \sqrt{(x_2 - x_1)^2 + (y_2 - y_1)^2}. \end{aligned}$$

Since $d(\mathbf{U}, \mathbf{T}) = |z_2 - z_1|$, substitution now yields the stated equation (1).

Example. Find the distance between $\mathbf{S}(2, 4, 1)$ and $\mathbf{T}(6, 7, 13)$.

Solution: By the distance formula (1), above, with $x_2 - x_1 = 6 - 2 = 4$, $y_2 - y_1 = 7 - 4 = 3$, and $z_2 - z_1 = 13 - 1 = 12$, you have

$$d(\mathbf{S}, \mathbf{T}) = \sqrt{4^2 + 3^2 + 12^2} = \sqrt{16 + 9 + 144} = \sqrt{169} = 13.$$

Exercises 8–1

In Exercises 1–8, sketch the graph of the point in space which has the given coordinates.

1. (1, 2, 1)
2. (3, 2, 1)
3. (0, 2, 3)
4. (4, 0, 2)
5. (4, 3, 0)
6. (0, 0, 5)
7. (3, 0, 0)
8. (0, 4, 0)

In Exercises 9–14, state the condition(s) that must be satisfied by the coordinates of every point (x, y, z) lying in the given plane(s).

9. xy-plane
10. yz-plane
11. xz-plane
12. xy- and yz-planes
13. xy- and xz-planes
14. yz- and xz-planes

In Exercises 15–20, find values for x, y, and z so that the two ordered triples are equal.

Example. $(x + y, x - y, z), (3, 1, 4)$

Solution: If the two ordered triples are equal, then

$$x + y = 3, \quad x - y = 1, \quad \text{and} \quad z = 4.$$

You can find values for x and y by solving the first two of these equations simultaneously. The required values are

$$x = 2, \quad y = 1, \quad \text{and} \quad z = 4.$$

15. $(3, y, z + 1), (3, 4, 5)$
16. $(x, 5, z - 2), (4, 5, 2)$
17. $(x + 1, y - 3, z + 4), (-2, 3, 6)$
18. $(x - 2, y + 2, z - 4), (0, 0, 0)$
19. $(x + 4, 3, z + 5), (4, y + 3, 2)$
20. $(1, y - 3, z + 2), (x - 5, 4, 6)$

In Exercises 21–28, find the distance between the points **S** and **T**.

21. $\mathbf{S}(1, 1, 2), \mathbf{T}(2, 3, 4)$
22. $\mathbf{S}(-1, 1, 3), \mathbf{T}(0, -1, 1)$
23. $\mathbf{S}(2, -1, 5), \mathbf{T}(0, 2, -1)$
24. $\mathbf{S}(3, 1, -3), \mathbf{T}(5, 4, 3)$
25. $\mathbf{S}(2, 1, 0), \mathbf{T}(3, 5, 8)$
26. $\mathbf{S}(-2, 1, -3), \mathbf{T}(2, 2, 5)$
27. $\mathbf{S}(4, 5, 7), \mathbf{T}(6, 6, 5)$
28. $\mathbf{S}(5, 3, -6), \mathbf{T}(2, -1, 6)$

29. Use the distance formula and the converse of the Pythagorean theorem to show that the points $\mathbf{S}(3, 5, 2)$, $\mathbf{T}(2, 3, -1)$, and $\mathbf{U}(6, 1, -1)$ are the vertices of a right triangle.

30. Show that the points **S**(6, 3, 4), **T**(2, 1, −2), and **U**(4, −1, 10) are the vertices of an isosceles triangle.

31. Show that the points **S**(2, −1, 3), **T**(4, 2, 1), and **U**(−2, −7, 7) lie on the same line.

32. Show that the midpoint of the segment with endpoints $\mathbf{S}_1(x_1, y_1, z_1)$ and $\mathbf{S}_2(x_2, y_2, z_2)$ is the point $\mathbf{S_m}\left(\frac{x_1 + x_2}{2}, \frac{y_1 + y_2}{2}, \frac{z_1 + z_2}{2}\right)$.

33. Find an equation whose graph is the set of all points located a distance of 3 units from **S**(2, 3, 4).

34. Find an equation whose graph is the set of all points equidistant from **S**(−1, 4, 0) and **T**(2, −1, 1).

* **35.** Find an equation whose graph is the set of all points the sum of whose distances from **S**(0, 4, 0) and **T**(0, −4, 0) is 10 units.

* **36.** Show that an equation of the sphere with radius 3 and center **S**(1, −2, 2) is $x^2 + y^2 + z^2 - 2x + 4y - 4z = 0$.

* **37.** Show that an equation of the sphere with radius r and center $\mathbf{S}(x_1, y_1, z_1)$ is $(x - x_1)^2 + (y - y_1)^2 + (z - z_1)^2 = r^2$.

* **38.** By completing the squares in x, y, and z, find the radius and the coordinates of the center of the sphere with equation

$$x^2 + y^2 + z^2 - 2z + 8y - 8 = 0.$$

8–2 Vectors in Space

Each ordered triple (v_1, v_2, v_3) of real numbers can be associated with a translation in space, just as each ordered pair of real numbers can be associated with a translation in the plane (Figure 8–7). Accordingly, an ordered triple of real numbers is defined to be a (three-dimensional) **vector**.

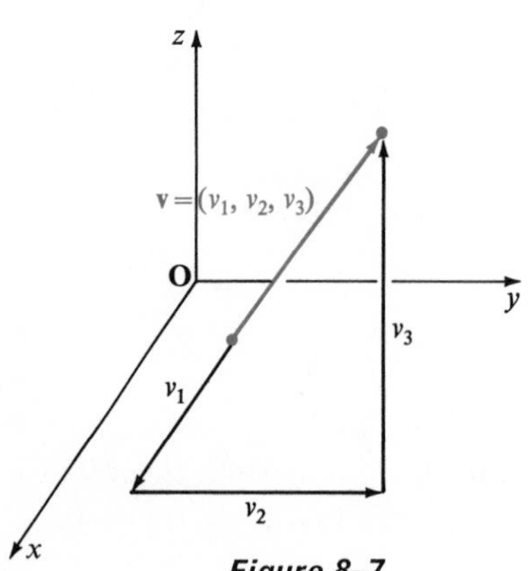

Figure 8–7

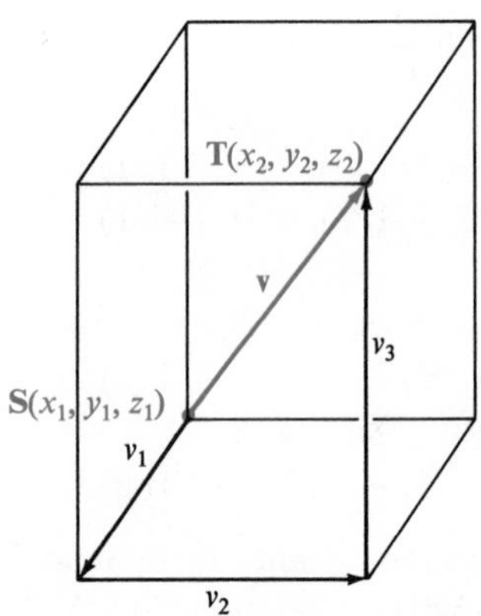

Figure 8–8

Figure 8–8 shows a directed line segment, or **geometric vector**, portraying the vector $\mathbf{v} = (v_1, v_2, v_3)$. This geometric vector represents a translation from the point $\mathbf{S}(x_1, y_1, z_1)$ to the point $\mathbf{T}(x_2, y_2, z_2)$. From the figure, you can see that $v_1 = x_2 - x_1, v_2 = y_2 - y_1$, and $v_3 = z_2 - z_1$. Thus, the geometric vector from the point with coordinates (x_1, y_1, z_1) to the point with coordinates (x_2, y_2, z_2) is a geometric representation of the vector

$$\mathbf{v} = (x_2 - x_1, y_2 - y_1, z_2 - z_1).$$

The point **S** in Figure 8–8 is said to be the **initial point** of the geometric vector, and **T** is said to be its **terminal point.** If the initial point of a geometric vector is the origin $\mathbf{O}(0, 0, 0)$, then the geometric vector is said to be in **standard position** and to be the **standard representation** of the corresponding vector.

The norm $\|\mathbf{v}\|$ of a vector $\mathbf{v} = (v_1, v_2, v_3)$ in $\mathcal{R}^3$ is defined by

$$\|\mathbf{v}\| = \sqrt{v_1^2 + v_2^2 + v_3^2}.$$

As in $\mathcal{R}^2$, the norm of a vector in $\mathcal{R}^3$ can be interpreted as the length of any one of its geometric representations. Thus, the norm of the vector $\mathbf{v} = (v_1, v_2, v_3)$ pictured in Figure 8–8 is equal to the length of $\overrightarrow{\mathbf{ST}}$, that is,

$$\sqrt{(x_2 - x_1)^2 + (y_2 - y_1)^2 + (z_2 - z_1)^2},$$

or

$$\sqrt{v_1^2 + v_2^2 + v_3^2}.$$

Other definitions applying to two-dimensional vectors also extend directly to three-dimensional vectors. In particular, if $\mathbf{v} = (v_1, v_2, v_3)$ and $\mathbf{u} = (u_1, u_2, u_3)$ are three-dimensional vectors, then:

$$\mathbf{v} + \mathbf{u} = (v_1 + u_1, v_2 + u_2, v_3 + u_3) \quad (1)$$

$$-\mathbf{v} = -(v_1, v_2, v_3) = (-v_1, -v_2, -v_3) \quad (2)$$

$$\mathbf{v} - \mathbf{u} = \mathbf{v} + (-\mathbf{u}) = (v_1 - u_1, v_2 - u_2, v_3 - u_3) \quad (3)$$

$$\mathbf{v} - \mathbf{v} = \mathbf{v} + (-\mathbf{v}) = (0, 0, 0) = \mathbf{0} \quad (4)$$

$\mathbf{v}$ is a unit vector if and only if $\|\mathbf{v}\| = 1$. (5)

If r is a scalar, then $r\mathbf{v} = r(v_1, v_2, v_3) = (rv_1, rv_2, rv_3)$. (6)

$$\mathbf{v} \cdot \mathbf{u} = v_1u_1 + v_2u_2 + v_3u_3 \quad (7)$$

Example 1. Show that if $\mathbf{v} = (v_1, v_2, v_3)$, then $\mathbf{v} \cdot \mathbf{v} = \|\mathbf{v}\|^2$.

Solution: From definition (7) above, you have

$$\mathbf{v} \cdot \mathbf{v} = (v_1, v_2, v_3) \cdot (v_1, v_2, v_3) = v_1^2 + v_2^2 + v_3^2.$$

Since $\|\mathbf{v}\|^2 = (\sqrt{v_1^2 + v_2^2 + v_3^2})^2 = v_1^2 + v_2^2 + v_3^2$, it follows that

$$\mathbf{v} \cdot \mathbf{v} = \|\mathbf{v}\|^2.$$

As in $\mathcal{R}^2$, a vector in $\mathcal{R}^3$ can be expressed as the sum of vector components parallel to the coordinate axes. Recall from Section 1–8 that **i** and **j** are unit vectors in the directions of the positive rays of the x- and y-axes, respectively, and that any vector can be written in one and only one way as a linear combination of **i** and **j**. For example, $\mathbf{v} = (2, 3)$ can be written as $\mathbf{v} = 2\mathbf{i} + 3\mathbf{j}$. In $\mathcal{R}^3$, **i** and **j** also represent unit vectors in the directions of the positive rays of the x- and y-axes, respectively, and **k** is defined to be the unit vector in the direction of the positive ray of the z-axis. Thus,

$$\mathbf{i} = (1, 0, 0), \mathbf{j} = (0, 1, 0), \mathbf{k} = (0, 0, 1).$$

Any vector in $\mathcal{R}^3$ can be written in one and only one way as a linear combination of **i**, **j**, and **k**. For example, for the vector $\mathbf{v} = (2, 3, 4)$, you have $\mathbf{v} = 2\mathbf{i} + 3\mathbf{j} + 4\mathbf{k}$ (see Figure 8–9).

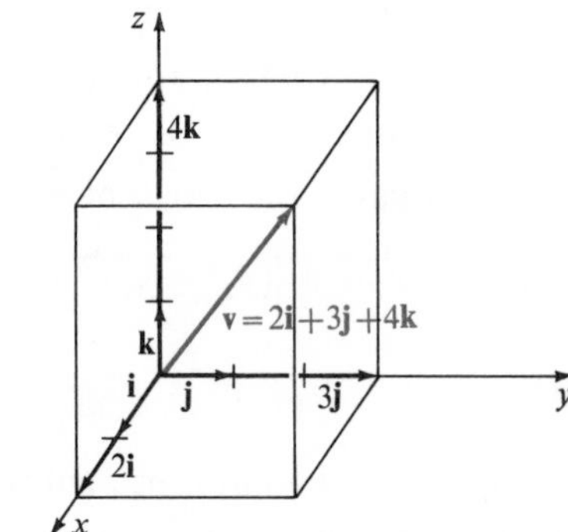

Figure 8–9

Exercises 8–2

In Exercises 1–8, the coordinates of two points, **S** and **T**, are given. Express the vector **v** representing a translation from **S** to **T** both (a) as an ordered triple and (b) in the form $a\mathbf{i} + b\mathbf{j} + c\mathbf{k}$. Then (c) find $\|\mathbf{v}\|$.

Example. S(3, 1, 5), T(−1, 4, 2)

Solution: **(a)** $\mathbf{v} = (-1 - 3, 4 - 1, 2 - 5) = (-4, 3, -3)$
(b) $\mathbf{v} = -4\mathbf{i} + 3\mathbf{j} - 3\mathbf{k}$
(c) $\|\mathbf{v}\| = \sqrt{(-4)^2 + 3^2 + (-3)^2} = \sqrt{16 + 9 + 9} = \sqrt{34}$

1. S(2, −1, 5), T(4, 3, 3)
2. S(2, 1, 5), T(−3, 2, −1)
3. S(0, 4, 5), T(5, −1, 0)
4. S(3, 0, −2), T(0, 4, 6)
5. S(−3, −3, −3), T(3, 3, 3)
6. S(0, 0, 3), T(−3, 0, 0)
7. S(5, −1, 0), T(0, 0, −2)
8. S(8, −6, −2), T(5, 1, −1)

In Exercises 9–16, $\mathbf{u} = (3, -1, 2)$ and $\mathbf{w} = (1, 7, -6)$. Express the vector $\mathbf{v}$ in the given equation as an ordered triple.

Example. $\mathbf{v} = \mathbf{u} - 2\mathbf{w}$

Solution: You have $2\mathbf{w} = 2(1, 7, -6) = (2, 14, -12)$. Then

$$\begin{aligned} \mathbf{v} &= (3, -1, 2) - (2, 14, -12) \\ &= (3 - 2, -1 - 14, 2 + 12) \\ &= (1, -15, 14). \end{aligned}$$

9. $\mathbf{v} = -3\mathbf{u}$

10. $\mathbf{v} = \frac{2}{3}\mathbf{w}$

11. $\mathbf{v} = \frac{1}{2}(\mathbf{u} + \mathbf{w})$

12. $\mathbf{v} = 2\mathbf{w} - \mathbf{u}$

13. $\mathbf{u} = \mathbf{v} + \mathbf{w}$

14. $\mathbf{w} = \mathbf{u} - 2\mathbf{v}$

15. $\mathbf{u} - \mathbf{v} = \mathbf{v} + 2\mathbf{w}$

16. $2(\mathbf{u} + \mathbf{v}) = 3(\mathbf{v} + \mathbf{u} + \mathbf{w})$

In Exercises 17–24, find $\mathbf{u} \cdot \mathbf{v}$ for the given vectors $\mathbf{u}$ and $\mathbf{v}$.

17. $\mathbf{u} = (-3, 2, 0), \mathbf{v} = (1, 1, 1)$

18. $\mathbf{u} = (1, -4, 5), \mathbf{v} = (-5, -2, 8)$

19. $\mathbf{u} = (-2, 3, -6), \mathbf{v} = (0, 4, -9)$

20. $\mathbf{u} = (7, 0, -3), \mathbf{v} = (-1, 6, 9)$

21. $\mathbf{u} = (-3, -4, -5), \mathbf{v} = (-6, 5, 7)$

22. $\mathbf{u} = (7, 2, -9), \mathbf{v} = (3, 1, 9)$

23. $\mathbf{u} = (9, 0, -2), \mathbf{v} = (\frac{1}{2}, 3, \frac{2}{3})$

24. $\mathbf{u} = (-\frac{1}{2}, 0, \frac{2}{3}), \mathbf{v} = (5, \frac{3}{4}, -2)$

In Exercises 25–35, let $\mathbf{u} = (u_1, u_2, u_3)$, $\mathbf{v} = (v_1, v_2, v_3)$, $\mathbf{w} = (w_1, w_2, w_3)$ be vectors in $\mathcal{R}^3$, and let r and s be scalars. Prove each statement.

* **25.** $\mathbf{u} + \mathbf{v} = \mathbf{v} + \mathbf{u}$

* **26.** $(\mathbf{u} + \mathbf{v}) + \mathbf{w} = \mathbf{u} + (\mathbf{v} + \mathbf{w})$

* **27.** $r(s\mathbf{v}) = (rs)\mathbf{v}$

* **28.** $r\mathbf{v} = \mathbf{0}$ if and only if either $r = 0$ or $\mathbf{v} = \mathbf{0}$

* **29.** $r(\mathbf{u} + \mathbf{v}) = r\mathbf{u} + r\mathbf{v}$

* **30.** $(r + s)\mathbf{u} = r\mathbf{u} + s\mathbf{u}$

* **31.** $\|r\mathbf{u}\| = |r|\,\|\mathbf{u}\|$

* **32.** $\mathbf{u} \cdot \mathbf{v} = \mathbf{v} \cdot \mathbf{u}$

* **33.** $r(\mathbf{u} \cdot \mathbf{v}) = (r\mathbf{u}) \cdot \mathbf{v}$

* **34.** $\mathbf{u} \cdot (\mathbf{v} + \mathbf{w}) = \mathbf{u} \cdot \mathbf{v} + \mathbf{u} \cdot \mathbf{w}$

* **35.** $\|\mathbf{u} + \mathbf{v}\|^2 = \|\mathbf{u}\|^2 + 2\mathbf{u} \cdot \mathbf{v} + \|\mathbf{v}\|^2$

Properties of Vectors in Space

8–3 Direction Cosines

As shown in Figure 8–10, the standard geometric representations of the vectors $(1, -2, 3)$ and $(2, -4, 6)$ lie on the same ray from the origin because $(2, -4, 6) = 2(1, -2, 3)$. Accordingly, you say that the vectors $(1, -2, 3)$ and $(2, -4, 6)$ have the *same* direction.

Figure 8–10

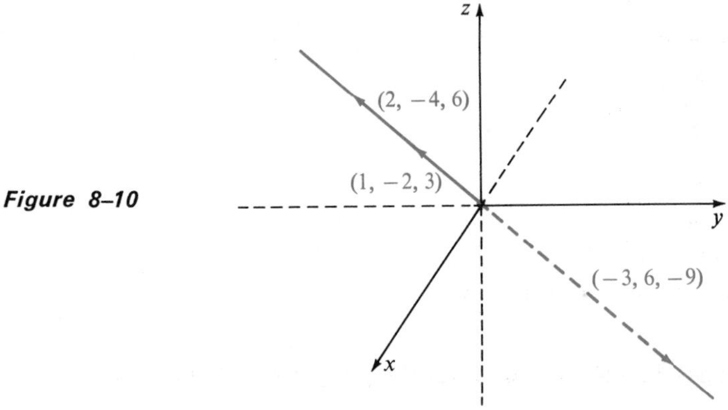

The standard representation of the vector $(-3, 6, -9)$, on the other hand, lies on the same line as the standard representation of $(1, -2, 3)$, since $(-3, 6, -9) = -3(1, -2, 3)$, but extends in the opposite direction from the origin. Accordingly, you say that the vectors $(1, -2, 3)$ and $(-3, 6, -9)$ have *opposite* directions.

In general, if $\mathbf{v} \in \mathcal{R}^3$ and k is a scalar, then the vectors $\mathbf{v}$ and $k\mathbf{v}$ are said to have the **same direction** if $k \geq 0$ and **opposite directions** if $k < 0$. In either case, $\mathbf{v}$ and $k\mathbf{v}$ are said to be **parallel vectors.**

You have seen that the direction of a nonzero two-dimensional vector is determined by the measure $m°(\theta)$, $0 \leq m°(\theta) < 360$, of the angle from the positive ray of the x-axis to the standard geometric representation of the vector. The direction of a nonzero three-dimensional vector is determined by three **direction angles**, one between each of the positive rays of the coordinate axes and the standard geometric representation of the vector. These three direction angles are usually named by Greek letters as follows: α is the direction angle with respect to the positive x-axis; β is the direction angle with respect to the positive y-axis; and γ is the direction angle with respect to the positive z-axis. Figure 8–11 illustrates the situation for the vector $\mathbf{v} = (v_1, v_2, v_3)$. The direction angles α, β, and γ are chosen so that

$$\begin{aligned} 0 \leq m°(\alpha) \leq 180, \\ 0 \leq m°(\beta) \leq 180, \\ 0 \leq m°(\gamma) \leq 180. \end{aligned} \tag{1}$$

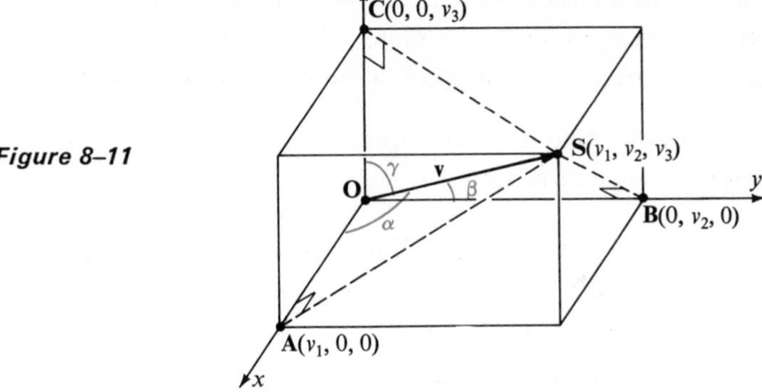

Figure 8–11

The cosines of the direction angles of a vector are called the **direction cosines** of the vector. In Figure 8–11, angles **OAS**, **OBS**, and **OCS** are right angles. Accordingly, the direction cosines of $\mathbf{v}$ are given by

$$\cos\alpha = \frac{v_1}{\|\mathbf{v}\|}, \cos\beta = \frac{v_2}{\|\mathbf{v}\|}, \cos\gamma = \frac{v_3}{\|\mathbf{v}\|}, \tag{2}$$

where $\|\mathbf{v}\| = \sqrt{v_1^2 + v_2^2 + v_3^2} \neq 0$. Since α, β, and γ are restricted to satisfy conditions (1), page 282, the direction angles are uniquely determined by the direction cosines. For example, if $\mathbf{v} = (1, -1, 0)$, then

$$\cos\alpha = \frac{1}{\sqrt{2}}, \cos\beta = \frac{-1}{\sqrt{2}}, \cos\gamma = 0,$$

and $$m^\circ(\alpha) = 45, m^\circ(\beta) = 135, m^\circ(\gamma) = 90.$$

(Note that the standard geometric representation of $\mathbf{v}$ lies in the xy-plane.)

If $\|\mathbf{v}\| = 0$, that is, if $\mathbf{v}$ is the zero vector, then the expressions $\frac{v_1}{\|\mathbf{v}\|}$, $\frac{v_2}{\|\mathbf{v}\|}$, and $\frac{v_3}{\|\mathbf{v}\|}$ for the direction cosines are meaningless, and the direction angles are undetermined. In this case, we shall agree that any desired direction may be assigned to $\mathbf{v}$. Again (see page 16), this seems geometrically reasonable since the standard geometric representation of the zero vector (the origin) lies on every geometric vector in standard position.

Now consider a relationship among the direction cosines of a nonzero vector $\mathbf{v} = (v_1, v_2, v_3)$. By Equation (2) above you have

$$\cos^2\alpha = \frac{v_1^2}{\|\mathbf{v}\|^2}, \cos^2\beta = \frac{v_2^2}{\|\mathbf{v}\|^2}, \text{ and } \cos^2\gamma = \frac{v_3^2}{\|\mathbf{v}\|^2}.$$

Hence, $\cos^2\alpha + \cos^2\beta + \cos^2\gamma = \dfrac{v_1^2 + v_2^2 + v_3^2}{\|\mathbf{v}\|^2} = \dfrac{v_1^2 + v_2^2 + v_3^2}{v_1^2 + v_2^2 + v_3^2}$, or

$$\cos^2\alpha + \cos^2\beta + \cos^2\gamma = 1. \tag{3}$$

Thus, the direction cosines of a vector are not independent; if you know any two of them, you can find the absolute value of the third. If $\cos\alpha$, $\cos\beta$, and $\cos\gamma$ are the direction cosines of a nonzero vector $\mathbf{v} = (v_1, v_2, v_3)$, then Equation (3) implies that $\mathbf{u} = (\cos\alpha, \cos\beta, \cos\gamma) = \frac{1}{\|\mathbf{v}\|}(v_1, v_2, v_3)$ is the unit vector with the same direction as $\mathbf{v}$.

Example 1. Verify that the sum of the squares of the direction cosines of the vector $\mathbf{v} = (1, -2, 3)$ equals 1, and find a unit vector with the same direction as $\mathbf{v}$.

Solution: You first compute $\|\mathbf{v}\|$:

$$\|\mathbf{v}\| = \sqrt{1^2 + (-2)^2 + 3^2} = \sqrt{14}\,.$$

Then

$$\cos\alpha = \frac{1}{\sqrt{14}},\ \cos\beta = \frac{-2}{\sqrt{14}},\ \cos\gamma = \frac{3}{\sqrt{14}}, \tag{4}$$

and

$$\cos^2\alpha = \frac{1}{14},\ \cos^2\beta = \frac{4}{14},\ \cos^2\gamma = \frac{9}{14}.$$

Then

$$\cos^2\alpha + \cos^2\beta + \cos^2\gamma = \frac{1}{14} + \frac{4}{14} + \frac{9}{14} = \frac{14}{14} = 1.$$

By (4), the required unit vector with the same direction as $\mathbf{v}$ is $\left(\frac{1}{\sqrt{14}}, \frac{-2}{\sqrt{14}}, \frac{3}{\sqrt{14}}\right)$.

If the nonzero vectors $\mathbf{u}$ and $\mathbf{v}$ are parallel, that is, if $\mathbf{u} = k\mathbf{v}$ for some nonzero scalar k, then either $\mathbf{u}$ and $\mathbf{v}$ have the same direction cosines or the direction cosines of $\mathbf{u}$ are the negatives of those of $\mathbf{v}$. Conversely, if the nonzero vectors $\mathbf{u}$ and $\mathbf{v}$ have equal direction cosines or direction cosines that are negatives of each other, then $\mathbf{u}$ is a nonzero scalar multiple of $\mathbf{v}$, that is, $\mathbf{u}$ and $\mathbf{v}$ are parallel.

To see that the first assertion given above is true, you can observe that for $\mathbf{u} = k\mathbf{v} = (kv_1, kv_2, kv_3)$, the direction cosines are given by

$$\cos\alpha = \frac{kv_1}{\|k\mathbf{v}\|} = \frac{kv_1}{|k|\,\|\mathbf{v}\|},\ \cos\beta = \frac{kv_2}{\|k\mathbf{v}\|} = \frac{kv_2}{|k|\,\|\mathbf{v}\|},\ \cos\gamma = \frac{kv_3}{\|k\mathbf{v}\|} = \frac{kv_3}{|k|\,\|\mathbf{v}\|},$$

from which you have

$$\cos\alpha = \pm\frac{v_1}{\|\mathbf{v}\|},\ \cos\beta = \pm\frac{v_2}{\|\mathbf{v}\|},\ \text{and } \cos\gamma = \pm\frac{v_3}{\|\mathbf{v}\|},$$

where the $+$ sign is to be taken throughout if k is positive and the $-$ sign is to be taken throughout if k is negative. But these are just the direction cosines, or the negatives of the direction cosines, of $\mathbf{v}$.

For the converse assertion, suppose that the nonzero vectors $\mathbf{u} = (u_1, u_2, u_3)$ and $\mathbf{v} = (v_1, v_2, v_3)$ have equal direction cosines or direction cosines that are negatives of each other. Then either

$$\frac{u_1}{\|\mathbf{u}\|} = \frac{v_1}{\|\mathbf{v}\|}, \quad \frac{u_2}{\|\mathbf{u}\|} = \frac{v_2}{\|\mathbf{v}\|}, \quad \frac{u_3}{\|\mathbf{u}\|} = \frac{v_3}{\|\mathbf{v}\|} \tag{5}$$

or

$$\frac{u_1}{\|\mathbf{u}\|} = -\frac{v_1}{\|\mathbf{v}\|}, \quad \frac{u_2}{\|\mathbf{u}\|} = -\frac{v_2}{\|\mathbf{v}\|}, \quad \frac{u_3}{\|\mathbf{u}\|} = -\frac{v_3}{\|\mathbf{v}\|}. \tag{6}$$

In the latter case, you have

$$u_1 = -\frac{\|\mathbf{u}\|}{\|\mathbf{v}\|}v_1, \quad u_2 = -\frac{\|\mathbf{u}\|}{\|\mathbf{v}\|}v_2, \quad u_3 = -\frac{\|\mathbf{u}\|}{\|\mathbf{v}\|}v_3,$$

so that

$$(u_1, u_2, u_3) = -\frac{\|\mathbf{u}\|}{\|\mathbf{v}\|}(v_1, v_2, v_3).$$

Since $\|\mathbf{u}\| \neq 0$ and $\|\mathbf{v}\| \neq 0$, $-\dfrac{\|\mathbf{u}\|}{\|\mathbf{v}\|}$ is a nonzero scalar, and hence $\mathbf{u}$ is a nonzero scalar multiple of $\mathbf{v}$.

Similar reasoning establishes the assertion, that $\mathbf{u}$ is a nonzero scalar multiple of $\mathbf{v}$, in the case in which the direction cosines are equal [Equation (5)].

Thus the nonzero vectors $\mathbf{u}$ and $\mathbf{v}$ are parallel if and only if either $\mathbf{u}$ and $\mathbf{v}$ have the same direction cosines or the direction cosines of $\mathbf{u}$ are the negatives of those of $\mathbf{v}$. If the direction cosines are the same, then $\mathbf{u}$ and $\mathbf{v}$ have the same direction; if the direction cosines of $\mathbf{u}$ are the negatives of those of $\mathbf{v}$, then the vectors have opposite directions.

Example 2. Find the vector $\mathbf{u}$ if $\|\mathbf{u}\| = 14$ and $\mathbf{u}$ has the direction opposite to that of the vector $\mathbf{v} = (2, 5, -3)$.

Solution: Direction cosines for $\mathbf{v}$ are

$$\cos\alpha = \frac{2}{\sqrt{2^2 + 5^2 + (-3)^2}} = \frac{2}{\sqrt{38}}, \quad \cos\beta = \frac{5}{\sqrt{38}},$$

and

$$\cos\gamma = \frac{-3}{\sqrt{38}}.$$

Hence, a unit vector in the direction opposite to that of $\mathbf{v}$ is $\left(\dfrac{-2}{\sqrt{38}}, \dfrac{-5}{\sqrt{38}}, \dfrac{3}{\sqrt{38}}\right)$. Since $\|\mathbf{u}\| = 14$, it follows that

$$\mathbf{u} = 14\left(\frac{-2}{\sqrt{38}}, \frac{-5}{\sqrt{38}}, \frac{3}{\sqrt{38}}\right) = \left(\frac{-28}{\sqrt{38}}, \frac{-70}{\sqrt{38}}, \frac{42}{\sqrt{38}}\right).$$

Exercises 8–3

In Exercises 1–8, find the direction cosines of the given vector.

1. $(-1, 2, 2)$
2. $(0, 3, -4)$
3. $(5, 0, 12)$
4. $(4, -4, 2)$
5. $(-3, -5, \sqrt{2})$
6. $(3, 2, 6)$
7. $(12, -3, 4)$
8. $(-8, -4, 1)$

In Exercises 9–14, find the unit vector in the direction of the vector with geometric representation from **S** to **T**.

9. $\mathbf{S}(1, -2, 5), \mathbf{T}(4, 0, 11)$
10. $\mathbf{S}(-3, 1, 0), \mathbf{T}(-2, 3, 2)$
11. $\mathbf{S}(2, -2, -1), \mathbf{T}(-4, -5, 1)$
12. $\mathbf{S}(9, 2, -1), \mathbf{T}(-3, 5, -5)$
13. $\mathbf{S}(10, 9, -2), \mathbf{T}(3, 4, -3)$
14. $\mathbf{S}(-3, -1, -2), \mathbf{T}(1, 3, -2)$

15. If, for a vector $\mathbf{v}$, $\cos \alpha = \dfrac{2}{11}$ and $\cos \beta = \dfrac{-6}{11}$, find $\cos \gamma$. (*Hint:* There are two correct answers.)
16. If, for a vector $\mathbf{v}$, $\cos \beta = \dfrac{3}{10}$ and $\cos \gamma = \dfrac{2}{5}$, find $\cos \alpha$.

In Exercises 17–22, find the vector **u** with the given norm and the same direction as the given vector **v**.

17. $\|\mathbf{u}\| = 8$; $\mathbf{v} = (1, 2, 5)$
18. $\|\mathbf{u}\| = 6$; $\mathbf{v} = (3, 3, 3)$
19. $\|\mathbf{u}\| = \frac{1}{2}$; $\mathbf{v} = (6, 12, 4)$
20. $\|\mathbf{u}\| = \frac{1}{4}$; $\mathbf{v} = (3, 0, 4)$
21. $\|\mathbf{u}\| = 14$; $\mathbf{v} = (6, 4, -2)$
22. $\|\mathbf{u}\| = 7\sqrt{2}$; $\mathbf{v} = (-3, 5, -4)$

23–28. Repeat Exercises 17–22 with **u** and **v** having opposite directions instead of the same direction.

In Exercises 29 and 30, state which of the given vectors are parallel.

29. $(3, 0, 2), (3, 1, 2), (-6, 0, -4)$
30. $(5, 2, 7), (3, 2, -1), (-3, -2, 1)$

In Exercises 31–34, α, β, and γ are direction angles of a vector.

31. Find $m°(\alpha)$ if $m°(\beta) = m°(\gamma) = 90$.
32. Find $m°(\alpha)$ if $m°(\beta) = m°(\gamma) = 45$.
33. If $m°(\alpha) = 120$ and $m°(\gamma) = 45$, find $m°(\beta)$.
34. If $m°(\alpha) = 135$ and $m°(\beta) = 60$, find $m°(\gamma)$.

* **35.** Show that if a vector $\mathbf{v}$ is parallel to the x-axis, then the direction cosines of $\mathbf{v}$ are either $(1, 0, 0)$ or $(-1, 0, 0)$.

* **36.** Show that if a vector $\mathbf{v}$ is parallel to a line in the yz-plane, then the direction cosines of $\mathbf{v}$ are of the form $(0, b, \pm\sqrt{1 - b^2})$.

8–4 Parallel and Perpendicular Vectors

If **u** and **v** are two nonzero vectors in $\mathcal{R}^3$, then the angle between them can be specified in exactly the same manner as the angle between two vectors in $\mathcal{R}^2$ (page 31). Namely, θ is the angle, $0 \leq m^\circ(\theta) \leq 180$, between the geometric vectors in standard position representing **u** and **v** [Figure 8–12(a)].

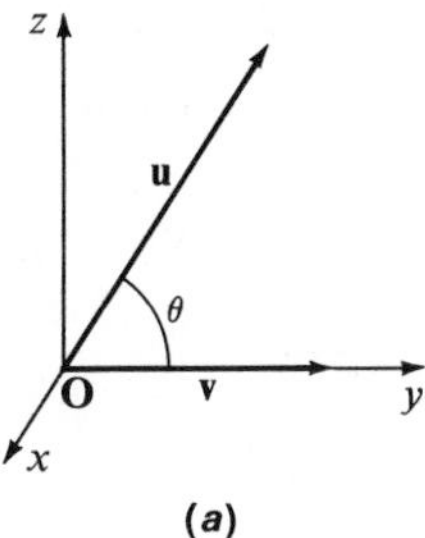

(a) (b)

Figure 8–12

You can see from Figure 8–12(b) that in case the given vectors **u** and **v** are not parallel, then the three vectors **u**, **v**, and **u** − **v** have geometric representations that form a triangle. By using the Law of Cosines, you can show by an argument directly paralleling that on page 31 (Exercise 30, page 292) that

$$\cos \theta = \frac{\mathbf{u} \cdot \mathbf{v}}{\|\mathbf{u}\| \, \|\mathbf{v}\|}. \tag{1}$$

Example 1. Find the cosine of the angle between the vectors $\mathbf{u} = (1, 2, 3)$ and $\mathbf{v} = (2, -1, 2)$.

Solution: From Equation (1), you have

$$\cos \theta = \frac{\mathbf{u} \cdot \mathbf{v}}{\|\mathbf{u}\| \, \|\mathbf{v}\|} = \frac{(1, 2, 3) \cdot (2, -1, 2)}{\sqrt{1^2 + 2^2 + 3^2}\,\sqrt{2^2 + (-1)^2 + 2^2}}$$

$$= \frac{2 - 2 + 6}{\sqrt{14}\,\sqrt{9}}$$

$$= \frac{6}{3\sqrt{14}} = \frac{2}{\sqrt{14}}.$$

Formula (1) holds also if the nonzero vectors **u** and **v** are parallel, since with $\mathbf{u} = k\mathbf{v}$ $(k \neq 0)$, either both members have the value 1 $(k > 0)$ or both members have the value -1 $(k < 0)$. This formula, however, does not hold if either of **u** and **v** is the zero vector; but in that case any value between -1 and 1, inclusive, may be assigned to $\cos \theta$ since any direction may be assigned to the zero vector.

Vectors **u** and **v** are parallel if and only if $m^\circ(\theta) = 0$ or 180, that is, if and only if $\cos\theta = \pm 1$. Thus, you can use Equation (1), page 287, as well as the scalar-multiple or the direction-cosine test (pages 282 and 285) to decide whether or not two nonzero vectors are parallel.

Example 2. Are $\mathbf{u} = (3, -1, 1)$ and $\mathbf{v} = (6, -2, 3)$ parallel vectors?

Solution: *Method 1.* Since the components of **u** and **v** are not proportional, that is, since $6 = 2(3)$, $-2 = 2(-1)$, but $3 \neq 2(1)$, the vectors are not parallel.

Method 2. You have

$$\|\mathbf{u}\| = \sqrt{3^2 + (-1)^2 + 1^2} = \sqrt{11}$$

and

$$\|\mathbf{v}\| = \sqrt{6^2 + (-2)^2 + 3^2} = \sqrt{49} = 7.$$

Let $\alpha_1, \beta_1, \gamma_1$, and $\alpha_2, \beta_2, \gamma_2$ be the direction angles of **u** and **v**, respectively. Then $\cos\alpha_1 = \frac{3}{\sqrt{11}}$ and $\cos\alpha_2 = \frac{6}{7} \neq \pm\frac{3}{\sqrt{11}}$. Thus **u** and **v** are not parallel.

Method 3. From Equation (1),

$$\begin{aligned}\cos\theta &= \frac{\mathbf{u}\cdot\mathbf{v}}{\|\mathbf{u}\|\,\|\mathbf{v}\|}\\ &= \frac{(3, -1, 1)\cdot(6, -2, 3)}{\sqrt{11}(7)}\\ &= \frac{18 + 2 + 3}{7\sqrt{11}} = \frac{23}{7\sqrt{11}}.\end{aligned}$$

Since $\cos\theta$ is neither 1 nor -1, the vectors are not parallel.

Two nonzero vectors **u** and **v** in $\mathcal{R}^3$ are perpendicular, or orthogonal, if and only if the measure of the angle θ between them is 90°, that is, if and only if $\cos\theta = 0$. It follows immediately from Equation (1) that nonzero vectors **u** and **v** in $\mathcal{R}^3$ are perpendicular if and only if $\mathbf{u}\cdot\mathbf{v} = 0$. We extend this notion to zero vectors and say that vectors **u** and **v** are perpendicular, or orthogonal, whenever $\mathbf{u}\cdot\mathbf{v} = 0$, whether or not **u** or **v** is the zero vector.

In Section 1–6, you saw that in $\mathcal{R}^2$ the vector $\mathbf{v_p} = (-v_2, v_1)$ is perpendicular to $\mathbf{v} = (v_1, v_2)$. If $\mathbf{v} \neq \mathbf{0}$, then every vector in $\mathcal{R}^2$ that is perpendicular to **v** must have either the same direction as $\mathbf{v_p}$ or the opposite direction; that is, every such vector must be parallel to (and, hence, a scalar multiple of) $\mathbf{v_p}$. (See Figure 8–13.) Hence all vectors perpendicular to $\mathbf{v}$ are parallel.

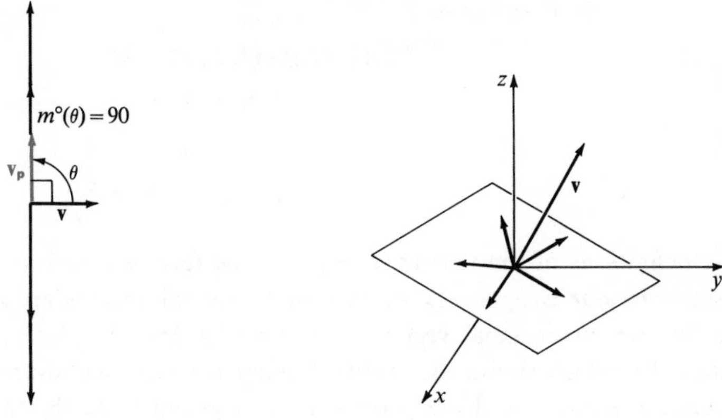

Figure 8–13 ***Figure 8–14***

In $\mathcal{R}^3$, the situation is different. Consider the vector $\mathbf{v} = (1, 2, 3)$. The dot product of $\mathbf{v}$ with each of the vectors $(0, -3, 2)$, $(3, 0, -1)$, and $(-2, 1, 0)$ is 0:

$$\begin{aligned}(1, 2, 3) \cdot (0, -3, 2) &= 0 - 6 + 6 = 0,\\ (1, 2, 3) \cdot (3, 0, -1) &= 3 + 0 - 3 = 0,\\ (1, 2, 3) \cdot (-2, 1, 0) &= -2 + 2 + 0 = 0.\end{aligned}$$

Hence $\mathbf{v}$ is perpendicular to all three. However, no two of these three vectors are parallel, since the direction cosines of $(0, -3, 2)$, $(3, 0, -1)$, and $(-2, 1, 0)$ are

$$0, -\frac{3}{\sqrt{13}}, \frac{2}{\sqrt{13}}$$

$$\frac{3}{\sqrt{10}}, 0, -\frac{1}{\sqrt{10}}$$

$$-\frac{2}{\sqrt{5}}, \frac{1}{\sqrt{5}}, 0$$

respectively.

Actually, you can find any number of nonparallel vectors each of which is perpendicular to $\mathbf{v}$, as indicated in Figure 8–14. This suggests that the set of geometric representations of all vectors perpendicular to $\mathbf{v}$ covers an entire plane, a fact which we shall use in the next chapter.

To find a vector $\mathbf{u}$ perpendicular to a given vector $\mathbf{v}$ in $\mathcal{R}^3$, you can assign two components to $\mathbf{u}$ and then use the fact that $\mathbf{u}$ is perpendicular to $\mathbf{v}$ if and only if $\mathbf{u} \cdot \mathbf{v} = 0$ to determine the third component of $\mathbf{u}$.

Example 3. Find a value for z so that $\mathbf{u} = (1, 4, z)$ and $\mathbf{v} = (1, 2, 3)$ are perpendicular.

Solution: The vector $\mathbf{u}$ is perpendicular to $\mathbf{v}$ if and only if $\mathbf{u} \cdot \mathbf{v} = 0$, that is, if and only if

$$\begin{aligned} (1, 4, z) \cdot (1, 2, 3) &= 0, \\ 1 + 8 + 3z &= 0, \\ 3z &= -9, \\ z &= -3. \end{aligned}$$

The definitions of the terms *vector projection* (*vector component*) and *scalar projection* (*scalar component*) for three-dimensional vectors are analogous to those for two-dimensional vectors (see pages 34–38). In particular, consider Figure 8–15, which shows the standard geometric representations of the nonzero vectors $\mathbf{u}$ and $\mathbf{v}$ and the perpendicular segment from the terminal point $V(x, y, z)$ of $\mathbf{v}$ to the line containing $\mathbf{u}$. The vector with geometric representation from the initial point of $\mathbf{v}$ to the foot of the perpendicular segment is called the **vector projection** of $\mathbf{v}$ on $\mathbf{u}$, or the **vector component** of $\mathbf{v}$ parallel to $\mathbf{u}$. The directed length of this vector projection is called the **scalar projection** of $\mathbf{v}$ on $\mathbf{u}$ or the **scalar component** of $\mathbf{v}$ parallel to $\mathbf{u}$, and it is designated by $\text{Comp}_{\mathbf{u}}\, \mathbf{v}$.

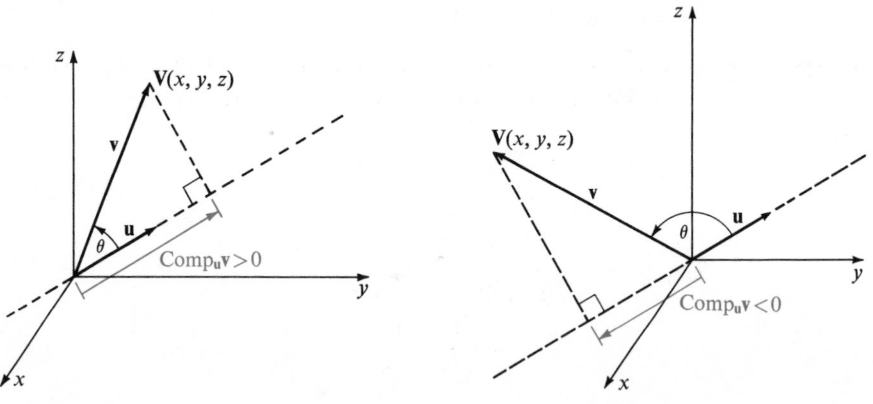

Figure 8–15 **Figure 8–16**

As suggested by Figure 8–16, if this scalar component is negative, then the vector component has the direction opposite to that of $\mathbf{u}$. In either case, the *undirected* length of the vector projection of $\mathbf{v}$ on $\mathbf{u}$ is just $|\text{Comp}_{\mathbf{u}}\, \mathbf{v}|$.

It is evident from Figures 8–15 and 8–16 that

$$\text{Comp}_{\mathbf{u}}\, \mathbf{v} = \|\mathbf{v}\| \cos \theta.$$

Then, from Equation (1), page 287, you have

$$\cos\theta = \frac{\mathbf{u}\cdot\mathbf{v}}{\|\mathbf{u}\|\,\|\mathbf{v}\|},$$

or

$$\|\mathbf{v}\|\cos\theta = \frac{\mathbf{u}\cdot\mathbf{v}}{\|\mathbf{u}\|}.$$

Thus

$$\text{Comp}_{\mathbf{u}}\,\mathbf{v} = \frac{\mathbf{u}\cdot\mathbf{v}}{\|\mathbf{u}\|}. \tag{2}$$

Example 4. Find the scalar component of $\mathbf{v} = (1, 3, 5)$ parallel to $\mathbf{u} = (1, -2, 2)$.

Solution: $\dfrac{\mathbf{u}\cdot\mathbf{v}}{\|\mathbf{u}\|} = \dfrac{1 - 6 + 10}{\sqrt{1 + 4 + 4}} = \dfrac{5}{3}.$

Exercises 8–4

In Exercises 1–6, find the cosine of the angle between the given vectors.

1. $(2, 1, -2), (1, 1, 0)$
2. $(2, 2, -1), (1, 1, 1)$
3. $(1, -1, -2), (-2, -1, 1)$
4. $(6, -2, 4), (5, -4, 3)$
5. $(2, 3, 5), (5, -2, 3)$
6. $(2, 1, 3), (-3, 3, -1)$

In Exercises 7–12, determine whether the given vectors are parallel, perpendicular, or neither.

7. $(6, -3, -9), (-2, 1, 3)$
8. $(3, 2, -5), (14, -6, 6)$
9. $(-2, 3, 4), (-6, -8, 3)$
10. $(1, 0, -2), (-2, -1, 1)$
11. $(2, 3, 0), (0, 0, -5)$
12. $(3, 0, 4), (0, 7, 0)$

In Exercises 13–18, determine a value for *x*, *y*, or *z* so that the second vector is perpendicular to the first.

13. $(3, -6, -2), (4, 2, z)$
14. $(5, 15, 7), (14, 7, z)$
15. $(2, 5, -6), (5, y, 10)$
16. $(8, -3, 6), (3, y, -1)$
17. $(2, -1, 3), (x, -9, -1)$
18. $(5, -4, 2), (x, 3, 1)$

In Exercises 19–24, find the scalar component of **v** on **u** for the given vectors **u** and **v**.

19. $\mathbf{u} = (2, 1, 2), \mathbf{v} = (2, 1, 1)$
20. $\mathbf{u} = (3, 0, -4), \mathbf{v} = (1, -3, 4)$
21. $\mathbf{u} = (0, -1, 0), \mathbf{v} = (-3, 4, -1)$
22. $\mathbf{u} = (12, 5, 0), \mathbf{v} = (3, 1, -2)$
23. $\mathbf{u} = (1, -3, 2), \mathbf{v} = (4, 3, 1)$
24. $\mathbf{u} = (-2, 3, 1), \mathbf{v} = (0, -2, 4)$

In Exercises 25–29, use Equation (1) on page 287.

25. Show that the triangle with vertices $\mathbf{R}(2, 7, 1)$, $\mathbf{S}(8, 5, 5)$, and $\mathbf{T}(7, 3, 4)$ is a right triangle.

26. Show that the triangle with vertices $\mathbf{R}(3, 1, -2)$, $\mathbf{S}(8, 4, 6)$, and $\mathbf{T}(6, 7, 0)$ is a right triangle.

27. Show that the triangle with vertices $\mathbf{R}(1, 3, -3)$, $\mathbf{S}(2, 2, -1)$, and $\mathbf{T}(3, 4, -2)$ is an equilateral triangle.

28. Show that the quadrilateral **QRST** with vertices $\mathbf{Q}(-1, 9, -2)$, $\mathbf{R}(-7, 1, 1)$, $\mathbf{S}(-9, 4, 5)$, and $\mathbf{T}(-3, 12, 2)$ is a rectangle.

* **29.** Show that if θ is the angle between vectors $\mathbf{v}_1$ and $\mathbf{v}_2$ in $\mathcal{R}^3$, then $\cos\theta = \cos\alpha_1 \cos\alpha_2 + \cos\beta_1 \cos\beta_2 + \cos\gamma_1 \cos\gamma_2$, where α_1, β_1, and γ_1 are the direction angles of $\mathbf{v}_1$, and α_2, β_2, and γ_2 are the direction angles of $\mathbf{v}_2$.

* **30.** Show that for nonzero vectors $\mathbf{u}$ and $\mathbf{v}$, the angle θ determined by their standard geometric representations satisfies the equation $\cos\theta = \dfrac{\mathbf{u}\cdot\mathbf{v}}{\|\mathbf{u}\|\,\|\mathbf{v}\|}$.

* **31.** Show that if $\mathbf{u} = (u_1, u_2, u_3)$ and $\mathbf{v} = (v_1, v_2, v_3)$ are nonzero vectors for which there is no scalar k such that $\mathbf{u} = k\mathbf{v}$, then the angle θ between them does not satisfy $\cos\theta = 1$ or $\cos\theta = -1$.

A Special Operation on Vectors in Space

8–5 The Cross Product of Two Vectors

The standard geometric representations of two nonparallel vectors $\mathbf{u}$ and $\mathbf{v}$ determine a plane through the origin. It should be possible, then, to find a nonzero vector $\mathbf{w}$ whose standard representation is perpendicular to this plane, that is, a vector $\mathbf{w}$ that is orthogonal to both $\mathbf{u}$ and $\mathbf{v}$.

Example 1. Find a nonzero vector $\mathbf{w}$ that is orthogonal to both the vector $\mathbf{u} = (0, 1, -2)$ and the vector $\mathbf{v} = (2, 0, 1)$.

Solution: You are asked to find a nonzero vector $\mathbf{w} = (x, y, z)$ for which both $\mathbf{u}\cdot\mathbf{w} = 0$ and $\mathbf{v}\cdot\mathbf{w} = 0$, that is, for which

$$(0, 1, -2)\cdot(x, y, z) = 0,$$
$$(2, 0, 1)\cdot(x, y, z) = 0,$$

or

$$y - 2z = 0,$$
$$2x + z = 0.$$

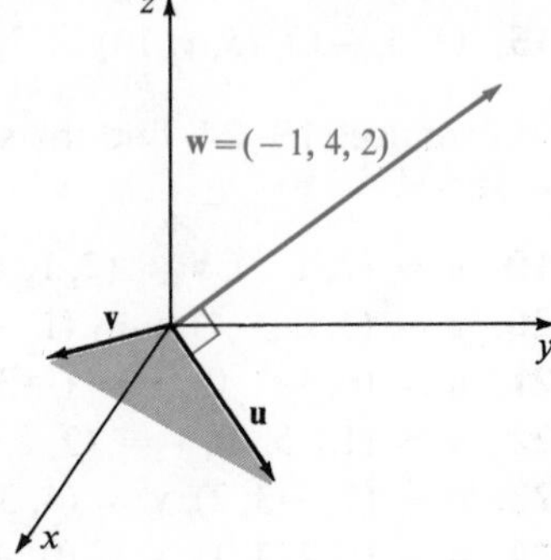

Setting $x = -1$, for example, you find that $z = 2$ and $y = 4$ satisfy these equations. Therefore the vector $\mathbf{w} = (x, y, z) = (-1, 4, 2)$ is perpendicular to both $\mathbf{u}$ and $\mathbf{v}$.

In general, for vectors $\mathbf{u} = (u_1, u_2, u_3)$ and $\mathbf{v} = (v_1, v_2, v_3)$, the vector $\mathbf{w} = (x, y, z)$ is orthogonal to both $\mathbf{u}$ and $\mathbf{v}$ if and only if $\mathbf{u} \cdot \mathbf{w} = 0$ and $\mathbf{v} \cdot \mathbf{w} = 0$, that is, if and only if

$$(u_1, u_2, u_3) \cdot (x, y, z) = 0,$$
$$(v_1, v_2, v_3) \cdot (x, y, z) = 0,$$

or

$$\begin{aligned} u_1x + u_2y + u_3z &= 0, \\ v_1x + v_2y + v_3z &= 0. \end{aligned} \tag{1}$$

If the determinant $\begin{vmatrix} u_1 & u_2 \\ v_1 & v_2 \end{vmatrix}$ is different from 0, you can apply Cramer's rule (page 106) to obtain unique solutions of Equations (1) for x and y in terms of z:

$$x = \frac{\begin{vmatrix} -u_3z & u_2 \\ -v_3z & v_2 \end{vmatrix}}{\begin{vmatrix} u_1 & u_2 \\ v_1 & v_2 \end{vmatrix}}, \qquad y = \frac{\begin{vmatrix} u_1 & -u_3z \\ v_1 & -v_3z \end{vmatrix}}{\begin{vmatrix} u_1 & u_2 \\ v_1 & v_2 \end{vmatrix}}. \tag{2}$$

Since multiplying each entry of a column (or a row) of a determinant by a number is equivalent to multiplying the determinant by that number (see Exercise 25, page 297), and since interchanging two columns of a determinant is equivalent to multiplying the determinant by -1 (see Exercise 26, page 297), Equations (2) can be written as

$$x = \frac{\begin{vmatrix} u_2 & u_3 \\ v_2 & v_3 \end{vmatrix}}{\begin{vmatrix} u_1 & u_2 \\ v_1 & v_2 \end{vmatrix}} z, \qquad y = \frac{\begin{vmatrix} u_3 & u_1 \\ v_3 & v_1 \end{vmatrix}}{\begin{vmatrix} u_1 & u_2 \\ v_1 & v_2 \end{vmatrix}} z.$$

By letting $z = k \begin{vmatrix} u_1 & u_2 \\ v_1 & v_2 \end{vmatrix}$, where $k \in \mathcal{R}$ is an arbitrary constant, you can write this solution in **symmetric form** as

$$(x, y, z) = k\left(\begin{vmatrix} u_2 & u_3 \\ v_2 & v_3 \end{vmatrix}, \begin{vmatrix} u_3 & u_1 \\ v_3 & v_1 \end{vmatrix}, \begin{vmatrix} u_1 & u_2 \\ v_1 & v_2 \end{vmatrix} \right). \tag{3}$$

If $\begin{vmatrix} u_1 & u_2 \\ v_1 & v_2 \end{vmatrix} = 0$ but one of the other determinants in Equation (3) is different from 0, you can solve Equations (1) either for x and z or for y and z in terms of the remaining variable, and the symmetric form (3) again results. Therefore:

If $\mathbf{u} = (u_1, u_2, u_3)$ and $\mathbf{v} = (v_1, v_2, v_3)$, and not all three of the determinants

$$\begin{vmatrix} u_2 & u_3 \\ v_2 & v_3 \end{vmatrix}, \quad \begin{vmatrix} u_3 & u_1 \\ v_3 & v_1 \end{vmatrix}, \quad \begin{vmatrix} u_1 & u_2 \\ v_1 & v_2 \end{vmatrix}$$

are 0, then the vector $\mathbf{w} = (x, y, z)$ is orthogonal to both $\mathbf{u}$ and $\mathbf{v}$ if and only if

$$(x, y, z) = k(\mathbf{u} \times \mathbf{v}),$$

where

$$\mathbf{u} \times \mathbf{v} = \left(\begin{vmatrix} u_2 & u_3 \\ v_2 & v_3 \end{vmatrix}, \begin{vmatrix} u_3 & u_1 \\ v_3 & v_1 \end{vmatrix}, \begin{vmatrix} u_1 & u_2 \\ v_1 & v_2 \end{vmatrix} \right). \tag{4}$$

(Notice that if the three determinants are all 0, then $\mathbf{u} \times \mathbf{v} = (0, 0, 0)$, and therefore $\mathbf{u} \times \mathbf{v}$ is still orthogonal to both $\mathbf{u}$ and $\mathbf{v}$; but in this case there are also vectors which are not of the form $k(\mathbf{u} \times \mathbf{v})$ that are orthogonal to both $\mathbf{u}$ and $\mathbf{v}$.)

For any two vectors $\mathbf{u} = (u_1, u_2, u_3)$ and $\mathbf{v} = (v_1, v_2, v_3)$, the vector $\mathbf{u} \times \mathbf{v}$ given by Equation (4) is called the **cross product** of $\mathbf{u}$ and $\mathbf{v}$. (Note that the cross product of two vectors is a *vector*, whereas the dot product of two vectors is a *scalar*.)

From Equation (4), you can see that $\mathbf{u} \times \mathbf{v}$ can be written as

$$\mathbf{u} \times \mathbf{v} = \begin{vmatrix} u_2 & u_3 \\ v_2 & v_3 \end{vmatrix} \mathbf{i} + \begin{vmatrix} u_3 & u_1 \\ v_3 & v_1 \end{vmatrix} \mathbf{j} + \begin{vmatrix} u_1 & u_2 \\ v_1 & v_2 \end{vmatrix} \mathbf{k}. \tag{5}$$

Although thus far we have considered only real numbers as entries in determinants, you will recognize the right-hand member of Equation (5) as the expansion of the vector determinant $\begin{vmatrix} u_1 & u_2 & u_3 \\ v_1 & v_2 & v_3 \\ \mathbf{i} & \mathbf{j} & \mathbf{k} \end{vmatrix}$ by minors of its third row (see Section 3–3).

We therefore write

$$\mathbf{u} \times \mathbf{v} = \begin{vmatrix} u_1 & u_2 & u_3 \\ v_1 & v_2 & v_3 \\ \mathbf{i} & \mathbf{j} & \mathbf{k} \end{vmatrix} = (u_2v_3 - v_2u_3)\mathbf{i} - (u_1v_3 - v_1u_3)\mathbf{j} + (u_1v_2 - v_1u_2)\mathbf{k}$$

$$= (u_2v_3 - v_2u_3)\mathbf{i} + (u_3v_1 - v_3u_1)\mathbf{j} + (u_1v_2 - v_1u_2)\mathbf{k} \tag{6}$$

Notice that you can readily obtain the expansion on the right in Equation (6) if you repeat part of the determinant:

$$\begin{matrix} u_1 & u_2 & u_3 & u_1 & u_2 \\ v_1 & v_2 & v_3 & v_1 & v_2 \\ \mathbf{i} & \mathbf{j} & \mathbf{k} & & \end{matrix}$$

The scalar coefficients of **i**, **j**, and **k** are then the second-order determinants appearing immediately above and to the right of **i**, **j**, and **k**, respectively. For example, **j** and its coefficient are shown in color.

Example 2. If $\mathbf{u} = (2, 2, 1)$ and $\mathbf{v} = (1, -1, 2)$, find $\mathbf{u} \times \mathbf{v}$ and verify that it is perpendicular to both $\mathbf{u}$ and $\mathbf{v}$. Is $\mathbf{u}$ perpendicular to $\mathbf{v}$?

Solution:
$$\mathbf{u} \times \mathbf{v} = \begin{vmatrix} 2 & 2 & 1 \\ 1 & -1 & 2 \\ \mathbf{i} & \mathbf{j} & \mathbf{k} \end{vmatrix} = [4 - (-1)]\mathbf{i} + [1 - 4]\mathbf{j} + [-2 - 2]\mathbf{k}$$
$$= 5\mathbf{i} - 3\mathbf{j} - 4\mathbf{k}$$
$$= (5, -3, -4).$$

$$\mathbf{u} \cdot (\mathbf{u} \times \mathbf{v}) = (2, 2, 1) \cdot (5, -3, -4) = 10 - 6 - 4 = 0,$$
$$\mathbf{v} \cdot (\mathbf{u} \times \mathbf{v}) = (1, -1, 2) \cdot (5, -3, -4) = 5 + 3 - 8 = 0.$$

Since both dot products are 0, $\mathbf{u} \times \mathbf{v}$ is perpendicular to both $\mathbf{u}$ and $\mathbf{v}$. Since $\mathbf{u} \cdot \mathbf{v} = (2, 2, 1) \cdot (1, -1, 2) = 2 - 2 + 2 = 2 \neq 0$, $\mathbf{u}$ is not perpendicular to $\mathbf{v}$.

The orientation of the vector $\mathbf{u} \times \mathbf{v}$ relative to the $\mathbf{u}$- and $\mathbf{v}$-directions is the same as the orientation of the z-axis relative to the positive x- and y-directions. (The vectors $\mathbf{u}$ and $\mathbf{v}$ are not necessarily perpendicular, of course.) Thus in a right-hand system (page 274), if you curl the fingers of your right hand from the $\mathbf{u}$-direction toward the $\mathbf{v}$-direction, then your thumb will point in the $\mathbf{u} \times \mathbf{v}$-direction, just as it points in the z-direction if you curl your fingers from the positive x-direction toward the positive y-direction. [See Figure 8–17(a).]

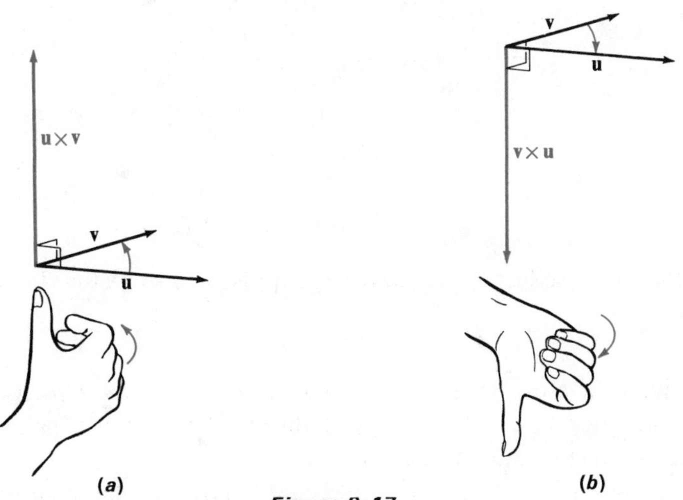

Figure 8–17

Notice from Figure 8–17(b) that the vectors $\mathbf{u} \times \mathbf{v}$ and $\mathbf{v} \times \mathbf{u}$ are oppositely directed. In fact, you have $\mathbf{u} \times \mathbf{v} = -(\mathbf{v} \times \mathbf{u})$, as you can readily verify. Thus the binary operation of taking the cross product of two vectors is *not* a commutative operation!

The norm of the vector $\mathbf{u} \times \mathbf{v}$ satisfies

$$\|\mathbf{u} \times \mathbf{v}\| = \|\mathbf{u}\| \, \|\mathbf{v}\| \sin \theta,$$

where θ is the angle between the vectors $\mathbf{u}$ and $\mathbf{v}$ (Exercise 36, page 298). Therefore, if $\mathbf{u}$ and $\mathbf{v}$ are not parallel, $\|\mathbf{u} \times \mathbf{v}\|$ has the physical interpretation of being the area of the region inside a parallelogram having geometric representations of $\mathbf{u}$ and $\mathbf{v}$ as two of its adjacent sides (Figure 8–18).

Figure 8–18

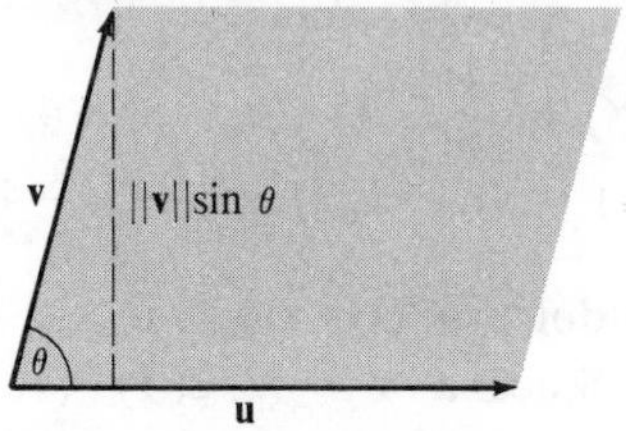

The fact that two vectors in $\mathcal{R}^3$ are parallel if and only if one is a scalar multiple of the other (page 282) has a special implication for the cross product of two such vectors. If $\mathbf{u} = (u_1, u_2, u_3)$ and $\mathbf{v} = (v_1, v_2, v_3)$ are parallel vectors with $\mathbf{u} = c\mathbf{v}$, then $u_1 = cv_1$, $u_2 = cv_2$, and $u_3 = cv_3$. Thus

$$\begin{aligned} \mathbf{u} \times \mathbf{v} &= \begin{vmatrix} u_1 & u_2 & u_3 \\ v_1 & v_2 & v_3 \\ \mathbf{i} & \mathbf{j} & \mathbf{k} \end{vmatrix} \\ &= \begin{vmatrix} cv_1 & cv_2 & cv_3 \\ v_1 & v_2 & v_3 \\ \mathbf{i} & \mathbf{j} & \mathbf{k} \end{vmatrix} \\ &= c \begin{vmatrix} v_1 & v_2 & v_3 \\ v_1 & v_2 & v_3 \\ \mathbf{i} & \mathbf{j} & \mathbf{k} \end{vmatrix} \\ &= c(0, 0, 0) = (0, 0, 0). \end{aligned}$$

That is, the cross product of two parallel vectors is the zero vector. In particular, you have

$$\mathbf{v} \times \mathbf{v} = \mathbf{0}.$$

Conversely, if the cross product of two vectors $\mathbf{u}$ and $\mathbf{v}$ is the zero vector, then one of $\mathbf{u}$ and $\mathbf{v}$ must be a scalar multiple of the other (Exercise 38, page 298), and hence the two vectors must be parallel.

The cross product of vectors in $\mathcal{R}^3$ has the following additional properties, which you should verify (Exercises 30 and 31, page 298):

If $\mathbf{u}, \mathbf{v}, \mathbf{w} \in \mathcal{R}^3$ and $r \in \mathcal{R}$, then

1. $\mathbf{u} \times (\mathbf{v} + \mathbf{w}) = (\mathbf{u} \times \mathbf{v}) + (\mathbf{u} \times \mathbf{w})$ Distributive property
2. $\mathbf{u} \times (r\mathbf{v}) = (r\mathbf{u}) \times \mathbf{v}$
   $= r(\mathbf{u} \times \mathbf{v})$ Scalar associative property

Exercises 8–5

In Exercises 1–10, find the cross product $\mathbf{u} \times \mathbf{v}$ of the vectors $\mathbf{u}$ and $\mathbf{v}$ and show that $\mathbf{u} \times \mathbf{v}$ is perpendicular to both $\mathbf{u}$ and $\mathbf{v}$.

1. $\mathbf{u} = (1, 2, 3), \mathbf{v} = (3, 0, -1)$

2. $\mathbf{u} = (1, -1, 1), \mathbf{v} = (2, 1, -2)$

3. $\mathbf{u} = (-2, 1, -1), \mathbf{v} = (2, 1, -1)$

4. $\mathbf{u} = (3, 2, -1), \mathbf{v} = (-1, 2, 2)$

5. $\mathbf{u} = (2, 0, 3), \mathbf{v} = (0, 5, 1)$

6. $\mathbf{u} = (-2, -3, -1)$, $\mathbf{v} = (-1, -2, -1)$

7. $\mathbf{u} = (3, 1, -2), \mathbf{v} = (4, -1, 3)$

8. $\mathbf{u} = (3, 5, -3), \mathbf{v} = (-2, 1, 7)$

9. $\mathbf{u} = (1, 2, a), \mathbf{v} = (-1, 1, b)$

10. $\mathbf{u} = (a, b, c), \mathbf{v} = (1, 2, 3)$

11. If $\mathbf{u} = (1, 2, 3)$ and $\mathbf{v} = (3, 0, -1)$, find $\mathbf{v} \times \mathbf{u}$. How does your answer compare with the vector $\mathbf{u} \times \mathbf{v}$ found in Exercise 1?

12. If $\mathbf{u} = (1, -1, 1)$ and $\mathbf{v} = (2, 1, -2)$, find $\mathbf{v} \times \mathbf{u}$. How does your answer compare with the vector $\mathbf{u} \times \mathbf{v}$ found in Exercise 2?

13. Show that if $\mathbf{u} = (1, 2, 3)$, $\mathbf{v} = (2, -1, 4)$, and $\mathbf{w} = (-1, 2, -3)$, then $\mathbf{u} \times (\mathbf{v} + \mathbf{w}) = (\mathbf{u} \times \mathbf{v}) + (\mathbf{u} \times \mathbf{w})$.

14. Show that if $\mathbf{u} = (2, 0, -1)$, $\mathbf{v} = (1, -1, 1)$, and $\mathbf{w} = (-1, 1, -2)$, then $\mathbf{u} \times (\mathbf{v} + \mathbf{w}) = (\mathbf{u} \times \mathbf{v}) + (\mathbf{u} \times \mathbf{w})$.

15. Show that if $r = 2$, $\mathbf{u} = (2, -3, 1)$, and $\mathbf{v} = (1, 0, -1)$, then $r\mathbf{u} \times \mathbf{v} = \mathbf{u} \times r\mathbf{v} = r(\mathbf{u} \times \mathbf{v})$.

16. Show that if $r = -3$, $\mathbf{u} = (1, 4, -1)$, and $\mathbf{v} = (1, -2, 3)$, then $r\mathbf{u} \times \mathbf{v} = \mathbf{u} \times r\mathbf{v} = r(\mathbf{u} \times \mathbf{v})$.

17. Show that for any vector $\mathbf{v}$ in $\mathcal{R}^3$, $\mathbf{v} \times \mathbf{v} = \mathbf{0}$.

18. Show that if $\mathbf{u}$ is a vector in $\mathcal{R}^3$, $\mathbf{v} = (2, -1, -2)$, and $\mathbf{u} \times \mathbf{v} = \mathbf{0}$, then $\mathbf{u}$ is a scalar multiple of $\mathbf{v}$.

In each of Exercises 19–22, show that the given statement is true.

19. $\mathbf{i} \times \mathbf{j} = \mathbf{k}$

20. $\mathbf{j} \times \mathbf{k} = \mathbf{i}$

21. $\mathbf{i} \times \mathbf{k} = -\mathbf{j}$

22. $-\mathbf{i} \times (\mathbf{j} + \mathbf{k}) = \mathbf{j} - \mathbf{k}$

23. Find values for a and b so that $(2, a, 1) \times (1, b, 2) = (3, -3, -1)$.

24. Find values for a and b so that $(1, 2, a) \times (3, -b, 1) = (10, 5, -10)$.

* **25.** Show that for all real numbers a, b, c, d, and k,

$$\begin{vmatrix} a & kb \\ c & kd \end{vmatrix} = k \begin{vmatrix} a & b \\ c & d \end{vmatrix}.$$

* **26.** Show that for all real numbers a, b, c, and d,

$$\begin{vmatrix} b & a \\ d & c \end{vmatrix} = - \begin{vmatrix} a & b \\ c & d \end{vmatrix}.$$

* **27.** Given that $\|\mathbf{u} \times \mathbf{v}\| = \|\mathbf{u}\| \, \|\mathbf{v}\| \sin \theta$, where θ is the angle between $\mathbf{u}$ and $\mathbf{v}$, find the area of a parallelogram with the geometric representations of the vectors $\mathbf{u} = (3, -3, 1)$ and $\mathbf{v} = (1, 2, -3)$ as adjacent sides.

* **28.** Use the fact stated in Exercise 27 to find the area of the parallelogram **ABCD** with vertices **A**$(-1, 3, 2)$, **B**$(3, -3, 4)$, **C**$(5, 0, 3)$, and **D**$(1, 6, 1)$.

* **29.** Show that if θ is the angle between the nonorthogonal vectors **u** and **v**, then

$$\tan\theta = \frac{\|\mathbf{u} \times \mathbf{v}\|}{\mathbf{u} \cdot \mathbf{v}}.$$

(*Hint:* Use the information given in Exercise 27.)

* **30.** Show that for any vectors **u**, **v**, and **w** in $\mathbb{R}^3$

$$\mathbf{u} \times (\mathbf{v} + \mathbf{w}) = (\mathbf{u} \times \mathbf{v}) + (\mathbf{u} \times \mathbf{w}).$$

* **31.** Show that for any $r \in \mathbb{R}$ and any vectors **u** and **v** in $\mathbb{R}^3$

$$\mathbf{u} \times (r\mathbf{v}) = (r\mathbf{u}) \times \mathbf{v} = r(\mathbf{u} \times \mathbf{v}).$$

* **32.** Show that $(\mathbf{i} \times \mathbf{j}) \cdot \mathbf{k} = 1$ and $\mathbf{j} \cdot (\mathbf{i} \times \mathbf{k}) = -1$.

* **33.** Show that for $\mathbf{u} = (u_1, u_2, u_3)$, $\mathbf{v} = (v_1, v_2, v_3)$, and $\mathbf{w} = (w_1, w_2, w_3)$,

$$\mathbf{u} \cdot (\mathbf{v} \times \mathbf{w}) = \begin{vmatrix} u_1 & u_2 & u_3 \\ v_1 & v_2 & v_3 \\ w_1 & w_2 & w_3 \end{vmatrix}.$$

This number is called the **scalar triple product** of **u**, **v**, and **w**.

* **34.** Use the result in Exercise 33 to show that $\mathbf{v} \times \mathbf{w}$ is orthogonal to both **v** and **w**.

* **35.** Show that $(\mathbf{u} \times \mathbf{v}) \cdot (\mathbf{u} \times \mathbf{v}) = (\mathbf{u} \cdot \mathbf{u})(\mathbf{v} \cdot \mathbf{v}) - (\mathbf{u} \cdot \mathbf{v})^2$.

* **36.** Use the formula in Exercise 35 and the identity $\cos^2\theta + \sin^2\theta = 1$ to obtain the formula in Exercise 27.

* **37.** Show that if **u** and **v** are orthogonal unit vectors, then $\mathbf{u} \times \mathbf{v}$ is a unit vector.

* **38.** Show that if $\mathbf{u} \times \mathbf{v} = \mathbf{0}$, then **u** and **v** are parallel vectors. (*Hint:* Use the formula in Exercise 27.)

* **39.** Given vectors **u**, **v**, **r**, and **s**, show that $(\mathbf{u} \times \mathbf{v}) \cdot (\mathbf{r} \times \mathbf{s}) = (\mathbf{u} \cdot \mathbf{r})(\mathbf{v} \cdot \mathbf{s}) - (\mathbf{u} \cdot \mathbf{s})(\mathbf{v} \cdot \mathbf{r})$. [Lagrange's identity.]

Chapter Summary

1. There is a one-to-one correspondence between the set of all points in three-dimensional space and the set $\mathbb{R}^3$ of all ordered triples (x, y, z) of real numbers. Each ordered triple (x, y, z) can be uniquely assigned to a point $\mathbf{S}(x, y, z)$ in space by means of a Cartesian coordinate system. In a right-hand Cartesian coordinate system, if you curl the fingers of your right hand from the positive x-direction toward the positive y-direction, then your thumb will point in the positive z-direction.
2. The distance between two points $\mathbf{S}(x_1, y_1, z_1)$ and $\mathbf{T}(x_2, y_2, z_2)$ in space is given by $d(\mathbf{S}, \mathbf{T}) = \sqrt{(x_2 - x_1)^2 + (y_2 - y_1)^2 + (z_2 - z_1)^2}$.

3. You can associate each ordered triple (v_1, v_2, v_3) of real numbers with a translation in space. Accordingly, an ordered triple is defined to be a **three-dimensional vector**. If the translation is from the point $\mathbf{S}(x_1, y_1, z_1)$ to the point $\mathbf{T}(x_2, y_2, z_2)$, then the components of the vector $\mathbf{v}$ are $v_1 = x_2 - x_1, v_2 = y_2 - y_1$, and $v_3 = z_2 - z_1$. The **norm** of $\mathbf{v} = (v_1, v_2, v_3)$ is given by

$$\|\mathbf{v}\| = \sqrt{v_1^2 + v_2^2 + v_3^2}.$$

4. The vector $\mathbf{v} = (v_1, v_2, v_3)$ can be expressed as the sum of vector components parallel to the coordinate axes. That is,

$$\mathbf{v} = v_1\mathbf{i} + v_2\mathbf{j} + v_3\mathbf{k},$$

where $\mathbf{i} = (1, 0, 0)$, $\mathbf{j} = (0, 1, 0)$, and $\mathbf{k} = (0, 0, 1)$ are unit vectors in the directions of the positive rays of the x-, y-, and z-axes, respectively.
5. The **direction** of a nonzero vector $\mathbf{v} = (v_1, v_2, v_3)$ is determined by angles α, β, and γ between the standard geometric representation of $\mathbf{v}$ and the positive rays of the x-, y-, and z-axes, respectively. The angles α, β, and γ are called the **direction angles** of $\mathbf{v}$. The measure of each direction angle is specified to be between 0° and 180°, inclusive.
6. The cosines of the direction angles of a nonzero vector $\mathbf{v} = (v_1, v_2, v_3)$ are called the **direction cosines** of $\mathbf{v}$ and are given by

$$\cos\alpha = \frac{v_1}{\|\mathbf{v}\|}, \cos\beta = \frac{v_2}{\|\mathbf{v}\|}, \cos\gamma = \frac{v_3}{\|\mathbf{v}\|}.$$

The direction cosines are related by

$$\cos^2\alpha + \cos^2\beta + \cos^2\gamma = 1.$$

7. Two vectors in $\mathcal{R}^3$ are parallel if and only if they have the same or opposite directions, that is, if and only if one is a scalar multiple of the other. Nonzero vectors are parallel if and only if they have the same direction cosines or direction cosines that are negatives of each other.
8. The cosine of the angle θ between two nonzero vectors $\mathbf{u}$ and $\mathbf{v}$ in $\mathcal{R}^3$ is given by

$$\cos\theta = \frac{\mathbf{u}\cdot\mathbf{v}}{\|\mathbf{u}\|\,\|\mathbf{v}\|}.$$

9. Two vectors $\mathbf{u}$ and $\mathbf{v}$ in $\mathcal{R}^3$ are parallel if and only if $\mathbf{u}\cdot\mathbf{v} = \pm\|\mathbf{u}\|\,\|\mathbf{v}\|$ and are perpendicular if and only if $\mathbf{u}\cdot\mathbf{v} = 0$. The zero vector is parallel to every vector and is also perpendicular to every vector.
10. The **scalar component** of the vector $\mathbf{v}$ parallel to the vector $\mathbf{u}$ is given by

$$\text{Comp}_{\mathbf{u}}\,\mathbf{v} = \frac{\mathbf{u}\cdot\mathbf{v}}{\|\mathbf{u}\|}.$$

11. The **cross product** of the vectors **u** and **v** is given by

$$\mathbf{u} \times \mathbf{v} = \begin{vmatrix} u_1 & u_2 & u_3 \\ v_1 & v_2 & v_3 \\ \mathbf{i} & \mathbf{j} & \mathbf{k} \end{vmatrix} = (u_2v_3 - v_2u_3)\mathbf{i} + (u_3v_1 - v_3u_1)\mathbf{j} + (u_1v_2 - v_1u_2)\mathbf{k};$$

$\mathbf{u} \times \mathbf{v}$ is a vector perpendicular to both **u** and **v**. If you curl the fingers of your right hand from the **u**-direction toward the **v**-direction, then in a right-hand coordinate system your thumb will point in the $\mathbf{u} \times \mathbf{v}$-direction. Since $\mathbf{u} \times \mathbf{v} = -(\mathbf{v} \times \mathbf{u})$, the binary operation of taking the cross product of two vectors is not commutative.

12. The norm of $\mathbf{u} \times \mathbf{v}$ is given by $\|\mathbf{u} \times \mathbf{v}\| = \|\mathbf{u}\|\,\|\mathbf{v}\| \sin \theta$, where θ is the angle between **u** and **v**. Thus $\|\mathbf{u} \times \mathbf{v}\|$ is the area of the region inside a parallelogram with geometric representations of **u** and **v** as two of its adjacent sides.

13. Vectors **u** and **v** are parallel if and only if $\mathbf{u} \times \mathbf{v} = \mathbf{0}$.

Chapter Review Exercises

1. Sketch the graph of the point in space with coordinates (2, 3, 4).
2. Find the distance between the points **S**(5, −1, −2) and **T**(7, 2, 4).
3. Express the vector **v** represented by the geometric vector from the point **S**(1, 0, −3) to the point **T**(3, 2, −5) first as an ordered triple and then in the form $a\mathbf{i} + b\mathbf{j} + c\mathbf{k}$. Also find $\|\mathbf{v}\|$.
4. If $\mathbf{u} = (2, -1, 4)$ and $\mathbf{v} = (5, 1, -3)$, find **w** such that

   (a) $\mathbf{w} = \mathbf{u} + 2\mathbf{v}$; (b) $\mathbf{w} = 3\mathbf{u} - 2\mathbf{v}$; (c) $\mathbf{u} + \mathbf{w} = \mathbf{v} - \mathbf{w}$.

5. If **u** and **v** are as in Exercise 4, find $\mathbf{u} \cdot \mathbf{v}$.
6. Find the direction cosines of the vector $\mathbf{v} = (2, -3, 6)$; find a unit vector with the same direction as **v**; and show that the sum of the squares of the direction cosines of **v** is 1.
7. Find the vector **v** if $\|\mathbf{v}\| = 5$ and **v** has the same direction as the vector $\mathbf{u} = (1, -1, 2)$.
8. Find the cosine of the angle between the vectors $\mathbf{u} = (1, 3, 2)$ and $\mathbf{v} = (3, -2, -1)$.
9. Find a value for y so that the vectors $\mathbf{u} = (1, 4, -2)$ and $\mathbf{v} = (-4, y, -3)$ are orthogonal.
10. Find $\text{Comp}_{\mathbf{u}}\, \mathbf{v}$ if $\mathbf{u} = (1, -4, 5)$ and $\mathbf{v} = (-4, 0, 3)$.
11. Find $\mathbf{u} \times \mathbf{v}$ if $\mathbf{u} = (2, 4, 1)$ and $\mathbf{v} = (3, -1, -2)$.
12. Find values for a and b so that $(1, a, 4) \times (2, b, 3) = (10, 5, -5)$.

Matrices and Homogeneous Affine Transformations

In the essay at the end of Chapter 7, you studied general affine transformations of the plane. A *homogeneous affine transformation* of the plane is an affine transformation T with equations of the form

$$\begin{aligned} x' &= ax + by, \\ y' &= cx + dy. \end{aligned} \tag{1}$$

Notice that the origin is a "fixed point" under such a transformation; that is, every homogeneous affine transformation maps the origin onto itself.

The transformation T, above, can be specified by stating the value of each of the four entries in its rectangular array, or *matrix*, of coefficients $\begin{bmatrix} a & b \\ c & d \end{bmatrix}$. (Recall Section 3–3.)

If the homogeneous affine transformation S has matrix of coefficients

$$\begin{bmatrix} e & f \\ g & h \end{bmatrix},$$

then you can readily verify that the affine transformation ST is homogeneous and has matrix of coefficients

$$\begin{bmatrix} ea + fc & eb + fd \\ ga + hc & gb + hd \end{bmatrix}.$$

In the algebra of matrices, the same letters T and S are used to denote matrices as are used in affine geometry to denote the corresponding homogeneous affine transformations. Thus you write

$$T = \begin{bmatrix} a & b \\ c & d \end{bmatrix}, \quad S = \begin{bmatrix} e & f \\ g & h \end{bmatrix}, \quad ST = \begin{bmatrix} ea + fc & eb + fd \\ ga + hc & gb + hd \end{bmatrix}.$$

Accordingly, *matrix multiplication* is defined in such a way that

$$ST = \begin{bmatrix} e & f \\ g & h \end{bmatrix}\begin{bmatrix} a & b \\ c & d \end{bmatrix} = \begin{bmatrix} ea + fc & eb + fd \\ ga + hc & gb + hd \end{bmatrix}. \tag{2}$$

By examining how the entries of the product matrix on the right in Equation (2) are obtained, you can see why matrix multiplication is called "row-by-column" multiplication.

If $X = \begin{bmatrix} x \\ y \end{bmatrix}$, then the same row-by-column definition gives

$$TX = \begin{bmatrix} a & b \\ c & d \end{bmatrix}\begin{bmatrix} x \\ y \end{bmatrix} = \begin{bmatrix} ax + by \\ cx + dy \end{bmatrix}. \tag{3}$$

Equations (1) can be written in matrix form as

$$\begin{bmatrix} x' \\ y' \end{bmatrix} = \begin{bmatrix} ax + by \\ cx + dy \end{bmatrix},$$

and therefore, by Equation (3), as

$$\begin{bmatrix} x' \\ y' \end{bmatrix} = \begin{bmatrix} a & b \\ c & d \end{bmatrix} \begin{bmatrix} x \\ y \end{bmatrix},$$

or simply, with $X' = \begin{bmatrix} x' \\ y' \end{bmatrix}$, as

$$X' = TX. \tag{4}$$

The identity matrix I, or matrix of coefficients of the identity transformation I, is $\begin{bmatrix} 1 & 0 \\ 0 & 1 \end{bmatrix}$, since

$$\begin{bmatrix} 1 & 0 \\ 0 & 1 \end{bmatrix} \begin{bmatrix} x \\ y \end{bmatrix} = \begin{bmatrix} x \\ y \end{bmatrix}.$$

By Equations (4) in the essay at the end of Chapter 7 (page 270), the inverse of the matrix T corresponding to the transformation T given by Equations (1) is

$$T^{-1} = \begin{bmatrix} \dfrac{d}{\begin{vmatrix} a & b \\ c & d \end{vmatrix}} & \dfrac{-b}{\begin{vmatrix} a & b \\ c & d \end{vmatrix}} \\ \dfrac{-c}{\begin{vmatrix} a & b \\ c & d \end{vmatrix}} & \dfrac{a}{\begin{vmatrix} a & b \\ c & d \end{vmatrix}} \end{bmatrix}.$$

You can readily verify that $T^{-1}T = I$ and that $TT^{-1} = I$.

To solve Equation (4) above for X in terms of X', you can simply "left-multiply" both members of this equation by T^{-1}:

$$T^{-1}X' = T^{-1}(TX) = (T^{-1}T)X = IX = X.$$

Thus,

$$X = T^{-1}X'. \tag{5}$$

For example, if you wish to determine the coordinates of the point R that the transformation T, defined by

$$x' = 2x - y,$$
$$y' = 4x + y,$$

maps onto the point $S(3, 9)$, you can set

$$\begin{aligned} 2x - y &= 3, \\ 4x + y &= 9, \end{aligned} \tag{6}$$

and solve for x and y as follows.

Writing these equations in matrix form as

$$TX = X',$$

with $T = \begin{bmatrix} 2 & -1 \\ 4 & 1 \end{bmatrix}$, $X = \begin{bmatrix} x \\ y \end{bmatrix}$, and $X' = \begin{bmatrix} 3 \\ 9 \end{bmatrix}$, you find that

$$T^{-1} = \begin{bmatrix} \frac{1}{6} & \frac{1}{6} \\ -\frac{4}{6} & \frac{2}{6} \end{bmatrix},$$

and therefore,

$$X = \begin{bmatrix} x \\ y \end{bmatrix} = T^{-1}X' = \begin{bmatrix} \frac{1}{6} & \frac{1}{6} \\ -\frac{4}{6} & \frac{2}{6} \end{bmatrix} \begin{bmatrix} 3 \\ 9 \end{bmatrix} = \begin{bmatrix} \frac{3}{6} + \frac{9}{6} \\ -\frac{12}{6} + \frac{18}{6} \end{bmatrix} = \begin{bmatrix} 2 \\ 1 \end{bmatrix}.$$

Hence the coordinates of R are $(2, 1)$.

Notice that another, quite different but perhaps more familiar, interpretation of the example given above is that Equations (6) are equations of lines which intersect at the point R whose coordinates are $(2, 1)$.

Since, for the homogeneous affine transformation T given by Equations (1), Equation (4) determines X' when X is given, and conversely Equation (5) determines X when X' is given, it follows that every homogeneous affine transformation is a one-to-one mapping of the coordinate plane onto itself. [The same is true of general affine transformations, of course, since such a transformation can be considered as a homogeneous affine transformation followed by a translation (see Equations (3) and (4) on pages 268 and 270).]

The theory of matrices is used in the design and operation of vehicles for oceanic and space exploration. Submersibles such as this one, equipped with advanced navigation and tracking systems, are being used for archeological research missions and the investigation of ocean phenomena.

A transformation of the form (1) is called a *linear transformation*, whether or not the determinant of coefficients is different from 0. If the determinant of coefficients is equal to 0, then the linear transformation is said to be *singular;* otherwise, it is *nonsingular.* Thus another name for "homogeneous affine transformation" is "nonsingular linear transformation."

The matrix of coefficients of a singular linear transformation does *not* have an inverse, and such a transformation is *not* one-to-one.

You can see by examples what the map of the coordinate plane under a singular linear transformation might be.

In the very special case

$$\begin{bmatrix} a & b \\ c & d \end{bmatrix} = \begin{bmatrix} 0 & 0 \\ 0 & 0 \end{bmatrix},$$

you have

$$\begin{bmatrix} x' \\ y' \end{bmatrix} = \begin{bmatrix} 0 & 0 \\ 0 & 0 \end{bmatrix}\begin{bmatrix} x \\ y \end{bmatrix} = \begin{bmatrix} 0 \\ 0 \end{bmatrix}.$$

Then the map of each point $P(x, y)$ is the origin $O(0, 0)$, and the entire coordinate plane is mapped onto a single point.

For another example, if

$$\begin{bmatrix} a & b \\ c & d \end{bmatrix} = \begin{bmatrix} 1 & 0 \\ 0 & 0 \end{bmatrix},$$

then

$$\begin{bmatrix} x' \\ y' \end{bmatrix} = \begin{bmatrix} 1 & 0 \\ 0 & 0 \end{bmatrix}\begin{bmatrix} x \\ y \end{bmatrix} = \begin{bmatrix} x \\ 0 \end{bmatrix},$$

and each point $P(x, y)$ is mapped onto $Q(x, 0)$, its projection on the x-axis; thus the entire coordinate plane is mapped onto an entire line through the origin.

Similarly, if

$$\begin{bmatrix} a & b \\ c & d \end{bmatrix} = \begin{bmatrix} 2 & 1 \\ 6 & 3 \end{bmatrix},$$

then

$$\begin{bmatrix} x' \\ y' \end{bmatrix} = \begin{bmatrix} 2 & 1 \\ 6 & 3 \end{bmatrix}\begin{bmatrix} x \\ y \end{bmatrix} = \begin{bmatrix} 2x + y \\ 6x + 3y \end{bmatrix}$$

and

$$x' = 2x + y,$$
$$y' = 6x + 3y,$$

whence

$$y' = 3x',$$

and again the entire plane is mapped onto a line through the origin.

These examples illustrate the fact that a singular linear transformation maps the entire plane either onto the origin or onto a line through the origin.

Matrix theory applies as well to nonhomogeneous affine transformations as to homogeneous ones. Thus, with matrix addition defined by the addition of corresponding entries of the matrices being added, you can write Equations (3) on page 268 in matrix form as

$$\begin{bmatrix} x' \\ y' \end{bmatrix} = \begin{bmatrix} a & b \\ c & d \end{bmatrix} \begin{bmatrix} x \\ y \end{bmatrix} + \begin{bmatrix} h \\ k \end{bmatrix},$$

or

$$X' = TX + H, \text{ where } H = \begin{bmatrix} h \\ k \end{bmatrix}.$$

Multiplication by T^{-1} now yields

$$T^{-1}X' = T^{-1}TX + T^{-1}H,$$

or

$$X = T^{-1}X' - T^{-1}H.$$

You might verify that this is the matrix form of Equations (4) in the essay at the end of Chapter 7 (page 270).

The word "matrix" (plural "matrices") was introduced into mathematics by the nineteenth-century mathematician James Joseph Sylvester (1814–1897). The theory of matrices today has many diverse applications in pure and applied mathematics. In particular, it is used in such new branches as the theory of games and linear programming. Much of the scientific and industrial computation now being performed on fast electronic digital computing machines for the design and use of vehicles for space and oceanic exploration, and for other environmental studies, involves matrices. The theory of matrices was first developed, however, in relation to geometric transformations such as those discussed above. Much of the early work in this development was done by the British mathematician William Rowan Hamilton (1806–1865), but the systematic development of the basic theory was carried out by another British mathematician, Arthur Cayley (1821–1895).

Chapter 9

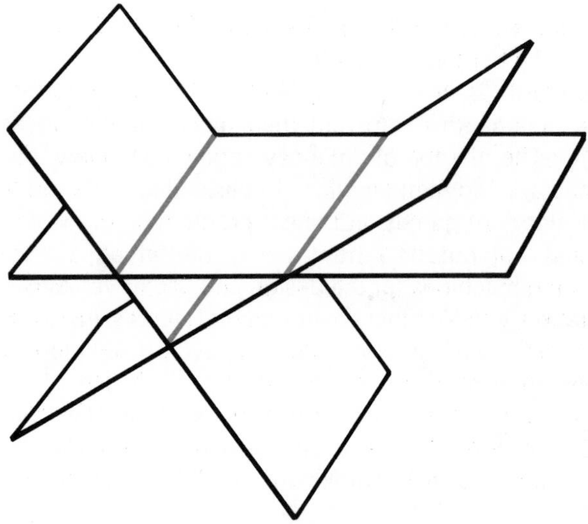

In this chapter, vector and Cartesian equations for lines and planes in space are derived. Intersections of lines and planes are discussed and some useful distance formulas presented.

Lines and Planes in Space

Equations for Lines and Planes

9–1 Lines in Space

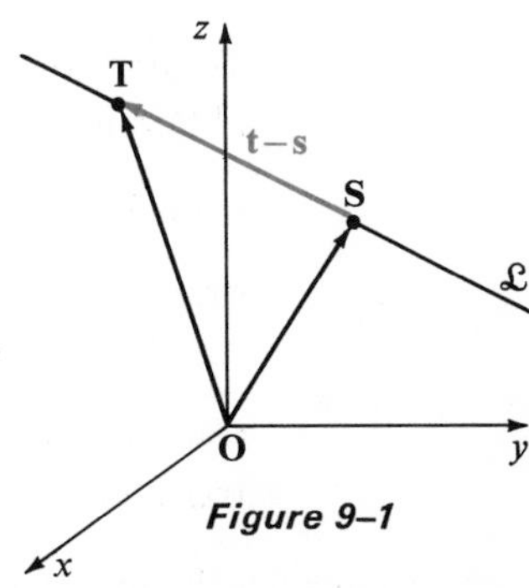

Figure 9–1

Many of the facts that you learned in Chapters 1 and 2 about vectors in $\mathcal{R}^2$ have analogues for vectors in $\mathcal{R}^3$. For example, if $\mathbf{S}(2, 4, 5)$ and $\mathbf{T}(0, -2, 6)$ are two points on a line $\mathcal{L}$ in space (Figure 9–1), then the vector $\mathbf{t} - \mathbf{s} = (0, -2, 6) - (2, 4, 5) = (-2, -6, 1)$ has a geometric representation lying on $\mathcal{L}$ and hence is parallel to $\mathcal{L}$. Thus the vector $(-2, -6, 1)$ is a direction vector (see page 53) for $\mathcal{L}$.

By reasoning similar to that in Section 2–1, you can show that if $\mathbf{U}(x, y, z)$ represents a point in space, then

$$\mathbf{u} = \mathbf{s} + r(\mathbf{t} - \mathbf{s}), \quad r \in \mathcal{R}, \tag{1}$$

is a parametric vector equation for $\mathcal{L}$. Thus $\mathcal{L}$ can be specified as

$$\{\mathbf{U}\colon \mathbf{u} = \mathbf{s} + r(\mathbf{t} - \mathbf{s}), r \in \mathcal{R}\}.$$

Example 1. Find a parametric vector equation for the line $\mathcal{L}$ through the points $\mathbf{S}(3, -1, 0)$ and $\mathbf{T}(4, 1, 3)$.

Solution: The point $\mathbf{S}(3, -1, 0)$ lies on $\mathcal{L}$, and $\mathbf{t} - \mathbf{s} = (4, 1, 3) - (3, -1, 0) = (1, 2, 3)$ is a direction vector for $\mathcal{L}$. Thus if $\mathbf{U}(x, y, z)$ represents a point in space, then a parametric vector equation for $\mathcal{L}$ is

$$\mathbf{u} = \mathbf{s} + r(\mathbf{t} - \mathbf{s}),$$

$$(x, y, z) = (3, -1, 0) + r(1, 2, 3),$$

or

$$(x, y, z) = (3 + r, -1 + 2r, 3r).$$

As in the case of two-dimensional vectors, if the domain of r in Equation (1) is restricted to a closed interval, then the graph of the equation is a line segment. In particular, if $0 \leq r \leq 1$, then the graph is the segment $\overline{\mathbf{ST}}$. You can identify points that lie a specified part of the way from **S** to **T** by appropriate choices of the parameter r, as illustrated in the following example.

Example 2. Find the coordinates of the points that trisect the line segment with endpoints $\mathbf{S}(2, 1, 8)$ and $\mathbf{T}(-1, 3, 7)$.

Solution: The vector $\mathbf{t} - \mathbf{s} = (-1 - 2, 3 - 1, 7 - 8) = (-3, 2, -1)$ is a direction vector for the line containing the given points **S** and **T**, and $\mathbf{t} - \mathbf{s}$ is directed from **S** to **T**. Therefore a parametric vector equation for the segment $\overline{\mathbf{ST}}$ is

$$\mathbf{u} = (2, 1, 8) + r(-3, 2, -1),\ 0 \leq r \leq 1.$$

To find the points of trisection, you set $r = \frac{1}{3}$ and $r = \frac{2}{3}$, obtaining

$$(2, 1, 8) + \tfrac{1}{3}(-3, 2, -1) \quad \text{and} \quad (2, 1, 8) + \tfrac{2}{3}(-3, 2, -1),$$

or

$$(2 - 1, 1 + \tfrac{2}{3}, 8 - \tfrac{1}{3}) \quad \text{and} \quad (2 - 2, 1 + \tfrac{4}{3}, 8 - \tfrac{2}{3}),$$

respectively. Therefore the coordinates of the required points of trisection are $(1, \frac{5}{3}, \frac{23}{3})$ and $(0, \frac{7}{3}, \frac{22}{3})$.

Now consider the line $\mathcal{L}$ in Example 1 again. The parametric vector equation of $\mathcal{L}$ given in the solution,

$$(x, y, z) = (3 + r, -1 + 2r, 3r),$$

is equivalent to the three Cartesian equations

$$x = 3 + r,\ y = -1 + 2r, \quad \text{and} \quad z = 3r.$$

These three equations are called *parametric Cartesian equations* of $\mathcal{L}$, with scalar parameter r. If you solve each of these equations for r, you have

$$r = x - 3,\ r = \frac{y - (-1)}{2}, \quad \text{and} \quad r = \frac{z}{3},$$

or

$$r = \frac{x - 3}{1},\ r = \frac{y - (-1)}{2}, \quad \text{and} \quad r = \frac{z - 0}{3}.$$

Since the right-hand member in each of these equations is equal to r, you can equate the right-hand members to obtain

$$\frac{x - 3}{1} = \frac{y - (-1)}{2} = \frac{z - 0}{3}. \tag{2}$$

Equations (2) are called *symmetric equations* for $\mathcal{L}$. The numbers 1, 2, and 3 in the denominators are called *direction numbers* for $\mathcal{L}$ because they are the components of a direction vector of $\mathcal{L}$. Since $\|(1, 2, 3)\| = \sqrt{1^2 + 2^2 + 3^2} = \sqrt{14}$, a set of direction cosines for $\mathcal{L}$ is $\left(\frac{1}{\sqrt{14}}, \frac{2}{\sqrt{14}}, \frac{3}{\sqrt{14}}\right)$. Note that the numbers 3, -1, and 0 in the numerators of Equations (2) are the coordinates of the point $\mathbf{S}(3, -1, 0)$ on $\mathcal{L}$.

More generally, if $\mathbf{S}(x_1, y_1, z_1)$ and $\mathbf{T}(x_2, y_2, z_2)$ are two points in space, then a **parametric vector equation** for the line $\mathcal{L}$ containing $\mathbf{S}$ and $\mathbf{T}$ is

$$(x, y, z) = (x_1, y_1, z_1) + r(x_2 - x_1, y_2 - y_1, z_2 - z_1),$$

or

$$(x, y, z) = (x_1 + r(x_2 - x_1), y_1 + r(y_2 - y_1), z_1 + r(z_2 - z_1)). \quad (3)$$

Parametric Cartesian equations for $\mathcal{L}$ are then

$$x = x_1 + r(x_2 - x_1), y = y_1 + r(y_2 - y_1), z = z_1 + r(z_2 - z_1). \quad (4)$$

Provided the differences $x_2 - x_1$, $y_2 - y_1$, and $z_2 - z_1$ are not all zero, $(x_2 - x_1, y_2 - y_1, z_2 - z_1)$ is a direction vector of $\mathcal{L}$ and hence $x_2 - x_1$, $y_2 - y_1$, and $z_2 - z_1$ are **direction numbers** for $\mathcal{L}$. If $x_2 - x_1$, $y_2 - y_1$, and $z_2 - z_1$ are all unequal to zero, then

$$\frac{x - x_1}{x_2 - x_1} = \frac{y - y_1}{y_2 - y_1} = \frac{z - z_1}{z_2 - z_1} \quad (5)$$

are **symmetric equations** for the given line $\mathcal{L}$. If the direction vector $\mathbf{v} = (x_2 - x_1, y_2 - y_1, z_2 - z_1)$ is written in the form (v_1, v_2, v_3), then Equations (5) become

$$\frac{x - x_1}{v_1} = \frac{y - y_1}{v_2} = \frac{z - z_1}{v_3}. \quad (6)$$

Example 3. Find symmetric equations for the line $\mathcal{L}$ containing the point $\mathbf{S}(2, 5, -1)$ and having direction vector $\mathbf{v} = (3, -2, 2)$.

Solution: Since $\mathbf{S}(2, 5, -1) \in \mathcal{L}$ and $\mathbf{v} = (3, -2, 2)$ is a direction vector of $\mathcal{L}$, symmetric equations for $\mathcal{L}$ are

$$\frac{x - 2}{3} = \frac{y - 5}{-2} = \frac{z + 1}{2}.$$

In geometry you learned that lines in space are parallel if they are coplanar and do not intersect. We shall define parallel lines in space in terms of their direction vectors. Thus, as in the plane (page 60), we shall say that two lines in space are parallel if and only if their direction vectors are parallel. (Note that coincident lines are again considered to be parallel.)

Example 4. Find symmetric equations for the line $\mathcal{L}$ that contains the point $\mathbf{S}(3, -1, 4)$ and is parallel to the line with symmetric equations

$$\frac{x-2}{1} = \frac{y+3}{-2} = \frac{z}{4}.$$

Solution: Direction numbers for $\mathcal{L}$ are 1, -2, and 4. Then, using Equation (6), you have

$$\frac{x-3}{1} = \frac{y+1}{-2} = \frac{z-4}{4}.$$

In geometry you learned that in space two planes, or a line and a plane, are parallel if they do not intersect. We shall adopt similar definitions, except that we shall consider coincident planes to be parallel and we shall consider lines lying in a plane to be parallel to that plane. In Sections 9–2 through 9–4, these concepts will be discussed in more detail in terms of vectors.

If a line is parallel to a coordinate plane, then one of its direction numbers is 0. Hence it does not have symmetric equations of the form (6), since one of the denominators would be 0. For example, if a line $\mathcal{L}$ is parallel to the xy-plane but not to the x- or y-axis [Figure 9–2(a)], then it has a direction vector of the form $(v_1, v_2, 0)$, where $v_1, v_2 \neq 0$. Although $\mathcal{L}$ does not have equations of the form (6), if it contains the point $\mathbf{S}(x_1, y_1, z_1)$ then it can be specified by the equations

$$\frac{x-x_1}{v_1} = \frac{y-y_1}{v_2} \quad \text{and} \quad z = z_1.$$

If a line is parallel to a coordinate axis, then two of its direction numbers are 0, and in place of symmetric equations you simply have equations expressing the two constant coordinates of each point on the line. Thus, the line $\mathcal{L}$ that is parallel to the z-axis and contains the point $\mathbf{S}(x_1, y_1, z_1)$ can be specified by the equations

$$x = x_1 \quad \text{and} \quad y = y_1$$

[see Figure 9–2(b)]. As illustrated, $\mathcal{L}$ intersects the xy-plane in the point $\mathbf{T}(x_1, y_1, 0)$.

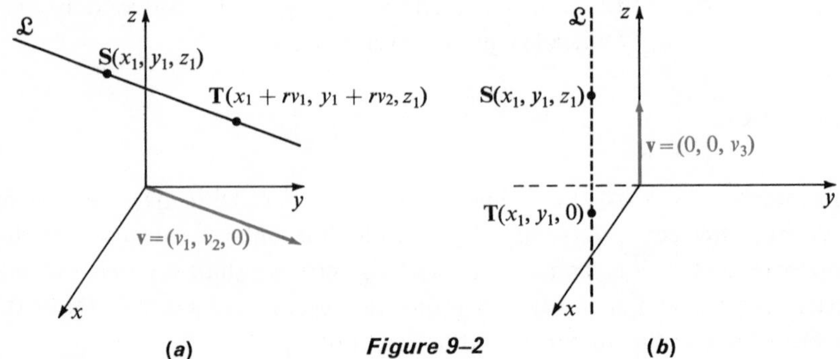

(*a*) ***Figure 9–2*** (*b*)

Just as in the plane, if a vector $\mathbf{v} \in \mathcal{R}^3$ is a direction vector of a line $\mathcal{L}$, then the vector $-\mathbf{v}$ is also a direction vector of $\mathcal{L}$ and the direction of $\mathcal{L}$ may be considered to be either that of $\mathbf{v}$ or that of $-\mathbf{v}$. When a particular direction vector $\mathbf{v}$ is associated with a line $\mathcal{L}$, then $\mathcal{L}$ is said to be a **directed line**, and its direction is taken to be that of $\mathbf{v}$. For example, the line $\mathcal{L}$ containing the point $\mathbf{S}(2, 4, 3)$ and having direction vector $\mathbf{v} = (3, -1, -2)$ is the same as the line containing $\mathbf{S}$ and having direction vector $-\mathbf{v} = (-3, 1, 2)$, but these two descriptions of $\mathcal{L}$ specify different *directed* lines since the two are oppositely directed.

The angle between two directed lines in space is defined to be the angle ϕ, $0 \leq m^\circ(\phi) \leq 180$, between their direction vectors. Notice that this definition applies to all directed lines in space, whether or not they intersect.

Example 5. Show that for the angle ϕ between directed lines $\mathcal{L}_1$ and $\mathcal{L}_2$ with direction vectors $\mathbf{v}_1 = (-1, -1, 0)$ and $\mathbf{v}_2 = (1, 1, \sqrt{6})$, respectively, you have $m^\circ(\phi) = 120$.

Solution: Since the angle ϕ satisfies the equation

$$\begin{aligned}\cos\phi &= \frac{\mathbf{v}_1 \cdot \mathbf{v}_2}{\|\mathbf{v}_1\|\,\|\mathbf{v}_2\|}\\ &= \frac{(-1, -1, 0)\cdot(1, 1, \sqrt{6})}{(\sqrt{2})(\sqrt{8})}\\ &= \frac{-2}{4} = -\frac{1}{2},\end{aligned}$$

you have $m^\circ(\phi) = 120$.

Exercises 9–1

In Exercises 1–6, find a parametric vector equation for the line through the given points **S** and **T**. State two sets of direction cosines for the line.

1. $\mathbf{S}(-1, 5, 7)$, $\mathbf{T}(-4, 1, 3)$

2. $\mathbf{S}(2, 3, -1)$, $\mathbf{T}(5, -3, 1)$

3. $\mathbf{S}(1, -2, -3)$, $\mathbf{T}(2, -3, 2)$

4. $\mathbf{S}(1, 2, 3)$, $\mathbf{T}(-2, 3, 4)$

5. $\mathbf{S}(2, -3, 4)$, $\mathbf{T}(5, 2, -1)$

6. $\mathbf{S}(1, 0, 3)$, $\mathbf{T}(2, 0, 3)$

In Exercises 7–12, find the coordinates of the point(s) which separate(s) the segment with given endpoints **S** and **T** in the given way.

7. $\mathbf{S}(-1, 3, 6)$, $\mathbf{T}(7, 5, -2)$; bisects

8. $\mathbf{S}(-4, -3, 0)$, $\mathbf{T}(8, 7, 12)$; bisects

9. $\mathbf{S}(-6, 1, 5)$, $\mathbf{T}(3, 13, -1)$; trisect

10. $\mathbf{S}(6, 0, -3)$, $\mathbf{T}(-6, 9, -12)$; trisect

11. $\mathbf{S}(-1, 2, 1)$, $\mathbf{T}(7, 6, -11)$; separate in fourths

12. $\mathbf{S}(-8, 2, 3)$, $\mathbf{T}(0, 6, 11)$; separate in fourths

In Exercises 13–18, determine the cosine of the angle ϕ between a directed line $\mathcal{L}_1$ with the given direction vector $\mathbf{v}_1$ and a directed line $\mathcal{L}_2$ with the given direction vector $\mathbf{v}_2$.

13. $\mathbf{v}_1 = (1, 2, 2), \mathbf{v}_2 = (3, 2, 6)$
14. $\mathbf{v}_1 = (3, 0, 4), \mathbf{v}_2 = (-5, 12, 0)$
15. $\mathbf{v}_1 = (-4, 8, 1), \mathbf{v}_2 = (3, 4, -12)$
16. $\mathbf{v}_1 = (9, 6, 2), \mathbf{v}_2 = (-2, 6, -9)$
17. $\mathbf{v}_1 = (9, -6, 2), \mathbf{v}_2 = (2, 6, 9)$
18. $\mathbf{v}_1 = (0, 0, 1), \mathbf{v}_2 = (0, 1, 0)$

In Exercises 19–40, find symmetric equations for the line through

19. $\mathbf{S}(0, 2, -1)$, with direction vector $(1, 3, 4)$.
20. $\mathbf{S}(-1, 1, -3)$, with direction vector $(2, 1, -3)$.
21. $\mathbf{S}(0, 0, 0)$, with direction vector $(1, 1, 1)$.
22. $\mathbf{S}(-2, 3, 2)$, with direction vector $(2, 1, -5)$.
23. $\mathbf{S}(3, -1, 4)$ and $\mathbf{T}(2, -3, 2)$.
24. $\mathbf{S}(-4, 1, 3)$ and $\mathbf{T}(2, -3, 0)$.
25. $\mathbf{S}(2, -1, 3)$ and $\mathbf{T}(-2, 1, -3)$.
26. $\mathbf{S}(1, 0, 1)$ and $\mathbf{T}(14, 7, 3)$.
27. $\mathbf{S}(2, 1, -4)$ and $\mathbf{T}(5, 3, -4)$.
28. $\mathbf{S}(-3, 2, 0)$ and $\mathbf{T}(6, 2, -1)$.
29. $\mathbf{S}(5, -1, 4)$ and $\mathbf{T}(5, -1, 2)$.
30. $\mathbf{S}(3, 5, -4)$ and $\mathbf{T}(-2, 5, -4)$.
31. $\mathbf{S}(1, 4, -2)$, and parallel to the line with symmetric equations

$$\frac{x+3}{1} = \frac{y+7}{3} = \frac{z-5}{-6}.$$

32. $\mathbf{S}(3, -4, 6)$, and parallel to the line with symmetric equations

$$\frac{x-5}{2} = \frac{y}{8} = \frac{z+3}{4}.$$

33. $\mathbf{S}(-2, -4, -1)$, and parallel to the line with symmetric equations

$$x + 2 = 3 - y = 4 - z.$$

34. $\mathbf{S}(5, -\sqrt{2}, \sqrt{3})$, and parallel to the line with symmetric equations

$$\frac{x+\sqrt{2}}{\sqrt{3}} = \frac{y-\sqrt{2}}{\sqrt{5}} = \frac{z+4}{\sqrt{7}}.$$

35. $\mathbf{S}(1, 2, 3)$, and parallel to the line containing $\mathbf{Q}(2, 1, -4)$ and $\mathbf{T}(1, 2, 2)$.
36. $\mathbf{S}(2, -1, -4)$, and parallel to the line containing $\mathbf{Q}(1, 1, 3)$ and $\mathbf{T}(0, 3, -2)$.
37. $\mathbf{S}(1, 5, -7)$, and parallel to the line containing $\mathbf{Q}(2, -1, 3)$ and $\mathbf{T}(5, 1, 3)$.
38. $\mathbf{S}(-5, 1, 0)$, and parallel to the line containing $\mathbf{Q}(5, -1, 4)$ and $\mathbf{T}(5, 1, 2)$.
39. $\mathbf{S}(6, -1, 4)$, and parallel to the line containing $\mathbf{Q}(3, 2, 5)$ and $\mathbf{T}(3, 2, 2)$.
40. $\mathbf{S}(1, 3, 1)$, and parallel to the line containing $\mathbf{Q}(4, 7, -6)$ and $\mathbf{T}(4, 3, -6)$.

9–2 Planes in Space

In Section 8–4, you learned that it is possible to find any number of nonparallel vectors perpendicular to a given vector in $\mathcal{R}^3$ and that the standard geometric representations of all such vectors lie in the same plane. These facts can be used to specify a plane $\mathcal{P}$ in space, since, as suggested by Figure 9–3(a), $\mathcal{P}$ is made up

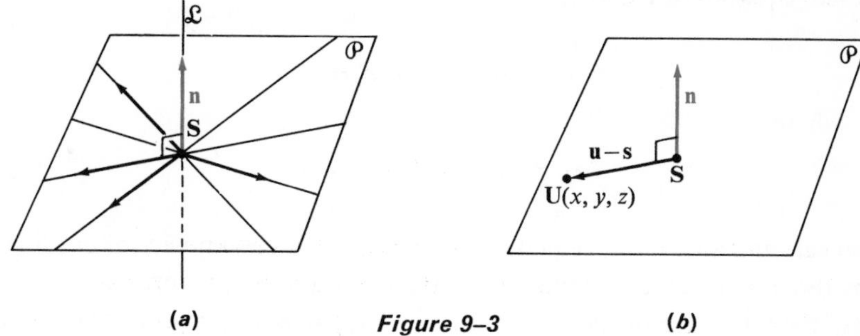

(a) *Figure 9–3* *(b)*

of the points on all lines perpendicular to the line $\mathcal{L}$ at the point **S** on $\mathcal{L}$. Thus, as illustrated in Figure 9–3(b):

■ If $\mathcal{P}$ is a plane and **S** is a point on $\mathcal{P}$, and if **n** is a nonzero vector having a geometric representation perpendicular to $\mathcal{P}$, then a point $\mathbf{U}(x, y, z)$ lies on $\mathcal{P}$ if and only if

$$(\mathbf{u} - \mathbf{s}) \cdot \mathbf{n} = 0. \tag{1}$$

Hence Equation (1) is an equation of $\mathcal{P}$.

Since $\mathbf{U}(x, y, z)$ is any (arbitrary) point on $\mathcal{P}$, Equation (1) is simply a statement that the vector **n** has a geometric representation which is perpendicular to *every* geometric vector having initial point at **S** and lying in $\mathcal{P}$. A vector having a geometric representation perpendicular to a plane $\mathcal{P}$ is said to be **perpendicular** to $\mathcal{P}$. A nonzero vector perpendicular to a plane $\mathcal{P}$ is called a **normal vector**, or simply a **normal**, to $\mathcal{P}$.

Because the dot product of two vectors is a scalar, Equation (1) can be used to find a Cartesian equation of a plane.

Example 1. Find a Cartesian equation for the plane $\mathcal{P}$ containing the point $\mathbf{S}(2, 3, -1)$ and having normal vector $\mathbf{n} = (1, 3, 2)$.

Solution: Let $\mathbf{U}(x, y, z)$ be any point on $\mathcal{P}$. Then from Equation (1) above, you have

$$\begin{aligned}
(\mathbf{u} - \mathbf{s}) \cdot \mathbf{n} &= 0,\\
\mathbf{u} \cdot \mathbf{n} - \mathbf{s} \cdot \mathbf{n} &= 0,\\
(x, y, z) \cdot (1, 3, 2) - (2, 3, -1) \cdot (1, 3, 2) &= 0,\\
x + 3y + 2z - (2 + 9 - 2) &= 0,\\
x + 3y + 2z - 9 &= 0.
\end{aligned}$$

Note that the Cartesian equation for the plane in Example 1 is of first degree with respect to the variables x, y, and z, and that the coefficients of the variables are the respective components of the normal vector $\mathbf{n}$. This suggests (Exercise 29, page 317) that if $\mathbf{U}(x, y, z)$ represents a point in space and $\mathcal{P}$ is a plane containing the point $\mathbf{S}(x_1, y_1, z_1)$ and having normal $\mathbf{n} = (a, b, c)$, then a Cartesian equation for $\mathcal{P}$ is

$$ax + by + cz + d = 0, \qquad (2)$$

where

$$d = -(ax_1 + by_1 + cz_1).$$

You can find a Cartesian equation for a plane $\mathcal{P}$ if you know the coordinates of any three noncollinear points on $\mathcal{P}$. If $\mathbf{S}_1$, $\mathbf{S}_2$, and $\mathbf{S}_3$ are three noncollinear points, then the vector $(\mathbf{s}_2 - \mathbf{s}_1) \times (\mathbf{s}_3 - \mathbf{s}_1) = \mathbf{n}$ is perpendicular to the plane $\mathcal{P}$ determined by $\mathbf{S}_1$, $\mathbf{S}_2$, and $\mathbf{S}_3$, and $\mathbf{n}$ is nonzero since the vectors $\mathbf{s}_2 - \mathbf{s}_1$ and $\mathbf{s}_3 - \mathbf{s}_1$ are not parallel. Hence $\mathbf{n}$ is a normal to $\mathcal{P}$ and you can find an equation for $\mathcal{P}$ by substituting the coordinates of $\mathbf{n}$ and those of $\mathbf{S}_1$ (or $\mathbf{S}_2$ or $\mathbf{S}_3$) in Equation (2) above.

Example 2. Find a Cartesian equation for the plane containing the points $\mathbf{S}_1(1, 2, 3)$, $\mathbf{S}_2(3, -1, 0)$, and $\mathbf{S}_3(2, 2, -1)$.

Solution: $\mathbf{s}_2 - \mathbf{s}_1 = (3, -1, 0) - (1, 2, 3) = (2, -3, -3)$, and $\mathbf{s}_3 - \mathbf{s}_1 = (2, 2, -1) - (1, 2, 3) = (1, 0, -4)$. Then the vector $(\mathbf{s}_2 - \mathbf{s}_1) \times (\mathbf{s}_3 - \mathbf{s}_1) = \mathbf{n}$ is a normal to the plane determined by the three points. You have

$$\begin{aligned} \mathbf{n} &= (\mathbf{s}_2 - \mathbf{s}_1) \times (\mathbf{s}_3 - \mathbf{s}_1) \\ &= \begin{vmatrix} 2 & -3 & -3 \\ 1 & 0 & -4 \\ \mathbf{i} & \mathbf{j} & \mathbf{k} \end{vmatrix} \\ &= 12\mathbf{i} + 5\mathbf{j} + 3\mathbf{k}, \end{aligned}$$

or

$$\mathbf{n} = (12, 5, 3).$$

Since $\mathbf{S}(x_1, y_1, z_1) = (1, 2, 3)$, by Equation (2) you have

$$12x + 5y + 3z - [12(1) + 5(2) + 3(3)] = 0,$$

or

$$12x + 5y + 3z - 31 = 0,$$

as an equation of the plane.

Notice in the solution of Example 2 that if you had used $\mathbf{S}_2(3, -1, 0)$ rather than $\mathbf{S}_1(1, 2, 3)$ in Equation (2), then you would have obtained

$$d = -[12(3) + 5(-1) + 3(0)] = -31;$$

or if you had used $\mathbf{S}_3(2, 2, -1)$, then you would have obtained

$$d = -[12(2) + 5(2) + 3(-1)] = -31,$$

so that the result is the same in each case.

If one of the components of a normal vector $\mathbf{n} = (a, b, c)$ to a plane $\mathcal{P}$ is zero, then a geometric representation of $\mathbf{n}$ is perpendicular to the corresponding axis, and $\mathcal{P}$ is parallel to that axis. This is clearly evident from a consideration of the associated direction cosine. For example, if $a = 0$, then

$$\cos \alpha = \frac{0}{\sqrt{b^2 + c^2}} = 0,$$

and a geometric representation of $\mathbf{n}$ is perpendicular to the x-axis. If two of the components of $\mathbf{n}$ are 0, then a geometric representation of $\mathbf{n}$ is perpendicular to both of the corresponding axes, and thus is parallel to the third axis. In such a case, a plane with $\mathbf{n}$ as normal vector is parallel to the coordinate plane containing the axes corresponding to the zero components. For example, any plane with a normal vector $\mathbf{n} = (0, 0, c)$ is parallel to the xy-plane and is perpendicular to the z-axis; such a plane can be represented by an equation of the form $z =$ *constant.*

A line $\mathcal{L}$ is **parallel** to a plane $\mathcal{P}$ if and only if a direction vector of $\mathcal{L}$ is perpendicular to a normal vector to $\mathcal{P}$. (Note that $\mathcal{L}$ could lie wholly in $\mathcal{P}$.) A line $\mathcal{L}$ is **perpendicular** to a plane $\mathcal{P}$ if and only if a direction vector of $\mathcal{L}$ is parallel to a normal vector to $\mathcal{P}$. Thus, if $\mathbf{v}$ is a direction vector of the line $\mathcal{L}$ and $\mathbf{n}$ is a normal to the plane $\mathcal{P}$, then:

■ $\mathcal{L}$ is parallel to $\mathcal{P}$ if and only if $\mathbf{v} \cdot \mathbf{n} = 0$.
$\mathcal{L}$ is perpendicular to $\mathcal{P}$ if and only if $\mathbf{v} \times \mathbf{n} = \mathbf{0}$.

Similarly, two planes are parallel or perpendicular if and only if their respective normals are parallel or perpendicular. That is, if $\mathcal{P}_1$ is a plane with normal vector $\mathbf{n}_1$ and $\mathcal{P}_2$ is a plane with normal vector $\mathbf{n}_2$, then:

■ $\mathcal{P}_1$ and $\mathcal{P}_2$ are parallel if and only if $\mathbf{n}_1 \times \mathbf{n}_2 = \mathbf{0}$.
$\mathcal{P}_1$ and $\mathcal{P}_2$ are perpendicular if and only if $\mathbf{n}_1 \cdot \mathbf{n}_2 = 0$.

Notice from this definition of parallel planes that any plane $\mathcal{P}$ is parallel to itself, since for any normal vector $\mathbf{n}$ to $\mathcal{P}$, you have $\mathbf{n} \times \mathbf{n} = \mathbf{0}$.

Exercises 9–2

In Exercises 1–6, find a Cartesian equation for the plane through the given point **S** with the given normal vector **n**.

1. $\mathbf{S}(1, 2, 3)$; $\mathbf{n} = (-1, 3, 5)$
2. $\mathbf{S}(2, -1, 4)$; $\mathbf{n} = (2, 3, -1)$
3. $\mathbf{S}(-1, -1, -1)$; $\mathbf{n} = (3, 4, -5)$
4. $\mathbf{S}(2, -1, 0)$; $\mathbf{n} = (2, -1, -2)$
5. $\mathbf{S}(0, 0, 0)$; $\mathbf{n} = (1, 0, 0)$
6. $\mathbf{S}(2, 2, -2)$; $\mathbf{n} = (0, 1, 2)$

In Exercises 7–12, find a Cartesian equation for the plane containing the given points **A**, **B**, and **C**.

7. $\mathbf{A}(3, 4, 1), \mathbf{B}(-1, -2, 5), \mathbf{C}(1, 7, 1)$
8. $\mathbf{A}(3, 1, 4)$, $\mathbf{B}(2, 1, 6)$, $\mathbf{C}(3, 2, 4)$
9. $\mathbf{A}(2, 1, 3)$, $\mathbf{B}(-1, -2, 4)$, $\mathbf{C}(4, 2, 1)$
10. $\mathbf{A}(3, 2, 1)$, $\mathbf{B}(1, 3, 2)$, $\mathbf{C}(1, -2, 3)$
11. $\mathbf{A}(4, 2, 1)$, $\mathbf{B}(-1, -2, 2)$, $\mathbf{C}(0, 4, -5)$
12. $\mathbf{A}(-1, -2, -1)$, $\mathbf{B}(-3, -1, -4)$, $\mathbf{C}(1, 2, 3)$

In Exercises 13–18, find symmetric equations for the line which contains the given point **S** and is perpendicular to the plane with the given equation.

13. $\mathbf{S}(1, -3, 4)$; $x - 3y + 2z = 4$
14. $\mathbf{S}(-2, 1, 3)$; $x + 2y - 2z - 5 = 0$
15. $\mathbf{S}(1, -1, 2)$; $4x - 3y + 2z - 7 = 0$
16. $\mathbf{S}(1, -2, -3)$; $x - 3y + 2z + 4 = 0$
17. $\mathbf{S}(3, -1, 4)$; $2x + 2y - z = 4$
18. $\mathbf{S}(-6, 4, 1)$; $3x - 2y + 5z + 8 = 0$
19. Find an equation for the plane that bisects perpendicularly the segment with endpoints $\mathbf{S}(4, 7, -1)$ and $\mathbf{T}(6, -1, 5)$.
20. Find an equation for the plane that bisects perpendicularly the segment with endpoints $\mathbf{S}(0, -1, -4)$ and $\mathbf{T}(-2, 3, 0)$.
21. Find an equation for the locus of the third vertex of all isosceles triangles having base with endpoints $\mathbf{S}(3, 1, -2)$ and $\mathbf{T}(-1, 3, 0)$. What is the locus?
22. Find an equation for the locus of the third vertex of all isosceles triangles having base with endpoints $\mathbf{S}(3, 6, -2)$ and $\mathbf{T}(5, -2, 4)$. What is the locus?
23. Equations for the intersections of a plane $\mathcal{P}$ with the xy-plane and the yz-plane are $2x - y = 7$, $z = 0$, and $y + 3z = -7$, $x = 0$, respectively. Find an equation for $\mathcal{P}$.
24. Equations for the intersections of a plane $\mathcal{P}$ with the xy-plane and the yz-plane are $x - 4y = 12$, $z = 0$, and $2y + 5z = -6$, $x = 0$, respectively. Find an equation for $\mathcal{P}$.
25. Find an equation for the plane containing the intersecting lines with equations
$$\frac{x-1}{2} = \frac{y+3}{4} = \frac{z}{7} \quad \text{and} \quad \frac{x-1}{-1} = \frac{y+3}{5} = \frac{z}{-2}.$$

26. Find an equation for the plane containing the intersecting lines with equations

$$\frac{x}{3} = \frac{y-1}{-1} = \frac{z+2}{3} \quad \text{and} \quad \frac{x}{-1} = \frac{y-1}{2} = \frac{z+2}{-5}.$$

27. Show that the line with symmetric equations

$$\frac{x-5}{3} = \frac{y+1}{-4} = \frac{z}{2}$$

is perpendicular to the plane with Cartesian equation

$$3x - 4y + 2z = 7.$$

28. Show that the line with parametric Cartesian equations

$$x = -3 - 2r,\ y = -4 - 7r, \quad \text{and} \quad z = 3r$$

is parallel to the plane with Cartesian equation

$$4x - 2y - 2z = 9.$$

* **29.** Show that if $\mathbf{U}(x, y, z)$ represents a point in space, and $\mathcal{P}$ is a plane containing the point $\mathbf{S}(x_1, y_1, z_1)$ and having normal $\mathbf{n} = (a, b, c)$, then a Cartesian equation for $\mathcal{P}$ is

$$ax + by + cz + d = 0,$$

where

$$d = -(ax_1 + by_1 + cz_1).$$

* **30.** Show that the plane with equation $ax + by + d = 0$ is parallel to the z-axis.

* **31.** Find an equation of the plane containing the point $\mathbf{S}(0, 2, 1)$ and also the line with symmetric equations

$$\frac{x-1}{3} = \frac{y+2}{5} = \frac{z-3}{2}.$$

* **32.** Find an equation of the plane containing the point $\mathbf{S}(3, -2, 1)$ and also the line with symmetric equations

$$\frac{x+2}{1} = \frac{y-5}{-1} = \frac{z}{6}.$$

* **33.** Show that the plane with vector equation $\mathbf{n} \cdot \mathbf{r} = d$, where $\mathbf{n} = (a, b, c)$ and $\mathbf{r} = (x, y, z)$, is also defined by the vector equation $\mathbf{N} \cdot \mathbf{r} = p$, where $\mathbf{N}$ is the unit vector in the direction of $\mathbf{n}$, and $p = \dfrac{d}{\|\mathbf{n}\|}$. Explain the geometric significance of p.

* **34.** Let α, β, and γ be the direction angles of the vector **n** in Exercise 33. Then use the result of Exercise 33 to derive the *normal form*,

$$x \cos \alpha + y \cos \beta + z \cos \gamma = p,$$

of the equation of a plane.

* **35.** Show that the locus of all points equidistant from the points $\mathbf{S}(x_1, y_1, z_1)$ and $\mathbf{T}(x_2, y_2, z_2)$ is a plane.

* **36.** Show that the plane which intersects the x-, y-, and z-axes at the points $(a, 0, 0)$, $(0, b, 0)$, and $(0, 0, c)$, respectively, where a, b, and c are different from 0, has an equation of the form

$$\frac{x}{a} + \frac{y}{b} + \frac{z}{c} = 1.$$

* **37.** Show that for any three noncollinear points $\mathbf{Q}(x_1, y_1, z_1)$, $\mathbf{S}(x_2, y_2, z_2)$, and $\mathbf{T}(x_3, y_3, z_3)$, an equation of the plane containing **Q**, **S**, and **T** is

$$\begin{vmatrix} x & y & z & 1 \\ x_1 & y_1 & z_1 & 1 \\ x_2 & y_2 & z_2 & 1 \\ x_3 & y_3 & z_3 & 1 \end{vmatrix} = 0.$$

Relationships involving Lines and Planes

9–3 Intersection of Planes

You learned in geometry that two given planes either are parallel or coincident, or they intersect in a line. For example, in three-dimensional Cartesian coordinate space, the intersection of the xy-plane and the xz-plane is the x-axis. Since it was observed in Section 9–2 that parallel (including coincident) planes have parallel normal vectors, you can draw the conclusion that:

Two planes whose normal vectors are not parallel intersect in a line.

This line is called the **line of intersection** of the planes.

If **v** is a direction vector of any line $\mathcal{L}$ in a plane $\mathcal{P}$ and **n** is a normal vector to $\mathcal{P}$, then **v** is perpendicular to **n**. Hence, a direction vector **v** of the line $\mathcal{L}$ of intersection of two nonparallel planes $\mathcal{P}_1$ and $\mathcal{P}_2$ must be perpendicular to the normal vectors to both planes (Figure 9–4).

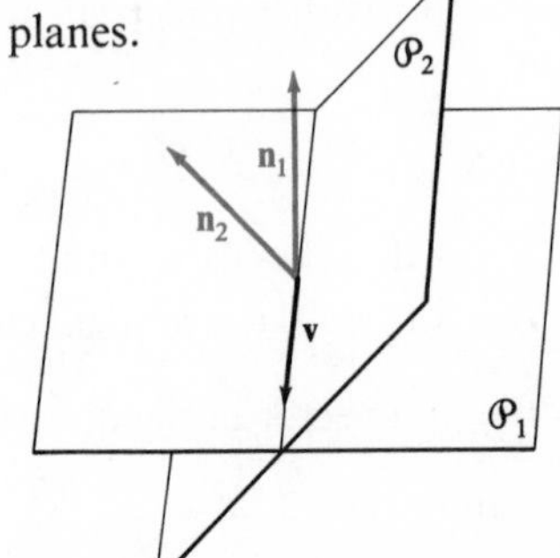

Figure 9–4

Thus, since the cross product of two given vectors is a vector perpendicular to each of the given vectors:

■ If $\mathbf{n}_1$ is a normal vector to the plane $\mathcal{P}_1$ and $\mathbf{n}_2$ is a normal vector to the plane $\mathcal{P}_2$, and if $\mathcal{P}_1$ and $\mathcal{P}_2$ intersect in the line $\mathcal{L}$, then $\mathbf{v} = \mathbf{n}_1 \times \mathbf{n}_2$ is a direction vector of $\mathcal{L}$.

If $\mathcal{L}$ itself is to be determined, then you must find the coordinates of at least one point $\mathbf{S}$ on $\mathcal{L}$. Since $\mathcal{L}$ lies in both $\mathcal{P}_1$ and $\mathcal{P}_2$, such a point $\mathbf{S}$ must lie in both planes. The fact that every line in space must intersect at least one of the coordinate planes can be used to identify a point $\mathbf{S}$ on $\mathcal{L}$ that lies in a coordinate plane. Knowing the coordinates of this point, you can then find an equation for $\mathcal{L}$.

Example 1. Find a parametric vector equation for the line $\mathcal{L}$ of intersection of the planes with Cartesian equations $2x + 3y + z = 8$ and $x - 3y - 2z = 1$.

Solution: By inspection, normal vectors to the two planes are $\mathbf{n}_1 = (2, 3, 1)$ and $\mathbf{n}_2 = (1, -3, -2)$. Then a direction vector for the line $\mathcal{L}$ of intersection is

$$\mathbf{v} = \begin{vmatrix} 2 & 3 & 1 \\ 1 & -3 & -2 \\ \mathbf{i} & \mathbf{j} & \mathbf{k} \end{vmatrix} = -3\mathbf{i} + 5\mathbf{j} - 9\mathbf{k} = (-3, 5, -9).$$

Since the z-coordinate of $\mathbf{v}$ is not zero, $\mathcal{L}$ is not parallel to the xy-plane, and you can replace z with 0 in the equations of the planes to find the point $\mathbf{S}$ of intersection of $\mathcal{L}$ and the xy-plane. Thus, the x- and y-coordinates of $\mathbf{S}$ must satisfy

$$\begin{aligned} 2x + 3y &= 8, \\ x - 3y &= 1. \end{aligned}$$

When solved simultaneously, this system yields

$$x = 3 \quad \text{and} \quad y = \tfrac{2}{3}.$$

Therefore, $\mathbf{S}(3, \frac{2}{3}, 0)$ is a point on each of the two planes, and hence $\mathbf{S}$ is on $\mathcal{L}$. Thus a parametric vector equation for $\mathcal{L}$ is

$$(x, y, z) = (3, \tfrac{2}{3}, 0) + r(-3, 5, -9), \quad r \in \mathcal{R}. \qquad (1)$$

You can use the vector equation (1) for $\mathcal{L}$ given above to determine parametric and symmetric Cartesian equations for $\mathcal{L}$.

Parametric form: $x = 3 - 3r,\ y = \frac{2}{3} + 5r,\ z = -9r.$

Symmetric form: $\dfrac{x - 3}{-3} = \dfrac{y - \frac{2}{3}}{5} = \dfrac{z}{-9}.$

In general, if $\mathbf{S}(x_1, y_1, z_1)$ is a point on the line $\mathcal{L}$ of intersection of non-parallel planes $\mathcal{P}_1$ and $\mathcal{P}_2$ with normal vectors $\mathbf{n}_1$ and $\mathbf{n}_2$, respectively, and if $\mathbf{U}(x, y, z)$ represents a point in space, then

$$\mathbf{u} = \mathbf{s} + r(\mathbf{n}_1 \times \mathbf{n}_2), \quad r \in \mathcal{R}$$

is a parametric vector equation for $\mathcal{L}$.

Example 2. Find parametric vector, parametric Cartesian, and symmetric equations for the line $\mathcal{L}$ that contains $\mathbf{S}(3, -1, 5)$ and is parallel to the line of intersection of the planes with equations $x - 2y + z = 2$ and $2x + y - z = 6$.

Solution: By inspection, normal vectors to the planes are $\mathbf{n}_1 = (1, -2, 1)$ and $\mathbf{n}_2 = (2, 1, -1)$, so that a direction vector for the line $\mathcal{M}$ of intersection is

$$\mathbf{n}_1 \times \mathbf{n}_2 = \begin{vmatrix} 1 & -2 & 1 \\ 2 & 1 & -1 \\ \mathbf{i} & \mathbf{j} & \mathbf{k} \end{vmatrix} = \mathbf{i} + 3\mathbf{j} + 5\mathbf{k} = (1, 3, 5).$$

Since $\mathcal{L}$ is parallel to $\mathcal{M}$ and contains $\mathbf{S}$, a parametric vector equation for $\mathcal{L}$ is

$$\mathbf{u} = (3, -1, 5) + r(1, 3, 5), \quad r \in \mathcal{R}.$$

Parametric Cartesian equations are

$$x = 3 + r, y = -1 + 3r, z = 5 + 5r;$$

and symmetric equations are

$$x - 3 = \frac{y + 1}{3} = \frac{z - 5}{5}.$$

Several possible relationships exist for the relative positions and the common intersection of three planes in space.

1. If all three of the planes are parallel [Figure 9–5(a), (b), (c)], then there is no common intersection unless all of the planes are coincident, in which case the common intersection is the entire plane.

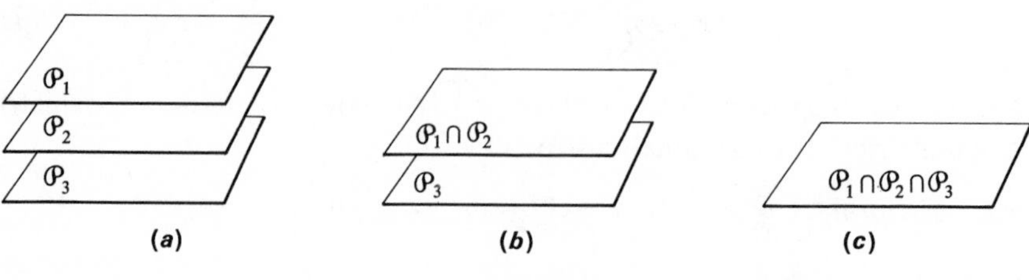

Figure 9–5

2. If two, but not all three, of the planes are parallel [Figure 9–6(a), (b)], then there is no common intersection unless the two parallel planes are coincident, in which case the common intersection is a line.

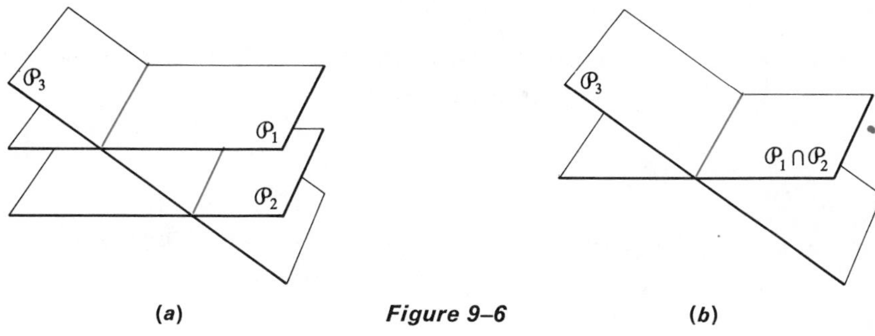

(*a*) **Figure 9–6** (*b*)

3. If no two of the planes are parallel, but their lines of intersection are parallel [Figure 9–7(a), (b)], then there is no common intersection unless the lines of intersection are coincident, in which case the common intersection is a line.

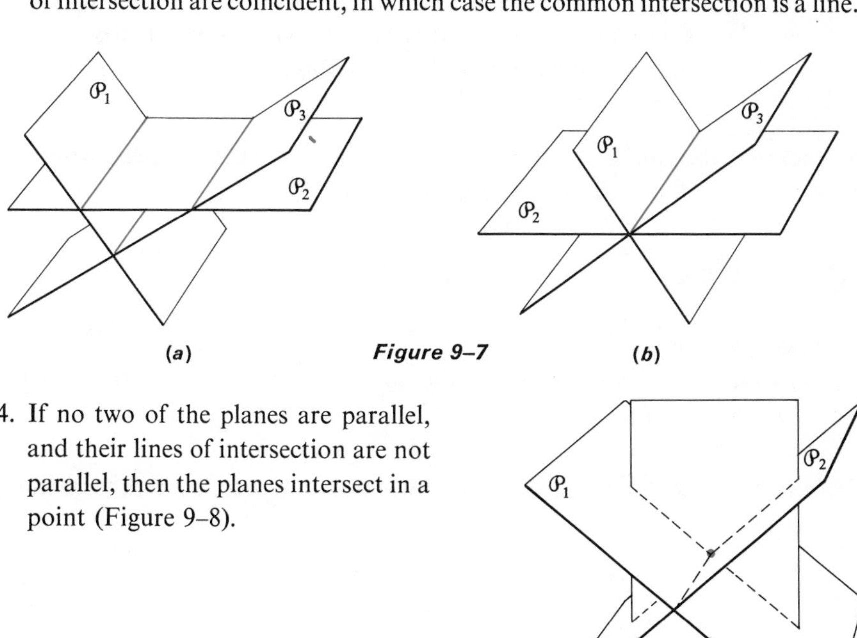

(*a*) **Figure 9–7** (*b*)

4. If no two of the planes are parallel, and their lines of intersection are not parallel, then the planes intersect in a point (Figure 9–8).

Figure 9–8

If three planes intersect in a single point, then the coordinates of the point can be found by solving three simultaneous linear equations representing the planes. If the equations do not have a unique solution, then the planes do not intersect in a single point.

Example 3. Determine the common intersection of the planes with equations

$$\begin{aligned} x + y + 4z &= -1, \\ -2x - y + z &= -5, \\ 3x - 2y + 3z &= -4. \end{aligned}$$

Solution: If you eliminate y between the first two equations and between the first and third equations, you obtain the system

$$\begin{aligned} -x + 5z &= -6, \\ 5x + 11z &= -6. \end{aligned}$$

Then, by eliminating x between these two equations, you obtain

$$36z = -36,$$

or

$$z = -1.$$

By substitution, you find that $x = 1$ and $y = 2$. Thus there can be no common solution other than $(x, y, z) = (1, 2, -1)$. Checking, you find that $(x, y, z) = (1, 2, -1)$ satisfies all three of the original equations. Therefore, the planes intersect in the single point $\mathbf{S}(1, 2, -1)$.

Note that Example 3 could also have been solved using Cramer's Rule (see page 106) since the coefficient determinant

$$D = \begin{vmatrix} 1 & 1 & 4 \\ -2 & -1 & 1 \\ 3 & -2 & 3 \end{vmatrix}$$

is not equal to zero.

Sometimes, as mentioned above, the common intersection of three planes is a plane, a line, or $\emptyset$.

Example 4. Determine the common intersection of the planes with equations

$$\begin{aligned} x + y + z &= 0, \\ 2x + y - z &= 3, \\ 4x + 3y + z &= 5. \end{aligned}$$

Solution: By eliminating z between the first two equations and between the second and third equations, you obtain the system

$$\begin{aligned} 3x + 2y &= 3, \\ 6x + 4y &= 8. \end{aligned}$$

By inspection, these two equations are inconsistent. Hence, there is no triple of values (x, y, z) that satisfies all three equations, and the solution set is $\emptyset$.

The intersection of a plane $\mathcal{P}$ in space with one of the coordinate planes is called the **trace** of $\mathcal{P}$ in that coordinate plane. You can often use the traces of a plane to help picture its graph. Figure 9–9 shows the part of the plane with equation $2x + 6y + 3z - 12 = 0$ that lies in the first octant.

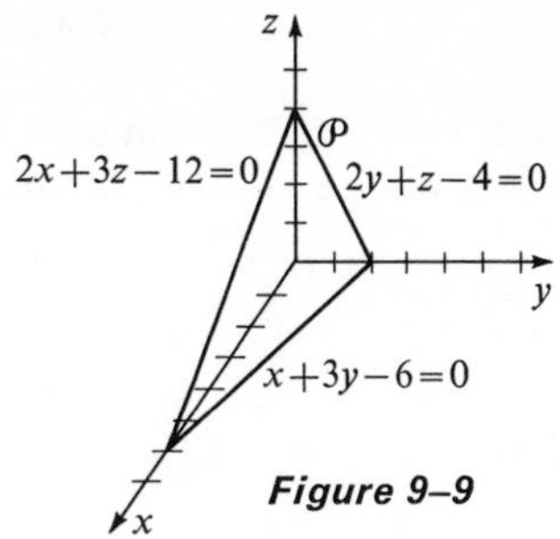

Figure 9–9

Since all points in the xy-plane have z-coordinate 0, you can find an equation for the trace of $\mathcal{P}$ in the xy-plane by setting $z = 0$ in the equation for $\mathcal{P}$. You have

$$2x + 6y + 3(0) - 12 = 0,$$
$$2x + 6y - 12 = 0,$$

or

$$x + 3y - 6 = 0.$$

Similarly, you can find an equation for the yz-trace by setting $x = 0$ in the equation for $\mathcal{P}$. You have

$$2y + z - 4 = 0.$$

Finally, by setting $y = 0$ in the equation for $\mathcal{P}$, you obtain an equation for the xz-trace:

$$2x + 3z - 12 = 0.$$

Example 5. Sketch the graph of $5x + 4z - 2y = 4$ in the first octant.

Solution: Determine and sketch the traces of the plane in the first octant.

xy-trace: $5x - 2y = 4$
xz-trace: $5x + 4z = 4$
yz-trace: $4z - 2y = 4$

Then indicate the portion of the plane in the first octant, as shown at the right.

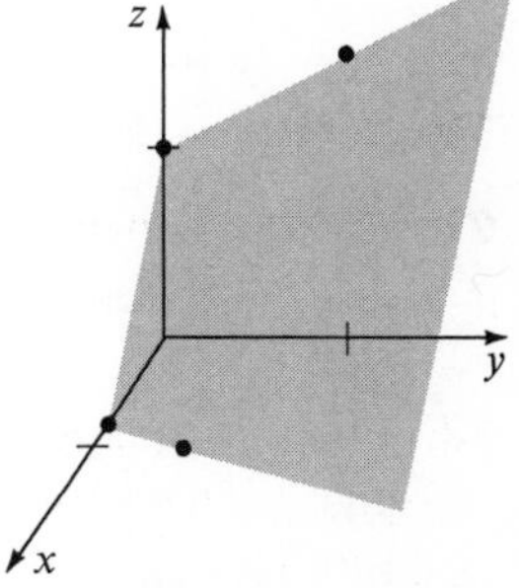

The **intercepts** of a plane, that is, the points where the plane intersects the coordinate axes, can also help you to sketch the plane. The x-, y-, and z-intercepts of a plane have coordinates of the form $(a, 0, 0)$, $(0, b, 0)$, and $(0, 0, c)$, respectively. Since these intercepts are also the points in which the *traces* intersect the coordinate axes, they can sometimes be used to provide immediate sketches of the traces.

Exercises 9–3

In Exercises 1–4, find a parametric vector equation for the line of intersection of the pair of planes whose equations are given.

1. $x - 2y + z = 0, 3x + y + 2z - 7 = 0$
2. $2x - y + 3z + 1 = 0$, $5x + 4y - z - 17 = 0$
3. $2x + 3y - 2 = 0, y - 3z + 4 = 0$
4. $x + 2y - 6 = 0, z = 4$

In Exercises 5–8, find (a) parametric Cartesian equations and (b) symmetric Cartesian equations for the line of intersection of the pair of planes whose equations are given.

5. $3x + y - z - 6 = 0, 4x - 2y - 3z + 2 = 0$
6. $2x - 3y + z - 5 = 0$, $2x - y + 8z - 3 = 0$
7. $x + y + 3z - 1 = 0$, $2x - 3y + z - 7 = 0$
8. $2x + 3y - 2z + 2 = 0$, $3x - y + 3z - 19 = 0$

In Exercises 9–14, determine the common intersection of the three planes whose equations are given.

9. $x + y + z - 6 = 0$, $2x + 2y - 3z + 3 = 0$, $2x - 3y + 2z - 2 = 0$
10. $2x + 5y + 2z - 1 = 0$, $x - 4y + 3z - 7 = 0$, $5x + 10y - 4z - 5 = 0$
11. $2x - y + 3z - 4 = 0$, $x + 4y - 5z = 0$, $13x - 2y + 13z - 4 = 0$
12. $x + y + 2z - 5 = 0$, $2x - y + 3z + 7 = 0$, $x - 2y + z + 13 = 0$
13. $2x + y + z - 7 = 0$, $x - 2y + z - 2 = 0$, $3x + 4y + z - 12 = 0$
14. $3x + 2y - 2z + 2 = 0$, $x - y + 2z - 11 = 0$, $9x + y + 2z - 20 = 0$

In Exercises 15–20, sketch the graph in the first octant of the plane with the given equation.

15. $x + 2y + 3z = 6$
16. $x + 2y = 5$
17. $y + 3z = 6$
18. $y = 4$
19. $x = 3$
20. $x + y - z = 1$

21. Find an equation for the plane which is parallel to the plane with x-, y-, and z-intercepts 3, -1, and 2, respectively, and which contains the point $S(5, -8, 3)$.
22. Find an equation for the plane which is parallel to the plane with x-, y-, and z-intercepts -1, 3, and 5, respectively, and which contains the point $S(0, 1, -1)$.
23. Find symmetric equations for the line which contains the point $\mathbf{S}(-3, 1, 6)$ and which is parallel to the line of intersection of the planes with equations $3x + 2y - z = 4$ and $x - y + 2z = 6$.
24. Find symmetric equations for the line which contains the point $\mathbf{S}(-1, 6, 2)$ and which is parallel to the line of intersection of the planes with equations $4x - y + 2z = 7$ and $x + 3y + z = 2$.

* **25.** A line perpendicular to a plane might be assigned either of two opposite directions. When the assignment has been made, the plane is **oriented**. If the equation $ax + by + cz + d = 0$ of the plane is given, then $\mathbf{n} = (a, b, c)$ [not $-\mathbf{n} = (-a, -b, -c)$] will be understood to be its normal vector unless a statement to the contrary is made. The dihedral angle θ, $0 \le m^\circ(\theta) \le 180$, between two oriented planes is defined to have the same measure as the angle between the normals to the planes. (See diagram.) Show that for the intersecting planes with equations

$$a_1x + b_1y + c_1z + d_1 = 0$$

and

$$a_2x + b_2y + c_2z + d_2 = 0,$$

$$\cos\theta = \frac{a_1a_2 + b_1b_2 + c_1c_2}{\sqrt{a_1^2 + b_1^2 + c_1^2}\sqrt{a_2^2 + b_2^2 + c_2^2}}.$$

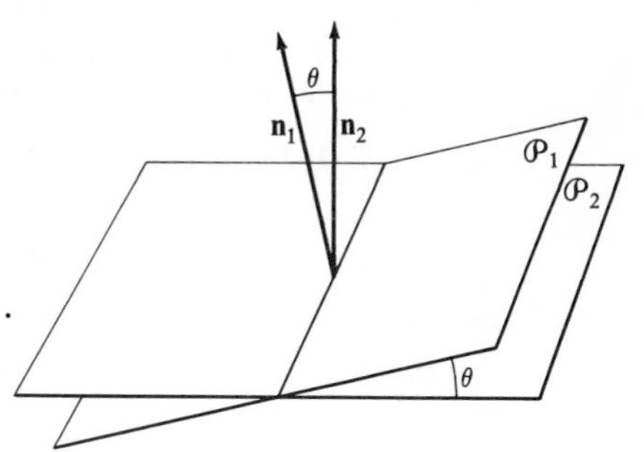

In Exercises 26–30, use the result in Exercise 25 to determine the cosine of the dihedral angle between the oriented planes with the given equations.

* **26.** $4x - 2y + z = 0$ and $2x + 4y - z + 2 = 0$

* **27.** $x - 2y + 2z + 3 = 0$ and $2x - y - 2z - 1 = 0$

* **28.** $4x + 2y - 6z + 3 = 0$ and $2x - y + 3z + 5 = 0$

* **29.** $x + 3y + 4z = 5$ and $z = 0$

* **30.** $2x + 3y + 2z = 3$ and $y = 0$

* **31.** Show that if $\mathbf{n}_1 = (a_1, b_1, c_1)$, $\mathbf{n}_2 = (a_2, b_2, c_2)$, and $\mathbf{n}_3 = (a_3, b_3, c_3)$, then the determinant

$$D = \begin{vmatrix} a_1 & b_1 & c_1 \\ a_2 & b_2 & c_2 \\ a_3 & b_3 & c_3 \end{vmatrix}$$

satisfies the equation

$$D = \mathbf{n}_1 \cdot (\mathbf{n}_2 \times \mathbf{n}_3) = (\mathbf{n}_1 \times \mathbf{n}_2) \cdot \mathbf{n}_3 = (\mathbf{n}_3 \times \mathbf{n}_1) \cdot \mathbf{n}_2.$$

In Exercises 32 and 33, $\mathcal{P}_1$, $\mathcal{P}_2$, and $\mathcal{P}_3$ are planes with normal vectors $\mathbf{n}_1 = (a_1, b_1, c_1)$, $\mathbf{n}_2 = (a_2, b_2, c_2)$, and $\mathbf{n}_3 = (a_3, b_3, c_3)$, respectively.

* **32.** Use the result in Exercise 31 to show that $D = 0$ if and only if either $\mathcal{P}_1$ is parallel to the line of intersection of $\mathcal{P}_2$ and $\mathcal{P}_3$, or $\mathcal{P}_2$ and $\mathcal{P}_3$ are parallel to each other.

* **33.** Use the result in Exercise 31 to show that if no two of the planes $\mathcal{P}_1$, $\mathcal{P}_2$, and $\mathcal{P}_3$ are parallel and if $\mathcal{P}_1$ is parallel to the intersection of $\mathcal{P}_2$ and $\mathcal{P}_3$, then the three lines of intersection of the planes are parallel.

* **34.** Show that the planes with equations

$$a_1x + b_1y + c_1z + d_1 = 0$$
$$a_2x + b_2y + c_2z + d_2 = 0$$

are parallel and distinct if and only if there is a constant k such that $a_2 = ka_1$, $b_2 = kb_1$, $c_2 = kc_1$, but $d_2 \neq kd_1$.

In Exercises 35–42, for the planes $\mathcal{P}_1$, $\mathcal{P}_2$, and $\mathcal{P}_3$ represented by the given equations (a), (b), and (c), respectively, find equations for the intersection of $\mathcal{P}_1$ and $\mathcal{P}_2$, of $\mathcal{P}_1$ and $\mathcal{P}_3$, and of $\mathcal{P}_2$ and $\mathcal{P}_3$, and then find the coordinates of the common intersection of the three planes.

* **35.** (a) $-2x + y - z + 2 = 0$
(b) $x + y - 2z + 2 = 0$
(c) $2x - y + z + 2 = 0$

* **36.** (a) $2x - y + z + 2 = 0$
(b) $x + 3y - 2z - 3 = 0$
(c) $3x + 2y - z + 1 = 0$

* **37.** (a) $x - y + z + 3 = 0$
(b) $2x - 2y + 2z + 7 = 0$
(c) $-x + y - z - 7 = 0$

* **38.** (a) $x + 2y + 3z = 0$
(b) $2x + 3y + 4z = 0$
(c) $3x + 4y + 4z - 1 = 0$

* **39.** (a) $2x - y + 3z + 4 = 0$
(b) $-2x + y - 3z - 4 = 0$
(c) $8x - 4y + 12z + 16 = 0$

* **40.** (a) $-2x - 6y + 2z - 2 = 0$
(b) $x + 3y - z + 1 = 0$
(c) $x + y - z = 0$

* **41.** (a) $x - y + 3z + 4 = 0$
(b) $2x - 2y + 6z + 8 = 0$
(c) $-3x + 3y - 9z + 12 = 0$

* **42.** (a) $7x - 4y + z + 1 = 0$
(b) $4x - y + 2z - 2 = 0$
(c) $x + 2y + 3z - 5 = 0$

9–4 Intersections of Lines and Planes

For a line and a plane in space, there are three possible configurations [Figures 9–10 (a), (b), (c)]: either the line is parallel to but does not intersect the plane, or it is parallel to and lies wholly in the plane, or it intersects the plane in a single point.

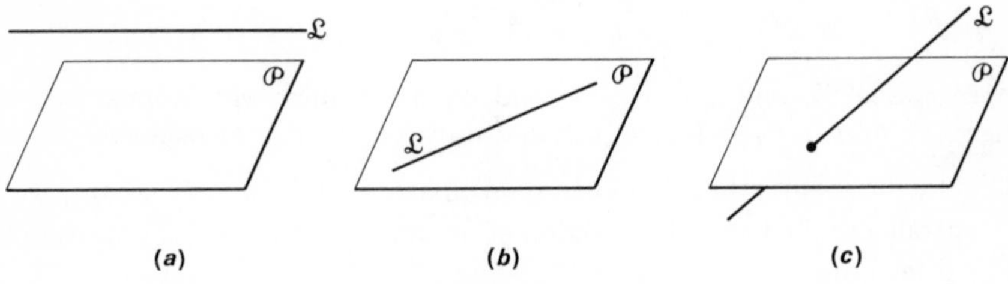

Figure 9–10

As you saw on page 315, a line $\mathcal{L}$ is parallel to a plane $\mathcal{P}$ if and only if a direction vector of $\mathcal{L}$ is perpendicular to a normal vector to $\mathcal{P}$.

If $\mathcal{L}$ is not parallel to $\mathcal{P}$, then $\mathcal{L}$ intersects $\mathcal{P}$ in a single point (see Exercises 25 and 26, page 330). You can use equations for $\mathcal{L}$ and $\mathcal{P}$ to identify the coordinates of this point.

Example 1. Find the coordinates of the point **S** of intersection of the plane with Cartesian equation

$$2x + 3y - z = 1$$

and the line with symmetric equations

$$\frac{x-1}{3} = \frac{y+2}{1} = \frac{z-2}{2}.$$

Solution: Observe first from the given direction numbers for the line and the given equation of the plane that the line is not parallel to the plane. That is,

$$(3, 1, 2) \cdot (2, 3, -1) = 6 + 3 - 2 \neq 0.$$

Next, if you solve the equation $\frac{x-1}{3} = \frac{y+2}{1}$ for x in terms of y, you have $x = 3y + 7$; and if you solve the equation $\frac{y+2}{1} = \frac{z-2}{2}$ for z in terms of y, you have $z = 2y + 6$. These equations express relationships between x and y, and between z and y, for every point $\mathbf{U}(x, y, z)$ on the line. In particular, they hold for the point **S** of intersection of the line and the plane.

Upon substituting the values $3y + 7$ and $2y + 6$ for x and z, respectively, in the equation for the plane, you obtain

$$\begin{aligned} 2(3y + 7) + 3y - (2y + 6) &= 1, \\ 6y + 14 + 3y - 2y - 6 &= 1, \\ 7y &= -7, \\ y &= -1. \end{aligned}$$

Thus -1 is the y-coordinate of the desired point of intersection. Finally, substituting -1 for y in the equations $x = 3y + 7$ and $z = 2y + 6$, you have $x = 4$ and $z = 4$. Hence, the point $\mathbf{S}(4, -1, 4)$ is the only point that can lie both on the line and in the plane, and you can verify that these coordinates do satisfy the given equations. Hence, the point $\mathbf{S}(4, -1, 4)$ is the unique point of intersection.

On page 315, you saw that a line $\mathcal{L}$ is perpendicular to a plane $\mathcal{P}$ if and only if a direction vector of $\mathcal{L}$ is parallel to a normal vector to $\mathcal{P}$. You can use this fact to help you determine equations for some planes.

Example 2. Find a Cartesian equation for the plane $\mathcal{P}$ that contains the point $\mathbf{Q}(2, -1, 0)$ and is perpendicular to the line containing the points $\mathbf{S}(3, -1, 2)$ and $\mathbf{T}(5, 7, -3)$.

Solution: First, find a direction vector $\mathbf{v}$ for the line containing $\mathbf{S}$ and $\mathbf{T}$. You have

$$\begin{aligned}\mathbf{v} &= (5, 7, -3) - (3, -1, 2)\\ &= (5 - 3, 7 - (-1), -3 - 2) = (2, 8, -5).\end{aligned}$$

Since the vector $\mathbf{v}$ is a normal to the plane $\mathcal{P}$, a Cartesian equation for $\mathcal{P}$ is of the form

$$2x + 8y - 5z + d = 0.$$

Then, because $\mathbf{Q}$ is contained in $\mathcal{P}$, its coordinates must satisfy this equation, and you have $2(2) + 8(-1) - 5(0) + d = 0$, or $d = 4$. Therefore a Cartesian equation of the plane $\mathcal{P}$ is $2x + 8y - 5z + 4 = 0$.

Exercises 9–4

In Exercises 1–6, find the coordinates of the point **S** of intersection of the line and the plane having the given equations.

1. $\dfrac{x-2}{1} = \dfrac{y+1}{-1} = \dfrac{z+6}{-1}$; $3x - 2y + 3z + 16 = 0$
2. $\dfrac{x-1}{1} = \dfrac{y+2}{2} = \dfrac{z-3}{4}$; $x + 4y - z + 5 = 0$
3. $\dfrac{x-3}{1} = \dfrac{y+8}{-2} = \dfrac{z+6}{-11}$; $3x + y + 2z - 10 = 0$
4. $\dfrac{x-2}{5} = \dfrac{y-1}{2} = \dfrac{z-5}{7}$; $2x + 3y - 3z + 3 = 0$
5. $\dfrac{x+3}{2} = \dfrac{y-5}{3} = \dfrac{z+4}{2}$; $5x - 2y + 3z + 17 = 0$
6. $\dfrac{x-4}{3} = \dfrac{y+2}{-2} = \dfrac{z-1}{4}$; $4x + 3y + 3z + 5 = 0$
7. Find an equation for the plane that contains the point $\mathbf{S}(1, 1, 1)$ and is perpendicular to the line of intersection of the planes with equations $2x - y + z - 5 = 0$ and $x + 2y + 2z - 5 = 0$.
8. Find an equation for the plane that contains the point $\mathbf{S}(2, 1, -4)$ and is perpendicular to the line of intersection of the planes with equations $4x - 3y + 2z - 7 = 0$ and $x + 4y - z - 5 = 0$.
9. Find a Cartesian equation for the plane that contains the point $\mathbf{Q}(3, -3, -4)$ and is perpendicular to the line containing the points $\mathbf{S}(0, 6, -6)$ and $\mathbf{T}(5, 2, 2)$.

10. Find a Cartesian equation for the plane that contains the point **Q**$(-4, 0, 3)$ and is perpendicular to the line containing the points **S**$(4, 10, -3)$ and **T**$(4, 14, -8)$.

11. Find an equation for the plane that contains the points **S**$(1, 2, 3)$ and **T**$(3, -1, 0)$ and is parallel to the line of intersection of the planes with equations $x + y + z - 3 = 0$ and $x + 2y - 3z + 5 = 0$.

12. Find an equation for the plane that contains the points **S**$(3, 0, 2)$ and **T**$(4, 1, -1)$ and is parallel to the line of intersection of the planes with equations $x - 2y + z - 2 = 0$ and $2x + 3y - 2z - 3 = 0$.

13. Find a Cartesian equation for the plane that contains the point **S**$(-6, 1, -3)$ and is perpendicular to a line whose three direction cosines are all equal.

14. Find a Cartesian equation for the plane which contains the line of intersection of the planes with equations $x + 3y - z = 6$ and $x + y = 4$ and also contains the origin.

15. Find a Cartesian equation for the plane that contains the point **S**$(5, -2, 3)$ and is parallel both to the x-axis and to the z-axis.

16. Find a Cartesian equation for the plane that contains the point **S**$(-3, 5, 1)$ and is parallel both to the y-axis and to the z-axis.

17. Find the length of the segment cut off on the line with symmetric equations $\frac{x - 3}{1} = \frac{y + 2}{-1} = \frac{z + 7}{-1}$ by the planes with Cartesian equations $3x - 2y + 3z + 16 = 0$ and $3x - 2y + 3z = 0$.

18. Find the length of the segment cut off on the line with symmetric equations $\frac{x - 1}{5} = \frac{y}{2} = \frac{z - 4}{7}$ by the planes with Cartesian equations $2x + 3y - 3z + 5 = 0$ and $2x + 3y - 3z - 5 = 0$.

19. A line $\mathcal{L}$ containing the point **S**$(2, -5, 8)$ is perpendicular to the plane $\mathcal{P}$ with Cartesian equation $x - 2y + 3z - 8 = 0$. Find the coordinates of the point of intersection of $\mathcal{L}$ and $\mathcal{P}$.

20. A line $\mathcal{L}$ containing the point **S**$(7, 3, -5)$ is perpendicular to the plane $\mathcal{P}$ with Cartesian equation $x - y - z - 3 = 0$. Find the coordinates of the point of intersection of $\mathcal{L}$ and $\mathcal{P}$.

21. Find the coordinates of the point of intersection of the plane $\mathcal{P}$ with Cartesian equation $2x + y + z = 6$ and the line through the origin perpendicular to $\mathcal{P}$.

22. Find the coordinates of the point of intersection of the plane $\mathcal{P}$ with Cartesian equation $x - 2y - 2z = 3$ and the line through the origin perpendicular to $\mathcal{P}$.

* **23.** Show that the coordinates of the point of intersection of the plane $\mathcal{P}$ with Cartesian equation $ax + by + cz = d$ and the line through the origin perpendicular to $\mathcal{P}$ are

$$\left(\frac{ad}{a^2 + b^2 + c^2}, \frac{bd}{a^2 + b^2 + c^2}, \frac{cd}{a^2 + b^2 + c^2}\right).$$

* **24.** Use the result in Exercise 23, together with the distance formula, to show that the distance from the origin to the plane $\mathcal{P}$ with Cartesian equation $ax + by + cz = d$ is given by

$$\frac{|d|}{\sqrt{a^2 + b^2 + c^2}}.$$

* **25.** Let $\mathbf{S}(x_1, y_1, z_1)$ be a point on the line $\mathcal{L}$ with direction vector $\mathbf{v} = (a_1, b_1, c_1)$. Parametric Cartesian equations for $\mathcal{L}$ are

$$x = x_1 + ra_1, y = y_1 + rb_1, z = z_1 + rc_1.$$

Let $\mathcal{P}$ be a plane with Cartesian equation $ax + by + cz = d$ which is not parallel to $\mathcal{L}$. Show that for any point of intersection of $\mathcal{L}$ with the plane $\mathcal{P}$, you have

$$r = \frac{d - ax_1 - by_1 - cz_1}{aa_1 + bb_1 + cc_1}.$$

* **26.** In Exercise 25, show that the point with the given value of r actually is on $\mathcal{P}$.

Distance Formulas

9–5 Distance between a Point and a Plane

From Equation (1) on page 287, it follows that for nonzero vectors $\mathbf{u}$ and $\mathbf{v}$ in $\mathcal{R}^3$, you have

$$\|\mathbf{v}\| \cos \theta = \frac{\mathbf{u} \cdot \mathbf{v}}{\|\mathbf{u}\|},$$

where θ is the angle between $\mathbf{u}$ and $\mathbf{v}$. As you have seen (pages 290 and 291), the number $\frac{\mathbf{u} \cdot \mathbf{v}}{\|\mathbf{u}\|}$ represents the scalar component of $\mathbf{v}$ parallel to $\mathbf{u}$. You can use this fact to find the (perpendicular) distance between a point $\mathbf{S}$ and a plane $\mathcal{P}$ in space in the same way that you find the distance between a point $\mathbf{S}$ and a line $\mathcal{L}$ in the plane (page 89). Thus, if $\mathbf{S}$ is a point and $\mathcal{P}$ is a plane, and if $\mathbf{T}$ is any point on $\mathcal{P}$ and $\mathbf{n}$ is a normal vector to $\mathcal{P}$, then the distance between $\mathbf{S}$ and $\mathcal{P}$ is equal to the absolute value of the scalar component of $\mathbf{s} - \mathbf{t}$ parallel to $\mathbf{n}$. That is,

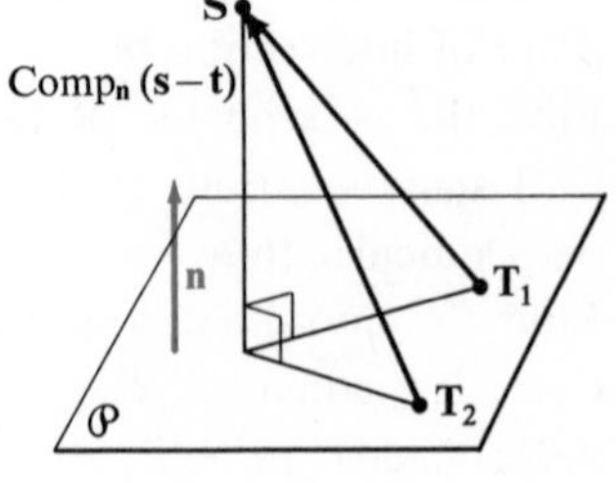

Figure 9–11

$$d(\mathbf{S}, \mathcal{P}) = \frac{|(\mathbf{s} - \mathbf{t}) \cdot \mathbf{n}|}{\|\mathbf{n}\|}. \tag{1}$$

Figure 9–11 illustrates the fact that $d(\mathbf{S}, \mathcal{P})$ does not depend on the selection of a specific point $\mathbf{T}$ on $\mathcal{P}$. The scalar component of $\mathbf{s} - \mathbf{t}$ parallel to $\mathbf{n}$ is the same for all points $\mathbf{T}$ on $\mathcal{P}$. That is, for any points $\mathbf{T}_1$ and $\mathbf{T}_2$ on $\mathcal{P}$, you have $\text{Comp}_{\mathbf{n}}(\mathbf{s} - \mathbf{t}_1) = \text{Comp}_{\mathbf{n}}(\mathbf{s} - \mathbf{t}_2)$.

Example 1. Find the distance between the point $\mathbf{S}(3, 2, 4)$ and the plane $\mathcal{P}$ with equation $4x - 3y + z = 2$.

Solution: You can find a point $\mathbf{T}$ on $\mathcal{P}$ by arbitrarily assigning two components to $\mathbf{T}$ and then substituting these values in the equation for $\mathcal{P}$ to determine the third component. If you replace both x and y in the given equation with 0, you obtain $z = 2$, so that the point $\mathbf{T}(0, 0, 2)$ lies on $\mathcal{P}$. Then the vector $\mathbf{s} - \mathbf{t}$ is $(3, 2, 4) - (0, 0, 2) = (3, 2, 2)$. By inspection of the given equation for $\mathcal{P}$, a normal to $\mathcal{P}$ is $\mathbf{n} = (4, -3, 1)$. Thus from Equation (1),

$$\begin{aligned} d(\mathbf{S}, \mathcal{P}) &= \frac{|(3, 2, 2) \cdot (4, -3, 1)|}{\sqrt{4^2 + (-3)^2 + 1^2}} \\ &= \frac{|12 - 6 + 2|}{\sqrt{16 + 9 + 1}} \\ &= \frac{8}{\sqrt{26}}. \end{aligned}$$

To find a Cartesian expression for $d(\mathbf{S}, \mathcal{P})$, observe that the Cartesian equation $ax + by + cz + d = 0$ for $\mathcal{P}$ can be written as $(x, y, z) \cdot (a, b, c) + d = 0$; that is, for any point $\mathbf{T}$ on $\mathcal{P}$, you have $\mathbf{t} \cdot \mathbf{n} + d = 0$, or $\mathbf{t} \cdot \mathbf{n} = -d$. With this in mind, let the coordinates of $\mathbf{S}$ be (x_1, y_1, z_1) and substitute in Equation (1) on page 330. You obtain

$$\begin{aligned} d(\mathbf{S}, \mathcal{P}) &= \frac{|(\mathbf{s} - \mathbf{t}) \cdot \mathbf{n}|}{\|\mathbf{n}\|} = \frac{|\mathbf{s} \cdot \mathbf{n} - \mathbf{t} \cdot \mathbf{n}|}{\|\mathbf{n}\|} \\ &= \frac{|(x_1, y_1, z_1) \cdot (a, b, c) + d|}{\sqrt{a^2 + b^2 + c^2}}, \end{aligned}$$

or

■ $$d(\mathbf{S}, \mathcal{P}) = \frac{|ax_1 + by_1 + cz_1 + d|}{\sqrt{a^2 + b^2 + c^2}}. \tag{2}$$

Example 2. Find the distance between the point $\mathbf{S}(3, -1, 2)$ and the plane $\mathcal{P}$ with equation $x + 2y - 2z = 6$.

Solution: Using Equation (2), you have

$$d(\mathbf{S}, \mathcal{P}) = \frac{|1(3) + 2(-1) - 2(2) - 6|}{\sqrt{1^2 + 2^2 + (-2)^2}} = \frac{|-9|}{\sqrt{9}} = \frac{9}{3} = 3.$$

You can see from Equation (2) that the distance between the origin $\mathbf{O}(0, 0, 0)$ and the plane $\mathcal{P}$ with equation $ax + by + cz + d = 0$ is just

$$d(\mathbf{O}, \mathcal{P}) = \frac{|a(0) + b(0) + c(0) + d|}{\sqrt{a^2 + b^2 + c^2}} = \frac{|d|}{\sqrt{a^2 + b^2 + c^2}}.$$

Now if you divide both the left-hand and right-hand members of the equation $ax + by + cz + d = 0$ by $\sqrt{a^2 + b^2 + c^2}$ or $-\sqrt{a^2 + b^2 + c^2}$, you have

$$\frac{a}{\sqrt{a^2 + b^2 + c^2}}x + \frac{b}{\sqrt{a^2 + b^2 + c^2}}y + \frac{c}{\sqrt{a^2 + b^2 + c^2}}z + \frac{d}{\sqrt{a^2 + b^2 + c^2}} = 0$$

or

$$\frac{a}{-\sqrt{a^2 + b^2 + c^2}}x + \frac{b}{-\sqrt{a^2 + b^2 + c^2}}y + \frac{c}{-\sqrt{a^2 + b^2 + c^2}}z + \frac{d}{-\sqrt{a^2 + b^2 + c^2}} = 0, \tag{3}$$

respectively. Notice that the coefficients of x, y, and z in Equations (3) are a set of direction cosines for $\mathbf{n} = (a, b, c)$ or $-\mathbf{n} = (-a, -b, -c)$, respectively, and that the absolute value of the constant term is the distance between the origin and the plane $\mathcal{P}$. This permits you to write Equations (3) as

$$x \cos \alpha + y \cos \beta + z \cos \gamma - p = 0, \tag{4}$$

where $p = \pm \dfrac{d}{\sqrt{a^2 + b^2 + c^2}}$ is the *directed* distance from the origin to the plane $\mathcal{P}$ (and $|p|$ is the distance between the origin and $\mathcal{P}$). The form (4) is called the **normal form** of the Cartesian equation of a plane. Since either sign may be chosen in determining p, each plane not containing the origin has two equations in normal form.

You can use Equation (4), as an alternative to the distance formula, to determine the distance between a given point and a given plane.

Example 3. For each of the points $\mathbf{Q}(1, 2, 6)$, $\mathbf{R}(1, 2, -1)$, $\mathbf{S}(2, 4, -2)$, and $\mathbf{T}(5, 1, -8)$, determine the distance between the point and the plane $\mathcal{P}$ with Cartesian equation $3x - 2y + 6z + 14 = 0$.

Solution: *Plan:* First find the directed distance from the origin to the plane $\mathcal{P}$. Then find the directed distance from the origin to the plane that is parallel to $\mathcal{P}$ and contains the given point. Finally, find the absolute value of the difference of these two directed distances.

In the equation for $\mathcal{P}$, you have $a = 3$, $b = -2$, and $c = 6$. Thus $\sqrt{a^2 + b^2 + c^2} = 7$, and you can write the equation for $\mathcal{P}$ in normal form by dividing by 7:

$$\tfrac{3}{7}x - \tfrac{2}{7}y + \tfrac{6}{7}z + 2 = 0.$$

Then, from Equation (4), the directed distance p from the origin to the plane $\mathcal{P}$ is -2.

The normal form of the equation of a plane parallel to $\mathcal{P}$ is

$$\tfrac{3}{7}x - \tfrac{2}{7}y + \tfrac{6}{7}z - p = 0.$$

For the plane parallel to $\mathcal{P}$ and containing $\mathbf{Q}(1, 2, 6)$, you have $\tfrac{3}{7}(1) - \tfrac{2}{7}(2) + \tfrac{6}{7}(6) - p = 0$, or $p = 5$. Similarly, the directed distance p from the origin to the plane parallel to $\mathcal{P}$ and containing

$$\begin{aligned} \mathbf{R} \text{ is } & \tfrac{3}{7}(1) - \tfrac{2}{7}(2) + \tfrac{6}{7}(-1) = -1, \\ \mathbf{S} \text{ is } & \tfrac{3}{7}(2) - \tfrac{2}{7}(4) + \tfrac{6}{7}(-2) = -2, \\ \mathbf{T} \text{ is } & \tfrac{3}{7}(5) - \tfrac{2}{7}(1) + \tfrac{6}{7}(-8) = -5. \end{aligned}$$

You therefore have

$$\begin{aligned} d(\mathbf{Q}, \mathcal{P}) &= |5 - (-2)| = 7, \\ d(\mathbf{R}, \mathcal{P}) &= |-1 - (-2)| = 1, \\ d(\mathbf{S}, \mathcal{P}) &= |-2 - (-2)| = 0, \\ d(\mathbf{T}, \mathcal{P}) &= |-5 - (-2)| = 3. \end{aligned}$$

Exercises 9–5

In Exercises 1–6, find the distance between the given point **S** and the plane $\mathcal{P}$ with the given equation.

1. $\mathbf{S}(2, -1, 4)$; $2x - 2y + z - 3 = 0$
2. $\mathbf{S}(2, 3, 1)$; $4x + 4y - 7z + 14 = 0$
3. $\mathbf{S}(3, -5, 9)$; $3x - 6y + 2z - 8 = 0$
4. $\mathbf{S}(5, -7, -1)$; $x - 2y - 2z + 6 = 0$
5. $\mathbf{S}(2, -1, 7)$; $4x - 8y + z - 5 = 0$
6. $\mathbf{S}(1, -2, 1)$; $14x - 5y - 2z + 4 = 0$

In Exercises 7–12, transform the given Cartesian equation of a plane to normal form, and then find the distance between the origin and the plane.

7. $x - 2y + 2z - 27 = 0$
8. $5x - 4y + 20z + 42 = 0$
9. $7x + 4y - 4z + 45 = 0$
10. $12x - 3y + 4z - 91 = 0$
11. $6x + 3y - 2z - 84 = 0$
12. $5x + 14y + 2z - 150 = 0$

In Exercises 13–18, find the distance between the parallel planes with the given equations.

13. $2x + y - 2z = 12$
$2x + y - 2z = 18$

14. $6x - 2y + 3z = 14$
$6x - 2y + 3z = 35$

15. $x + 8y + 4z = 36$
$x + 8y + 4z = -18$

16. $x - 2y + 2z = -12$
$x - 2y + 2z = 6$

17. $x - 2y + 3z + 6 = 0$
$3x - 6y + 9z - 8 = 0$

18. $-x + y - 6z - 3 = 0$
$2x - 2y + 12z - 11 = 0$

19. Find the distance between the plane with Cartesian equation $4x - 8y + z = 1$ and the line with symmetric equations $\frac{x-2}{15} = \frac{y-10}{9} = \frac{z-1}{12}$.

20. Find the distance between the plane with equation $x + 3y - 4z - 2 = 0$ and the line of intersection of the planes with equations $x - 2y + 3z = 5$ and $2x + y - z = 3$.

21. Find the locus of all points in space equidistant from the planes with equations $2x + 2y - z - 1 = 0$ and $x - 2y + 2z + 1 = 0$.

22. Find the locus of all points in space equidistant from the planes with equations $2x - y + 2z = 0$ and $x + 2y - 2z + 3 = 0$.

23. Find an equation of a plane $\mathcal{P}$ which is parallel to the plane with equation $3x - 2y + 6z = 9$ and is at a distance of 7 units from the origin. (*Hint:* There are two correct answers.)

24. Find an equation of a plane $\mathcal{P}$ which is parallel to the plane with equation $x - 3y + 5z = 8$ and is at a distance of 3 units from the origin. (*Hint:* There are two correct answers.)

9–6 Distance between a Point and a Line (Optional)

The distance between a point and a line in space is defined to be the length of the perpendicular segment joining the point and the line. To find the distance $d(\mathbf{S}, \mathcal{L})$ between a point **S** and a line $\mathcal{L}$ with vector equation

$$\mathbf{u} = \mathbf{t} + r\mathbf{v}, \; r \in \mathcal{R},$$

you can proceed as follows.

Note in Figure 9–12 that the distance $d(\mathbf{S}, \mathbf{T})$ between **S** and the point **T** on $\mathcal{L}$ is $\|\mathbf{s} - \mathbf{t}\|$, and accordingly that

$$d(\mathbf{S}, \mathcal{L}) = \|\mathbf{s} - \mathbf{t}\| \sin \theta. \qquad (1)$$

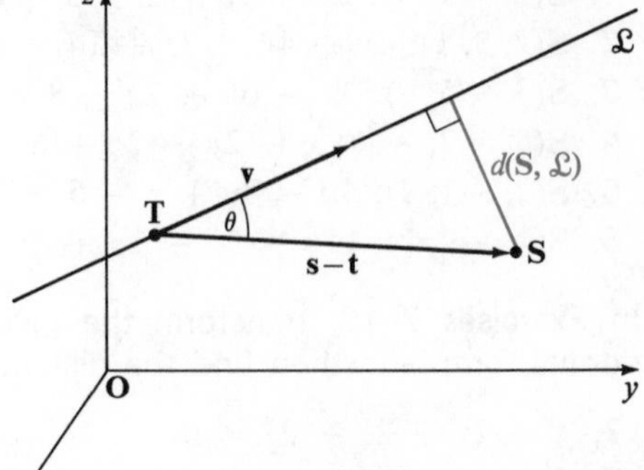

Figure 9–12

By a direct computation (Exercise 35, page 298), you can verify that for any vectors **u** and **v**,

$$(\mathbf{u} \times \mathbf{v}) \cdot (\mathbf{u} \times \mathbf{v}) = (\mathbf{u} \cdot \mathbf{u})(\mathbf{v} \cdot \mathbf{v}) - (\mathbf{u} \cdot \mathbf{v})^2. \tag{2}$$

Since $(\mathbf{u} \cdot \mathbf{v})^2 = (\mathbf{u} \cdot \mathbf{u})(\mathbf{v} \cdot \mathbf{v}) \cos^2 \theta$, where θ is the angle between **u** and **v**, you can write (2) equivalently in the form

$$\begin{aligned} \|\mathbf{u} \times \mathbf{v}\|^2 &= \|\mathbf{u}\|^2\|\mathbf{v}\|^2 (1 - \cos^2 \theta) \\ &= \|\mathbf{u}\|^2\|\mathbf{v}\|^2 \sin^2 \theta. \end{aligned}$$

Thus, if $\|\mathbf{v}\| \neq 0$,

$$\|\mathbf{u}\| \sin \theta = \frac{\|\mathbf{u} \times \mathbf{v}\|}{\|\mathbf{v}\|}. \tag{3}$$

Substituting from Equation (3), with $\mathbf{u} = \mathbf{s} - \mathbf{t}$, into Equation (1), you have

■ $$d(\mathbf{S}, \mathcal{L}) = \frac{\|(\mathbf{s} - \mathbf{t}) \times \mathbf{v}\|}{\|\mathbf{v}\|}. \tag{4}$$

Example. Find the distance between the point $\mathbf{S}(3, -1, 2)$ and the line $\mathcal{L}$ with symmetric equations

$$\frac{x - 2}{1} = \frac{y - 1}{2} = \frac{z}{-1}.$$

Solution: First, by inspection, a point on $\mathcal{L}$ is $\mathbf{T}(2, 1, 0)$ and a direction vector for $\mathcal{L}$ is $\mathbf{v} = (1, 2, -1)$. Then

$$\begin{aligned} \mathbf{s} - \mathbf{t} &= (3, -1, 2) - (2, 1, 0) \\ &= (1, -2, 2), \end{aligned}$$

and

$$\begin{aligned} (\mathbf{s} - \mathbf{t}) \times \mathbf{v} &= (1, -2, 2) \times (1, 2, -1) \\ &= (-2, 3, 4). \end{aligned}$$

Next, you find that

$$\|(\mathbf{s} - \mathbf{t}) \times \mathbf{v}\| = \sqrt{4 + 9 + 16} = \sqrt{29}$$

and

$$\|\mathbf{v}\| = \sqrt{1 + 4 + 1} = \sqrt{6}.$$

Finally, from Equation (4), you have

$$d(\mathbf{S}, \mathcal{L}) = \frac{\sqrt{29}}{\sqrt{6}} = \sqrt{\frac{29}{6}}.$$

Exercises 9–6

In Exercises 1–14, find the distance between the given point **S** and the line with the given equations.

1. $\mathbf{S}(3, 0, 2)$; $x - 1 = y - 1 = z$
2. $\mathbf{S}(2, 3, 1)$; $\dfrac{x}{-1} = \dfrac{y-2}{-2} = \dfrac{z+1}{2}$
3. $\mathbf{S}(7, 6, 5)$; $\dfrac{x-1}{6} = \dfrac{y-1}{7} = \dfrac{z-1}{8}$
4. $\mathbf{S}(3, 1, -2)$; $\dfrac{x-1}{1} = \dfrac{y-2}{-1} = \dfrac{z-1}{2}$
5. $\mathbf{S}(11, 2, 4)$; $\dfrac{x-2}{1} = \dfrac{y-1}{1} = \dfrac{z+2}{-1}$
6. $\mathbf{S}(9, 2, 3)$; $\dfrac{x-3}{-2} = \dfrac{y+2}{1} = \dfrac{z-1}{2}$
7. $\mathbf{S}(1, -1, 2)$; $\dfrac{x-3}{2} = \dfrac{y-2}{-1} = \dfrac{z+3}{3}$
8. $\mathbf{S}(3, -1, -3)$; $\dfrac{x-2}{4} = \dfrac{y+2}{-2} = \dfrac{z+1}{5}$
9. $\mathbf{S}(3, 1, 4)$; $\dfrac{x-3}{4} = \dfrac{y+2}{2} = \dfrac{z-2}{-3}$
10. $\mathbf{S}(-2, 2, 3)$; $\dfrac{x+2}{2} = \dfrac{y-5}{3} = \dfrac{z+2}{-4}$
11. $\mathbf{S}(5, -3, 6)$; $\dfrac{x-3}{3} = \dfrac{y+2}{-4}$ and $z = 4$
12. $\mathbf{S}(1, 5, 15)$; $\dfrac{y+4}{5} = \dfrac{z-3}{12}$ and $x = -3$
13. $\mathbf{S}(0, 0, 0)$; $\dfrac{x-1}{3} = \dfrac{y+1}{-2} = \dfrac{z+3}{-6}$
14. $\mathbf{S}(0, 0, 0)$; $\dfrac{x+3}{-1} = \dfrac{y}{-2} = \dfrac{z}{2}$

* **15.** Let $\mathcal{L}_1$ be a line in space having direction vector $\mathbf{v}_1$ and containing point $\mathbf{T}_1$, and $\mathcal{L}_2$ be a line **skew** to $\mathcal{L}_1$ (nonparallel and nonintersecting) having direction vector $\mathbf{v}_2$ and containing point $\mathbf{T}_2$. Use the fact that $\mathbf{v}_1 \times \mathbf{v}_2$ is a vector perpendicular to both $\mathcal{L}_1$ and $\mathcal{L}_2$ to show that the (perpendicular) distance between $\mathcal{L}_1$ and $\mathcal{L}_2$ is

$$d(\mathcal{L}_1, \mathcal{L}_2) = \frac{|(\mathbf{t}_2 - \mathbf{t}_1) \cdot (\mathbf{v}_1 \times \mathbf{v}_2)|}{\|\mathbf{v}_1 \times \mathbf{v}_2\|}.$$

In Exercises 16 and 17, use the result in Exercise 15 to find the distance between the lines with the given symmetric equations.

* **16.** $\dfrac{x}{2} = \dfrac{y+2}{3} = \dfrac{z-1}{1}$; $\dfrac{x+1}{3} = \dfrac{y-1}{4} = \dfrac{z+1}{-1}$

* **17.** $\dfrac{x-1}{3} = \dfrac{y}{4} = \dfrac{z-5}{-1}$; $\dfrac{x}{2} = \dfrac{y+1}{-1} = \dfrac{z-4}{1}$

Chapter Summary

1. For the line $\mathcal{L}$ containing the (distinct) points $\mathbf{S}(x_1, y_1, z_1)$ and $\mathbf{T}(x_2, y_2, z_2)$, a parametric vector equation is
$$(x, y, z) = (x_1, y_1, z_1) + r(x_2 - x_1, y_2 - y_1, z_2 - z_1);$$
parametric Cartesian equations are
$$x = x_1 + r(x_2 - x_1), y = y_1 + r(y_2 - y_1), z = z_1 + r(z_2 - z_1);$$
and if $\mathcal{L}$ is not parallel to a coordinate plane, symmetric equations are
$$\frac{x - x_1}{x_2 - x_1} = \frac{y - y_1}{y_2 - y_1} = \frac{z - z_1}{z_2 - z_1}.$$
2. If $\mathbf{S}$ is a point on a plane $\mathcal{P}$ and $\mathbf{n}$ is a normal vector to $\mathcal{P}$, then a point $\mathbf{U}$ lies on $\mathcal{P}$ if and only if $(\mathbf{u} - \mathbf{s}) \cdot \mathbf{n} = 0$.
3. The graph of the first-degree Cartesian equation $ax + by + cz + d = 0$, where a, b, and c are not all 0, is a plane $\mathcal{P}$. The vector $\mathbf{n} = (a, b, c)$ is a normal vector to $\mathcal{P}$. If $\mathbf{v} = (v_1, v_2, v_3)$ is a direction vector for a line $\mathcal{L}$, then $\mathcal{L}$ is parallel to $\mathcal{P}$ if and only if $\mathbf{v} \cdot \mathbf{n} = 0$, and $\mathcal{L}$ is perpendicular to $\mathcal{P}$ if and only if $\mathbf{v} \times \mathbf{n} = \mathbf{0}$. Planes $\mathcal{P}_1$ and $\mathcal{P}_2$ with normal vectors $\mathbf{n}_1$ and $\mathbf{n}_2$, respectively, are parallel if and only if $\mathbf{n}_1 \times \mathbf{n}_2 = \mathbf{0}$; they are perpendicular if and only if $\mathbf{n}_1 \cdot \mathbf{n}_2 = 0$.
4. Two planes whose normal vectors are not parallel intersect in a line. If $\mathbf{n}_1$ and $\mathbf{n}_2$ are respective normal vectors for two planes that intersect in a line, then $\mathbf{n}_1 \times \mathbf{n}_2$ is a direction vector for the line of intersection. The intersection of a plane $\mathcal{P}$ with one of the coordinate planes is called the **trace** of $\mathcal{P}$ in that coordinate plane.
5. A line not parallel to a plane intersects the plane in one and only one point.
6. The distance between a point $\mathbf{S}$ and a plane $\mathcal{P}$ containing a point $\mathbf{T}$ is given by
$$d(\mathbf{S}, \mathcal{P}) = \frac{|(\mathbf{s} - \mathbf{t}) \cdot \mathbf{n}|}{\|\mathbf{n}\|},$$
where $\mathbf{n}$ is a normal vector to $\mathcal{P}$. In Cartesian form, the distance is given by
$$d(\mathbf{S}, \mathcal{P}) = \frac{|ax_1 + by_1 + cz_1 + d|}{\sqrt{a^2 + b^2 + c^2}},$$
where the coordinates of $\mathbf{S}$ are (x_1, y_1, z_1), and a Cartesian equation for $\mathcal{P}$ is $ax + by + cz + d = 0$.
7. The **normal form** of the equation of a plane $\mathcal{P}$ with Cartesian equation $ax + by + cz + d = 0$ is $x \cos \alpha + y \cos \beta + z \cos \gamma - p = 0$, where $\cos \alpha$, $\cos \beta$, $\cos \gamma$ are the direction cosines of $\mathbf{n} = (a, b, c)$ or $-\mathbf{n} = (-a, -b, -c)$, and $p = \pm \dfrac{d}{\sqrt{a^2 + b^2 + c^2}}$ is the directed distance from $\mathbf{O}$ to $\mathcal{P}$.

8. (Optional) The distance between a point **S** and a line $\mathcal{L}$ is given by

$$d(\mathbf{S}, \mathcal{L}) = \frac{\|(\mathbf{s} - \mathbf{t}) \times \mathbf{v}\|}{\|\mathbf{v}\|},$$

where **T** is any point on $\mathcal{L}$ and **v** is a direction vector of $\mathcal{L}$.

Chapter Review Exercises

1. Find **(a)** parametric vector, **(b)** parametric Cartesian, and **(c)** symmetric equations for the line containing the points **S**$(3, -1, 2)$ and **T**$(1, 5, 5)$.

2. Find the coordinates of the midpoint of the segment with endpoints **S**$(5, -3, 7)$ and **T**$(3, 5, -3)$.

3. Find a Cartesian equation for the plane containing the point **S**$(3, -1, 5)$ and having normal vector $\mathbf{n} = (-1, 2, 1)$.

4. Find symmetric equations for the line which contains the point **S**$(-1, 5, 3)$ and is perpendicular to the plane with Cartesian equation

$$x - 2y + z - 3 = 0.$$

5. Find a parametric vector equation for the line which contains the origin and is parallel to the line of intersection of the planes with Cartesian equations $x - 2y + z - 4 = 0$ and $2x + y - z - 3 = 0$.

6. Find symmetric equations for the line of intersection of the planes with Cartesian equations $x + y - 2z - 3 = 0$ and $2x - 3y + 2z - 5 = 0$.

7. Find the coordinates of the point of intersection of the line with symmetric equations

$$\frac{x + 2}{3} = \frac{y - 4}{1} = \frac{z}{-3}$$

and the plane with Cartesian equation $2x - y + z - 4 = 0$.

8. Find a Cartesian equation for the plane which contains **S**$(3, 2, -1)$ and is perpendicular to the line of intersection of the planes with Cartesian equations $x - 2y + z - 3 = 0$ and $x + y + 2 = 0$.

9. Find the distance between the point **S**$(1, 2, -1)$ and the plane with Cartesian equation $3x + 4y - 12z = 3$.

10. Find the distance between the origin and the plane with Cartesian equation $2x - y - 2z - 30 = 0$.

11. (Optional) Find the distance between the point **S**$(2, -1, 1)$ and the line with symmetric equations $x - 2 = y - 2 = z - 1$.

12. (Optional) Find the distance between the origin and the line with symmetric equations $\dfrac{x + 1}{2} = \dfrac{y - 1}{-2} = z + 2.$

Projective Geometry

When an artist paints a landscape, he maps what he sees—that is, he maps the image that the lens of his eye focuses on the retina of the eye—on a plane sheet of canvas. When you look at the result—that is, when you form an image of the painting on the retina of your own eye—you recognize the scene that the artist has painted because of the invariance of geometric properties under all these transformations.

Painting and photography involve perspective transformations. Historically, perspective painting developed only gradually, as you will realize when you think of pictures you have seen of the flat-appearing paintings on the walls of ancient Egyptian and Mayan tombs, and on the beautiful vases of the early Greeks. The subject of perspective images was investigated at length by the great artists Leonardo da Vinci (1452–1519) and Albrecht Dürer (1471–1528).

Physically, a perspective plane transformation can be described more readily using two planes rather than one. For a given point V in space, corresponding, say, to the eye of the beholder, and for two given planes $\mathcal{P}_1$ and $\mathcal{P}_2$

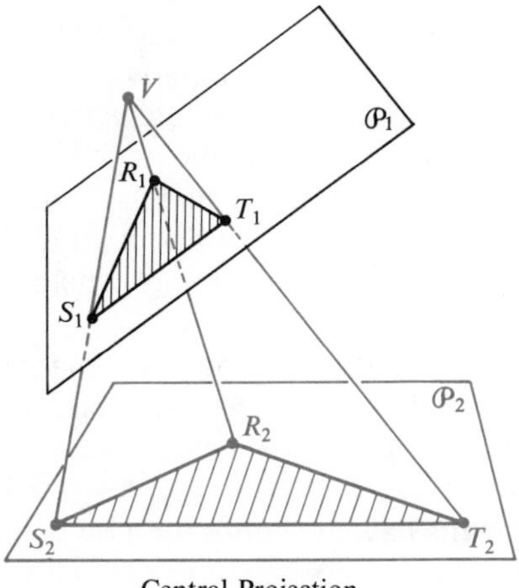

Central Projection

not containing V, consider the line joining V and any point S_1 of $\mathcal{P}_1$. The point S_2 where the line intersects $\mathcal{P}_2$ is called the *central projection* of S_1 on $\mathcal{P}_2$, with V as center of projection, as illustrated above. Another kindred type of projection is *parallel projection,* in which the mapping is made by lines parallel to a given line $\mathcal{L}$, not parallel to either $\mathcal{P}_1$ or $\mathcal{P}_2$, as illustrated on the top of page 340.

When each point of a configuration $\mathcal{C}_1$, such as the triangle $S_1T_1R_1$, on $\mathcal{P}_1$ is thus projected, from V or parallel to $\mathcal{L}$, into a point of $\mathcal{P}_2$, the resulting configuration $\mathcal{C}_2$ is called a central or parallel projection, respectively, of $\mathcal{C}_1$ on $\mathcal{P}_2$.

Now $\mathcal{C}_2$ can be projected, by a central or parallel projection, onto a configuration $\mathcal{C}_3$ on a plane $\mathcal{P}_3$, yielding an indirect map of $\mathcal{C}_1$ onto $\mathcal{C}_3$, and so on.

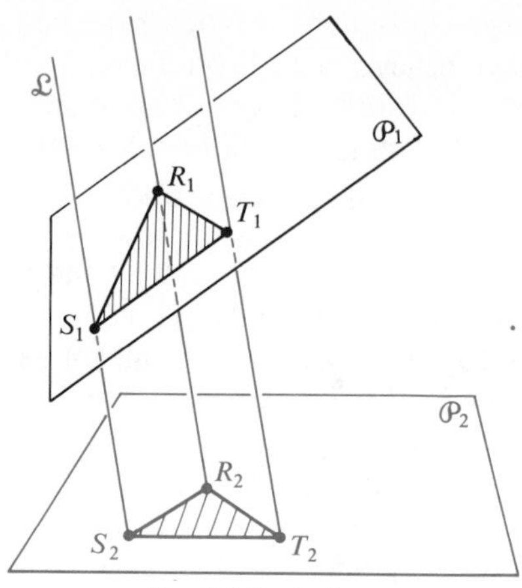

Parallel Projection

Any plane transformation that is either a central or parallel projection, or that can be obtained by a succession of a finite number of such transformations, is called a *projective transformation*. A single central or parallel projection is sometimes called a *perspective transformation*. *Projective geometry* is the study of properties that are invariant under projective transformations.

If, in the central projection discussed above, the planes $\mathcal{P}_1$ and $\mathcal{P}_2$ are not parallel, then the plane $\mathcal{P}$ that contains V and is parallel to $\mathcal{P}_2$ intersects $\mathcal{P}_1$ in a line $\mathcal{L}_1$, as in the figure on page 341. If the point Q_1 is on $\mathcal{L}_1$, then the line VQ_1 is parallel to the plane $\mathcal{P}_2$, and accordingly there is no point of $\mathcal{P}_2$ onto which Q_1 is projected.

Similarly, the plane that contains V and is parallel to $\mathcal{P}_1$ intersects $\mathcal{P}_2$ in a line $\mathcal{L}_2$. If the point Q_2 is on $\mathcal{L}_2$, then the line VQ_2 is parallel to the plane $\mathcal{P}_1$, and accordingly there is no point of $\mathcal{P}_1$ which projects onto Q_2. Thus on the plane $\mathcal{P}_1$ there is an exceptional line $\mathcal{L}_1$, and on the plane $\mathcal{P}_2$ there is an exceptional line $\mathcal{L}_2$, whose points have no corresponding points on the other plane.

Since there is no point on $\mathcal{P}_2$ corresponding to a point Q_1 on the line $\mathcal{L}_1$, it appears that even the elementary property of "being a point" is not an invariant under general projective transformations. This defect, however, can readily be removed by a simple, consistent extension of the notions concerning points and lines. (In geometry, you will recall, "point" and "line" are axiomatic entities, not the marks you make on paper.)

For a point Q_1 on the line $\mathcal{L}_1$ in the figure on page 341, think of a nearby point Q_1' on $\mathcal{P}_1$ but not on $\mathcal{L}_1$, and of the corresponding point Q_2' on $\mathcal{P}_2$. If Q_1' is quite near to Q_1, then the line VQ_1' is nearly parallel to VQ_1, and Q_2' is at a remote position on the line VQ_1', in one direction or the other according as Q_1' lies on one side or the other of $\mathcal{L}_1$ on $\mathcal{P}_1$.

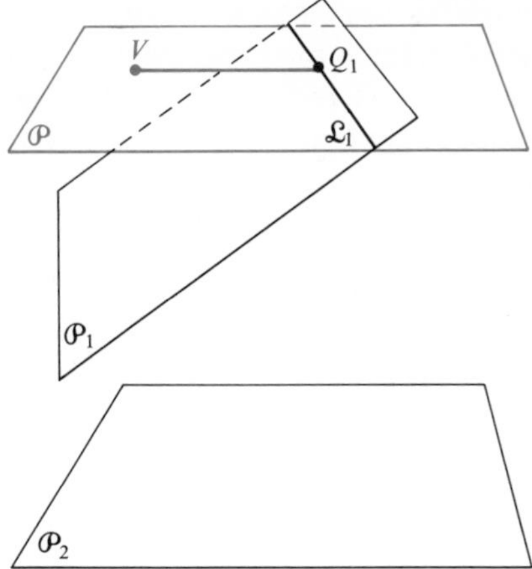

Thinking of the limiting process as Q_1' comes toward coincidence with Q_1, you are thus led to extend your notions concerning points, lines, and planes in space as follows: On each line $\mathcal{L}$, there is a single "ideal" *point at infinity*, which is on every line parallel to $\mathcal{L}$ but on no other line. On each plane $\mathcal{P}$ there is a single ideal *line at infinity*, which contains every ideal point on $\mathcal{P}$ but no other point. There is a single ideal *plane at infinity*, which contains every ideal point and every ideal line but no other point or line.

With these extensions, you can readily see, for example, that there is exactly one line containing any two given points in the plane (whether or not these are ideal points), that every pair of given lines in a plane (whether or not these are parallel or ideal lines) have exactly one point in common, and that every pair of given planes in space (whether or not these are parallel planes or one of them is the ideal plane) have exactly one line in common.

Further, with these extensions, a projective transformation of plane $\mathcal{P}_1$ onto plane $\mathcal{P}_2$ gives a one-to-one correspondence between the set of points of $\mathcal{P}_1$ and the set of points of $\mathcal{P}_2$. Points are mapped onto points, lines onto lines, and the point of intersection of two lines on $\mathcal{P}_1$ onto the point of intersection of the lines on $\mathcal{P}_2$ onto which the given lines are mapped.

Thus, as illustrated in the figures on pages 339 and 340, the property of being a triangle is invariant under projective transformations, with the understanding that any of the vertices of a triangle might very well be ideal points at infinity. On the other hand, although the property of being a pair of *congruent* triangles is invariant under rigid-motion transformations, this property is not invariant under projective transformations.

Projective geometry was developed by French mathematicians during the 17th and 19th centuries. Jean Victor Poncelet (1788–1867) is generally considered to have initiated the systematic development of the subject. He spent the winter of 1813–14 as a prisoner of war in Russia, and the geometric thoughts he had there were published in 1822 in his *Traité des propriétés projectives des figures* (*Treatise on the projective properties of figures*).

Since congruent triangles are not, in general, mapped onto congruent triangles under projective transformations, theorems concerning congruent triangles have no place in projective geometry. By contrast, the configuration involved in the following theorem of Pappus of Alexandria (3rd and 4th centuries A.D.) is mapped onto a configuration of the same sort, and accordingly this is considered to be a theorem of projective geometry:

> If points A, B, C and A', B', C' are on intersecting lines $\mathfrak{L}$ and $\mathfrak{L}'$, respectively, and if the lines AB' and $A'B$ intersect in point R, BC' and $B'C$ in point S, and CA' and $C'A$ in point T, then R, S, and T are collinear.

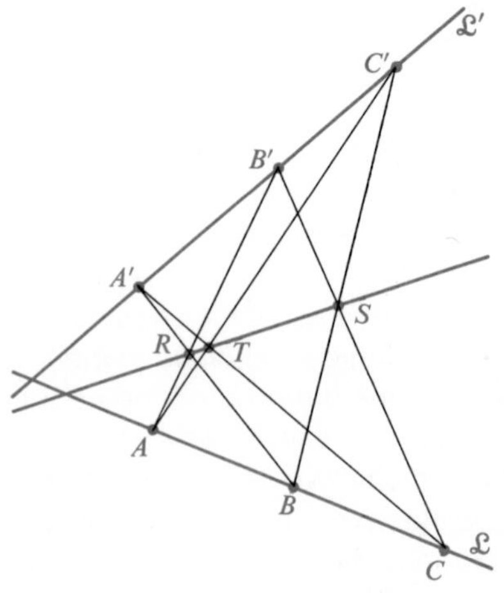

For the same reason, the following result of Gérard Desargues (1591–1661) is a theorem of projective geometry:

> In a plane, if two triangles ABC and $A'B'C'$ are so situated that the lines joining corresponding vertices are concurrent in a point O, then the lines containing corresponding sides intersect in collinear points.

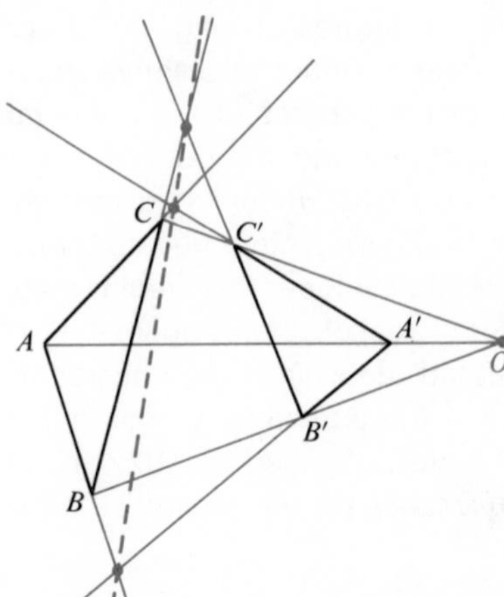

If projective geometry were concerned only with theorems about intersecting lines and collinear points, then it would be interesting but not particularly rich. Actually, its most beautiful results are concerned with conic sections.

Consider the figure on page 166 in the essay on Dandelin spheres. Do you see that the ellipse $\mathcal{E}$ on the plane $\mathcal{P}$ and the circle $\mathcal{C}_1$ on the plane $\mathcal{P}_1$ are central projections of each other, with the vertex V of the cone $\mathcal{K}$ as center of projection? As stated later in that essay, there are positions of the plane $\mathcal{P}$ for which the projection on $\mathcal{P}$ of the circle $\mathcal{C}_1$ would be a parabola or a hyperbola. In general:

> The property of being a conic section is invariant under projective transformations.

You may recall from page 271, in the essay on affine transformations, that the property of being a circle or an ellipse, the property of being a parabola, and the property of being a hyperbola are invariant under affine transformations. Affine transformations, in fact, are just those projective transformations for which the ideal points at infinity are mapped on the ideal points at infinity.

One of the most significant results of projective geometry, discovered by the brilliant prodigy Blaise Pascal (1623–1662) before he was 17 years old, is the following theorem:

> If a hexagon is inscribed in a conic section, then the three pairs of opposite sides intersect in collinear points.

Notice that the theorem of Pappus (page 342) is a limiting case of Pascal's theorem, in which the conic section has degenerated into a pair of straight lines.

Chapter 10

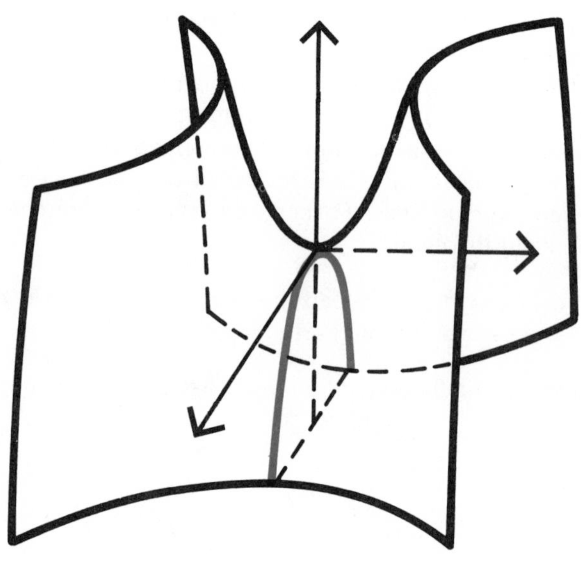

In this chapter, surfaces, such as the hyperbolic paraboloid shown above, and their equations are studied. The chapter ends with a discussion of cylindrical and spherical coordinates.

Surfaces and Transformations of Coordinates in Space

Surfaces

10–1 Spheres

Recall from geometry that a **sphere** is defined to be the locus of all points in space at a given distance (the **radius** of the sphere) from a given point (the **center** of the sphere). Thus (Figure 10–1) if the given point (the center) is $\mathbf{C}(x_0, y_0, z_0)$ and the given distance (the radius) is r, then the point $\mathbf{U}(x, y, z)$ is on the sphere if and only if

$$\|\mathbf{u} - \mathbf{c}\| = r.$$

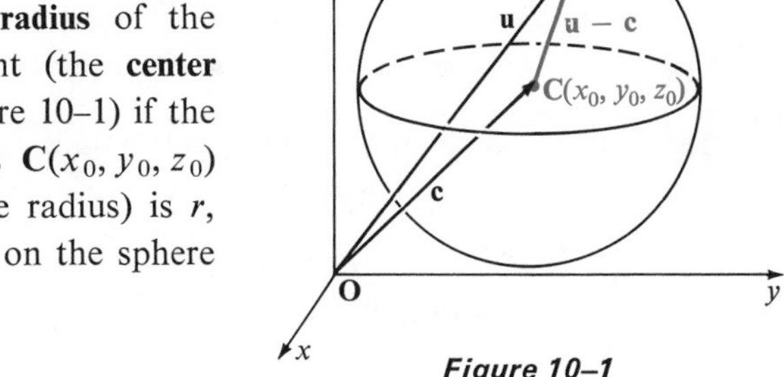

Figure 10–1

In terms of Cartesian coordinates, $\mathbf{U}$ is on the sphere if and only if

$$\sqrt{(x - x_0)^2 + (y - y_0)^2 + (z - z_0)^2} = r,$$

or, equivalently,

$$(x - x_0)^2 + (y - y_0)^2 + (z - z_0)^2 = r^2. \tag{1}$$

Thus Equation (1) is an equation of the sphere with center $\mathbf{C}(x_0, y_0, z_0)$ and radius r.

Example 1. Find an equation of the sphere with center $\mathbf{C}(1, -2, 3)$ and radius 4.

Solution: Substituting 1, -2, and 3 for x_0, y_0, and z_0, respectively, and 4 for r in Equation (1), you have

$$(x - 1)^2 + (y + 2)^2 + (z - 3)^2 = 16.$$

Expanding the squares in the left-hand member of Equation (1) and rearranging terms, you obtain

$$x^2 + y^2 + z^2 - 2x_0x - 2y_0y - 2z_0z + (x_0^2 + y_0^2 + z_0^2 - r^2) = 0$$

as an equation of the sphere. This equation is of the form

$$x^2 + y^2 + z^2 + Gx + Hy + Iz + J = 0. \tag{2}$$

Working in the other direction, by completing squares you can write a given equation of the form (2) equivalently in the form

$$(x - x_0)^2 + (y - y_0)^2 + (z - z_0)^2 = k, \tag{3}$$

where k is a constant. If $k > 0$, then Equation (3) is an equation of a sphere with radius $\sqrt{k}$. If, however, $k = 0$ or $k < 0$, then the graph of Equation (3) is the point $\mathbf{C}(x_0, y_0, z_0)$ or the empty set $\emptyset$, respectively.

Example 2. Identify the graph of each equation.

(a) $x^2 + y^2 + z^2 + 6x + 4y - 2z + 10 = 0$

(b) $x^2 + y^2 + z^2 + 6x + 4y - 2z + 14 = 0$

(c) $x^2 + y^2 + z^2 + 6x + 4y - 2z + 18 = 0$

Solution: Completing squares, you obtain

(a) $(x + 3)^2 + (y + 2)^2 + (z - 1)^2 = -10 + 9 + 4 + 1 = 4;$

(b) $(x + 3)^2 + (y + 2)^2 + (z - 1)^2 = -14 + 9 + 4 + 1 = 0;$

(c) $(x + 3)^2 + (y + 2)^2 + (z - 1)^2 = -18 + 9 + 4 + 1 = -4.$

Thus the loci are

(a) a sphere with center $\mathbf{C}(-3, -2, 1)$ and radius 2;

(b) the point $\mathbf{C}(-3, -2, 1)$;

(c) the empty set $\emptyset$.

A sphere is determined by any four noncoplanar points through which the sphere passes. By solving a set of linear equations for G, H, I, and J, you can determine an equation of the form (2) for the sphere through four given noncoplanar points. You can then proceed as in Example 2 to determine the radius and the coordinates of the center of the sphere.

Example 3. Determine an equation, the radius, and the coordinates of the center, of the sphere through the points $\mathbf{Q}(0, 0, 0)$, $\mathbf{R}(0, 0, 2)$, $\mathbf{S}(0, -1, 3)$, and $\mathbf{T}(1, -4, 3)$.

Solution: The sphere has an equation of the form (2). Thus the coordinates of **Q**, **R**, **S**, and **T** satisfy this equation and you have:

$$0^2 + 0^2 + 0^2 + G(0) + H(0) + I(0) + J = 0$$
$$0^2 + 0^2 + 2^2 + G(0) + H(0) + I(2) + J = 0$$
$$0^2 + (-1)^2 + 3^2 + G(0) + H(-1) + I(3) + J = 0$$
$$1^2 + (-4)^2 + 3^2 + G(1) + H(-4) + I(3) + J = 0$$

or

$$\begin{aligned} J &= 0 \\ 2I + J &= -4 \\ -H + 3I + J &= -10 \\ G - 4H + 3I + J &= -26 \end{aligned}$$

Solving this set of equations for G, H, I, and J, you obtain $G = -4$, $H = 4$, $I = -2$, $J = 0$. Thus an equation of the sphere is $x^2 + y^2 + z^2 - 4x + 4y - 2z = 0$. Completing the squares, you have

$$(x - 2)^2 + (y + 2)^2 + (z - 1)^2 = 9 = 3^2.$$

Therefore, the radius is 3 and the coordinates of the center are $(2, -2, 1)$.

Through a given point **T** on a sphere $\mathcal{S}$ there is a unique plane $\mathcal{P}$ tangent to $\mathcal{S}$. You can use the fact that the radius vector **v** from the center **C** of $\mathcal{S}$ to **T** is perpendicular to $\mathcal{P}$ (Figure 10–2) to determine an equation of $\mathcal{P}$. Thus, if the co-

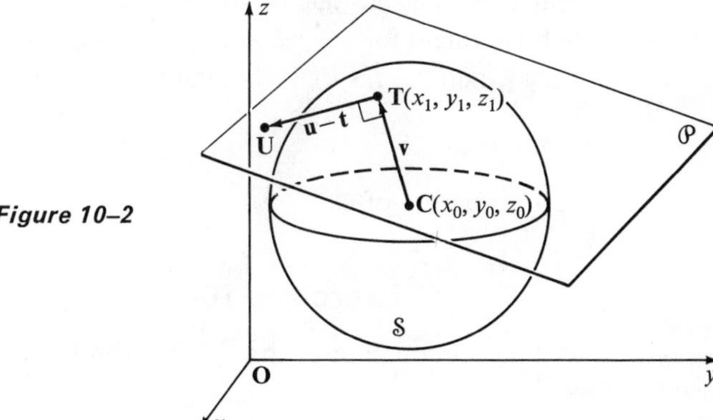

Figure 10–2

ordinates of **T** are (x_1, y_1, z_1) and an equation of $\mathcal{S}$ is $(x - x_0)^2 + (y - y_0)^2 + (z - z_0)^2 = r^2$, then $\mathbf{v} = (x_1 - x_0, y_1 - y_0, z_1 - z_0)$ is the radius vector to **T**. Since $\mathcal{P}$ contains **T** and is perpendicular to **v**, a point $\mathbf{U}(x, y, z)$ is on $\mathcal{P}$ if and only if

$$\mathbf{v} \cdot (\mathbf{u} - \mathbf{t}) = 0,$$

that is, if and only if

$$(x_1 - x_0)(x - x_1) + (y_1 - y_0)(y - y_1) + (z_1 - z_0)(z - z_1) = 0. \quad (4)$$

Accordingly, Equation (4) is an equation of the tangent plane $\mathcal{P}$.

An alternative form of the equation for the tangent plane $\mathcal{P}$ can be obtained as follows. Since **T** is on $\mathcal{S}$, you have

$$(x_1 - x_0)^2 + (y_1 - y_0)^2 + (z_1 - z_0)^2 = r^2.$$

Adding the left-hand member of this equation to the left-hand member of Equation (4), and the right-hand member to the right-hand member of (4), you obtain

$$(x_1 - x_0)(x - x_1 + x_1 - x_0) + (y_1 - y_0)(y - y_1 + y_1 - y_0) + (z_1 - z_0)(z - z_1 + z_1 - z_0) = r^2,$$

or

$$(x_1 - x_0)(x - x_0) + (y_1 - y_0)(y - y_0) + (z_1 - z_0)(z - z_0) = r^2. \quad (5)$$

Since the steps are reversible, Equation (5) is also an equation of the tangent plane $\mathcal{P}$.

Example 4. Find an equation of the tangent plane $\mathcal{P}$ to the sphere $\mathcal{S}$ with equation

$$x^2 + y^2 + z^2 = 9$$

at the point $\mathbf{T}(1, -2, 2)$ on $\mathcal{S}$.

Solution: Notice first that **T** is on $\mathcal{S}$ since $1^2 + (-2)^2 + 2^2 = 9$. The center of $\mathcal{S}$ is at the origin, so that $(x_0, y_0, z_0) = (0, 0, 0)$. Substituting 0 for x_0, y_0, and z_0, respectively, 1, -2, and 2 for x_1, y_1, and z_1, respectively, and 9 for r^2 in Equation (5), you have

$$x - 2y + 2z = 9$$

as an equation of $\mathcal{P}$.

Exercises 10–1

In Exercises 1–8, find an equation of the sphere with the given radius r and the given center **C**.

1. $\mathbf{C}(5, 0, -7)$; $r = 2$
2. $\mathbf{C}(-2, 0, 3)$; $r = 4$
3. $\mathbf{C}(0, 0, -1)$; $r = 5$
4. $\mathbf{C}(-2, 0, 0)$; $r = 3$
5. $\mathbf{C}(-1, -2, 2)$; $r = 7$
6. $\mathbf{C}(3, -4, -3)$; $r = 6$
7. $\mathbf{C}(\frac{3}{2}, \frac{2}{3}, -\frac{1}{2})$; $r = \frac{1}{4}$
8. $\mathbf{C}(\frac{1}{4}, -\frac{1}{4}, \frac{3}{4})$; $r = \frac{1}{5}$

In Exercises 9–16, identify the graph of the given equation.

9. $x^2 + y^2 + z^2 - 2x + 4y + 6z + 5 = 0$
10. $x^2 + y^2 + z^2 + 8x - 4y - 4z + 8 = 0$
11. $x^2 + y^2 + z^2 + x + 3y = 0$
12. $x^2 + y^2 + z^2 - 2x + z = 0$

13. $x^2 + y^2 + z^2 - 8x - 6y + 4z + 29 = 0$
14. $x^2 + y^2 + z^2 + 2x - 2y - 4z + 6 = 0$
15. $x^2 + y^2 + z^2 + 2x - 2y - 4z + 7 = 0$
16. $x^2 + y^2 + z^2 + 4x + 3y - 2z + 7 = 0$

In Exercises 17–24, determine an equation, the coordinates of the center, and the radius, of the sphere through the points **Q**, **R**, **S**, and **T** whose coordinates are given.

17. $\mathbf{Q}(0, 0, 0)$, $\mathbf{R}(1, 0, 0)$, $\mathbf{S}(0, 1, 0)$, $\mathbf{T}(0, 0, 1)$
18. $\mathbf{Q}(0, 0, 0)$, $\mathbf{R}(1, 1, 0)$, $\mathbf{S}(1, 0, 1)$, $\mathbf{T}(0, 1, 1)$
19. $\mathbf{Q}(1, 1, 1)$, $\mathbf{R}(1, 1, -1)$, $\mathbf{S}(1, -1, 1)$, $\mathbf{T}(-1, 1, 1)$
20. $\mathbf{Q}(1, 1, 1)$, $\mathbf{R}(1, -1, -1)$, $\mathbf{S}(-1, 1, -1)$, $\mathbf{T}(-1, -1, 1)$
21. $\mathbf{Q}(2, 0, 5)$, $\mathbf{R}(0, 0, 1)$, $\mathbf{S}(3, -1, 5)$, $\mathbf{T}(-1, -4, 2)$
22. $\mathbf{Q}(-5, -4, 1)$, $\mathbf{R}(3, 4, -5)$, $\mathbf{S}(0, 0, 4)$, $\mathbf{T}(0, 8, 0)$
23. $\mathbf{Q}(2, 1, -3)$, $\mathbf{R}(1, 2, 0)$, $\mathbf{S}(0, 2, -2)$, $\mathbf{T}(0, 0, 3)$
24. $\mathbf{Q}(5, -3, 7)$, $\mathbf{R}(2, 0, 1)$, $\mathbf{S}(0, -2, 0)$, $\mathbf{T}(-2, 1, 2)$

In Exercises 25–32, find an equation of the plane $\mathcal{P}$ that is tangent at the given point **T** to the sphere $\mathcal{S}$ whose equation is given.

25. $x^2 + y^2 + z^2 = 49$; $\mathbf{T}(6, 2, -3)$
26. $x^2 + y^2 + z^2 = 169$; $\mathbf{T}(-12, 3, 4)$
27. $(x - 4)^2 + (y + 2)^2 + z^2 = 121$; $\mathbf{T}(-5, 4, 2)$
28. $x^2 + (y - 1)^2 + (z + 3)^2 = 25$; $\mathbf{T}(3, 1, -7)$
29. $(x + 2)^2 + (y + 3)^2 + (z - 1)^2 = 10$; $\mathbf{T}(-3, -3, 4)$
30. $(x - 1)^2 + (y + 1)^2 + (z + 3)^2 = 11$; $\mathbf{T}(0, 0, 0)$
31. $x^2 + y^2 + z^2 + 2x - 4y - 2z - 15 = 0$; $\mathbf{T}(3, 0, 2)$
32. $x^2 + y^2 + z^2 - 4x + 6y + 4z + 3 = 0$; $\mathbf{T}(0, -2, 1)$

* 33. Show that the tangent plane $\mathcal{P}$ to the sphere $\mathcal{S}$ with equation

$$x^2 + y^2 + z^2 + Gx + Hy + Iz + J = 0$$

at the point $\mathbf{T}(x_1, y_1, z_1)$ on $\mathcal{S}$ has equation

$$x_1x + y_1y + z_1z + \frac{G}{2}(x + x_1) + \frac{H}{2}(y + y_1) + \frac{I}{2}(z + z_1) + J = 0.$$

* 34. Show that the sphere through the noncoplanar points $\mathbf{Q}(x_1, y_1, z_1)$, $\mathbf{R}(x_2, y_2, z_2)$, $\mathbf{S}(x_3, y_3, z_3)$, and $\mathbf{T}(x_4, y_4, z_4)$ has equation

$$\begin{vmatrix} x^2 + y^2 + z^2 & x & y & z & 1 \\ x_1^2 + y_1^2 + z_1^2 & x_1 & y_1 & z_1 & 1 \\ x_2^2 + y_2^2 + z_2^2 & x_2 & y_2 & z_2 & 1 \\ x_3^2 + y_3^2 + z_3^2 & x_3 & y_3 & z_3 & 1 \\ x_4^2 + y_4^2 + z_4^2 & x_4 & y_4 & z_4 & 1 \end{vmatrix} = 0.$$

In Exercises 35 and 36, identify the graph of the equation obtained by substituting the coordinates of the given points in the equation of Exercise 34. If the graph is a sphere, find its radius and the coordinates of its center.

* **35.** **Q**(0, 0, 0), **R**(1, 0, 0), **S**(2, 1, 3), **T**(5, 2, 6)

* **36.** **Q**(0, 0, 0), **R**(0, 1, 0), **S**(4, 0, 2), **T**(−1, 2, 3)

In Exercises 37 and 38, determine whether or not the five points whose coordinates are given lie on a sphere.

* **37.** **Q**(0, 0, 0), **R**(2, −4, 6), **S**(3, −1, 0), **T**(4, −3, 1), **K**(−2, −1, 5)

* **38.** **Q**(0, 0, 0), **R**(5, 5, 0), **S**(−1, 0, −1), **T**(1, 1, 1), **K**(4, 6, −6)

In Exercises 39–42, identify the intersection of the two graphs whose equations are given.

Example. $x^2 + y^2 + z^2 - 169 = 0;\ z - 12 = 0$

Solution: Solve the two equations simultaneously by substituting 12 for z in the first equation. You obtain $x^2 + y^2 - 25 = 0$. Thus the intersection is a circle, with center **C**(0, 0, 12) and radius 5, lying in the plane $z = 12$.

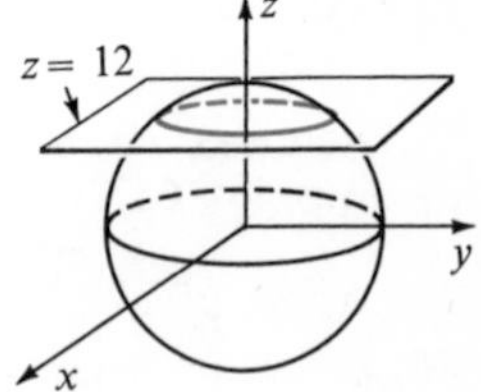

* **39.** $x^2 + y^2 + z^2 - 25 = 0;\ y - 3 = 0$

* **40.** $x^2 + y^2 + z^2 - 25 = 0;\ x^2 + y^2 + (z - 2)^2 - 16 = 0$

* **41.** $x^2 + y^2 + z^2 - 25 = 0;\ x^2 + y^2 + (z - 1)^2 - 16 = 0$

* **42.** $x^2 + y^2 + z^2 - 25 = 0;\ x^2 + y^2 + (z - 1)^2 - 9 = 0$

10–2 Cylinders and Surfaces of Revolution

You are familiar with such physical surfaces in space as tin cans and firecrackers (Figure 10–3). Such surfaces are called *cylinders* and are said to be *cylindrical.* In analytic geometry, a cylinder, or a cylindrical surface, can be defined as follows: Imagine a plane $\mathcal{P}$ and a line $\mathcal{L}$ not parallel to $\mathcal{P}$. If $\mathcal{L}$ moves in such a way that a point **T** of $\mathcal{L}$ traverses a path $\mathcal{C}$ in $\mathcal{P}$ while the direction of $\mathcal{L}$ remains unchanged, then $\mathcal{L}$ sweeps out a surface $\mathcal{S}$, as indicated in Figure 10–4. You call $\mathcal{S}$ a **cylinder**, or **cylindrical surface**. The curve $\mathcal{C}$ in $\mathcal{P}$ is said to be the **directrix** of the cylinder; the line $\mathcal{L}$ and all other lines on the surface parallel to $\mathcal{L}$ are called **elements**, or **generators**, of the surface.

Figure 10–3

Notice that in analytic geometry a cylinder is considered to extend indefinitely in both directions of an element; Figure 10–4 illustrates only a section of a cylinder.

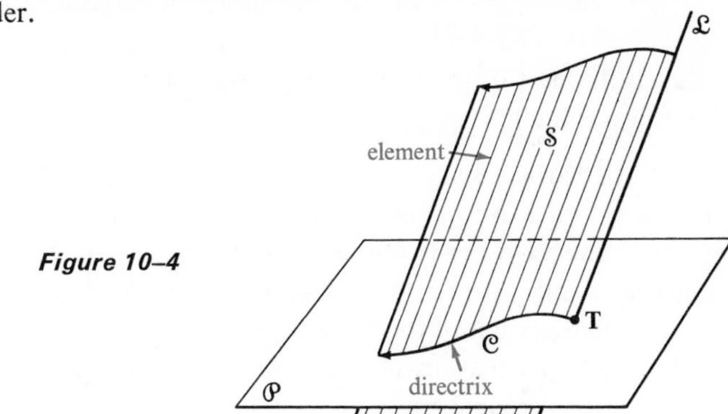

Figure 10–4

If for each point on a surface, there is a line through that point which lies entirely on the surface, then the surface is said to be **ruled** and the lines are said to be **rulings of the surface**. Thus a cylinder is a ruled surface, with the elements as rulings.

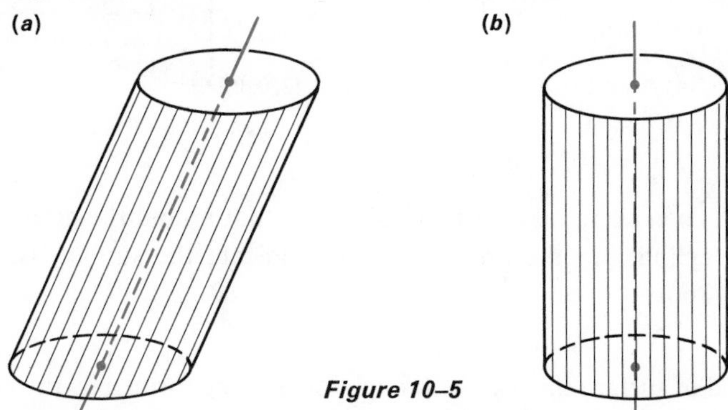

Figure 10–5

Cylinders are classified according to the nature of the directrix. For example, if the directrix is a circle, then the cylinder is a **circular cylinder** [Figure 10–5(a)]; the line that is parallel to an element and contains the center of the circle is called the **axis** of the circular cylinder. If the elements of a circular cylinder are perpendicular to the plane of the circle, then the cylinder is called a **right circular cylinder** [Figure 10–5(b)]. Note that a plane can be considered to be a cylinder whose directrix is a straight line.

A Cartesian equation in two variables can be considered to be an equation of a cylinder whose elements are parallel to the coordinate axis associated with the missing variable. For example, $3x^2 + 4y^2 = 12$ is an equation of an elliptical cylinder with an ellipse in the xy-plane as directrix, and with elements parallel to the z-axis.

The intersection of any surface with a coordinate plane is called the *trace* of the surface in that coordinate plane. As you saw in Section 9–3 and as is illustrated in Example 1 below, the traces are often useful in graphing.

Example 1. Sketch the part of the cylinder with equation $3x^2 + 4y^2 = 12$ that lies in the first octant.

Solution: As noted earlier, since the variable z is missing, the elements of the cylinder are parallel to the z-axis. An equation of the trace in the xy-plane is found by setting $z = 0$ in the original equation; in this case z does not appear in the original equation, so that an equation of the trace is just $3x^2 + 4y^2 = 12$.

To find an equation of the trace in the xz-plane, you set $y = 0$ in the original equation to obtain

$$3x^2 = 12,$$
$$x^2 = 4,$$
$$x = 2 \quad \text{or} \quad x = -2.$$

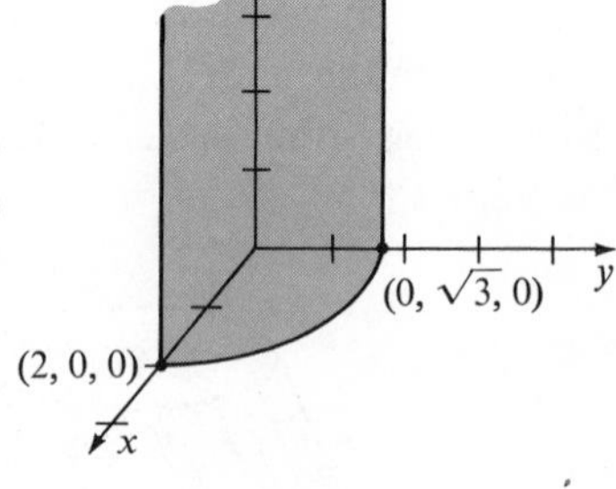

To find an equation of the trace in the yz-plane, you set $x = 0$ to obtain

$$4y^2 = 12,$$
$$y^2 = 3,$$
$$y = \sqrt{3} \quad \text{or} \quad y = -\sqrt{3}.$$

Then sketch the parts of the traces that bound the graph of the cylinder in the first octant and indicate the surface as shown above.

Another kind of surface that can be thought of as being formed by the motion of a set of points is the following: If a plane curve $\mathcal{C}$ is revolved about a line $\mathcal{L}$ in its plane so that each point on the curve describes a circle, then the resulting surface is called a **surface of revolution**. The plane curve $\mathcal{C}$ is called the **generating curve** of the surface, and the line about which the curve is revolved is called the **axis of revolution** (Figure 10–6). In this book, we shall ordinarily

Figure 10–6

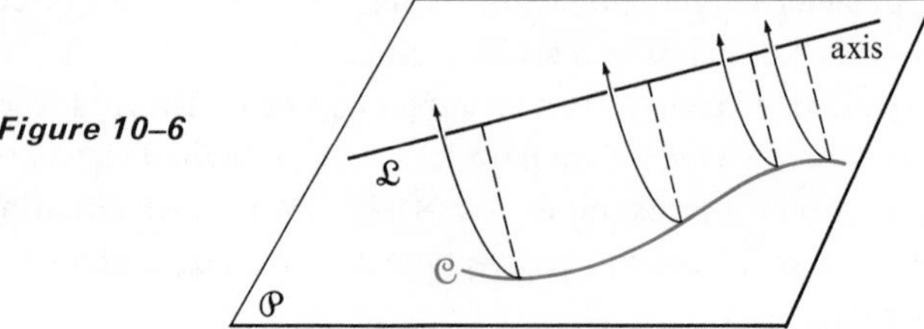

consider only surfaces of revolution having one of the coordinate axes as axis of revolution. The following example illustrates how you can find an equation for such a surface.

Example 2. Find an equation for the surface of revolution generated by revolving the parabolic arc with equation $z = \sqrt{y}$ about the y-axis.

Solution: Choose any point $\mathbf{S}_1$ on the surface, as shown, and then consider the cross section lying in the plane perpendicular to the y-axis and containing this point. Let $\mathbf{C}$ be the center of this

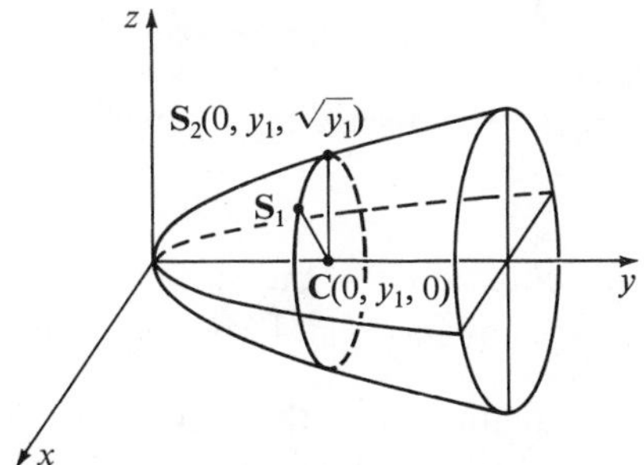

cross section and $\mathbf{S}_2$ be the point in the cross section lying on the generating curve. If the coordinates of $\mathbf{S}_1$ are (x_1, y_1, z_1), then the coordinates of $\mathbf{C}$ are $(0, y_1, 0)$ and the length of $\overline{\mathbf{CS}_1}$ is $\sqrt{x_1^2 + z_1^2}$. Since $\mathbf{S}_2$ lies on the parabolic arc with equation $z = \sqrt{y}$, the coordinates of $\mathbf{S}_2$ are $(0, y_1, \sqrt{y_1})$. Thus the length of $\overline{\mathbf{CS}_2}$ is $\sqrt{y_1}$. Since $\overline{\mathbf{CS}_1}$ and $\overline{\mathbf{CS}_2}$ are equal in length, being radial segments of the same circle, you have

$$\sqrt{y_1} = \sqrt{x_1^2 + z_1^2}.$$

Upon squaring both members of this equation, you obtain

$$y_1 = x_1^2 + z_1^2.$$

Since $\mathbf{S}_1$ is any point on the surface, you can write

$$y = x^2 + z^2 \tag{1}$$

as an equation satisfied by all points on the locus. Conversely, since the steps are reversible, any point whose coordinates satisfy Equation (1) is on the locus. Therefore Equation (1) is an equation of the surface.

You can use the method of Example 2 to derive an equation for any surface of revolution generated by revolving a curve in one of the coordinate planes about one of the coordinate axes. The results can be summarized as follows:

■ To obtain an equation for the surface of revolution generated by revolving a given curve $\mathcal{C}$ lying in a coordinate plane about a given coordinate axis lying in that plane: In the equation for $\mathcal{C}$, if the axis of revolution is

The x-axis:

Replace y or z, whichever appears in the equation, with $\pm\sqrt{y^2 + z^2}$;

The y-axis:

Replace x or z, whichever appears in the equation, with $\pm\sqrt{x^2 + z^2}$;

The z-axis:

Replace x or y, whichever appears in the equation, with $\pm\sqrt{x^2 + y^2}$.

Example 3. Find an equation for the surface generated by revolving the ellipse with equation $x^2 + 4y^2 = 1$ about the x-axis.

Solution: In the equation for the generating curve, $x^2 + 4y^2 = 1$, replace y with $\pm\sqrt{y^2 + z^2}$ to obtain

$$x^2 + 4(\pm\sqrt{y^2 + z^2})^2 = 1,$$

or

$$x^2 + 4y^2 + 4z^2 = 1,$$

which is the required equation.

Exercises 10–2

In Exercises 1–6, sketch the portion in the first octant of the cylinder whose equation in $\mathcal{R}^3$ is as given.

1. $y^2 + z^2 = 4$
2. $x^2 + z^2 = 16$
3. $y = x^2$
4. $x^2 - z^2 = 4$
5. $4y^2 + 9z^2 = 36$
6. $z = x^2 + 1$

In Exercises 7–14, find an equation of the surface obtained by revolving the curve with the given equation about the indicated axis.

7. $x^2 + 2y^2 = 1$; the x-axis
8. $x^2 + 2y^2 = 1$; the y-axis
9. $2x^2 - z^2 = 1$; the x-axis
10. $2x^2 - z^2 = 1$; the z-axis
11. $y^2 + x - 4 = 0$; the x-axis
12. $x^2 - 2y + 3 = 0$; the y-axis
13. $x^2 - 4x + y^2 - 21 = 0$; the x-axis
14. $x^2 + y^2 + 6y - 7 = 0$; the x-axis

15. Find an equation of the curve which is the intersection of the surface in Exercise 3 with the plane with equation $y = 4$.

16. Find an equation of the intersection of the surface in Exercise 4 with the plane with equation $z = 3$.

17. Find an equation of the intersection of the surface in Exercise 7 with the plane with equation $x = 0$.

18. Find an equation of the intersection of the surface in Exercise 8 with the xz-plane.

10–3 Quadric Surfaces

In earlier chapters, you saw that in the plane the graph of a first-degree Cartesian equation in two variables is a line and the graph of a second-degree Cartesian equation in two variables is a conic section (which might be a "degenerate" conic section, that is, one or two lines, a single point, or the empty set $\emptyset$). In Chapter 9, you saw that in space the graph of a first-degree Cartesian equation in three variables is a plane, and that to describe a line in space in terms of Cartesian coordinates you can use two first-degree equations whose graphs (planes) have the desired line as their intersection.

It seems reasonable to inquire about the graphs in space of second-degree Cartesian equations in three variables,

$$Ax^2 + By^2 + Cz^2 + Dxy + Exz + Fyz + Gx + Hy + Iz + J = 0,$$

where the coefficients are real-number constants and A, B, C, D, E, and F are not all 0. You might expect that the graph of such an equation would have some relationship with the conic sections. To see that this is so, consider the intersection of a plane $\mathcal{P}$ parallel to one of the coordinate planes and the graph of a second-degree equation in x, y, and z. On $\mathcal{P}$, one of the variables is constant. When this constant is substituted for the variable in the second-degree equation in x, y, and z, there results an equation of at most second degree in the other two variables. Thus the intersection ordinarily is a conic section.

For example, for the intersection of the graph of the equation

$$x^2 + 2y^2 + z^2 - 3z - 5 = 0$$

with the plane $z = 4$, you have

$$x^2 + 2y^2 + 4^2 - 3(4) - 5 = 0,$$
$$x^2 + 2y^2 - 1 = 0.$$

Thus the intersection is an ellipse. (The intersection of $\mathcal{P}$ and the graph of a second-degree equation in x, y, and z might be the plane $\mathcal{P}$ itself. This is the case, for example, when the equation is $x^2 = 0$ and $\mathcal{P}$ is the yz-plane.)

The graph of a second-degree Cartesian equation in three variables is called a **quadric surface**. In this section, you will study the general types of quadric surfaces. Their equations will be given in terms of conveniently chosen axes and positive constants a, b, and c. The axes are chosen, in particular, so that there are no xy-, xz-, or yz-terms in the equations. (See Section 10–4.)

1. The **ellipsoid** (Figure 10–7):

$$\frac{x^2}{a^2} + \frac{y^2}{b^2} + \frac{z^2}{c^2} = 1$$

The intercepts on the x-, y-, and z-axes are, respectively, $\pm a$, $\pm b$, and $\pm c$. The intersection (trace) of this surface with each of the coordinate planes is an ellipse or a circle.

If two of a, b, and c are equal, then the surface is called a **spheroid**; if all three are equal, then the surface is a sphere.

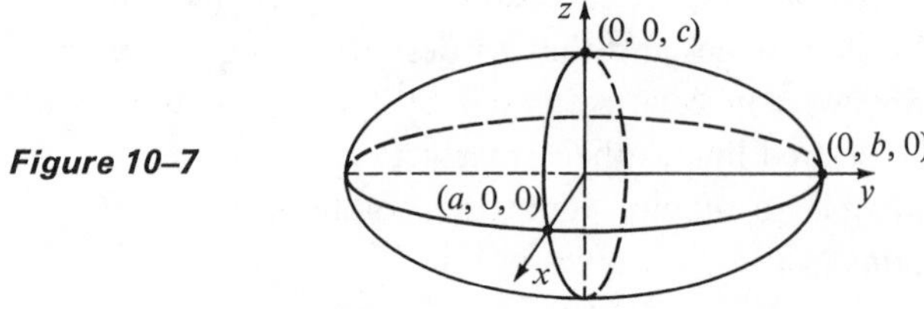

Figure 10–7

A spheroid is a surface of revolution obtained by revolving an ellipse about one of its axes. It is a **prolate spheroid**, like a football, if the axis of revolution is the major axis of the ellipse [Figure 10–8(a)]; and it is an **oblate spheroid**, like a doorknob, if the axis of revolution is the minor axis of the ellipse [Figure 10–8(b)].

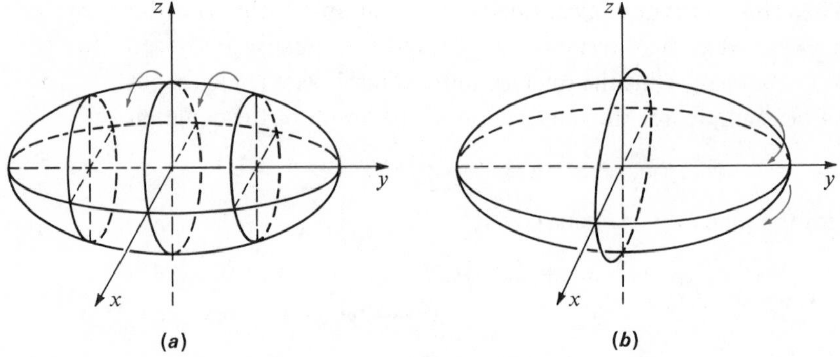

Figure 10–8

2. The **hyperboloid of one sheet** (Figure 10–9):

$$\frac{x^2}{a^2} + \frac{y^2}{b^2} - \frac{z^2}{c^2} = 1$$

The surface has cross sections in planes parallel to the xy-plane which are ellipses or, if $a = b$, circles. Cross sections in planes parallel to the other coordinate planes are hyperbolas. If $a = b$, the surface can be thought of as being generated by revolving a hyperbola about the line containing its conjugate axis. The expression "of one sheet" refers to the fact that the surface is connected, or all in one piece.

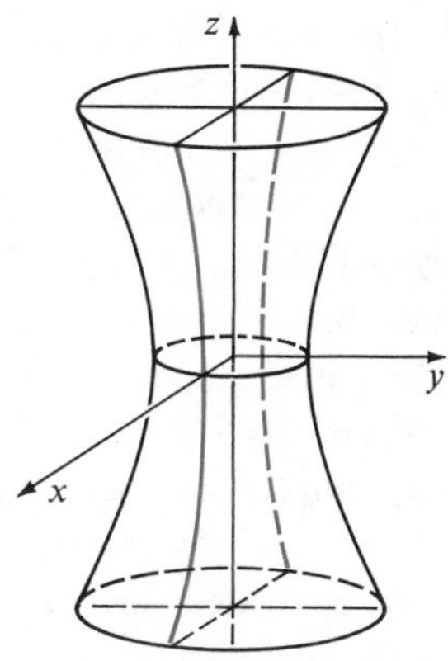

Figure 10–9

A perhaps surprising fact is that through each point on a hyperboloid of one sheet there are two straight lines that lie entirely on the surface! (See page 359 and Exercises 19–24, page 364.) As mentioned earlier, these lines are called **rulings**, and thus the hyperboloid of one sheet is said to be a **doubly ruled** surface.

3. The **hyperboloid of two sheets** (Figure 10–10):

$$-\frac{x^2}{a^2} - \frac{y^2}{b^2} + \frac{z^2}{c^2} = 1, \quad \text{or} \quad \frac{x^2}{a^2} + \frac{y^2}{b^2} - \frac{z^2}{c^2} = -1$$

The cross section of this surface in a plane parallel to the xy-plane is an ellipse or a circle if $z^2 > c^2$, is a point if $z^2 = c^2$, and is the empty set if $z^2 < c^2$. Cross sections in planes parallel to the other coordinate planes are hyperbolas. If $a = b$, the surface can be thought of as being generated by revolving a hyperbola about its principal axis. This hyperboloid is said to be "of two sheets" because the surface consists of two connected sets of points which are not connected to each other.

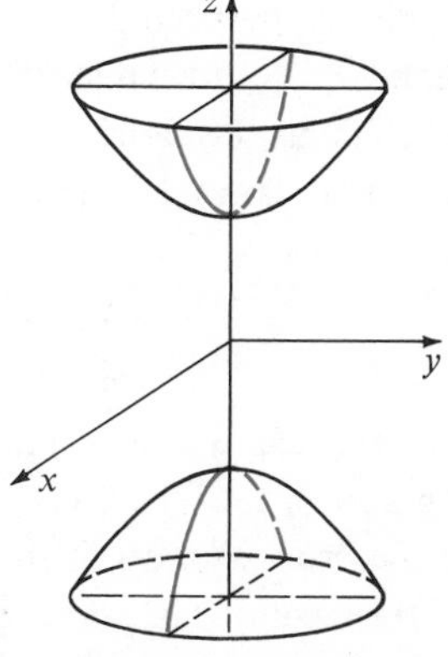

Figure 10–10

4. The **elliptic cone** (Figure 10–11):

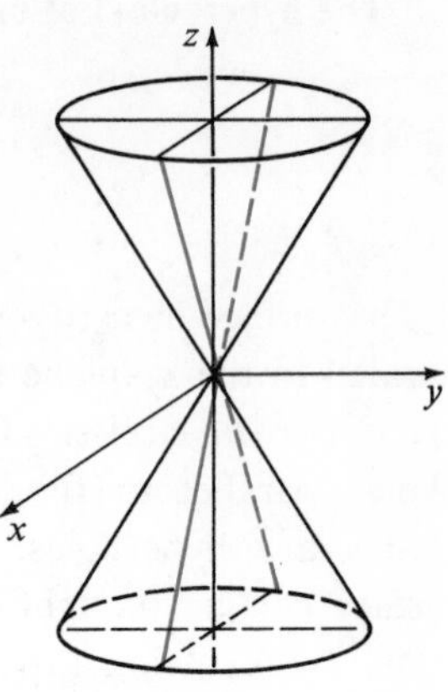

Figure 10–11

$$\frac{x^2}{a^2} + \frac{y^2}{b^2} - \frac{z^2}{c^2} = 0$$

This surface has cross sections in planes parallel to, and lying above or below, the xy-plane which are ellipses or, if $a = b$, circles; for the xy-plane itself, the ellipse or circle reduces to a point. Cross sections in planes parallel to, but not coincident with, the other coordinate planes are hyperbolas; for the coordinate planes themselves, the hyperbolas reduce to pairs of intersecting lines. The line joining each point of an elliptic cone with the origin lies entirely on the surface of the cone, and accordingly the elliptic cone is a ruled surface.

You can think of an elliptic cone as being a degenerate form of a hyperboloid of either one or two sheets. It is related to these hyperboloids in the same manner that the asymptotes of a hyperbola are related to the hyperbola. You say that the graph of

$$\frac{x^2}{a^2} + \frac{y^2}{b^2} - \frac{z^2}{c^2} = 0$$

is the **asymptotic cone** of the hyperboloid of one sheet

$$\frac{x^2}{a^2} + \frac{y^2}{b^2} - \frac{z^2}{c^2} = 1$$

and also of the hyperboloid of two sheets

$$\frac{x^2}{a^2} + \frac{y^2}{b^2} - \frac{z^2}{c^2} = -1.$$

Ellipsoids, hyperboloids of one sheet, hyperboloids of two sheets, and elliptic cones are **central quadric surfaces**. In the standard representations given above, each central quadric surface has its **center** at the origin.

5. The **elliptic paraboloid** (Figure 10–12):

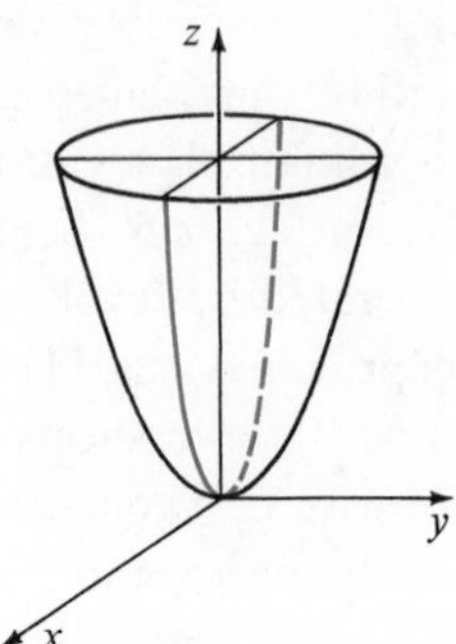

Figure 10–12

$$z = \frac{x^2}{a^2} + \frac{y^2}{b^2}$$

For this surface, a cross section in a plane parallel to, and lying above, the xy-plane is an ellipse or, if $a = b$, a circle; the cross section in the xy-plane is simply the origin. Cross sections in planes parallel to the other two coordinate planes are parabolas. If $a = b$, the surface can be thought of as being generated by revolving a parabola about its axis.

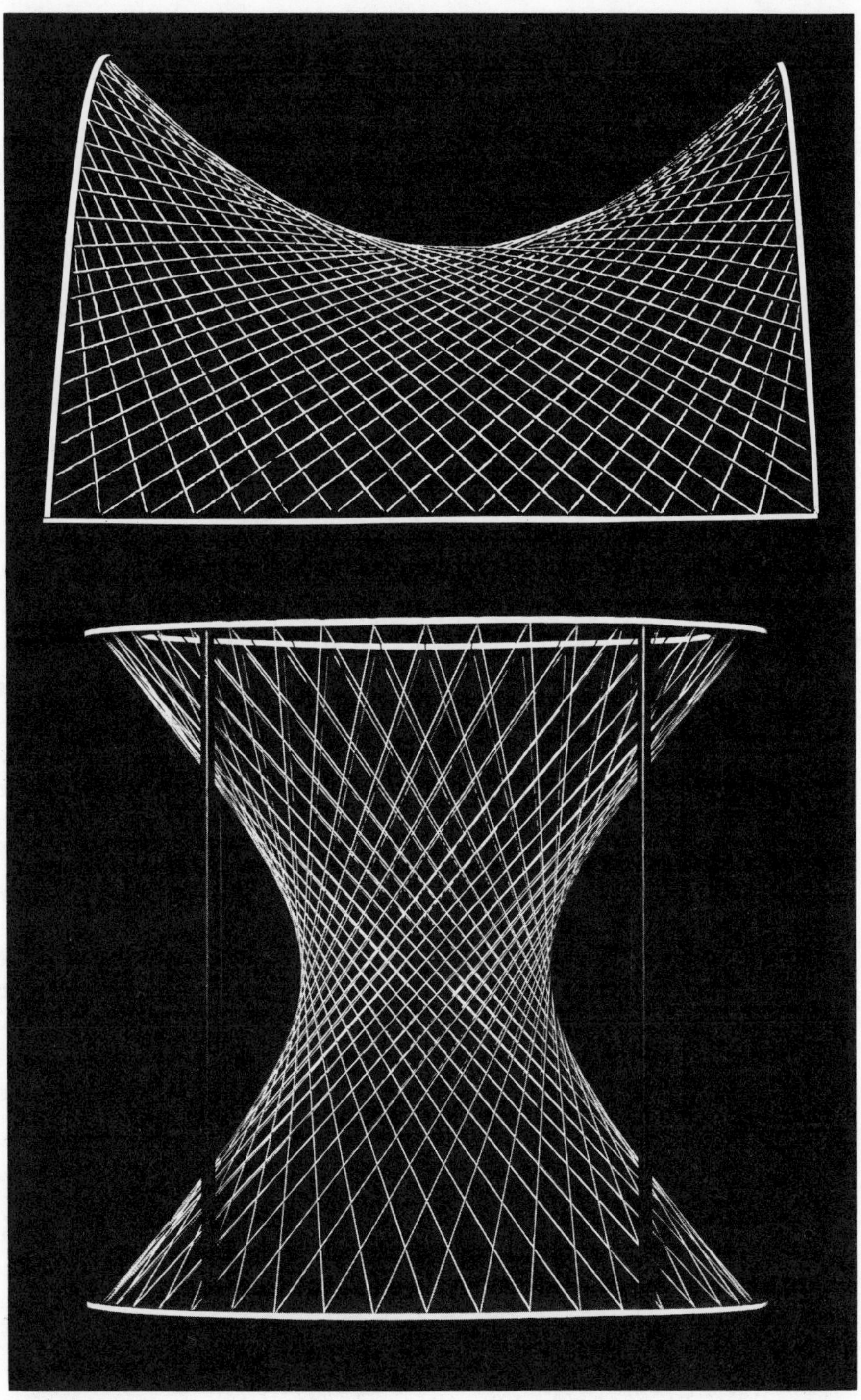

The string models of a hyperbolic paraboloid (top) and a hyperboloid of one sheet (bottom) illustrate the fact that both of these surfaces are doubly ruled.

6. The **hyperbolic paraboloid** (Figure 10–13):

$$z = \frac{y^2}{b^2} - \frac{x^2}{a^2}$$

A hyperbolic paraboloid is sometimes called a "saddle-shaped" surface since it is shaped rather like a saddle in that the cross sections which are parabolas have opposite orientations; the section (trace) on the yz-plane opens upward,

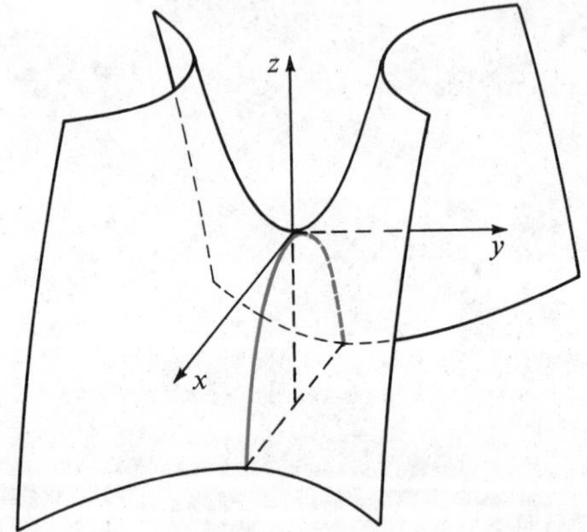

Figure 10–13

and the section (trace) on the xz-plane opens downward. The trace of this surface in the xy-plane is a pair of lines through the origin, but any other cross section in a plane parallel to the xy-plane is a hyperbola. If the plane lies above the xy-plane, the cross section is a hyperbola with principal axis parallel to the y-axis; if the plane lies below the xy-plane, the cross section is a hyperbola with principal axis parallel to the x-axis.

Like the hyperboloid of one sheet, the hyperbolic paraboloid is a doubly ruled surface (see page 359 and Exercises 25–28, page 365).

Elliptic paraboloids and hyperbolic paraboloids are **noncentral quadric surfaces.** In the standard representations given above, each noncentral quadric surface has its **vertex** at the origin.

In all of the foregoing equations of quadric surfaces, the roles of x, y, and z, and of their negatives, might, of course, be interchanged. For example, the graph of

$$-\frac{x^2}{a^2} + \frac{y^2}{b^2} - \frac{z^2}{c^2} = 1$$

is a hyperboloid of two sheets, and the graph of

$$-x = \frac{y^2}{a^2} + \frac{z^2}{b^2}$$

is an elliptic paraboloid.

As in the case of conic sections, there are degenerate quadric surfaces. For the graph of

$$Ax^2 + By^2 + Cz^2 + J = 0, \tag{1}$$

where A, B, and C are not all 0, there are the following degenerate graphs (in addition to the elliptic cone):

1. No points at all if A, B, and C all have the same sign (one or two of A, B, and C might be 0), and J has the same sign (and is not 0). For example, the equation $x^2 + y^2 + z^2 + 4 = 0$ has $\emptyset$ as its graph.

2. One point, the origin, if A, B, and C all have the same sign and $J = 0$. For example, the graph of

$$3x^2 + 2y^2 + z^2 = 0$$

is the origin.

3. A line, one of the coordinate axes, if two of A, B, and C have the same sign and the other coefficient and the constant J are 0. For example, the graph of $y^2 + z^2 = 0$ is the x-axis.

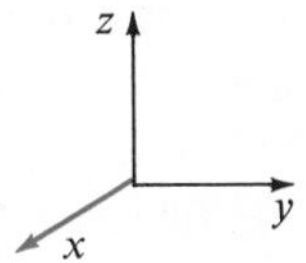

4. A plane, one of the coordinate planes, if two of A, B, and C as well as J are zero. For example, the graph of $x^2 = 0$ is the yz-plane.

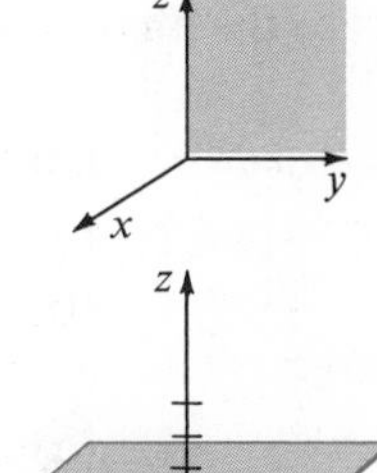

5. A pair of planes parallel to one of the coordinate planes if two of A, B, and C are 0, and the other coefficient and J are of opposite sign. For example, the graph of $z^2 - 4 = 0$ is the pair of planes whose equations are $z = 2$ and $z = -2$.

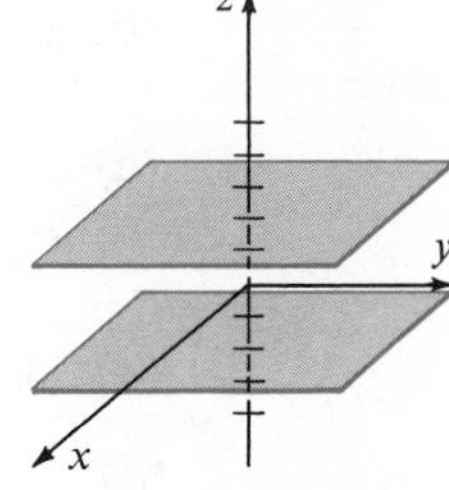

6. A pair of planes intersecting in a coordinate axis if one of A, B, and C is 0, the other two have opposite signs, and $J = 0$. For example, the graph of $4x^2 - y^2 = 0$ is the pair of planes whose equations are $2x + y = 0$ and $2x - y = 0$.

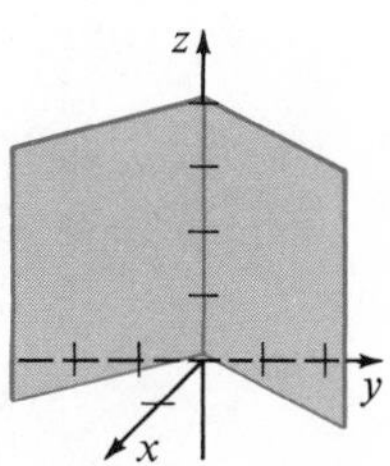

7. An elliptical or circular cylinder if one of A, B, and C is 0, the other two have the same sign, and J has the opposite sign. For example, the graph of $4x^2 + 9y^2 - 36 = 0$ is an elliptical cylinder with elements parallel to the z-axis.

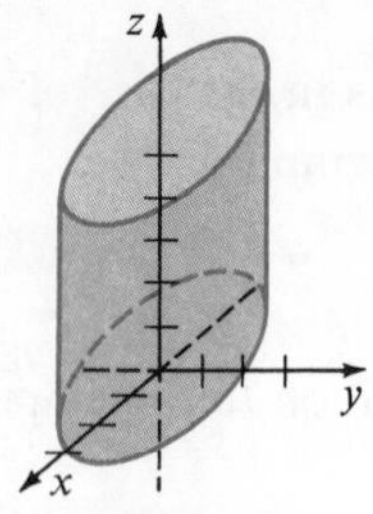

8. A hyperbolic cylinder if one of A, B, and C is 0, the other two have opposite signs, and $J \neq 0$. For example, the graph of $y^2 - 2z^2 - 1 = 0$ is a hyperbolic cylinder with elements parallel to the x-axis.

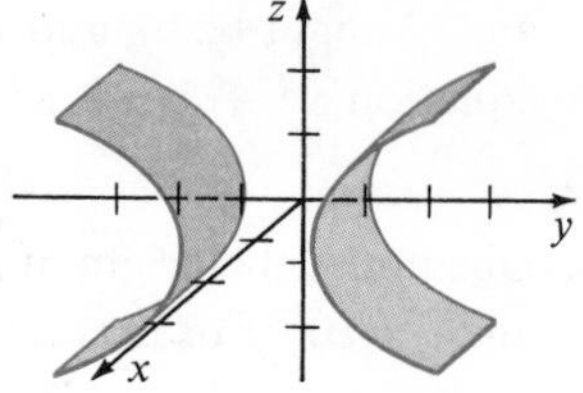

There is just one type of degenerate quadric surface that cannot be represented by an equation of the form (1), namely, the parabolic cylinder. For example, the graph of $z = y^2$ is a parabolic cylinder with elements parallel to the x-axis.

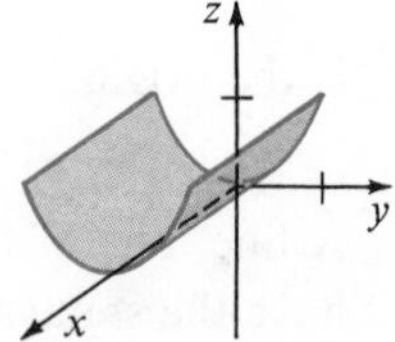

As illustrated by Example 1 on page 352, in sketching a surface you can use the traces of the surface in the coordinate planes, and cross sections of the surface in planes parallel to the coordinate planes, as a framework on which to base the final sketch.

Example. Let $\mathcal{S}$ be the surface with equation $4z = x^2 + y^2$.

(a) Name the graph $\mathcal{S}$.

(b) Find equations for the traces of $\mathcal{S}$ in the coordinate planes and name the graph of each trace.

(c) Find equations for the cross sections of $\mathcal{S}$ in the planes with equations $z = 1$ and $z = 4$, and name the graph of each cross section.

(d) Sketch the surface $\mathcal{S}$.

Solution: **(a)** Comparing the given equation with the standard equations for quadric surfaces, you can see that since the given equation can be written equivalently in the form

$$z = \frac{x^2}{a^2} + \frac{y^2}{b^2},$$

where $a = b = 2$, the graph is an elliptic paraboloid. (Since the coefficients of x^2 and y^2 are equal, it is also a surface of revolution.)

(b) Equations and names for the traces are the following:

in the xz-plane, $x^2 = 4z$, a parabola;
in the yz-plane, $y^2 = 4z$, a parabola;
in the xy-plane, $x^2 + y^2 = 0$, the origin.

The graphs of these traces are shown at the left below.

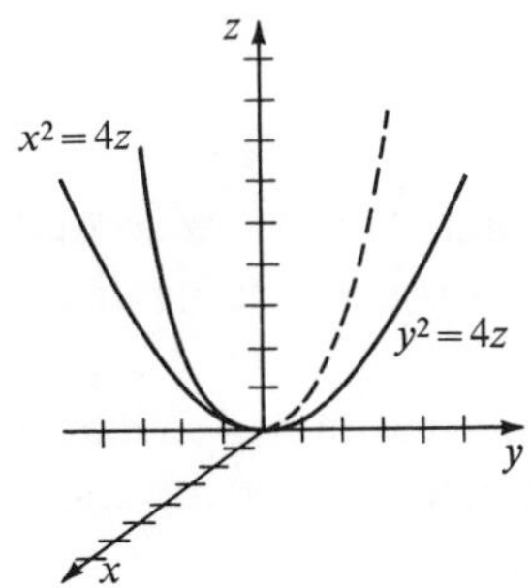

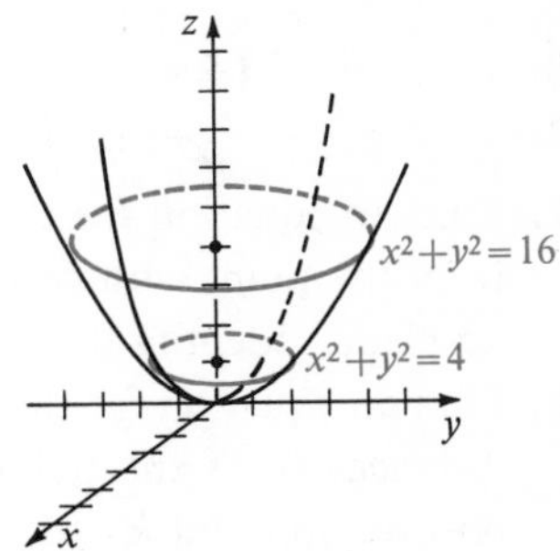

(c) Equations and names for the cross sections are the following:

in the plane $z = 1$, $x^2 + y^2 = 4$, a circle;
in the plane $z = 4$, $x^2 + y^2 = 16$, a circle.

(d) Graphing the cross sections, you have the sketch shown at the right above.

Exercises 10–3

In Exercises 1–8:

a. Name the graph of the given equation.
b. Find equations for the traces (if they exist) in the coordinate planes.
c. Find equations for the cross sections (if they exist) in the planes with equations $x = 4$, $y = 4$, and $z = 4$.
d. Draw a rough sketch of the surface.

1. $9x^2 + 4z^2 = 36y$

2. $4y^2 + 4z^2 - x^2 = 0$

3. $9x^2 - y^2 = 4z$

4. $2y^2 + 4z^2 = x^2$

5. $x^2 + \frac{y^2}{4} + \frac{z^2}{9} = 1$

6. $\frac{x^2}{9} + \frac{y^2}{16} - \frac{z^2}{4} = 1$

7. $\frac{-x^2}{4} - \frac{y^2}{9} + z^2 = 1$

8. $\frac{x^2}{4} - \frac{y^2}{9} = 0$

In Exercises 9–16, complete the square in x, y, and z as in Section 10–1. Then identify the surface.

9. $x^2 - y^2 - z^2 + 4x - 8y + 2z - 17 = 0$
10. $x^2 + y^2 + z^2 - 6x + 4y - 8 = 0$
11. $2x^2 + y^2 - 4z^2 + 4z - 6y - 2 = 0$
12. $y^2 - x^2 - 4z^2 = 2x + 8z$
13. $x^2 - 4y^2 + 2x - z + 8y - 3 = 0$
14. $x^2 + z^2 + 2x + \frac{3}{2}y + 2z - 3 = 0$
15. $x^2 + z^2 - 4x - y - 5 = 0$
16. $x^2 - z^2 - y^2 - 4x + 4z - 1 = 0$

17. Find an equation for the locus of a point that moves so that its distance from the point with coordinates $(0, c, 0)$ is equal to its distance from the xz-plane. Identify the locus.

18. Find an equation for the locus of a point that moves so that the sum of its distances from the points with coordinates $(-c, 0, 0)$ and $(c, 0, 0)$ is a positive constant k. Identify the locus.

Exercises 19–24 outline a proof for the statement in the text that a hyperboloid of one sheet is a doubly ruled surface.

* **19.** Show that if the point $\mathbf{U}(x_0, y_0, z_0)$, with $x_0 \neq -a$, is on the hyperboloid of one sheet,

$$\frac{x^2}{a^2} + \frac{y^2}{b^2} - \frac{z^2}{c^2} = 1, \tag{1}$$

then there is a unique real value r for which

$$\frac{y_0}{b} + \frac{z_0}{c} = r\left(1 + \frac{x_0}{a}\right) \quad \text{and} \quad r\left(\frac{y_0}{b} - \frac{z_0}{c}\right) = 1 - \frac{x_0}{a}. \tag{2}$$

[*Hint:* Rewrite Equation (1) equivalently in the form

$$\left(\frac{y}{b} + \frac{z}{c}\right)\left(\frac{y}{b} - \frac{z}{c}\right) = \left(1 + \frac{x}{a}\right)\left(1 - \frac{x}{a}\right).$$

Solve the first of Equations (2) for r, and show that since $\mathbf{U}$ is on the hyperboloid this value of r also satisfies the second of Equations (2).]

* **20.** Show that, for each fixed value of the parameter r, if $\mathbf{U}(x_0, y_0, z_0)$ is on the line of intersection of the planes with equations

$$\frac{y}{b} + \frac{z}{c} = r\left(1 + \frac{x}{a}\right) \quad \text{and} \quad r\left(\frac{y}{b} - \frac{z}{c}\right) = 1 - \frac{x}{a}, \tag{3}$$

then $\mathbf{U}$ is also on the hyperboloid of one sheet represented by Equation (1).

* **21.** Show that the line of intersection of the planes with equations

$$\frac{y}{b} - \frac{z}{c} = 0 \quad \text{and} \quad 1 + \frac{x}{a} = 0 \tag{4}$$

lies on the hyperboloid of one sheet represented by Equation (1).

*** 22.** Repeat Exercise 19 with

$$\frac{y_0}{b} - \frac{z_0}{c} = r\left(1 + \frac{x_0}{a}\right) \quad \text{and} \quad r\left(\frac{y_0}{b} + \frac{z_0}{c}\right) = 1 - \frac{x_0}{a}$$

in place of Equations (2).

*** 23.** Repeat Exercise 20 with

$$\frac{y}{b} - \frac{z}{c} = r\left(1 + \frac{x}{a}\right) \quad \text{and} \quad r\left(\frac{y}{b} + \frac{z}{c}\right) = 1 - \frac{x}{a}$$

in place of Equations (3).

*** 24.** Repeat Exercise 21 with

$$\frac{y}{b} + \frac{z}{c} = 0 \quad \text{and} \quad 1 + \frac{x}{a} = 0$$

in place of Equations (4).

Exercises 25–28 outline a proof of the fact that a hyperbolic paraboloid is a doubly ruled surface.

*** 25.** Show that if the point $\mathbf{U}(x_0, y_0, z_0)$ is on the hyperbolic paraboloid

$$\frac{y^2}{b^2} - \frac{x^2}{a^2} = z, \tag{5}$$

then there is a unique real number r for which

$$\frac{y_0}{b} + \frac{x_0}{a} = r \quad \text{and} \quad r\left(\frac{y_0}{b} - \frac{x_0}{a}\right) = z_0. \tag{6}$$

[*Hint:* Factor the left-hand member of Equation (5).]

*** 26.** Show that, for each fixed value of the parameter r, if $\mathbf{U}(x_0, y_0, z_0)$ is on the line of intersection of the planes with equations

$$\frac{y}{b} + \frac{x}{a} = r \quad \text{and} \quad r\left(\frac{y}{b} - \frac{x}{a}\right) = z, \tag{7}$$

then $\mathbf{U}$ is also on the hyperbolic paraboloid represented by Equation (5) of Exercise 25.

*** 27.** Repeat Exercise 25 with

$$\frac{y_0}{b} - \frac{x_0}{a} = r \quad \text{and} \quad r\left(\frac{y_0}{b} + \frac{x_0}{a}\right) = z_0$$

in place of Equations (6).

*** 28.** Repeat Exercise 26 with

$$\frac{y}{b} - \frac{x}{a} = r \quad \text{and} \quad r\left(\frac{y}{b} + \frac{x}{a}\right) = z$$

in place of Equations (7).

Transformation of Coordinates; Other Coordinate Systems

10–4 Translation and Rotation of Axes

In Section 10–1, you saw that an equation of the form

$$(x - x_0)^2 + (y - y_0)^2 + (z - z_0)^2 = r^2, \qquad r > 0, \tag{1}$$

represents a sphere with center $\mathbf{C}(x_0, y_0, z_0)$ and radius r. Making the substitutions

$$\begin{aligned} x' &= x - x_0, \\ y' &= y - y_0, \\ z' &= z - z_0, \end{aligned} \tag{2}$$

you can write Equation (1) in the form

$$x'^2 + y'^2 + z'^2 = r^2. \tag{3}$$

Equations (2), or the equivalent equations

$$\begin{aligned} x &= x' + x_0, \\ y &= y' + y_0, \\ z &= z' + z_0, \end{aligned} \tag{4}$$

are equations for a *translation of axes* (Figure 10–14). You can think of the x'-, y'-, and z'-axes as resulting from sliding, or translating, the x-, y-, and z-axes, while keeping them parallel to their original orientations, until the origin coincides with the point whose original coordinates were (x_0, y_0, z_0). Equations (1) and (3) both represent the sphere shown in Figure 10–14, but Equation (3) is simpler and it represents the sphere with its center at the origin of the $x'y'z'$-system.

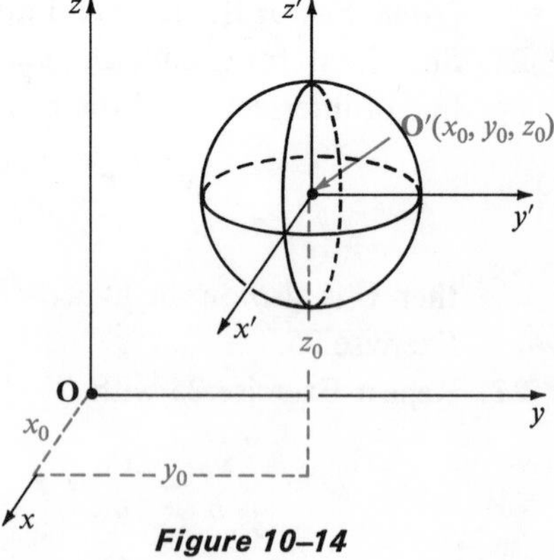

Figure 10–14

By a translation of axes, you can similarly transform any second-degree equation in x, y, and z in which there are no xy-, xz-, or yz-terms into one of the simple forms you studied in Section 10–3.

Example 1. By a translation of axes, transform each equation so that its graph has its center (if the graph is an ellipsoid or a hyperboloid) or its vertex (if the graph is a paraboloid) at the origin of the new coordinate system. Identify the graph and give the xyz-coordinates of its center or vertex.

(a) $36x^2 + 9y^2 - 4z^2 + 18y + 16z - 43 = 0$

(b) $y^2 + 5z^2 - 5x - 6y + 10z - 1 = 0$

Solution: **(a)** Collect terms, factor constants, complete squares, simplify:

$$36x^2 + 9(y^2 + 2y) - 4(z^2 - 4z) = 43,$$
$$36x^2 + 9(y^2 + 2y + 1) - 4(z^2 - 4z + 4) = 43 + 9 - 16,$$
$$36x^2 + 9(y + 1)^2 - 4(z - 2)^2 = 36.$$

Then let $x' = x$, $y' = y + 1$, $z' = z - 2$, and write the equation in standard form:

$$\frac{x'^2}{1^2} + \frac{y'^2}{2^2} - \frac{z'^2}{3^2} = 1.$$

This is an equation of a hyperboloid of one sheet (see page 357). The coordinates of its center are $(x', y', z') = (0, 0, 0)$. Therefore, since $x = x'$, $y = y' - 1$, and $z = z' + 2$, the xyz-coordinates of the center are $(0, -1, 2)$.

(b) Collect terms, factor constants, complete squares, simplify:

$$(y^2 - 6y) + 5(z^2 + 2z) = 5x + 1,$$
$$(y^2 - 6y + 9) + 5(z^2 + 2z + 1) = 5x + 1 + 9 + 5,$$
$$(y - 3)^2 + 5(z + 1)^2 = 5x + 15 = 5(x + 3).$$

Then let $x' = x + 3$, $y' = y - 3$, and $z' = z + 1$, and write the equation in standard form:

$$\frac{y'^2}{(\sqrt{5})^2} + \frac{z'^2}{1^2} = x'.$$

This is an equation of an elliptic paraboloid (see page 358). The coordinates of its vertex are $(x', y', z') = (0, 0, 0)$. Therefore, because $x = x' - 3$, $y = y' + 3$, and $z = z' - 1$, the xyz-coordinates of the vertex are $(-3, 3, -1)$.

As in the case of Cartesian axes in the plane, it is often useful to rotate Cartesian axes in space. This procedure can be used in particular to simplify a second-degree equation in x, y, and z which contains xy-, xz-, or yz-terms. The rotation *can* be made all at once (Exercise 13, page 369), but it can also always be achieved by a succession of at most three rotations in coordinate planes, with which you are already familiar.

Example 2. Identify the graph of the equation

$$x^2 - yz - y + 1 = 0.$$

Solution: To eliminate the yz-term (see Section 5–3), let

$$y = y' \cos\phi - z' \sin\phi,$$
$$z = y' \sin\phi + z' \cos\phi,$$

and

$$x = x'.$$

Then the given equation can be written equivalently in the form

$$x'^2 - (y' \cos\phi - z' \sin\phi)(y' \sin\phi + z' \cos\phi) - (y' \cos\phi - z' \sin\phi) + 1 = 0.$$

The coefficient B' of the $y'z'$-term is given by

$$B' = \sin^2\phi - \cos^2\phi;$$

$B' = 0$ if

$$m^\circ(\phi) = 45, \quad \cos\phi = \sin\phi = \frac{1}{\sqrt{2}}\cdot$$

Substituting these values for $\sin\phi$ and $\cos\phi$, you obtain

$$x'^2 - \left(\frac{y'}{\sqrt{2}} - \frac{z'}{\sqrt{2}}\right)\left(\frac{y'}{\sqrt{2}} + \frac{z'}{\sqrt{2}}\right) - \left(\frac{y'}{\sqrt{2}} - \frac{z'}{\sqrt{2}}\right) + 1 = 0,$$

$$x'^2 - \frac{y'^2}{2} + \frac{z'^2}{2} - \frac{y'}{\sqrt{2}} + \frac{z'}{\sqrt{2}} + 1 = 0.$$

Collecting terms, factoring, and completing squares, you have

$$x'^2 - \tfrac{1}{2}(y'^2 + \sqrt{2}y' + \tfrac{1}{2}) + \tfrac{1}{2}(z'^2 + \sqrt{2}z' + \tfrac{1}{2}) + 1 = 0 - \tfrac{1}{4} + \tfrac{1}{4},$$

$$x'^2 - \tfrac{1}{2}\left(y' + \frac{1}{\sqrt{2}}\right)^2 + \tfrac{1}{2}\left(z' + \frac{1}{\sqrt{2}}\right)^2 + 1 = 0.$$

You can then simplify the equation further by a translation of axes: Let $x' = x''$, $y' + \dfrac{1}{\sqrt{2}} = y''$, and $z' + \dfrac{1}{\sqrt{2}} = z''$. Making these substitutions and writing the equation in standard form, you have

$$\frac{y''^2}{(\sqrt{2})^2} - \frac{z''^2}{(\sqrt{2})^2} - \frac{x''^2}{1^2} = 1.$$

This is an equation of a hyperboloid of two sheets (see page 357).

Exercises 10–4

In Exercises 1–8, transform the given equation by a translation of axes so that its graph has its center or its vertex at the origin of the new coordinate system. Identify the graph and give the *xyz*-coordinates of its center or vertex.

1. $x^2 + 3y^2 + 4z^2 - 4x + 18y + 20 = 0$
2. $4x^2 - y^2 + 9z^2 - 2y + 54z + 44 = 0$
3. $2x^2 - 3y^2 - z^2 - 4x - 12y + 4z - 19 = 0$
4. $5x^2 + 4y^2 - 10x + 16y - 4z + 20 = 0$
5. $3x^2 - z^2 + 12x + 3y + 2z + 9 = 0$
6. $2x^2 + y^2 + 6z^2 - 2x - 2y + 18z + 9 = 0$
7. $9x^2 - 4y^2 - 8y - 5z - 14 = 0$
8. $x^2 + y^2 - 3z^3 - 8x + 6y + 12z + 13 = 0$

In Exercises 9–12, use an appropriate rotation of axes, and, if necessary, a translation of axes, to identify the graph of the given equation.

9. $xy + z^2 - 3z + 1 = 0$

10. $x^2 - yz + 4x - 3 = 0$

11. $\frac{2}{\sqrt{3}}x^2 + 2xy - 3z^2 + 4 = 0$

* 12. $xz + y - 4 = 0$

* 13. Suppose that you have two right-hand Cartesian coordinate systems in space, an *xyz*-system and an *x′y′z′*-system, and suppose that the *xyz*- and *x′y′z′*-axes have a common origin, as shown in the diagram at the right. Let the direction angles of the *x′*-axis (with respect to the *x*-, *y*-, and *z*-axes) be $\alpha_1, \beta_1, \gamma_1$, let those of the *y′*-axis be α_2, β_2, γ_2, and let those of the *z′*-axis be $\alpha_3, \beta_3, \gamma_3$.

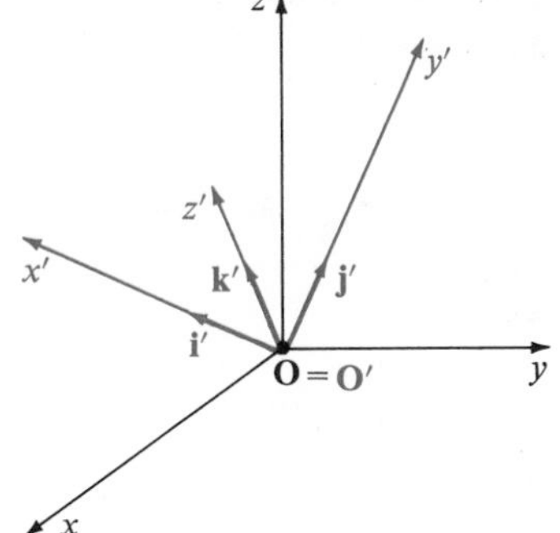

Show that if a point **S** has coordinates (x, y, z) with respect to the *xyz*-system and (x', y', z') with respect to the *x′y′z′*-system, then

$$\begin{aligned} x' &= x\cos\alpha_1 + y\cos\beta_1 + z\cos\gamma_1, \\ y' &= x\cos\alpha_2 + y\cos\beta_2 + z\cos\gamma_2, \\ z' &= x\cos\alpha_3 + y\cos\beta_3 + z\cos\gamma_3. \end{aligned}$$

Hint: Let $\mathbf{i}', \mathbf{j}'$, and $\mathbf{k}'$ be the unit vectors in the directions of the positive rays of the *x′*-, *y′*-, and *z′*-axes, respectively, and let **s** be the vector (x, y, z) with respect to the *xyz*-system and (x', y', z') with respect to the *x′y′z′*-system. Then, for example, you have

$$\begin{aligned} x' = \text{Comp}_{\mathbf{i}'}\,\mathbf{s} &= \text{Comp}_{\mathbf{i}'}\,(x\,\mathbf{i} + y\,\mathbf{j} + z\,\mathbf{k}) \\ &= x\,\text{Comp}_{\mathbf{i}'}\,\mathbf{i} + y\,\text{Comp}_{\mathbf{i}'}\,\mathbf{j} + z\,\text{Comp}_{\mathbf{i}'}\,\mathbf{k}. \end{aligned}$$

*** 14.** Show that, with the notation of Exercise 13, page 369,

$$\begin{aligned} x &= x' \cos \alpha_1 + y' \cos \alpha_2 + z' \cos \alpha_3, \\ y &= x' \cos \beta_1 + y' \cos \beta_2 + z' \cos \beta_3, \\ z &= x' \cos \gamma_1 + y' \cos \gamma_2 + z' \cos \gamma_3. \end{aligned}$$

*** 15.** Transform the equation

$$10x^2 + 13y^2 + 13z^2 - 4xy - 4xz + 8yz - 36 = 0$$

into an $x'y'z'$-equation, where the $x'y'z'$ direction-cosines are $(\frac{2}{3}, \frac{2}{3}, -\frac{1}{3})$, $(\frac{2}{3}, -\frac{1}{3}, \frac{2}{3})$, and $(-\frac{1}{3}, \frac{2}{3}, \frac{2}{3})$, respectively. Identify the graph.

10–5 Cylindrical and Spherical Coordinates

In the plane, as you saw in Chapter 7, polar coordinates are sometimes more useful than Cartesian coordinates. In space, there are also other useful coordinate systems. One of these is a *cylindrical coordinate system.*

To see how a cylindrical coordinate system can be established in space, consider a plane $\mathcal{P}$ and a line $\mathcal{L}$ perpendicular to $\mathcal{P}$ at a point **O** of $\mathcal{P}$ (Figure 10–15). Using **O** as the pole, establish a polar coordinate system in $\mathcal{P}$ (as discussed in Chapter 7), and make $\mathcal{L}$ into a number line (called the z-axis) having **O** as origin.

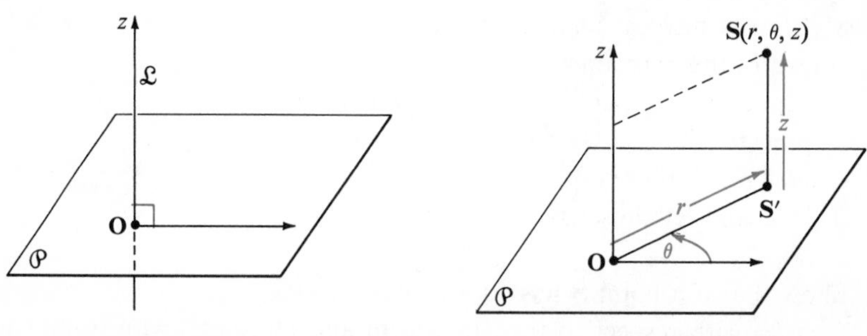

Figure 10–15

Figure 10–16

Thus, as illustrated in Figure 10–16, any point **S** in space can be assigned an ordered triple $(r, m(\theta), z)$, or, more briefly, (r, θ, z), of numbers, where z is the directed perpendicular distance from the polar-coordinate plane to **S**; the coordinates (r, θ, z) are called **cylindrical coordinates** of **S**. Because the polar coordinates of a point in the plane $\mathcal{P}$ are not unique, a cylindrical coordinate system does *not* assign unique coordinates to points in space. The term "cylindrical" is applied to this coordinate system because for each positive constant c the graph of $r = c$ is a right circular cylinder whose radius is c and whose axis is the z-axis (Figure 10–17).

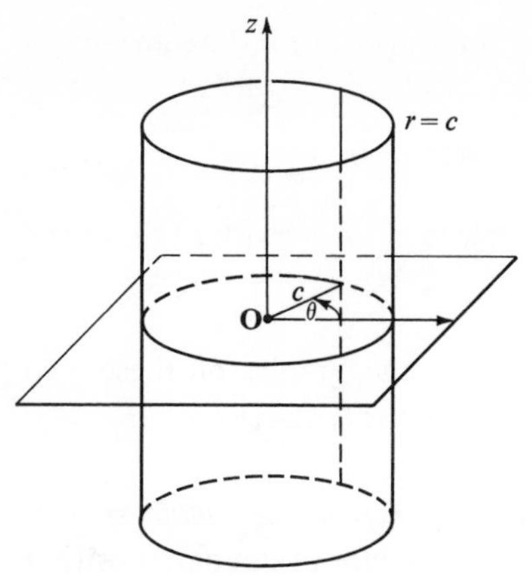

Figure 10–17

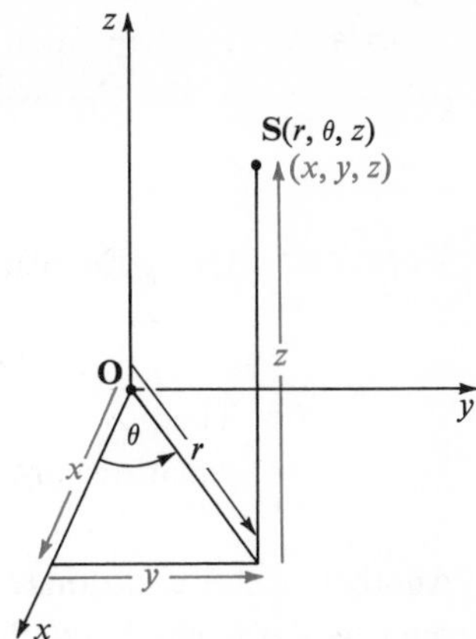

Figure 10–18

By orienting a cylindrical coordinate system so that the polar axis corresponds to the positive x-axis in a Cartesian coordinate system, the z-axis is the same for both systems, and the ray $r > 0$, $m°(\theta) = 90$, $z = 0$ corresponds to the positive y-axis (Figure 10–18), you can find relationships between the Cartesian and cylindrical coordinates of each point in space. (Note that the positive sense of rotation is the same in the polar coordinate plane and the xy-plane.) As Figure 10–18 suggests, for $x^2 + y^2 \neq 0$, you have

$$r = \pm\sqrt{x^2 + y^2}; \cos\theta = \frac{x}{\pm\sqrt{x^2 + y^2}}, \sin\theta = \frac{y}{\pm\sqrt{x^2 + y^2}};$$
$$z = z. \tag{1}$$

Also, whether or not $r = 0$,

$$x = r\cos\theta, y = r\sin\theta, \text{ and } z = z. \tag{2}$$

Example 1. Find cylindrical coordinates for the point **S** whose Cartesian coordinates are $(\sqrt{3}, -1, 3)$.

Solution: From Equations (1), using $x = \sqrt{3}$, $y = -1$, and $z = 3$, you have

$$r = \pm\sqrt{(\sqrt{3})^2 + (-1)^2} = \pm\sqrt{4} = \pm 2.$$

Suppose you select (arbitrarily) $r = 2$. Then $\cos\theta = \frac{\sqrt{3}}{2}$ and $\sin\theta = -\frac{1}{2}$, so that $m^R(\theta) = \frac{11\pi}{6} + 2k\pi$, where the integer k is arbitrary. Suppose you select $k = 0$; since $z = 3$, cylindrical coordinates of **S** are then $\left(2, \frac{11\pi}{6}, 3\right)$.

Example 2. Find an equation in cylindrical coordinates with the same graph as the Cartesian equation

$$x^2 + y^2 - 4z^2 = 0.$$

Solution: Using Equations (2), you can replace $x^2 + y^2$ with r^2 to obtain

$$r^2 - 4z^2 = 0.$$

Then $r = 2z$ or $r = -2z$. Since the graphs of these two equations are identical, you may arbitrarily select $r = 2z$.

Another useful coordinate system in space is a *spherical coordinate system.* From Chapter 8, you know that in space a geometric vector in standard position is determined by its length and its direction angles. Thus you can locate a point **S** in space by specifying the length and direction angles of the vector from the origin **O** in a Cartesian coordinate system to the point **S**. However, because the direction angles are not independent (see page 283), it is simpler to locate **S** by considering just two angles. This is what is done, for example, in locating a point on the surface of the Earth by giving its angles of longitude and latitude.

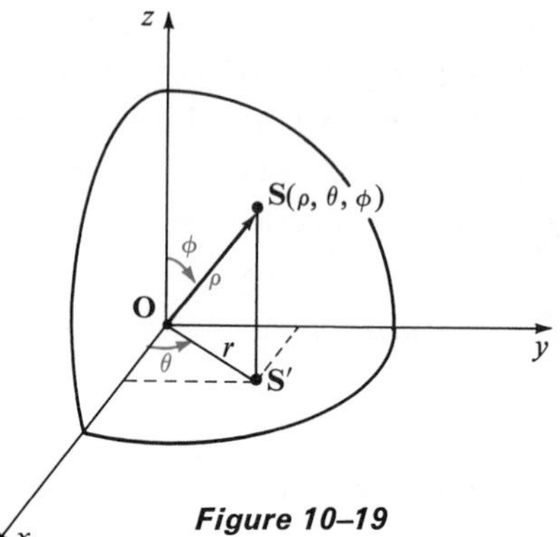

Figure 10–19

Consider the projection **S′**, on the xy-plane, of a given point **S**, as shown in Figure 10–19; by definition, **S′** is the point of intersection of the xy-plane and the line through **S** perpendicular to the plane. The polar coordinates of **S′** in the xy-plane are $(r, m(\theta))$, or (r, θ), as indicated.

If ρ is the distance from the origin **O** to the point **S** (the length of the vector $\overrightarrow{\mathbf{OS}}$) and ϕ is the angle between the positive ray of the z-axis and the ray $\overrightarrow{\mathbf{OS}}$, then $(\rho, m(\theta), m(\phi))$, or (ρ, θ, ϕ), is the ordered triple assigned to **S**; the coordinates (ρ, θ, ϕ) are called **spherical coordinates** of **S**. The term "spherical" refers to the fact that in a spherical coordinate system the graph of the equation $\rho = c$, $c > 0$, is a sphere with center at the origin and radius c. Like cylindrical coordinates, spherical coordinates are *not* unique. However, the spherical coordinates of all points other than those on the z-axis can be made unique by placing restrictions on ρ, θ, and ϕ. This is customarily done as follows:

$$\rho \geq 0, \quad 0 \leq \theta < 2\pi, \quad 0 \leq \phi \leq \pi.$$

The angle θ is then called the **longitude** of **S**, and ϕ is called its **colatitude**.

Figure 10–19 suggests that the relationships given by Equations (3) below exist between the Cartesian and spherical coordinates of a point. Notice first that the angle between $\overrightarrow{\mathbf{OS}}$ and $\overrightarrow{\mathbf{OS}}'$ in Figure 10–19 is $90 - \phi$, so that you have $r = \rho \sin \phi$, where $r \geq 0$ because of the restrictions placed on ρ and ϕ. Then you have

$$\begin{aligned} x &= r \cos \theta = \rho \sin \phi \cos \theta, \\ y &= r \sin \theta = \rho \sin \phi \sin \theta, \\ z &= \rho \cos \phi. \end{aligned} \tag{3}$$

Also,

$$\rho = \sqrt{x^2 + y^2 + z^2};$$

and, if $x^2 + y^2 \neq 0$, then

$$\begin{aligned} \cos \theta &= \frac{x}{\sqrt{x^2 + y^2}}, & \sin \theta &= \frac{y}{\sqrt{x^2 + y^2}}; \\ \cos \phi &= \frac{z}{\sqrt{x^2 + y^2 + z^2}}, & \sin \phi &= \frac{\sqrt{x^2 + y^2}}{\sqrt{x^2 + y^2 + z^2}}. \end{aligned} \tag{4}$$

Example 3. Write a Cartesian equation having the same graph as the spherical-coordinate equation

$$\rho = a \sin \phi \cos \theta.$$

Solution: Using Equations (4), you have, upon substitution in the given equation,

$$\sqrt{x^2 + y^2 + z^2} = a\left(\frac{\sqrt{x^2 + y^2}}{\sqrt{x^2 + y^2 + z^2}}\right)\left(\frac{x}{\sqrt{x^2 + y^2}}\right).$$

Then, upon simplifying, you find that a Cartesian equation of the graph is

$$x^2 + y^2 + z^2 = ax,$$

or

$$x^2 + y^2 + z^2 - ax = 0.$$

Note that this is an equation of the sphere with center $\left(\frac{a}{2}, 0, 0\right)$ and radius $\frac{a}{2}$.

Exercises 10–5

In Exercises 1–8, find cylindrical and spherical coordinates for the point whose Cartesian coordinates are given.

1. $(1, 1, 1)$
2. $(2, 1, 1)$
3. $(-2, \sqrt{3}, 2)$
4. $(\sqrt{5}, 2, 4)$
5. $(3, 4, 5)$
6. $(3, 3, -4)$
7. $(\sqrt{3}, -2, -1)$
8. $(-\sqrt{6}, -\sqrt{2}, -2\sqrt{2})$

In Exercises 9–12, find Cartesian coordinates for the point whose cylindrical coordinates are given.

9. $\left(4, \frac{\pi}{2}, 1\right)$
10. $\left(-3, \frac{\pi}{3}, 4\right)$
11. $\left(-3, \frac{\pi}{4}, -5\right)$
12. $\left(2, -\frac{\pi}{2}, 3\right)$

In Exercises 13–16, find Cartesian coordinates for the point whose spherical coordinates are given.

13. $\left(2, \frac{\pi}{2}, \frac{\pi}{6}\right)$
14. $\left(2, \frac{\pi}{2}, \frac{\pi}{2}\right)$
15. $\left(5, \frac{3\pi}{2}, \frac{\pi}{3}\right)$
16. $\left(3, \frac{\pi}{3}, \frac{3\pi}{4}\right)$

In Exercises 17–20, find a Cartesian equation for the surface whose equation in cylindrical coordinates is given.

17. $r^2 = 4 - z^2$
18. $r = 2$
19. $r = 9 \cos \theta$
20. $r^2 - z^2 = -4$

In Exercises 21–24, find a Cartesian equation for the surface whose equation in spherical coordinates is given.

21. $\rho = 2$
22. $\rho = 2 \sin \phi \cos \theta$
23. $\rho = 4 \sin \phi$
24. $\rho \cos \phi = 4$

In Exercises 25–32, find an equation in (a) cylindrical coordinates and (b) spherical coordinates for the surface whose Cartesian equation is given.

25. $x^2 + y^2 + z^2 = 9$
26. $x^2 + y^2 - 4z^2 = 16$
27. $x^2 + y^2 + 9z^2 = 36$
28. $x^2 + y^2 = 9z$
29. $x^2 + y^2 + z^2 - 2x = 0$
30. $y = 3$
31. $x = 4z^2$
32. $y = 2x^2 + z$

In Exercises 33–36, find an equation in (a) cylindrical coordinates and (b) spherical coordinates for the given surface.

33. A sphere with center at the origin and radius k.
34. A right circular cylinder with the z-axis as axis and radius k.
35. A plane parallel to the xy-plane and 4 units above it.
36. A plane parallel to the yz-plane and intersecting the x-axis at $x = 4$.

* **37.** Show that if $r \geq 0$ and $0 \leq \theta < 2\pi$ then the cylindrical and spherical coordinates of a point are related by

$$r = \rho \sin \phi,\ \theta = \theta,\ z = \rho \cos \phi.$$

* **38.** Use the result of Exercise 37 to find an equation in spherical coordinates for the surface having cylindrical-coordinate equation $r^2 + 4z^2 = 16$.

* **39.** Show that the distance between the points **S** and **T** whose cylindrical coordinates are (r_1, θ_1, z_1) and (r_2, θ_2, z_2) is given by

$$d(\mathbf{S}, \mathbf{T}) = \sqrt{r_2^2 - 2r_1r_2 \cos(\theta_2 - \theta_1) + r_1^2 + (z_2 - z_1)^2}.$$

* **40.** Show that the distance between the points **S** and **T** whose spherical coordinates are $(\rho_1, \theta_1, \phi_1)$ and $(\rho_2, \theta_2, \phi_2)$ is given by

$$d(\mathbf{S}, \mathbf{T}) = \sqrt{\rho_2^2 - 2\rho_1\rho_2 [\sin \phi_1 \sin \phi_2 \cos(\theta_2 - \theta_1) + \cos \phi_1 \cos \phi_2] + \rho_1^2}.$$

Chapter Summary

1. An equation of the **sphere** with center $\mathbf{C}(x_0, y_0, z_0)$ and radius r is

$$(x - x_0)^2 + (y - y_0)^2 + (z - z_0)^2 = r^2.$$

2. By completing squares, you can write an equation of the form

$$x^2 + y^2 + z^2 + Gx + Hy + Iz + J = 0 \qquad (1)$$

in the form

$$(x - x_0)^2 + (y - y_0)^2 + (z - z_0)^2 = k.$$

The locus is a sphere, a point, or $\emptyset$ according as $k > 0$, $k = 0$, or $k < 0$.

3. You can find an equation for the sphere through four given noncollinear points by substituting the coordinates of the points in Equation (1) above and then solving the resulting system of four linear equations for G, H, I, and J.
4. To determine an equation of the plane tangent to a given sphere at a given point on the sphere, you can use the fact that the plane is perpendicular to the vector from the center of the sphere to the point of contact.
5. The surface swept out by a line $\mathcal{L}$ that moves in such a way that a point **T** on $\mathcal{L}$ traces a curve $\mathcal{C}$ in a plane $\mathcal{P}$, and $\mathcal{L}$ always remains parallel to its original position, is called a **cylinder**. The curve $\mathcal{C}$ is called the **directrix** of the cylinder. The line $\mathcal{L}$ and all other lines on the surface parallel to $\mathcal{L}$ are called **elements**, or **generators**, of the surface.
6. If a curve $\mathcal{C}$ lying in a plane $\mathcal{P}$ is revolved about a line $\mathcal{L}$ in $\mathcal{P}$ so that each point on $\mathcal{C}$ describes a circle about $\mathcal{L}$, then the surface swept out is called a **surface of revolution**. The curve $\mathcal{C}$ is called the **generating curve**, and $\mathcal{L}$ is called the **axis of revolution**.

7. The graph in space of a second-degree Cartesian equation in three variables is called a **quadric surface**.
8. The central quadric surfaces are ellipsoids (including spheres and spheroids), hyperboloids of one sheet, hyperboloids of two sheets, and elliptic cones. In standard representation, a central quadric surface has its center at the origin.
9. The noncentral quadric surfaces are elliptic paraboloids and hyperbolic paraboloids. In standard representation, a noncentral quadric surface has its vertex at the origin.
10. Cylinders, hyperboloids of one sheet, elliptic cones, and hyperbolic paraboloids are **ruled surfaces**.
11. Just as there are degenerate conic sections, so are there degenerate quadric surfaces (see pages 361–362).
12. Equations for a translation of axes in space are:

$$\begin{aligned} x' &= x - x_0, & x &= x' + x_0,\\ y' &= y - y_0, & y &= y' + y_0,\\ z' &= z - z_0. & z &= z' + z_0. \end{aligned}$$

13. A rotation of axes in space can be made by a succession of at most three rotations in coordinate planes.
14. Cartesian coordinates (x, y, z) are related, for $x^2 + y^2 \neq 0$, to **cylindrical coordinates** (r, θ, z) by the equations

$$r = \pm\sqrt{x^2 + y^2},\ \cos\theta = \frac{x}{\pm\sqrt{x^2 + y^2}},\ \sin\theta = \frac{y}{\pm\sqrt{x^2 + y^2}},\ z = z;$$

and by the equations

$$x = r\cos\theta,\ y = r\sin\theta,\ z = z.$$

15. Cartesian coordinates (x, y, z) are related to **spherical coordinates** (ρ, θ, ϕ) by the equations

$$\begin{aligned} x &= \rho\sin\phi\cos\theta,\\ y &= \rho\sin\phi\sin\theta,\\ z &= \rho\cos\phi; \end{aligned}$$

and, for $x^2 + y^2 \neq 0$, by the equations

$$\rho = \sqrt{x^2 + y^2 + z^2},$$

$$\cos\theta = \frac{x}{\sqrt{x^2 + y^2}}, \qquad \sin\theta = \frac{y}{\sqrt{x^2 + y^2}},$$

$$\cos\phi = \frac{z}{\sqrt{x^2 + y^2 + z^2}},\ \sin\phi = \frac{\sqrt{x^2 + y^2}}{\sqrt{x^2 + y^2 + z^2}}.$$

Chapter Review Exercises

1. Determine the center and radius of the sphere with equation
$$x^2 + y^2 + z^2 - 8x + 6y = 0.$$
2. Determine the center and radius of the sphere through the points **Q**(0, 0, 0), **R**(2, 0, 0), **S**(0, 3, 0), and **T**(0, 0, 4).
3. Determine an equation of the plane that is tangent to the sphere $\mathcal{S}$ with equation
$$x^2 + y^2 + z^2 - 2x + 4y - 4 = 0$$
at the point (0, 0, 2) on $\mathcal{S}$.
4. Sketch the portion in the first octant of the cylinder with equation
$$x^2 + y^2 = 9.$$
5. Find an equation for the surface obtained by revolving the curve with equation
$$4x^2 + y^2 = 16$$
about the y-axis.
6. Name the graph of $x^2 + 4y^2 = z$ and sketch the surface.
7. For what value of J will the graph of $Ax^2 + By^2 + Cz^2 + J = 0$ be a single point?
8. Identify the graph of
$$x^2 - 2y^2 - z^2 - 2x - 12y - 4z - 24 = 0$$
and give the coordinates of its center.
9. Identify the graph of
$$x^2 + 2y^2 + 8y - z + 10 = 0$$
and give the coordinates of its vertex.
10. Identify the graph of $x^2 + yz - 2y - 2 = 0$.
11. Find cylindrical and spherical coordinates for the point with Cartesian coordinates $(1, -1, -1)$.
12. Find a Cartesian equation for the surface having cylindrical-coordinate equation $r^2 + z^2 = 9$.

Topology

In his inaugural address as Professor of Mathematics at the University of Erlangen, the German mathematician Christian Felix Klein (1849–1925) made a notable proposal. If you have read the essays at the end of earlier chapters in this book, then perhaps you will better understand the significance of his proposal. Namely, he suggested the classification of different branches of geometry according to the groups of transformations under which various theorems about geometrical figures remain true.

The suggestion given above was a dominant force in geometrical studies for almost fifty years, until considerations of relativity theory forced a change of emphasis to the nature of space itself, and in fact the suggestion still greatly influences geometrical thinking.

In the "Erlangen [or Erlanger (*adj.*)] program," a geometry is the system of definitions and theorems that remain invariant under a given group of transformations. In earlier essays of this book, you have encountered several such geometries. You saw that metric geometry is the geometry of rigid-motion transformations, and that Euclidean geometry includes metric geometry but is concerned also in part with invariants under similarity transformations. You saw further that rigid-motion and similarity transformations are special cases of the affine transformations of affine geometry, and that affine transformations in turn are special cases of the projective transformations of projective geometry.

Are there even more general transformations under which a significant set of definitions and theorems remains valid? If you crumple a sheet of paper

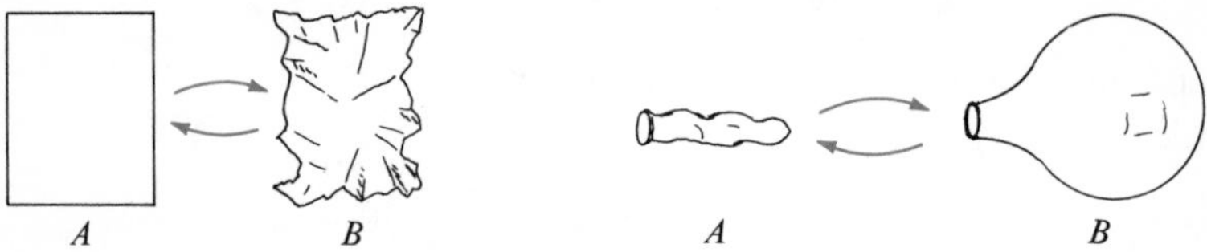

or inflate a toy balloon, then you effect a transformation from the original object A to the transformed object B that is

1. one-to-one

and

2. continuous from A to B and from B to A.

Such a transformation is called a *topological transformation*, and *topology* is the geometry of topological invariants.

The objects A and B above are *topologically equivalent;* in fact, all four objects are topologically equivalent figures in space. Can you see that a two-hole doughnut and a vase with two handles, as shown at the top of the next page, are topologically equivalent?

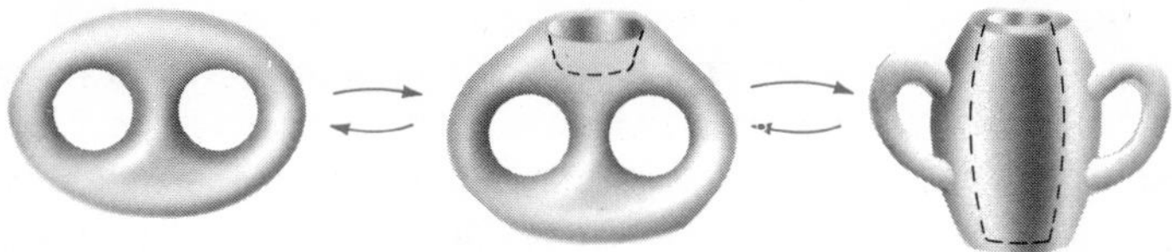

From the examples given above, you can see why topology is sometimes called "rubber geometry."

Notice that tearing a sheet of paper does not effect a continuous transformation, since points that are arbitrarily close are then mapped onto points that are not arbitrarily close. Notice also that gluing, so that two points are considered as being mapped onto one point, is not one-to-one. Thus tearing and gluing are not topological transformations.

Not much attention was given to topological problems before the middle of the nineteenth century, and in fact most of the topological developments have occurred during the present century.

One of the first topological problems ever to be studied had to do with the seven bridges over the Pregel River in the East Prussian city of Königsberg. The question asked was whether or not it would be possible for a burgher in an afternoon walk to pass exactly once over each of the seven bridges. You can see that this is a topological problem, because it is concerned not with the size or shape of things but rather with their arrangement relative to each other; the problem would be the same on a badly distorted map of the city.

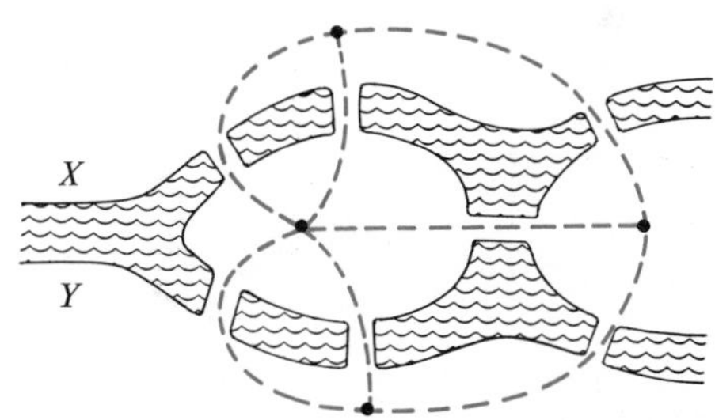

The bridge problem was solved by the great and versatile Swiss mathematician Leonhard Euler (1707–1783) in 1736. Choosing a point to represent each of the land regions involved, you can picture a walk over each bridge by means of a curved segment joining two of the points. For any such connected network of points (*nodes*) and segments (*arcs*), you can say that a node is *even* if it is on an even number of arcs, and that it is *odd* if it is on an odd number of arcs. Because each arc has two ends, there cannot be an odd number of odd nodes. Euler showed that:

1. If each node is even, then the walk is possible and it can be made starting and ending at the same point.
2. If exactly two nodes are odd, then the walk is possible but it cannot be made starting and ending at the same point.
3. If more than two nodes are odd, then the walk is impossible.

Since all four nodes in the Königsberg-bridge problem are odd, the walk is impossible for that configuration.

If there had been an eighth bridge, from *X* to *Y* in the figure (as indeed today there is), would a burgher have been able to make the walk? If so, would he have been able both to start and to end the walk at his own home? You can find the answer by applying Euler's criteria given on page 379.

For each of the following figures, is it possible to make a copy with one continuous, nonretracing mark on a sheet of paper? You can apply Euler's criteria to see that for one of the figures it is possible, beginning and ending

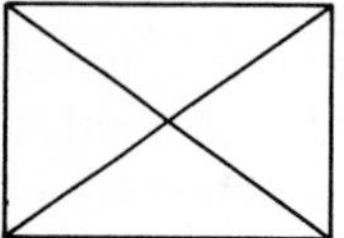
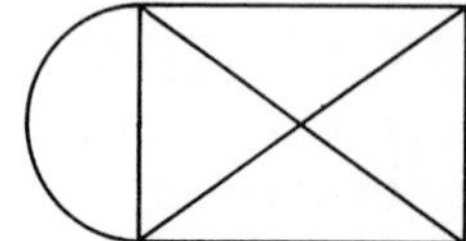

at the same point; for another it is possible, but not beginning and ending at the same point; and for the remaining one it is impossible.

For each of the following networks, is it possible to draw one continuous, nonretracing mark cutting each arc (*between* the nodes) exactly once? To

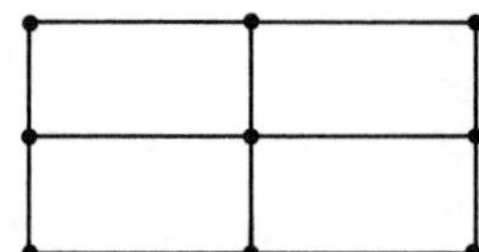
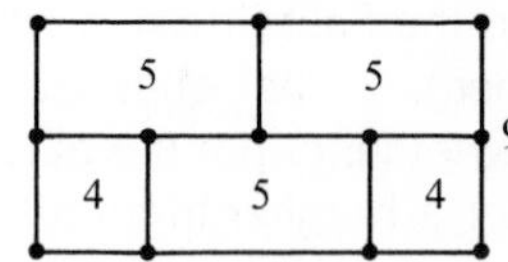

find the answer, you can construct an auxiliary network and apply Euler's criteria to it. It is not, however, actually necessary to construct the auxiliary network. You can simply count the arcs in the boundary of each region (including the *outside* one) and apply Euler's odd-and-even criteria to these numbers. The numbers are shown, for example, for the third network.

In topology, proofs of "intuitively obvious" facts sometimes are deceptively difficult. In the examples above, we have referred to regions and their boundaries. Each boundary is a simple closed curve; that is, it is topologically

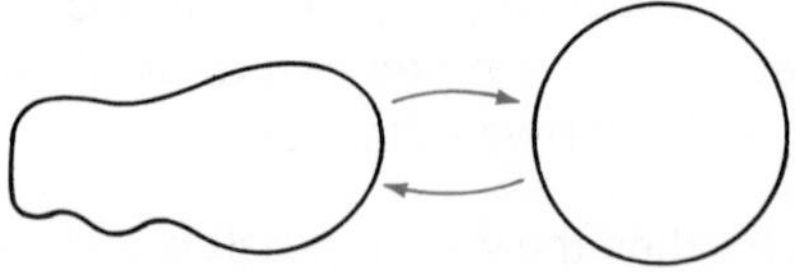

equivalent to a circle. The *Jordan curve theorem* (Camille Jordan; French, 1838–1922) is the statement that every simple closed curve in the plane divides the plane into exactly three sets: the curve itself and two regions, one inside the curve and the other outside. Jordan's own proof of his theorem was long, difficult—and erroneous.

In the figure at the left on the top of page 381, can you tell whether the point O is inside or outside the simple closed curve $\mathcal{C}$? It is inside if the ray shown in color cuts $\mathcal{C}$ an odd number of times; otherwise, it is outside.

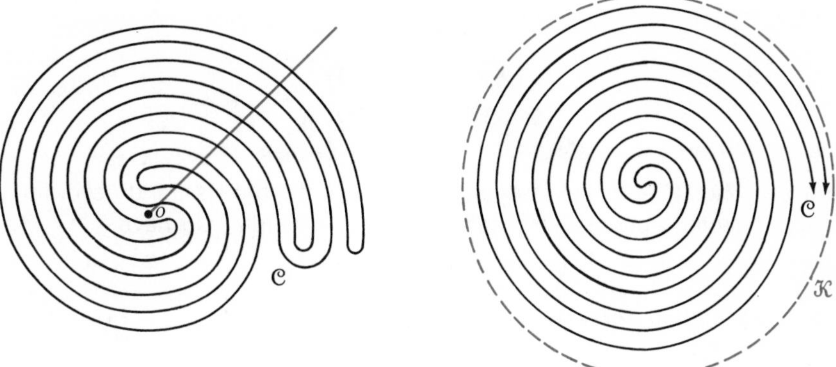

Somewhat surprisingly, the plane can be divided into more parts by an open arc (that is, by a set topologically equivalent to the set of points on the number line with x-coordinate satisfying $0 < x < 1$) than it can by a simple closed curve. In the figure at the right above, the open arc $\mathcal{C}$ determines three sets: two inside the circle $\mathcal{K}$ to which $\mathcal{C}$ converges, and the other the union of $\mathcal{K}$ and its exterior.

Another early topological result is the very important formula relating the number V of vertices, the number E of edges, and the number F of faces of a simple polyhedron:

$$V - E + F = 2.$$

For example, for the cube you have

$$V - E + F = 8 - 12 + 6 = 2.$$

Descartes found this formula in 1640, and Euler proved it in 1752; it is known as *Euler's formula for polyhedrons.* The result does not depend on the fact that the edges of a polyhedron are line segments or that the faces lie in planes; it holds equally well, for example, for connected networks on a sphere. It can be proved by starting with a network consisting of a single point, or node, on the sphere. For this, you have

$$V - E + F = 1 - 0 + 1 = 2,$$

as desired. You can then proceed inductively to build up any given connected

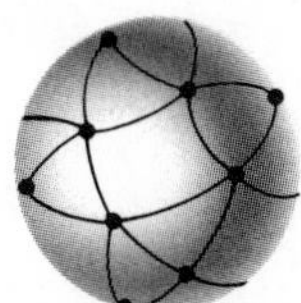

network by successively adjoining new arcs and nodes and noting that at each possible step both $V + F$ and E are increased equally.

As suggested by the figure below, you can add a "handle" to a sphere by removing 2 faces from a network on the sphere and replacing them with a handle having 3 new edges, 3 new faces without holes, and no new

vertices. Thus $V - E + F$ is decreased by 2 in the process, and accordingly for a connected network on a sphere with one handle you have

$$V - E + F = 2 - 2 = 0.$$

In general, then, for a sphere with p handles you have

$$V - E + F = 2 - 2p.$$

You say that p is the *genus* of the surface.

Since the surfaces of the figures shown on page 379 are topologically

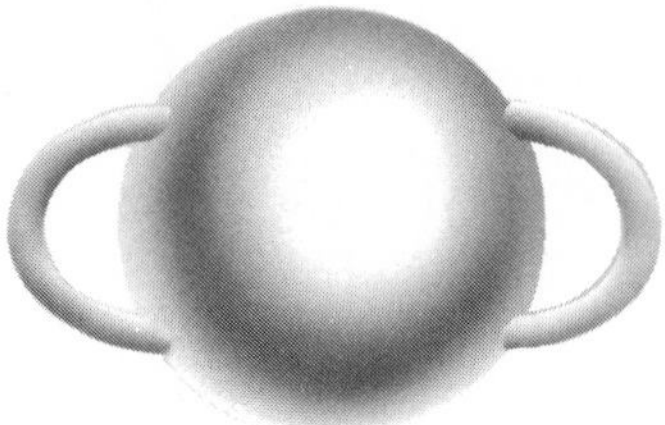

equivalent to a sphere with 2 handles, these surfaces also are of genus 2.

The surfaces we have been considering are *two-sided* surfaces. A bug crawling on one side could never meet a bug crawling on the other side. Even for a *piece* of one of the surfaces, the bugs could meet only at or over an edge.

The German mathematician August Ferdinand Möbius (1790–1868) noted that there also are *one-sided* surfaces, and he investigated some of their properties. One such surface is the Klein bottle, shown on the left below,

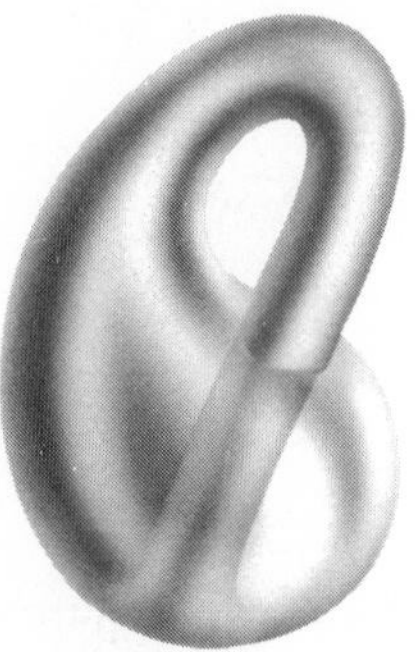

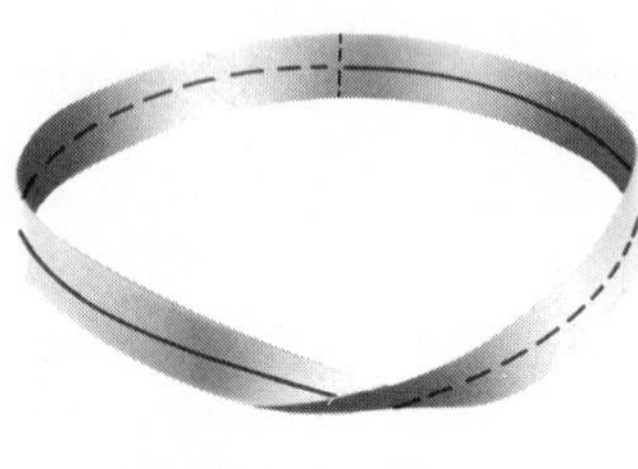

which is sometimes used as a flytrap. To its right is the amusing Möbius strip.

Such a strip can be made from a long rectangular strip of paper by giving it a half twist and gluing the ends together. You might experiment with such a strip by cutting the strip its entire length along a line parallel to the edge and one-half or one-third of the way across the strip. You might also experiment with other strips having different twists.

In the imaginary geographical map shown below, each of the four countries has a boundary arc in common with every other country. Accordingly, if

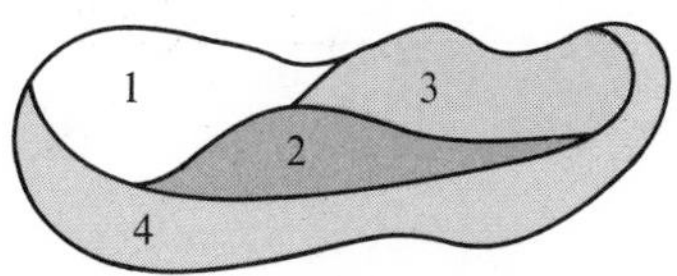

regions sharing a boundary arc are to be given different colors, then four colors are needed for this map.

The famous unsolved *four-color problem* can now be stated very simply: Determine the least number of colors that will suffice for *all possible* plane maps. The problem was posed by Möbius in 1840. It is now known that 5 colors will always suffice, and the example above shows that at least 4 are sometimes needed. The final solution, however, still remains to be found.

A spherical solid can be considered to be topologically mapped onto itself when it is rotated, say through an angle of 30°, about a polar axis (see the diagram on the left below). In this mapping, no point not on the polar axis is mapped onto itself. Is there a topological map of the spherical solid onto itself for which *every* point is mapped onto a point different from itself? The answer is "No." There are not necessarily *many* fixed points, as in this rotation, but for each topological map of a spherical solid onto itself there must be *at least one* point that is mapped onto itself.

Since each closed, two-sided surface $\mathcal{S}$ of genus $p = 0$ bounds a solid $\mathcal{V}$ that is topologically equivalent to a spherical solid, it follows that for each topological map of $\mathcal{V}$ onto itself there must be at least one point that is mapped onto itself.

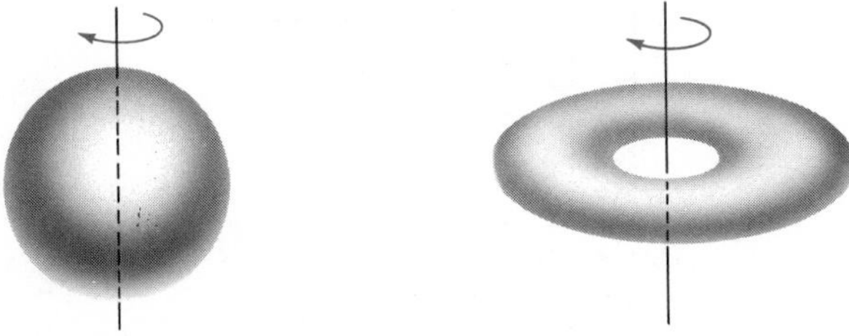

A result such as the one given above is called a *fixed-point theorem.* As suggested by the figure at the right above, there is no corresponding fixed-point theorem for topological maps of the solid *torus*, that is, the solid $\mathcal{V}$ bounded by a closed, two-sided surface of genus $p = 1$, onto itself.

Topological results are applied in many other branches of mathematics. Suppose, for example, that the function f, defined by $y = f(x)$, is continuous

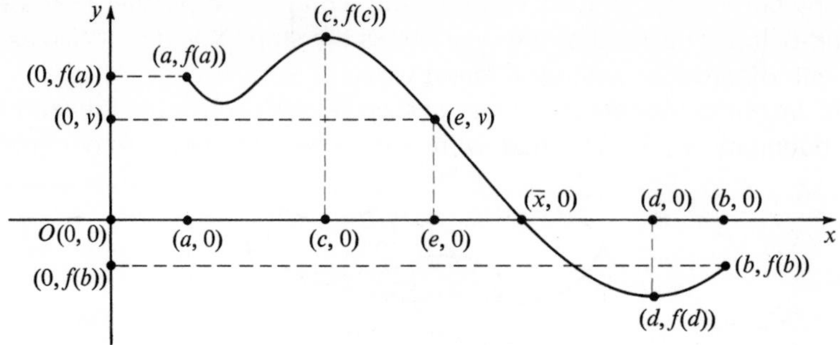

in the interval $a \leq x \leq b$. Then it can be shown, by considering the topology of sets of points, that:

1. There is at least one value c, $a \leq c \leq b$, at which $f(x)$ takes on a maximum value, and at least one value d, $a \leq d \leq b$, at which $f(x)$ takes on a minimum value.
2. For each value v between $f(a)$ and $f(b)$, there is at least one value e, $a \leq e \leq b$, for which $f(e) = v$.

Notice that the results given above concerning the function f are *existence theorems*. The actual values c, d, and e are not determined. It is asserted only that there *are* such values. The values would be different for different functions. Although precise values are not given, the results are nevertheless of great theoretical importance.

In particular, by the second of these results, if $f(a)$ and $f(b)$ are of opposite sign, then:

3. There is a value $\bar{x}$ between a and b for which $f(\bar{x}) = 0$.

This last result can be used to demonstrate another interesting fixed-point theorem. Suppose the function g, defined by $y = g(x)$, is continuous in the

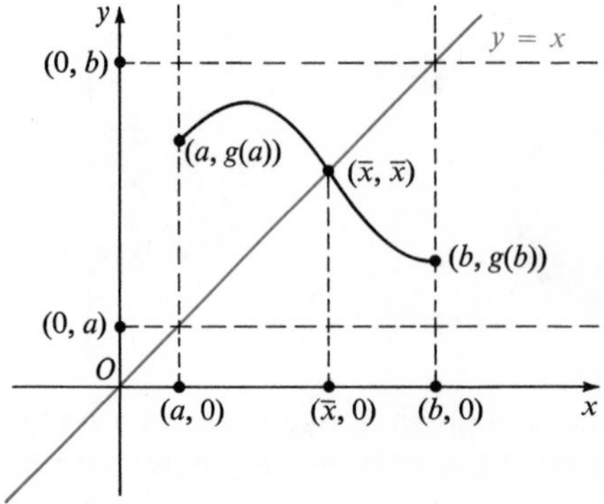

interval $a \leq x \leq b$. Suppose also that you have $a \leq g(x) \leq b$ in this interval;

then g maps the interval $a \leq x \leq b$ *into itself* (although not necessarily topologically, since the map might not be one-to-one). The theorem is that there must be some value $\bar{x}$, $a \leq \bar{x} \leq b$, which is mapped onto itself, that is, for which $g(\bar{x}) = \bar{x}$. To prove this, you should consider the function f defined by $f(x) = g(x) - x$. Perhaps you can supply the details.

Topology formerly was often called *analysis situs* (analysis of place, or of position). In 1895, the French mathematical universalist Jules Henri Poincaré (1854–1912) wrote a long research article entitled *Analysis situs*, and he followed this with additional articles devoted to the subject. He recognized the fundamental topological importance of Euler's formula and the basic significance of the topological work done in the second half of the nineteenth century by Möbius, Jordan, and Klein, mentioned above, and also by such German mathematicians as Carl Friedrich Gauss (1777–1855), Johann Benedict Listing (1808–1882), Georg Friedrich Bernhard Riemann (1826–1866), and Karl Theodor Wilhelm Weierstrass (1815–1897). Poincaré's work firmly established topology as a distinct branch of mathematics, and it has been systematically pursued as such since the start of this century.

Much of the recent development of geometry was observed and participated in by the American mathematician Oswald Veblen (1880–1960). His extensive writing contributed significantly, in succession, to the foundations of geometry, projective geometry, topology, differential geometry, and relativity theory. In 1905, he gave the first correct proof of the Jordan curve theorem (page 380). Even more important than his own research and expository writing, however, were his good judgment, his devotion to the development of mathematics, and his encouragement and kindly understanding of other mathematicians.

Oswald Veblen

Table 1 Squares and Square Roots

N	N^2	$\sqrt{N}$	$\sqrt{10N}$	N	N^2	$\sqrt{N}$	$\sqrt{10N}$
1.0	1.00	1.000	3.162	**5.5**	30.25	2.345	7.416
1.1	1.21	1.049	3.317	**5.6**	31.36	2.366	7.483
1.2	1.44	1.095	3.464	**5.7**	32.49	2.387	7.550
1.3	1.69	1.140	3.606	**5.8**	33.64	2.408	7.616
1.4	1.96	1.183	3.742	**5.9**	34.81	2.429	7.681
1.5	2.25	1.225	3.873	**6.0**	36.00	2.449	7.746
1.6	2.56	1.265	4.000	**6.1**	37.21	2.470	7.810
1.7	2.89	1.304	4.123	**6.2**	38.44	2.490	7.874
1.8	3.24	1.342	4.243	**6.3**	39.69	2.510	7.937
1.9	3.61	1.378	4.359	**6.4**	40.96	2.530	8.000
2.0	4.00	1.414	4.472	**6.5**	42.25	2.550	8.062
2.1	4.41	1.449	4.583	**6.6**	43.56	2.569	8.124
2.2	4.84	1.483	4.690	**6.7**	44.89	2.588	8.185
2.3	5.29	1.517	4.796	**6.8**	46.24	2.608	8.246
2.4	5.76	1.549	4.899	**6.9**	47.61	2.627	8.307
2.5	6.25	1.581	5.000	**7.0**	49.00	2.646	8.367
2.6	6.76	1.612	5.099	**7.1**	50.41	2.665	8.426
2.7	7.29	1.643	5.196	**7.2**	51.84	2.683	8.485
2.8	7.84	1.673	5.292	**7.3**	53.29	2.702	8.544
2.9	8.41	1.703	5.385	**7.4**	54.76	2.720	8.602
3.0	9.00	1.732	5.477	**7.5**	56.25	2.739	8.660
3.1	9.61	1.761	5.568	**7.6**	57.76	2.757	8.718
3.2	10.24	1.789	5.657	**7.7**	59.29	2.775	8.775
3.3	10.89	1.817	5.745	**7.8**	60.84	2.793	8.832
3.4	11.56	1.844	5.831	**7.9**	62.41	2.811	8.888
3.5	12.25	1.871	5.916	**8.0**	64.00	2.828	8.944
3.6	12.96	1.897	6.000	**8.1**	65.61	2.846	9.000
3.7	13.69	1.924	6.083	**8.2**	67.24	2.864	9.055
3.8	14.44	1.949	6.164	**8.3**	68.89	2.881	9.110
3.9	15.21	1.975	6.245	**8.4**	70.56	2.898	9.165
4.0	16.00	2.000	6.325	**8.5**	72.25	2.915	9.220
4.1	16.81	2.025	6.403	**8.6**	73.96	2.933	9.274
4.2	17.64	2.049	6.481	**8.7**	75.69	2.950	9.327
4.3	18.49	2.074	6.557	**8.8**	77.44	2.966	9.381
4.4	19.36	2.098	6.633	**8.9**	79.21	2.983	9.434
4.5	20.25	2.121	6.708	**9.0**	81.00	3.000	9.487
4.6	21.16	2.145	6.782	**9.1**	82.81	3.017	9.539
4.7	22.09	2.168	6.856	**9.2**	84.64	3.033	9.592
4.8	23.04	2.191	6.928	**9.3**	86.49	3.050	9.644
4.9	24.01	2.214	7.000	**9.4**	88.36	3.066	9.695
5.0	25.00	2.236	7.071	**9.5**	90.25	3.082	9.747
5.1	26.01	2.258	7.141	**9.6**	92.16	3.098	9.798
5.2	27.04	2.280	7.211	**9.7**	94.09	3.114	9.849
5.3	28.09	2.302	7.280	**9.8**	96.04	3.130	9.899
5.4	29.16	2.324	7.348	**9.9**	98.01	3.146	9.950
5.5	30.25	2.345	7.416	**10**	100.00	3.162	10.000

Table 2 Values of Trigonometric Functions

Angle	Sin	Cos	Tan	Cot	Sec	Csc	
0° 00′	.0000	1.0000	.0000	- - - -	1.000	- - - -	**90° 00′**
10′	.0029	1.0000	.0029	343.8	1.000	343.8	**50′**
20′	.0058	1.0000	.0058	171.9	1.000	171.9	**40′**
30′	.0087	1.0000	.0087	114.6	1.000	114.6	**30′**
40′	.0116	.9999	.0116	85.94	1.000	85.95	**20′**
50′	.0145	.9999	.0145	68.75	1.000	68.76	**10′**
1° 00′	.0175	.9998	.0175	57.29	1.000	57.30	**89° 00′**
10′	.0204	.9998	.0204	49.10	1.000	49.11	**50′**
20′	.0233	.9997	.0233	42.96	1.000	42.98	**40′**
30′	.0262	.9997	.0262	38.19	1.000	38.20	**30′**
40′	.0291	.9996	.0291	34.37	1.000	34.38	**20′**
50′	.0320	.9995	.0320	31.24	1.001	31.26	**10′**
2° 00′	.0349	.9994	.0349	28.64	1.001	28.65	**88° 00′**
10′	.0378	.9993	.0378	26.43	1.001	26.45	**50′**
20′	.0407	.9992	.0407	24.54	1.001	24.56	**40′**
30′	.0436	.9990	.0437	22.90	1.001	22.93	**30′**
40′	.0465	.9989	.0466	21.47	1.001	21.49	**20′**
50′	.0494	.9988	.0495	20.21	1.001	20.23	**10′**
3° 00′	.0523	.9986	.0524	19.08	1.001	19.11	**87° 00′**
10′	.0552	.9985	.0553	18.07	1.002	18.10	**50′**
20′	.0581	.9983	.0582	17.17	1.002	17.20	**40′**
30′	.0610	.9981	.0612	16.35	1.002	16.38	**30′**
40′	.0640	.9980	.0641	15.60	1.002	15.64	**20′**
50′	.0669	.9978	.0670	14.92	1.002	14.96	**10′**
4° 00′	.0698	.9976	.0699	14.30	1.002	14.34	**86° 00′**
10′	.0727	.9974	.0729	13.73	1.003	13.76	**50′**
20′	.0756	.9971	.0758	13.20	1.003	13.23	**40′**
30′	.0785	.9969	.0787	12.71	1.003	12.75	**30′**
40′	.0814	.9967	.0816	12.25	1.003	12.29	**20′**
50′	.0843	.9964	.0846	11.83	1.004	11.87	**10′**
5° 00′	.0872	.9962	.0875	11.43	1.004	11.47	**85° 00′**
10′	.0901	.9959	.0904	11.06	1.004	11.10	**50′**
20′	.0929	.9957	.0934	10.71	1.004	10.76	**40′**
30′	.0958	.9954	.0963	10.39	1.005	10.43	**30′**
40′	.0987	.9951	.0992	10.08	1.005	10.13	**20′**
50′	.1016	.9948	.1022	9.788	1.005	9.839	**10′**
6° 00′	.1045	.9945	.1051	9.514	1.006	9.567	**84° 00′**
10′	.1074	.9942	.1080	9.255	1.006	9.309	**50′**
20′	.1103	.9939	.1110	9.010	1.006	9.065	**40′**
30′	.1132	.9936	.1139	8.777	1.006	8.834	**30′**
40′	.1161	.9932	.1169	8.556	1.007	8.614	**20′**
50′	.1190	.9929	.1198	8.345	1.007	8.405	**10′**
7° 00′	.1219	.9925	.1228	8.144	1.008	8.206	**83° 00′**
10′	.1248	.9922	.1257	7.953	1.008	8.016	**50′**
20′	.1276	.9918	.1287	7.770	1.008	7.834	**40′**
30′	.1305	.9914	.1317	7.596	1.009	7.661	**30′**
40′	.1334	.9911	.1346	7.429	1.009	7.496	**20′**
50′	.1363	.9907	.1376	7.269	1.009	7.337	**10′**
8° 00′	.1392	.9903	.1405	7.115	1.010	7.185	**82° 00′**
10′	.1421	.9899	.1435	6.968	1.010	7.040	**50′**
20′	.1449	.9894	.1465	6.827	1.011	6.900	**40′**
30′	.1478	.9890	.1495	6.691	1.011	6.765	**30′**
40′	.1507	.9886	.1524	6.561	1.012	6.636	**20′**
50′	.1536	.9881	.1554	6.435	1.012	6.512	**10′**
9° 00′	.1564	.9877	.1584	6.314	1.012	6.392	**81° 00′**
	Cos	Sin	Cot	Tan	Csc	Sec	Angle

Table 2 Values of Trigonometric Functions

Angle	*Sin*	*Cos*	*Tan*	*Cot*	*Sec*	*Csc*	
9° 00′	.1564	.9877	.1584	6.314	1.012	6.392	**81° 00′**
10′	.1593	.9872	.1614	6.197	1.013	6.277	**50′**
20′	.1622	.9868	.1644	6.084	1.013	6.166	**40′**
30′	.1650	.9863	.1673	5.976	1.014	6.059	**30′**
40′	.1679	.9858	.1703	5.871	1.014	5.955	**20′**
50′	.1708	.9853	.1733	5.769	1.015	5.855	**10′**
10° 00′	.1736	.9848	.1763	5.671	1.015	5.759	**80° 00′**
10′	.1765	.9843	.1793	5.576	1.016	5.665	**50′**
20′	.1794	.9838	.1823	5.485	1.016	5.575	**40′**
30′	.1822	.9833	.1853	5.396	1.017	5.487	**30′**
40′	.1851	.9827	.1883	5.309	1.018	5.403	**20′**
50′	.1880	.9822	.1914	5.226	1.018	5.320	**10′**
11° 00′	.1908	.9816	.1944	5.145	1.019	5.241	**79° 00′**
10′	.1937	.9811	.1974	5.066	1.019	5.164	**50′**
20′	.1965	.9805	.2004	4.989	1.020	5.089	**40′**
30′	.1994	.9799	.2035	4.915	1.020	5.016	**30′**
40′	.2022	.9793	.2065	4.843	1.021	4.945	**20′**
50′	.2051	.9787	.2095	4.773	1.022	4.876	**10′**
12° 00′	.2079	.9781	.2126	4.705	1.022	4.810	**78° 00′**
10′	.2108	.9775	.2156	4.638	1.023	4.745	**50′**
20′	.2136	.9769	.2186	4.574	1.024	4.682	**40′**
30′	.2164	.9763	.2217	4.511	1.024	4.620	**30′**
40′	.2193	.9757	.2247	4.449	1.025	4.560	**20′**
50′	.2221	.9750	.2278	4.390	1.026	4.502	**10′**
13° 00′	.2250	.9744	.2309	4.331	1.026	4.445	**77° 00′**
10′	.2278	.9737	.2339	4.275	1.027	4.390	**50′**
20′	.2306	.9730	.2370	4.219	1.028	4.336	**40′**
30′	.2334	.9724	.2401	4.165	1.028	4.284	**30′**
40′	.2363	.9717	.2432	4.113	1.029	4.232	**20′**
50′	.2391	.9710	.2462	4.061	1.030	4.182	**10′**
14° 00′	.2419	.9703	.2493	4.011	1.031	4.134	**76° 00′**
10′	.2447	.9696	.2524	3.962	1.031	4.086	**50′**
20′	.2476	.9689	.2555	3.914	1.032	4.039	**40′**
30′	.2504	.9681	.2586	3.867	1.033	3.994	**30′**
40′	.2532	.9674	.2617	3.821	1.034	3.950	**20′**
50′	.2560	.9667	.2648	3.776	1.034	3.906	**10′**
15° 00′	.2588	.9659	.2679	3.732	1.035	3.864	**75° 00′**
10′	.2616	.9652	.2711	3.689	1.036	3.822	**50′**
20′	.2644	.9644	.2742	3.647	1.037	3.782	**40′**
30′	.2672	.9636	.2773	3.606	1.038	3.742	**30′**
40′	.2700	.9628	.2805	3.566	1.039	3.703	**20′**
50′	.2728	.9621	.2836	3.526	1.039	3.665	**10′**
16° 00′	.2756	.9613	.2867	3.487	1.040	3.628	**74° 00′**
10′	.2784	.9605	.2899	3.450	1.041	3.592	**50′**
20′	.2812	.9596	.2931	3.412	1.042	3.556	**40′**
30′	.2840	.9588	.2962	3.376	1.043	3.521	**30′**
40′	.2868	.9580	.2994	3.340	1.044	3.487	**20′**
50′	.2896	.9572	.3026	3.305	1.045	3.453	**10′**
17° 00′	.2924	.9563	.3057	3.271	1.046	3.420	**73° 00′**
10′	.2952	.9555	.3089	3.237	1.047	3.388	**50′**
20′	.2979	.9546	.3121	3.204	1.048	3.356	**40′**
30′	.3007	.9537	.3153	3.172	1.049	3.326	**30′**
40′	.3035	.9528	.3185	3.140	1.049	3.295	**20′**
50′	.3062	.9520	.3217	3.108	1.050	3.265	**10′**
18° 00′	.3090	.9511	.3249	3.078	1.051	3.236	**72° 00′**
	Cos	*Sin*	*Cot*	*Tan*	*Csc*	*Sec*	*Angle*

Table 2 Values of Trigonometric Functions

Angle	*Sin*	*Cos*	*Tan*	*Cot*	*Sec*	*Csc*	
18° 00′	.3090	.9511	.3249	3.078	1.051	3.236	**72° 00′**
10′	.3118	.9502	.3281	3.047	1.052	3.207	**50′**
20′	.3145	.9492	.3314	3.018	1.053	3.179	**40′**
30′	.3173	.9483	.3346	2.989	1.054	3.152	**30′**
40′	.3201	.9474	.3378	2.960	1.056	3.124	**20′**
50′	.3228	.9465	.3411	2.932	1.057	3.098	**10′**
19° 00′	.3256	.9455	.3443	2.904	1.058	3.072	**71° 00′**
10′	.3283	.9446	.3476	2.877	1.059	3.046	**50′**
20′	.3311	.9436	.3508	2.850	1.060	3.021	**40′**
30′	.3338	.9426	.3541	2.824	1.061	2.996	**30′**
40′	.3365	.9417	.3574	2.798	1.062	2.971	**20′**
50′	.3393	.9407	.3607	2.773	1.063	2.947	**10′**
20° 00′	.3420	.9397	.3640	2.747	1.064	2.924	**70° 00′**
10′	.3448	.9387	.3673	2.723	1.065	2.901	**50′**
20′	.3475	.9377	.3706	2.699	1.066	2.878	**40′**
30′	.3502	.9367	.3739	2.675	1.068	2.855	**30′**
40′	.3529	.9356	.3772	2.651	1.069	2.833	**20′**
50′	.3557	.9346	.3805	2.628	1.070	2.812	**10′**
21° 00′	.3584	.9336	.3839	2.605	1.071	2.790	**69° 00′**
10′	.3611	.9325	.3872	2.583	1.072	2.769	**50′**
20′	.3638	.9315	.3906	2.560	1.074	2.749	**40′**
30′	.3665	.9304	.3939	2.539	1.075	2.729	**30′**
40′	.3692	.9293	.3973	2.517	1.076	2.709	**20′**
50′	.3719	.9283	.4006	2.496	1.077	2.689	**10′**
22° 00′	.3746	.9272	.4040	2.475	1.079	2.669	**68° 00′**
10′	.3773	.9261	.4074	2.455	1.080	2.650	**50′**
20′	.3800	.9250	.4108	2.434	1.081	2.632	**40′**
30′	.3827	.9239	.4142	2.414	1.082	2.613	**30′**
40′	.3854	.9228	.4176	2.394	1.084	2.595	**20′**
50′	.3881	.9216	.4210	2.375	1.085	2.577	**10′**
23° 00′	.3907	.9205	.4245	2.356	1.086	2.559	**67° 00′**
10′	.3934	.9194	.4279	2.337	1.088	2.542	**50′**
20′	.3961	.9182	.4314	2.318	1.089	.2525	**40′**
30′	.3987	.9171	.4348	2.300	1.090	.2508	**30′**
40′	.4014	.9159	.4383	2.282	1.092	.2491	**20′**
50′	.4041	.9147	.4417	2.264	1.093	.2475	**10′**
24° 00′	.4067	.9135	.4452	2.246	1.095	2.459	**66° 00′**
10′	.4094	.9124	.4487	2.229	1.096	2.443	**50′**
20′	.4120	.9112	.4522	2.211	1.097	2.427	**40′**
30′	.4147	.9100	.4557	2.194	1.099	2.411	**30′**
40′	.4173	.9088	.4592	2.177	1.100	2.396	**20′**
50′	.4200	.9075	.4628	2.161	1.102	2.381	**10′**
25° 00′	.4226	.9063	.4663	2.145	1.103	2.366	**65° 00′**
10′	.4253	.9051	.4699	2.128	1.105	2.352	**50′**
20′	.4279	.9038	.4734	2.112	1.106	2.337	**40′**
30′	.4305	.9026	.4770	2.097	1.108	2.323	**30′**
40′	.4331	.9013	.4806	2.081	1.109	2.309	**20′**
50′	.4358	.9001	.4841	2.066	1.111	2.295	**10′**
26° 00′	.4384	.8988	.4877	2.050	1.113	2.281	**64° 00′**
10′	.4410	.8975	.4913	2.035	1.114	2.268	**50′**
20′	.4436	.8962	.4950	2.020	1.116	2.254	**40′**
30′	.4462	.8949	.4986	2.006	1.117	2.241	**30′**
40′	.4488	.8936	.5022	1.991	1.119	2.228	**20′**
50′	.4514	.8923	.5059	1.977	1.121	2.215	**10′**
27° 00′	.4540	.8910	.5095	1.963	1.122	2.203	**63° 00′**
	Cos	*Sin*	*Cot*	*Tan*	*Csc*	*Sec*	*Angle*

Table 2 Values of Trigonometric Functions

Angle	Sin	Cos	Tan	Cot	Sec	Csc	
27° 00′	.4540	.8910	.5095	1.963	1.122	2.203	**63° 00′**
10′	.4566	.8897	.5132	1.949	1.124	2.190	**50′**
20′	.4592	.8884	.5169	1.935	1.126	2.178	**40′**
30′	.4617	.8870	.5206	1.921	1.127	2.166	**30′**
40′	.4643	.8857	.5243	1.907	1.129	2.154	**20′**
50′	.4669	.8843	.5280	1.894	1.131	2.142	**10′**
28° 00′	.4695	.8829	.5317	1.881	1.133	2.130	**62° 00′**
10′	.4720	.8816	.5354	1.868	1.134	2.118	**50′**
20′	.4746	.8802	.5392	1.855	1.136	2.107	**40′**
30′	.4772	.8788	.5430	1.842	1.138	2.096	**30′**
40′	.4797	.8774	.5467	1.829	1.140	2.085	**20′**
50′	.4823	.8760	.5505	1.816	1.142	2.074	**10′**
29° 00′	.4848	.8746	.5543	1.804	1.143	2.063	**61° 00′**
10′	.4874	.8732	.5581	1.792	1.145	2.052	**50′**
20′	.4899	.8718	.5619	1.780	1.147	2.041	**40′**
30′	.4924	.8704	.5658	1.767	1.149	2.031	**30′**
40′	.4950	.8689	.5696	1.756	1.151	2.020	**20′**
50′	.4975	.8675	.5735	1.744	1.153	2.010	**10′**
30° 00′	.5000	.8660	.5774	1.732	1.155	2.000	**60° 00′**
10′	.5025	.8646	.5812	1.720	1.157	1.990	**50′**
20′	.5050	.8631	.5851	1.709	1.159	1.980	**40′**
30′	.5075	.8616	.5890	1.698	1.161	1.970	**30′**
40′	.5100	.8601	.5930	1.686	1.163	1.961	**20′**
50′	.5125	.8587	.5969	1.675	1.165	1.951	**10′**
31° 00′	.5150	.8572	.6009	1.664	1.167	1.942	**59° 00′**
10′	.5175	.8557	.6048	1.653	1.169	1.932	**50′**
20′	.5200	.8542	.6088	1.643	1.171	1.923	**40′**
30′	.5225	.8526	.6128	1.632	1.173	1.914	**30′**
40′	.5250	.8511	.6168	1.621	1.175	1.905	**20′**
50′	.5275	.8496	.6208	1.611	1.177	1.896	**10′**
32° 00′	.5299	.8480	.6249	1.600	1.179	1.887	**58° 00′**
10′	.5324	.8465	.6289	1.590	1.181	1.878	**50′**
20′	.5348	.8450	.6330	1.580	1.184	1.870	**40′**
30′	.5373	.8434	.6371	1.570	1.186	1.861	**30′**
40′	.5398	.8418	.6412	1.560	1.188	1.853	**20′**
50′	.5422	.8403	.6453	1.550	1.190	1.844	**10′**
33° 00′	.5446	.8387	.6494	1.540	1.192	1.836	**57° 00′**
10′	.5471	.8371	.6536	1.530	1.195	1.828	**50′**
20′	.5495	.8355	.6577	1.520	1.197	1.820	**40′**
30′	.5519	.8339	.6619	1.511	1.199	1.812	**30′**
40′	.5544	.8323	.6661	1.501	1.202	1.804	**20′**
50′	.5568	.8307	.6703	1.492	1.204	1.796	**10′**
34° 00′	.5592	.8290	.6745	1.483	1.206	1.788	**56° 00′**
10′	.5616	.8274	.6787	1.473	1.209	1.781	**50′**
20′	.5640	.8258	.6830	1.464	1.211	1.773	**40′**
30′	.5664	.8241	.6873	1.455	1.213	1.766	**30′**
40′	.5688	.8225	.6916	1.446	1.216	1.758	**20′**
50′	.5712	.8208	.6959	1.437	1.218	1.751	**10′**
35° 00′	.5736	.8192	.7002	1.428	1.221	1.743	**55° 00′**
10′	.5760	.8175	.7046	1.419	1.223	1.736	**50′**
20′	.5783	.8158	.7089	1.411	1.226	1.729	**40′**
30′	.5807	.8141	.7133	1.402	1.228	1.722	**30′**
40′	.5831	.8124	.7177	1.393	1.231	1.715	**20′**
50′	.5854	.8107	.7221	1.385	1.233	1.708	**10′**
36° 00′	.5878	.8090	.7265	1.376	1.236	1.701	**54° 00′**
	Cos	Sin	Cot	Tan	Csc	Sec	Angle

Table 2 Values of Trigonometric Functions

Angle	*Sin*	*Cos*	*Tan*	*Cot*	*Sec*	*Csc*	
36° 00′	.5878	.8090	.7265	1.376	1.236	1.701	**54° 00′**
10′	.5901	.8073	.7310	1.368	1.239	1.695	**50′**
20′	.5925	.8056	.7355	1.360	1.241	1.688	**40′**
30′	.5948	.8039	.7400	1.351	1.244	1.681	**30′**
40′	.5972	.8021	.7445	1.343	1.247	1.675	**20′**
50′	.5995	.8004	.7490	1.335	1.249	1.668	**10′**
37° 00′	.6018	.7986	.7536	1.327	1.252	1.662	**53° 00′**
10′	.6041	.7969	.7581	1.319	1.255	1.655	**50′**
20′	.6065	.7951	.7627	1.311	1.258	1.649	**40′**
30′	.6088	.7934	.7673	1.303	1.260	1.643	**30′**
40′	.6111	.7916	.7720	1.295	1.263	1.636	**20′**
50′	.6134	.7898	.7766	1.288	1.266	1.630	**10′**
38° 00′	.6157	.7880	.7813	1.280	1.269	1.624	**52° 00′**
10′	.6180	.7862	.7860	1.272	1.272	1.618	**50′**
20′	.6202	.7844	.7907	1.265	1.275	1.612	**40′**
30′	.6225	.7826	.7954	1.257	1.278	1.606	**30′**
40′	.6248	.7808	.8002	1.250	1.281	1.601	**20′**
50′	.6271	.7790	.8050	1.242	1.284	1.595	**10′**
39° 00′	.6293	.7771	.8098	1.235	1.287	1.589	**51° 00′**
10′	.6316	.7753	.8146	1.228	1.290	1.583	**50′**
20′	.6338	.7735	.8195	1.220	1.293	1.578	**40′**
30′	.6361	.7716	.8243	1.213	1.296	1.572	**30′**
40′	.6383	.7698	.8292	1.206	1.299	1.567	**20′**
50′	.6406	.7679	.8342	1.199	1.302	1.561	**10′**
40° 00′	.6428	.7660	.8391	1.192	1.305	1.556	**50° 00′**
10′	.6450	.7642	.8441	1.185	1.309	1.550	**50′**
20′	.6472	.7623	.8491	1.178	1.312	1.545	**40′**
30′	.6494	.7604	.8541	1.171	1.315	1.540	**30′**
40′	.6517	.7585	.8591	1.164	1.318	1.535	**20′**
50′	.6539	.7566	.8642	1.157	1.322	1.529	**10′**
41° 00′	.6561	.7547	.8693	1.150	1.325	1.524	**49° 00′**
10′	.6583	.7528	.8744	1.144	1.328	1.519	**50′**
20′	.6604	.7509	.8796	1.137	1.332	1.514	**40′**
30′	.6626	.7490	.8847	1.130	1.335	1.509	**30′**
40′	.6648	.7470	.8899	1.124	1.339	1.504	**20′**
50′	.6670	.7451	.8952	1.117	1.342	1.499	**10′**
42° 00′	.6691	.7431	.9004	1.111	1.346	1.494	**48° 00′**
10′	.6713	.7412	.9057	1.104	1.349	1.490	**50′**
20′	.6734	.7392	.9110	1.098	1.353	1.485	**40′**
30′	.6756	.7373	.9163	1.091	1.356	1.480	**30′**
40′	.6777	.7353	.9217	1.085	1.360	1.476	**20′**
50′	.6799	.7333	.9271	1.079	1.364	1.471	**10′**
43° 00′	.6820	.7314	.9325	1.072	1.367	1.466	**47° 00′**
10′	.6841	.7294	.9380	1.066	1.371	1.462	**50′**
20′	.6862	.7274	.9435	1.060	1.375	1.457	**40′**
30′	.6884	.7254	.9490	1.054	1.379	1.453	**30′**
40′	.6905	.7234	.9545	1.048	1.382	1.448	**20′**
50′	.6926	.7214	.9601	1.042	1.386	1.444	**10′**
44° 00′	.6947	.7193	.9657	1.036	1.390	1.440	**46° 00′**
10′	.6967	.7173	.9713	1.030	1.394	1.435	**50′**
20′	.6988	.7153	.9770	1.024	1.398	1.431	**40′**
30′	.7009	.7133	.9827	1.018	1.402	1.427	**30′**
40′	.7030	.7112	.9884	1.012	1.406	1.423	**20′**
50′	.7050	.7092	.9942	1.006	1.410	1.418	**10′**
45° 00′	.7071	.7071	1.000	1.000	1.414	1.414	**45° 00′**
	Cos	*Sin*	*Cot*	*Tan*	*Csc*	*Sec*	*Angle*

Properties of Real Numbers

In the following list, a, b, and c represent real numbers.

Properties of Equality

Reflexive For all real numbers a, $a = a$.

Symmetric If $a = b$, then $b = a$.

Transitive If $a = b$ and $b = c$, then $a = c$.

Properties of Order

Axiom of Comparison One and only one of the following statements is true:

$$a < b, a = b, a > b.$$

Transitive property If $a < b$ and $b < c$, then $a < c$.
If $a > b$ and $b > c$, then $a > c$.

Properties of Addition and Multiplication

	Addition	*Multiplication*
Closure	$a + b$ represents a unique real number.	ab represents a unique real number.
Commutativity	$a + b = b + a$	$ab = ba$
Associativity	$(a + b) + c = a + (b + c)$	$(ab)c = a(bc)$
Identity	0 is the unique real number satisfying $0 + a = a$ and $a + 0 = a$.	1 is the unique real number satisfying $1(a) = a$ and $a(1) = a$.
Inverses	For each real number a, there is a unique real number $-a$ such that $a + (-a) = (-a) + a = 0$.	For each real number $a \neq 0$, there is a unique real number $\frac{1}{a}$ such that $a\left(\frac{1}{a}\right) = \frac{1}{a}(a) = 1.$

Distributivity $a(b + c) = ab + ac$ and $(b + c)a = ba + ca$

Substitution Changing the symbol by which a real number is named in a mathematical expression does not change the meaning of the expression.

Completeness Every nonempty subset of the set $\mathcal{R}$ of real numbers that has an upper bound in $\mathcal{R}$ has a least upper bound in $\mathcal{R}$.

Trigonometric Identities

For all x and y for which both members of the equation are defined:

1. $\tan x = \dfrac{\sin x}{\cos x}$

2. $\cot x = \dfrac{\cos x}{\sin x}$

3. $\sec x = \dfrac{1}{\cos x}$

4. $\csc x = \dfrac{1}{\sin x}$

5. $\cot x = \dfrac{1}{\tan x}$

6. $\sin^2 x + \cos^2 x = 1$

7. $1 + \tan^2 x = \sec^2 x$

8. $1 + \cot^2 x = \csc^2 x$

9. $\cos(x \pm y) = \cos x \cos y \mp \sin x \sin y$

10. $\cos(-x) = \cos x$

11. $\cos\left(\dfrac{\pi}{2} - x\right) = \sin x$

12. $\cos 2x = \cos^2 x - \sin^2 x = 2\cos^2 x - 1 = 1 - 2\sin^2 x$

13. $\cos\dfrac{x}{2} = \pm\sqrt{\tfrac{1}{2}(1 + \cos x)}$

14. $\sin(x \pm y) = \sin x \cos y \pm \cos x \sin y$

15. $\sin(-x) = -\sin x$

16. $\sin\left(\dfrac{\pi}{2} - x\right) = \cos x$

17. $\sin 2x = 2\sin x \cos x$

18. $\sin\dfrac{x}{2} = \pm\sqrt{\tfrac{1}{2}(1 - \cos x)}$

19. $\tan(x \pm y) = \dfrac{\tan x \pm \tan y}{1 \mp \tan x \tan y}$

20. $\tan(-x) = -\tan x$

21. $\tan 2x = \dfrac{2\tan x}{1 - \tan^2 x}$

22. $\tan\dfrac{x}{2} = \dfrac{\sin x}{1 + \cos x}$

23. $\cos(\pi \pm x) = -\cos x$

24. $\sin(\pi \pm x) = \mp\sin x$

25. $\tan(\pi \pm x) = \pm\tan x$

Summary of Formulas

Page

Distance between points $\mathbf{A}(x_1, y_1)$ and $\mathbf{B}(x_2, y_2)$

$d(\mathbf{A}, \mathbf{B}) = \sqrt{(x_2 - x_1)^2 + (y_2 - y_1)^2}$ **3**

Operations in $\mathcal{R}^2$ with vectors $\mathbf{u}$, $\mathbf{v} = (x, y)$, $\mathbf{v}_1 = (x_1, y_1)$, $\mathbf{v}_2 = (x_2, y_2)$

(With the exception of the starred ones, these formulas can be extended to 3-dimensional vectors.)

$\mathbf{v}_1 + \mathbf{v}_2 = (x_1 + x_2, y_1 + y_2)$ (sum of vectors) **8**

$-\mathbf{v} = -(x, y) = (-x, -y)$ (negative of a vector) **10**

$\mathbf{v}_1 - \mathbf{v}_2 = \mathbf{v}_1 + (-\mathbf{v}_2) = (x_1 - x_2, y_1 - y_2)$ **10**
(difference of vectors)

$\|\mathbf{v}\| = \sqrt{x^2 + y^2}$ (norm of a vector) **13**

*$\sin\theta = \dfrac{y}{\|\mathbf{v}\|}$, $\cos\theta = \dfrac{x}{\|\mathbf{v}\|}$ (direction angle θ of the vector $\mathbf{v} \neq \mathbf{0}$) **14**

$r\mathbf{v} = r(x, y) = (rx, ry)$, $r \in \mathcal{R}$ (product of scalar and vector) **20**

$\mathbf{u}$ and $\mathbf{v}$ ($\mathbf{v} \neq \mathbf{0}$) are parallel if and only if $\mathbf{u} = r\mathbf{v}$ **21**

If $\mathbf{u} = (x_1, y_1)$ and $\mathbf{v} = (x_2, y_2)$, then $\mathbf{u} \cdot \mathbf{v} = x_1x_2 + y_1y_2$ **27**
(inner, dot, or scalar, product of vectors $\mathbf{u}$ and $\mathbf{v}$)

$\mathbf{u} \perp \mathbf{v}$ if and only if $\mathbf{u} \cdot \mathbf{v} = 0$ **27**

*If $\mathbf{v} = (x, y)$, then $\mathbf{v}_\mathbf{p} = (-y, x)$ **27**

$\cos\alpha = \dfrac{\mathbf{u} \cdot \mathbf{v}}{\|\mathbf{u}\|\,\|\mathbf{v}\|}$ (angle α between vectors $\mathbf{u} \neq \mathbf{0}$ and $\mathbf{v} \neq \mathbf{0}$) **31**

*$\mathbf{v} = (\mathbf{u} \cdot \mathbf{v})\mathbf{u} + (\mathbf{u}_\mathbf{p} \cdot \mathbf{v})\mathbf{u}_\mathbf{p}$, where $\|\mathbf{u}\| = 1$ **35**
(resolution of $\mathbf{v}$ into components)

Equations of lines in the plane

$\mathbf{u} = \mathbf{s} + r(\mathbf{t} - \mathbf{s})$, $r \in \mathcal{R}$ **49**
(standard parametric vector equation of the line through $\mathbf{S}$ and $\mathbf{T}$)

$x = x_1 + r(x_2 - x_1)$, $y = y_1 + r(y_2 - y_1)$, $r \in \mathcal{R}$ **51**
(parametric Cartesian equations for the line through $\mathbf{S}(x_1, y_1)$ and $\mathbf{T}(x_2, y_2)$)

$\mathbf{u} = \mathbf{s} + r\mathbf{v}$ (standard parametric vector equation of the line through $\mathbf{S}$ parallel to $\mathbf{v}$) **53**

$m = \dfrac{k}{h}$ (slope m of a line with direction vector (h, k), $h \neq 0$) **59**

$\mathcal{L}_1 \perp \mathcal{L}_2$ if and only if $m_1 = -\dfrac{1}{m_2}$, or one line is vertical (no slope) and one horizontal (slope 0) **62**

$Ax + By + C = 0$ (standard Cartesian, or scalar, equation) **65**

$y - y_1 = m(x - x_1)$ (point-slope form) **70**

$y - y_1 = \dfrac{y_2 - y_1}{x_2 - x_1}(x - x_1)$, $x_1 \neq x_2$ (two-point form) **70**

$y = mx + b$ (slope-intercept form) **74**

$\dfrac{x}{a} + \dfrac{y}{b} = 1$, $a \neq 0, b \neq 0$ (intercept form) **75**

$\dfrac{x - x_1}{h} = \dfrac{y - y_1}{k}$, $h \neq 0$, $k \neq 0$ (symmetric form, (h, k) a direction vector) **77**

$x \cos \alpha + y \sin \alpha - p = 0$ (normal form) **91**

or

$$\frac{A}{\pm\sqrt{A^2 + B^2}}x + \frac{B}{\pm\sqrt{A^2 + B^2}}y + \frac{C}{\pm\sqrt{A^2 + B^2}} = 0, \quad A^2 + B^2 \neq 0$$ **92**

Distance from point $S(x_1, y_1)$ to line $\mathcal{L}$: $Ax + By + C = 0$

$$d(\mathbf{S}, \mathcal{L}) = \frac{|(\mathbf{s} - \mathbf{t}) \cdot \mathbf{v_P}|}{\|\mathbf{v_P}\|} \quad \mathbf{T} \text{ on } \mathcal{L}, \quad \mathbf{v} \parallel \mathcal{L}$$ **89**

or

$$d(\mathbf{S}, \mathcal{L}) = \left|\frac{Ax_1 + By_1 + C}{\sqrt{A^2 + B^2}}\right|$$ **94**

Determinants

$$\begin{vmatrix} a_1 & b_1 \\ a_2 & b_2 \end{vmatrix} = a_1b_2 - a_2b_1$$ **104**

$$\begin{vmatrix} a_1 & b_1 & c_1 \\ a_2 & b_2 & c_2 \\ a_3 & b_3 & c_3 \end{vmatrix} = a_1b_2c_3 + a_2b_3c_1 + a_3b_1c_2 - a_1b_3c_2 - a_2b_1c_3 - a_3b_2c_1$$ **104**

Standard equations of conic sections

$$\left.\begin{array}{l}(x-h)^2+(y-k)^2=r^2 \\ x^2+y^2+Dx+Ey+F=0\end{array}\right\}\text{ circle}$$ 125, 126

$$\left.\begin{array}{l}x^2=4py \\ y^2=4px\end{array}\right\}\text{ parabola}$$ 131

$$\left.\begin{array}{l}\dfrac{x^2}{a^2}+\dfrac{y^2}{b^2}=1 \\ \dfrac{x^2}{b^2}+\dfrac{y^2}{a^2}=1\end{array}\right\}\text{ ellipse}$$ 140, 141

$$\left.\begin{array}{l}\dfrac{x^2}{a^2}-\dfrac{y^2}{b^2}=1 \\ \dfrac{y^2}{a^2}-\dfrac{x^2}{b^2}=1\end{array}\right\}\text{ hyperbola}$$ 146, 147

Translation of axes in the plane

$x = x' + h$ and $y = y' + k$ 171

$x' = x - h$ and $y' = y - k$ 171

Tables of equations of conic sections 174, 177

Rotation of axes

$x' = x \cos \phi + y \sin \phi$ and $y' = -x \sin \phi + y \cos \phi$ 181

$x = x' \cos \phi - y' \sin \phi$ and $y = x' \sin \phi + y' \cos \phi$ 181

$$\tan 2\phi = \frac{B}{A - C}$$ 188

Equation of line tangent at the origin to graph of $Ax^2 + Bxy + Cy^2 + Dx + Ey = 0$

$Dx + Ey = 0$, D and E not both zero 193

Polar coordinates

$x = r \cos \theta$ and $y = r \sin \theta$ 246

$r = \pm\sqrt{x^2 + y^2}$, $\cos \theta = \dfrac{x}{\pm\sqrt{x^2 + y^2}}$, $\sin \theta = \dfrac{y}{\pm\sqrt{x^2 + y^2}}$ 246

$$d = \sqrt{r_2^2 + r_1^2 - 2r_2r_1 \cos(\theta_2 - \theta_1)}$$

(distance between points with coordinates (r_1, θ_1) and (r_2, θ_2)) 250

Polar equations of conic sections

$r = \dfrac{ep}{1 \pm \cos\theta}$ (focus at pole, directrix $\perp$ to polar axis) **261**

$r = \dfrac{ep}{1 \pm \sin\theta}$ (focus at pole, directrix $\|$ to polar axis) **261**

Distance between points $\mathbf{S}(x_1, y_1, z_1)$ and $\mathbf{T}(x_2, y_2, z_2)$ in space

$d(\mathbf{S}, \mathbf{T}) = \sqrt{(x_2 - x_1)^2 + (y_2 - y_1)^2 + (z_2 - z_1)^2}$ **276**

Direction cosines of a vector $\mathbf{v} = (v_1, v_2, v_3)$

$\cos\alpha = \dfrac{v_1}{\|\mathbf{v}\|}, \quad \cos\beta = \dfrac{v_2}{\|\mathbf{v}\|}, \quad \cos\gamma = \dfrac{v_3}{\|\mathbf{v}\|}, \quad \text{where } \|\mathbf{v}\| \neq \mathbf{0}$ **283**

$\cos^2\alpha + \cos^2\beta + \cos^2\gamma = 1$ **283**

Cross product of vectors $\mathbf{u}(u_1, u_2, u_3)$ and $\mathbf{v}(v_1, v_2, v_3)$

$$\mathbf{u} \times \mathbf{v} = \begin{vmatrix} u_1 & u_2 & u_3 \\ v_1 & v_2 & v_3 \\ \mathbf{i} & \mathbf{j} & \mathbf{k} \end{vmatrix}$$ **294**

Equations of lines in space

$\mathbf{u} = \mathbf{s} + r(\mathbf{t} - \mathbf{s}),\ r \in \mathcal{R}$ (standard parametric vector equation of line through $\mathbf{S}$ and $\mathbf{T}$) **307**

$(x, y, z) = (x_1, y_1, z_1) + r(x_2 - x_1, y_2 - y_1, z_2 - z_1)$ **309**

(parametric vector equation of line through $\mathbf{S}(x_1, y_1, z_1)$ and $\mathbf{T}(x_2, y_2, z_2)$)

$x = x_1 + r(x_2 - x_1),\ y = y_1 + r(y_2 - y_1),\ z = z_1 + r(z_2 - z_1)$ **309**

(parametric Cartesian equations of line through $\mathbf{S}(x_1, y_1, z_1)$ and $\mathbf{T}(x_2, y_2, z_2)$)

$\dfrac{x - x_1}{x_2 - x_1} = \dfrac{y - y_1}{y_2 - y_1} = \dfrac{z - z_1}{z_2 - z_1}$ **309**

(symmetric equations of line through $\mathbf{S}(x_1, y_1, z_1)$ and $\mathbf{T}(x_2, y_2, z_2)$)

Equations of planes in space

$ax + by + cz + d = 0$ (Cartesian equation) **314**

$x\cos\alpha + y\cos\beta + z\cos\gamma - p = 0$ (normal form) **332**

$\mathbf{u} = \mathbf{s} + r(\mathbf{n}_1 \times \mathbf{n}_2),\ r \in \mathcal{R}$ (parametric vector equation for line of intersection of planes with normal vectors $\mathbf{n}_1$ and $\mathbf{n}_2$, $\mathbf{S}$ on line of intersection) **320**

Distance from point $S(x_1, y_1, z_1)$ to plane $\mathcal{P}$: $ax + by + cz + d = 0$

$(a^2 + b^2 + c^2 \neq 0)$

$$d(\mathbf{S}, \mathcal{P}) = \frac{|(\mathbf{s} - \mathbf{t}) \cdot \mathbf{n}|}{\|\mathbf{n}\|}, \quad \mathbf{T} \text{ on } \mathcal{P}, \quad \mathbf{n} \perp \mathcal{P}, \quad \|\mathbf{n}\| \neq 0$$ **330**

or

$$d(\mathbf{S}, \mathcal{P}) = \frac{|ax_1 + by_1 + cz_1 + d|}{\sqrt{a^2 + b^2 + c^2}}$$ **331**

Distance from point S to line $\mathcal{L}$ in space

$$d(\mathbf{S}, \mathcal{L}) = \frac{\|(\mathbf{s} - \mathbf{t}) \times \mathbf{v}\|}{\|\mathbf{v}\|}, \quad \mathbf{T} \text{ on } \mathcal{L}, \quad \mathbf{v} \parallel \mathcal{L}$$ **335**

Sphere with center $C(x_0, y_0, z_0)$, radius r

$$(x - x_0)^2 + (y - y_0)^2 + (z - z_0)^2 = r^2 \quad \text{(Cartesian equation)}$$ **345**

Vector equation of plane tangent to sphere at point T

$$\mathbf{v} \cdot (\mathbf{u} - \mathbf{t}) = 0, \quad \mathbf{v} \text{ the radius vector from center of sphere to } \mathbf{T}$$ **347**

Ellipsoid

$$\frac{x^2}{a^2} + \frac{y^2}{b^2} + \frac{z^2}{c^2} = 1$$ **356**

Hyperboloid of one sheet

$$\frac{x^2}{a^2} + \frac{y^2}{b^2} - \frac{z^2}{c^2} = 1$$ **357**

Hyperboloid of two sheets

$$-\frac{x^2}{a^2} - \frac{y^2}{b^2} + \frac{z^2}{c^2} = 1 \quad \text{or} \quad \frac{x^2}{a^2} + \frac{y^2}{b^2} - \frac{z^2}{c^2} = -1$$ **357**

Elliptic Cone

$$\frac{x^2}{a^2} + \frac{y^2}{b^2} - \frac{z^2}{c^2} = 0$$ **358**

Elliptic paraboloid

$$z = \frac{x^2}{a^2} + \frac{y^2}{b^2}$$ **358**

Hyperbolic paraboloid

$$z = \frac{y^2}{b^2} - \frac{x^2}{a^2}$$ **360**

Translation of axes in space

$$x' = x - x_0,\ y' = y - y_0,\ z' = z - z_0$$ **366**

$$x = x' + x_0,\ y = y' + y_0,\ z = z' + z_0$$ **366**

Cylindrical coordinates

$$x = r\cos\theta,\ y = r\sin\theta,\ z = z$$ **371**

$$r = \pm\sqrt{x^2 + y^2}, \quad \cos\theta = \frac{x}{\pm\sqrt{x^2 + y^2}}, \quad \sin\theta = \frac{y}{\pm\sqrt{x^2 + y^2}},$$ **371**

$$z = z$$

Spherical coordinates

$$x = \rho\sin\phi\cos\theta$$ **373**

$$y = \rho\sin\phi\sin\theta$$

$$z = \rho\cos\phi$$

$$\rho = \sqrt{x^2 + y^2 + z^2}$$

$$\cos\theta = \frac{x}{\sqrt{x^2 + y^2}}, \quad \sin\theta = \frac{y}{\sqrt{x^2 + y^2}}$$

$$\cos\phi = \frac{z}{\sqrt{x^2 + y^2 + z^2}}, \quad \sin\phi = \frac{\sqrt{x^2 + y^2}}{\sqrt{x^2 + y^2 + z^2}}$$

Index

ABCDEFGHIJ-RM-7987654321

Answers for Odd-Numbered Exercises

Exercises 1–1, page 4

1. -4 **3.** 1 **5.** $x = 1, y = 1$ **7.** 3
9. $3\sqrt{2}$ **11.** 4 **13.** $\sqrt{8 + 2\sqrt{3}}$ **25.** $(3, 4)$

Exercises 1–2, pages 7–8

1.

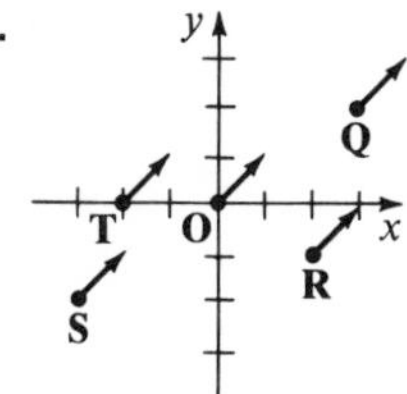

3.

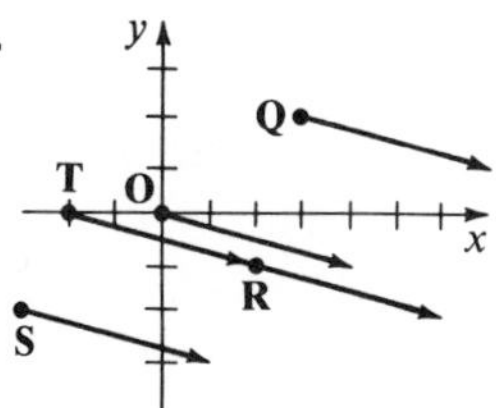

5.

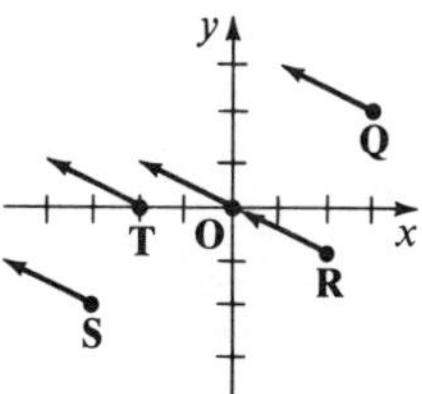

7.

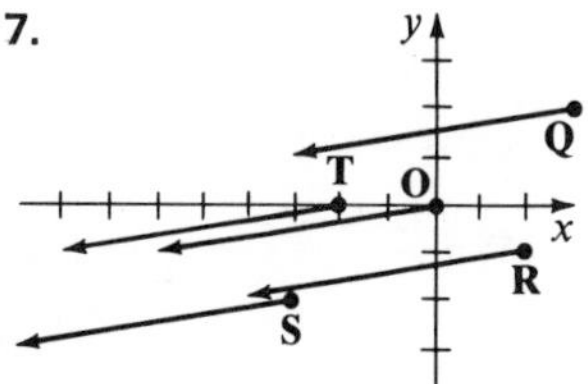

9.

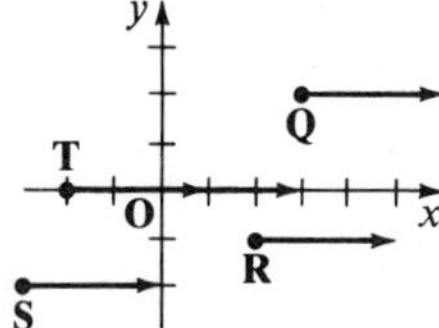

11. **(a)** $(1, 6)$
(b) $(4, 8)$
(c) $(5, 5)$
(d) $(-2, 13)$
(e) $(-5, 1)$

13. **(a)** $(-3, 5)$ **(b)** $(0, 7)$ **(c)** $(1, 4)$ **(d)** $(-6, 12)$ **(e)** $(-9, 0)$
15. **(a)** $(7, -8)$ **(b)** $(10, -6)$ **(c)** $(11, -9)$ **(d)** $(4, -1)$ **(e)** $(1, -13)$
17. **(a)** $(-3, -4)$ **(b)** $(0, -2)$ **(c)** $(1, -5)$ **(d)** $(-6, 3)$ **(e)** $(-9, -9)$
19. **(a)** $(-2, 0)$ **(b)** $(1, 2)$ **(c)** $(2, -1)$ **(d)** $(-5, 7)$ **(e)** $(-8, -5)$
21. $(4, 2)$ **23.** $(-5, 3)$ **25.** $(-15, -9)$ **27.** $(6, 0)$
29. $(\frac{7}{2}, \frac{3}{2})$ **31.** $(\frac{11}{2}, 3)$ **33.** $(7, -10)$

Exercises 1–3, pages 12–13

1. $(6, -2)$ **3.** $(-1, 6)$ **5.** $(5, -3)$
7. $(4, 3)$ **9.** $(2, -4)$ **11.** $(6, 1)$
13. $(-4, 2)$ **15.** $(-2, 5)$ **17.** $(1, -4)$

Exercises 1–4, pages 18–20

1. $3\sqrt{2}$; $45°$ **3.** 2; $30°$ **5.** 5; $126°50'$
7. 2; $180°$ **9.** 10; $223°10'$ **11.** 5; $53°10'$
13. $2\sqrt{5}$; $26°30'$ **15.** $3\sqrt{2}$; $45°$ **17.** $(4.35, 2.5)$
19. $(0, 6)$ **21.** $(1, -1.73)$ **23.** $(-6.06, -3.5)$

25. (a) 3, (b) $-\frac{16}{3}$ **27.** (a) $-\frac{9}{2}$, (b) 2 **29.** (a) $\frac{6}{5}$, (b) $-\frac{10}{3}$

31. $\left(\frac{15\sqrt{58}}{58}, \frac{35\sqrt{58}}{58}\right)$ **33.** $\left(-\frac{6\sqrt{29}}{29}, \frac{15\sqrt{29}}{29}\right)$ **35.** $\left(\frac{5}{2}, \frac{5\sqrt{3}}{2}\right)$

Exercises 1–5, pages 25–26

1. $(7, -1)$
3. $(-1, -3)$
5. $5\sqrt{2}$
7. $4\sqrt{29} - 4$
9. $\left(-\frac{\sqrt{5}}{5}, \frac{2\sqrt{5}}{5}\right)$
11. $\left(\frac{\sqrt{5}}{5}, \frac{2\sqrt{5}}{5}\right)$
13. $\left(-\frac{2\sqrt{5}}{5}, -\frac{\sqrt{5}}{5}\right)$
15. yes
17. no
19. yes
21. $2(\cos 30°, \sin 30°)$
23. $\sqrt{34}\,(\cos 301°, \sin 301°)$
25. $\sqrt{53}\,(\cos 254°, \sin 254°)$
27. $\left(-\frac{\sqrt{2}}{2}, \frac{\sqrt{2}}{2}\right)$
29. $\left(-\frac{7}{\sqrt{58}}, \frac{3}{\sqrt{58}}\right)$
31. $(0, 1)$
33. $\left(-\frac{b}{\sqrt{a^2 + b^2}}, \frac{a}{\sqrt{a^2 + b^2}}\right)$

Exercises 1–6, pages 29–30

1. -1 **3.** 0 **5.** $-\sqrt{2} + 15$
7. 16 **9.** $(-\frac{4}{5}, \frac{3}{5})$ **11.** $\left(-\frac{\sqrt{7}}{4}, -\frac{3}{4}\right)$
13. $\left(\frac{1}{\sqrt{10}}, \frac{3}{\sqrt{10}}\right)$ **15.** $(0, -1)$ **17.** $-\frac{5}{3}$
19. 6 **21.** 3 **23.** $-\frac{3}{2}$
25. $4\sqrt{3}$ **27.** 0 **29.** $6\sqrt{3}$

Exercises 1–7, page 33

1. perpendicular **3.** parallel **5.** neither, $m°(\alpha) \doteq 143$
7. neither, $m°(\alpha) = 30$ **9.** perpendicular **11.** neither, $m°(\alpha) \doteq 25$
13. $m°(\angle A) = 60$, $m°(\angle B) = 60$, $m°(\angle C) = 60$
15. $m°(\angle A) = 90$, $m°(\angle B) \doteq 53$, $m°(\angle C) \doteq 37$

Exercises 1–8, pages 39–40

1. (a) $\mathbf{i} + 2\mathbf{j}$ (b) $\frac{3}{\sqrt{2}}\mathbf{u} + \frac{1}{\sqrt{2}}\mathbf{u_p}$ (c) $\frac{11}{25}\mathbf{t} + \frac{2}{25}\mathbf{t_p}$

3. (a) $-\mathbf{i}$ (b) $-\frac{1}{\sqrt{2}}\mathbf{u} + \frac{1}{\sqrt{2}}\mathbf{u_p}$ (c) $-\frac{3}{25}\mathbf{t} + \frac{4}{25}\mathbf{t_p}$

5. (a) $-3\mathbf{i} - 4\mathbf{j}$ (b) $-\frac{7}{\sqrt{2}}\mathbf{u} - \frac{1}{\sqrt{2}}\mathbf{u_p}$ (c) $-\mathbf{t}$

7. (a) $5\mathbf{i} - 6\mathbf{j}$ (b) $-\frac{1}{\sqrt{2}}\mathbf{u} - \frac{11}{\sqrt{2}}\mathbf{u_p}$ (c) $-\frac{9}{25}\mathbf{t} - \frac{38}{25}\mathbf{t_p}$

9. 90° **11.** 120° **13.** $(-\frac{3}{13}, -\frac{2}{13})$; $\frac{-1}{\sqrt{13}}$

15. $(-\frac{4}{5}, \frac{2}{5})$; $\frac{-2}{\sqrt{5}}$ **17.** 0 **19.** $(\frac{3}{5}, \frac{4}{5})$

21. $\left(\frac{\sqrt{5}}{5}, \frac{2\sqrt{5}}{5}\right)$

Chapter Review Exercises, page 42

1. $x = 3$, $y = 3$ **3.** (2, 1) **5.** norm = 2; $\tan\alpha = \frac{1}{\sqrt{3}}$

7. −6 **9.** $\frac{26}{5}\mathbf{u} + \frac{7}{5}\mathbf{u_p}$

Exercises 2–1, pages 51–52

1. $\mathbf{u} = (2, 1) + r(-2, -1)$; $x = 2 - 2r$, $y = 1 - r$
3. $\mathbf{u} = (4, -2) + r(0, 5)$; $x = 4$, $y = -2 + 5r$
5. $\mathbf{u} = (-7, 2) + r(4, -3)$; $x = -7 + 4r$, $y = 2 - 3r$
7. $\mathbf{u} = (-6, -3) + r(2, 1)$; $x = -6 + 2r$, $y = -3 + r$
9. $\mathbf{u} = (a, b) + r(b - a, a - b)$; $x = a + r(b - a)$; $y = b + r(a - b)$
11. (a) $(\frac{1}{2}, \frac{1}{2})$ (b) (−1, 1), (2, 0)
13. (a) $(0, \frac{5}{2})$ (b) (−1, 0), (1, 5)
15. (a) (−4, 2) (b) (−6, 1), (−2, 3)
17. (a) $(\frac{1}{2}, 4)$ (b) $(-\frac{2}{3}, 5)$, $(\frac{5}{3}, 3)$
19. $\mathbf{v} = (2, 5) + r(4, -6)$, $0 \leq r \leq 1$
21. $\mathbf{v} = (-2, 4) + r(1, 3)$, $0 \leq r \leq 1$

Exercises 2–2, pages 56–57

1. $\mathbf{u} = (3, 2) + r(1, 1)$; $x = 3 + r$, $y = 2 + r$
3. $\mathbf{u} = (4, -3) + r(-1, 2)$; $x = 4 - r$, $y = -3 + 2r$
5. $\mathbf{u} = (-2, 4) + r(-1, 2)$; $x = -2 - r$, $y = 4 - 2r$
7. $\mathbf{u} = (-6, -4) + r(-3, -2)$; $x = -6 - 3r$, $y = -4 - 2r$
9. yes **11.** no
13. yes **15.** no
17. $\mathbf{u} = (2, 3) + r(3, 4)$ **19.** $\mathbf{u} = (3, -1) + r(5, 2)$
21. $\mathbf{u} = (-6, 2) + r(3, 1)$ **23.** $\mathbf{u} = (-1, -1) + r(-2, 7)$
25. yes **27.** yes
29. no **31.** $\mathbf{U}_1(7, 1)$, $\mathbf{U}_2(1, -5)$
33. $\mathbf{U}_1(15, 40)$, $\mathbf{U}_2(-5, -8)$

Exercises 2–3, pages 62–63

1. $m = -\frac{1}{9}$; $\mathbf{u} = (5, 1) + r(1, -\frac{1}{9})$
3. $m = -1$; $\mathbf{u} = (3, -4) + r(1, -1)$

5. $m = 0$; $\mathbf{u} = (2, -3) + r(1, 0)$
7. $m = -4$; $\mathbf{u} = (-2, -3) + r(1, -4)$
9. $\mathbf{u} = (3, -4) + r(1, 2)$
11. $\mathbf{u} = (0, -5) + r(1, 0)$
13. $\mathbf{u} = (-2, -3) + r(1, \frac{2}{3})$
15. $\mathbf{u} = (1, 0) + r(0, 1)$
17. parallel
19. parallel
21. perpendicular
23. collinear
25. not collinear
27. not collinear
29. (a) $\mathbf{u} = (-2, 1) + r(2, -1)$ (b) $\mathbf{u} = (-2, 1) + r(1, 2)$
31. (a) $\mathbf{u} = (3, 4) + r(5, -2)$ (b) $\mathbf{u} = (3, 4) + r(2, 5)$
33. (a) $\mathbf{u} = (0, 4) + r(-3, 3)$ (b) $\mathbf{u} = (0, 4) + r(-3, -3)$
35. (a) $\mathbf{u} = (-2, -3) + r(5, 6)$ (b) $\mathbf{u} = (-2, -3) + r(-6, 5)$
37. parallel
39. perpendicular
41. neither
43. $\mathbf{u} = \left(\frac{x_1 + x_2}{2}, \frac{y_1 + y_2}{2}\right) + r(c, d)$

Exercises 2–4, pages 67–69

1. $3x + 5y - 22 = 0$
3. $x - 7y - 9 = 0$
5. $-x - 2y - 1 = 0$
7. $x + y + 4 = 0$
9. $2x + 5y - 41 = 0$
11. $x + 2y - 6 = 0$
13. $y + 5 = 0$
15. parallel
17. perpendicular
19. parallel
21. (a) $5x + 2y - 11 = 0$ (b) $2x - 5y + 13 = 0$
23. (a) $7x + 5y - 11 = 0$ (b) $5x - 7y - 29 = 0$
25. (a) $-6x + 5y - 46 = 0$ (b) $5x + 6y + 18 = 0$
27. (a) $2x + y + 3 = 0$ (b) $x - 2y - 1 = 0$
29. $x + y - 8 = 0$
31. $4x + 3y - 1 = 0$
33. $y + 1 = 0$

Exercises 2–5, pages 72–73

1. $2x - y + 1 = 0$
3. $x + y + 1 = 0$
5. $3x + y - 2 = 0$
7. $x - 2y - 1 = 0$
9. $3x + 5y + 24 = 0$
11. $3x + 2y - 13 = 0$
13. $2x + y = 0$
15. $4x + 13y - 2 = 0$
17. $5x + 2y + 7 = 0$
19. $y + 3 = 0$
21. $x - 3 = 0$
23. $x - 2y = 0$; $3x + y = 0$; $2x + 3y - 14 = 0$
25. $x - y + 4 = 0$; $x + y - 10 = 0$; $y - 3 = 0$
27. $5x + 4y - 14 = 0$; $4x - y = 0$; $x + 5y - 14 = 0$
29. $x + 3y - 16 = 0$; $x - 3y + 10 = 0$; $x - 3 = 0$
31. $2x + y - 2 = 0$; $3x - 2y = 0$; $x - 3y + 2 = 0$
33. $x - y + 4 = 0$; $x - 3 = 0$; $x + y - 10 = 0$
35. $10x + y - 26 = 0$; $x + 4y - 13 = 0$; $8x - 7y = 0$
37. $(3, -4)$, $(-3, 4)$, $(7, 8)$
39. $3x - 4y + 11 = 0$; $x + 3y - 9 = 0$; $5x + 2y - 7 = 0$

Exercises 2-6, pages 76–77

1. $2x - y + 5 = 0$
3. $4x - y - 4 = 0$
5. $2x - 3y = 0$
7. $4x + 5y - 20 = 0$
9. $m = \frac{3}{4}$; $b = 2$
11. $m = -\frac{5}{4}$; $b = 4$
13. $m = \frac{7}{2}$; $b = -2$
15. $m = 0$; $b = -2$
17. $3x + 2y - 6 = 0$
19. $2x - 3y + 6 = 0$
21. $2x - 7y - 14 = 0$
23. $2x + y + 2 = 0$
25. $a = 4, b = 3$
27. $a = -3, b = 5$
29. $a = -2, b = 5$
31. $a = -\frac{8}{3}, b = \frac{8}{5}$
33. $15x + 2y = 0$
35. $22x + 6y - 33 = 0$
37. $x - y + 2 = 0$
39. $x + 3y - 6 = 0$
41. $a = 1$

Exercises 2–7, page 80

1. $\frac{x-5}{2} = \frac{y-2}{-4}$; $m°(\theta) \doteq 117$
3. $\frac{x-5}{2} = \frac{y-9}{11}$; $m°(\theta) \doteq 80$
5. $\frac{x+3}{-2} = \frac{y-4}{7}$; $m°(\theta) \doteq 106$
7. $\frac{x-4}{2} = \frac{y+5}{-4}$; $m°(\theta) \doteq 117$
9. $\frac{x+1}{-3} = \frac{y+3}{-6}$; $m°(\theta) \doteq 63$
11. $x + 2 = \frac{y+3}{2}$; $m°(\theta) \doteq 63$
13. $x = \frac{y-4}{2}$
15. $\frac{x}{-5} = \frac{y-2}{2}$
17. $\frac{x-1}{-3} = \frac{y-\frac{5}{3}}{2}$
19. $\frac{x-1}{7} = \frac{y-\frac{5}{7}}{2}$
21. $\sqrt{3}x - y + (3 + 5\sqrt{3}) = 0$
23. $3x + \sqrt{3}y + (2\sqrt{3} - 9) = 0$

Chapter Review Exercises, page 82

1. $\mathbf{u} = (3, -2) + r(-4, 7)$; $x = 3 - 4r, y = -2 + 7r$
3. $\mathbf{u} = (5, -1) + r(3, 1)$
5. $\mathbf{u} = (2, -5) + r(-3, 2)$
7. $-3x + y + 13 = 0$
9. $m = 2$; $b = \frac{7}{2}$

Exercises 3–1, pages 95–97

1. $\frac{5}{\sqrt{5}}$
3. $\frac{10}{\sqrt{2}}$
5. $\frac{11}{\sqrt{13}}$
7. $\frac{7}{\sqrt{5}}$
9. $\frac{3}{\sqrt{5}}$
11. $\frac{54}{5\sqrt{2}}$
13. 11
15. $\frac{36}{\sqrt{41}}$
17. $\frac{33}{\sqrt{65}}$
19. $\frac{62}{13}$
21. $\frac{33}{\sqrt{65}}, \frac{33}{\sqrt{65}}, \frac{11}{\sqrt{2}}$
23. $\frac{8}{\sqrt{5}}, \frac{8}{\sqrt{37}}, 1$
25. $\frac{33}{2}$
27. 4
29. $\frac{7}{\sqrt{10}}$
31. $k = \frac{1}{3}$ or $k = 2$
33. $x - y + 4 = 0$ and $x + y = 0$
35. $4y - 3 = 0$ and $4x + 1 = 0$

37. $x - 8y + 10 = 0, 9x - 2y + 10 = 0, 10x + 11y - 4 = 0$
39. $3x - 4y + 35 = 0$ and $3x - 4y - 15 = 0$
41. $\left|\dfrac{b^2 + a^2 - ab}{\sqrt{a^2 + b^2}}\right|$

Exercises 3–2, pages 101–103

1. $(1, 2)$ **3.** $(1, -1)$ **5.** $(-1, -1)$
7. $(-2, 3)$ **9.** the lines coincide **11.** $(4, 1)$
13. $(4, 1)$ **15.** $\emptyset$ **17.** $(-2, 3)$
19. the lines coincide **21.** $(4, 1)$ **23.** $(4, 1)$
25. $\emptyset$ **27.** $(1, 2)$, $(2, 0)$, and $(7, 4)$
29. $(-1, -1)$, $(1, 5)$, and $(2, 3)$ **31.** $12x - 15y - 8 = 0$
33. $x - 3y - 2 = 0$ **35.** $x - y - 9 = 0$
37. $x + 7y - 17 = 0, x - y + 1 = 0$

Exercises 3–3, pages 107–109

1. -26 **3.** 0 **5.** 0 **7.** 1
9. $a + b$ **11.** $a^2 + b^2$ **13.** 0 **15.** 0
17. -19 **19.** -19 **21.** $(2, 1)$ **23.** $(-2, 1, 3)$
25. $x - y - 3 = 0$ **29.** $\begin{vmatrix} x & y & 1 \\ a & 0 & 1 \\ 0 & b & 1 \end{vmatrix} = 0$

Chapter Review Exercises, page 115

1. $\frac{6}{13}$ **3.** $(2, 4)$ **5.** $(\frac{27}{26}, \frac{15}{26})$
7. 66 **9.** 9

Exercises 4–1, pages 123–124

1. $4x + 6y - 41 = 0$
3. $x^2 + y^2 - 4x - 8y - 16 = 0$
5. $x^2 + y^2 = 9$
7. $xy + 2x + y - 14 = 0 \quad (x \neq 1, x \neq 3)$
9. $2x^2 + x - 3y + 5 = 0 \quad (x \neq -1, x \neq 2)$
11. $y^2 - 12x + 36 = 0$
13. $x^2 + y^2 + 6x - 18y + 2xy - 15 = 0$
15. $3x^2 - 48x + 4y^2 = 0$ **17.** $16x^2 + 25y^2 = 400$
19. $x^2 - 2y = 0$ **21.** $y = \frac{9}{4}x^2$

Exercises 4–2, pages 128–129

1. $(x - 3)^2 + (y - 4)^2 = 4$ **3.** $(x - 2)^2 + (y + 3)^2 = 49$
5. $(x + 2)^2 + (y + 4)^2 = 9$ **7.** $(x - 1)^2 + (y - 2)^2 = 25$
9. $(x + 5)^2 + (y - 3)^2 = 49$ **11.** $(x + 7)^2 + (y + 8)^2 = 100$
13. $(x + \frac{3}{2})^2 + (y - \frac{5}{2})^2 = 5^2$ **15.** $x^2 + y^2 - 2x - 2y - 23 = 0$
17. $x^2 + y^2 + 8x + 6y - 144 = 0$ **19.** $4x - 3y = 0$
21. $x - 2y = 0$ **23.** $x + 3 = 0$
25. $(x - 3)^2 + (y - 4)^2 = 25$ **27.** $(x - 1)^2 + (y + 5)^2 = \frac{81}{2}$

29. $(x + 2)^2 + (y + 4)^2 = \frac{16}{5}$

31. $(x + 6)^2 + (y + 2)^2 = 26$

33. $(x - 4)^2 + (y - 5)^2 = 25$

35. $(x + 1)^2 + (y - 4)^2 = 16$

37. $(x + 7)^2 + (y - 7)^2 = 45$ or $(x - 5)^2 + (y - 1)^2 = 45$

Exercises 4–3, pages 133–136

1. $x^2 = 12y$; directrix $y = -3$

3. $y^2 = 8x$; directrix $x = -2$

5. $y^2 = -4x$; directrix $x = 1$

7. $x^2 = -20y$; directrix $y = 5$

9. focus $(0, 2)$; directrix $y = -2$

11. focus $(2, 0)$; directrix $x = -2$

13. $y^2 = 8x$

15. $x^2 = y$

17. $y^2 - 4y - 8x + 28 = 0$

19. $x^2 - 8x + 12y + 40 = 0$

21. $y^2 - 14x + 63 = 0$

23. $y = 2x^2 + 3x + 1$

25. $y = -x^2 + x + 3$

27. $y = 3x^2 - 5$

33. $x^2 + y^2 - 5py = 0$

Applied Problems 4–3, page 136

1. 75 m. above ground

3. 48 inches

5. 160 feet

Exercises 4–4, pages 141–143

1. $\frac{x^2}{25} + \frac{y^2}{9} = 1$
major axis = 10
minor axis = 6

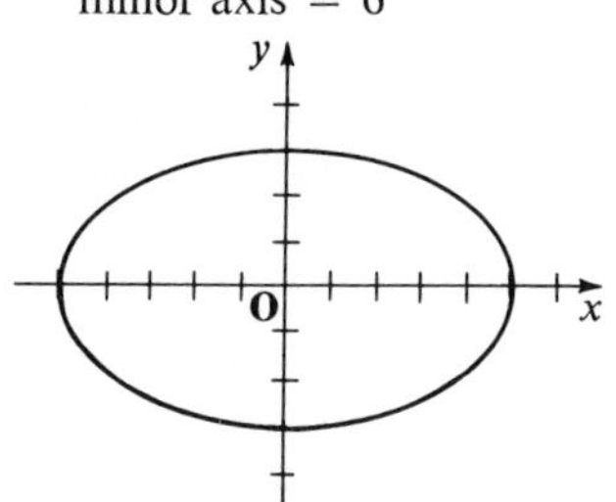

3. $\frac{x^2}{9} + \frac{y^2}{4} = 1$
major axis = 6
minor axis = 4

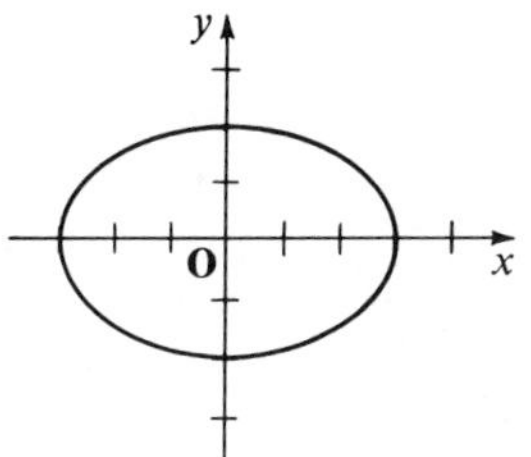

5. $\frac{x^2}{9} + \frac{y^2}{25} = 1$
major axis = 10
minor axis = 6

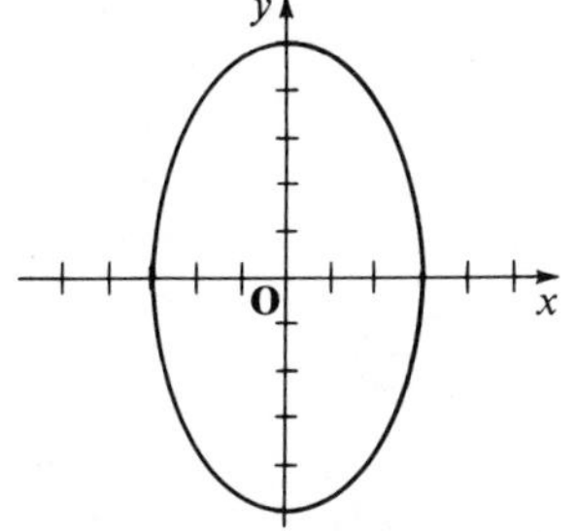

7. $\frac{x^2}{4} + \frac{y^2}{7} = 1$
major axis = $2\sqrt{7}$
minor axis = 4

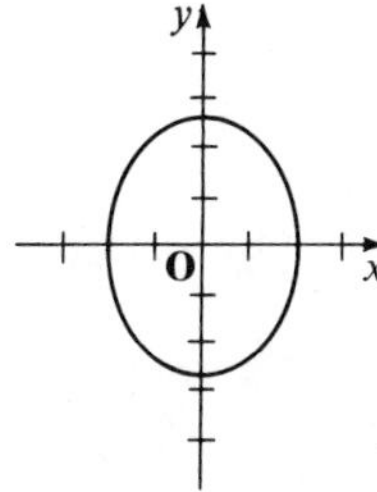

9. $\frac{x^2}{25} + \frac{y^2}{9} = 1$; $\mathbf{F}_1(-4, 0), \mathbf{F}_2(4, 0)$

11. $\frac{x^2}{9} + \frac{y^2}{13} = 1$; $\mathbf{F}_1(0, 2), \mathbf{F}_2(0, -2)$

13. $\mathbf{V}_1(-3, 0)$, $\mathbf{V}_2(3, 0)$; $\mathbf{F}_1(-\sqrt{5}, 0)$, $\mathbf{F}_2(\sqrt{5}, 0)$
15. $\mathbf{V}_1(0, 8)$, $\mathbf{V}_2(0, -8)$; $\mathbf{F}_1(0, 2\sqrt{15})$, $\mathbf{F}_2(0, -2\sqrt{15})$
17. $\mathbf{V}_1(-3, 0)$, $\mathbf{V}_2(3, 0)$; $\mathbf{F}_1(-2\sqrt{2}, 0)$, $\mathbf{F}_2(2\sqrt{2}, 0)$
19. $\mathbf{V}_1(0, \sqrt{7})$, $\mathbf{V}_2(0, -\sqrt{7})$; $\mathbf{F}_1(0, 2)$, $\mathbf{F}_2(0, -2)$

21. $\dfrac{x^2}{28} + \dfrac{3y^2}{28} = 1$

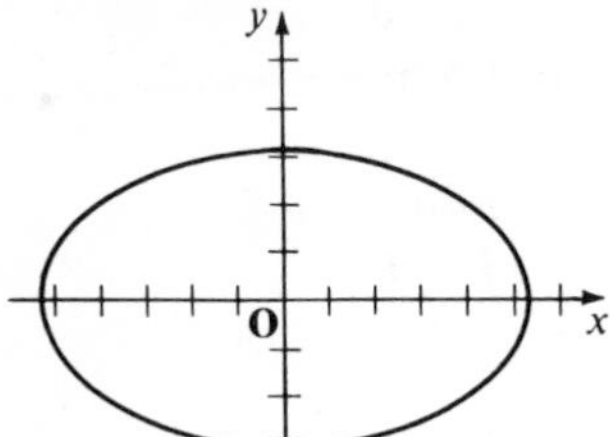

23. $\dfrac{x^2}{25} + \dfrac{y^2}{9} = 1$

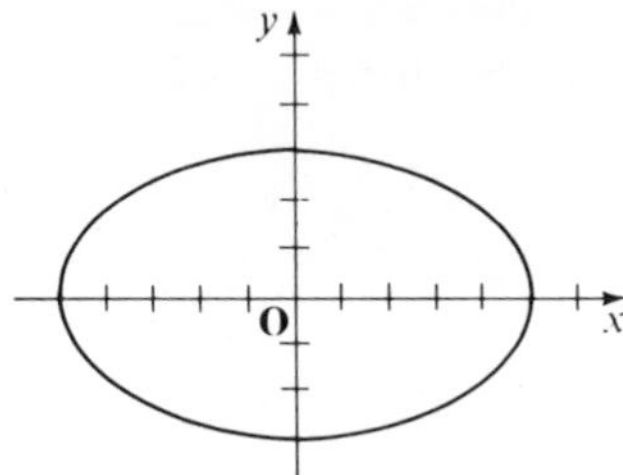

25. $5x^2 + 9y^2 - 72y + 99 = 0$

27. $4x^2 + 3y^2 - 48x + 132 = 0$

Applied Problems 4–4, page 144

3. 3,110,000 miles

Exercises 4–5, pages 150–153

1. $\dfrac{x^2}{9} - \dfrac{y^2}{16} = 1$

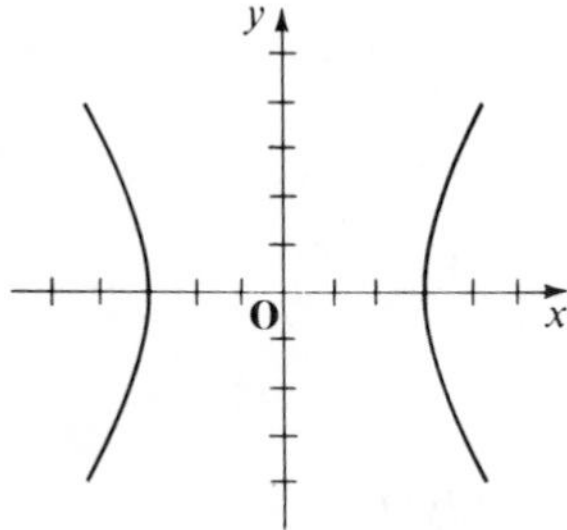

3. $\dfrac{x^2}{9} - \dfrac{y^2}{4} = 1$

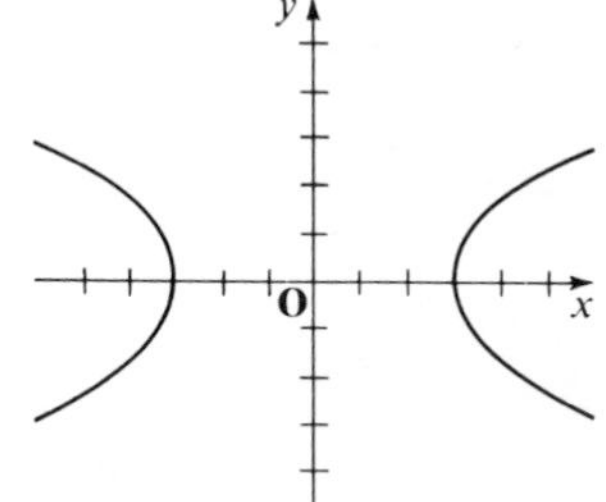

5. $\dfrac{y^2}{16} - \dfrac{x^2}{9} = 1$

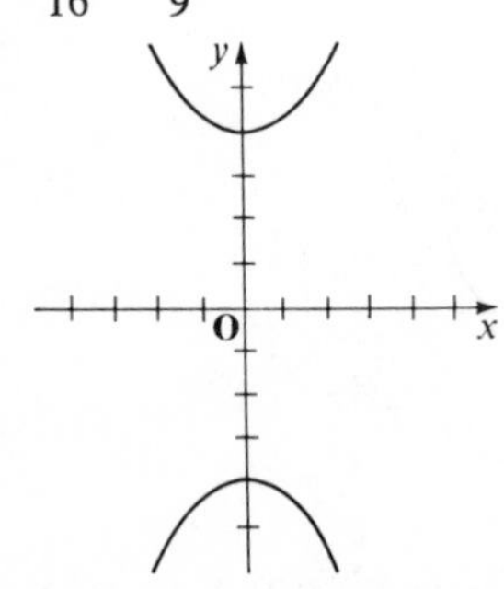

7. $\dfrac{y^2}{3} - \dfrac{x^2}{4} = 1$

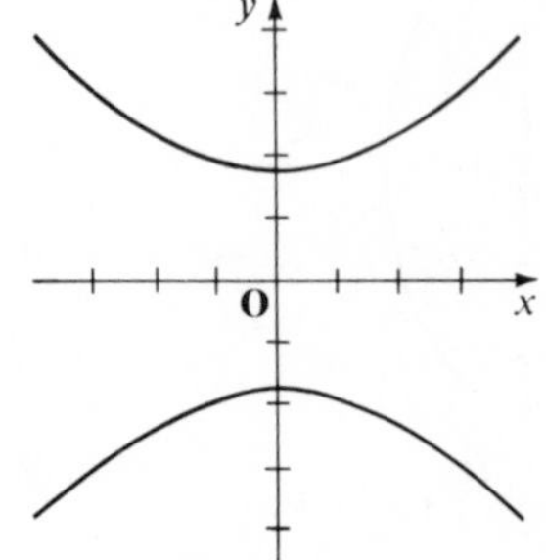

9. $\frac{x^2}{25} - \frac{y^2}{16} = 1$

11. $\frac{y^2}{169} - \frac{x^2}{25} = 1$

13. $V_1(-5, 0)$, $V_2(5, 0)$; $F_1(-\sqrt{41}, 0)$, $F_2(\sqrt{41}, 0)$; $y = \pm\frac{4}{5}x$

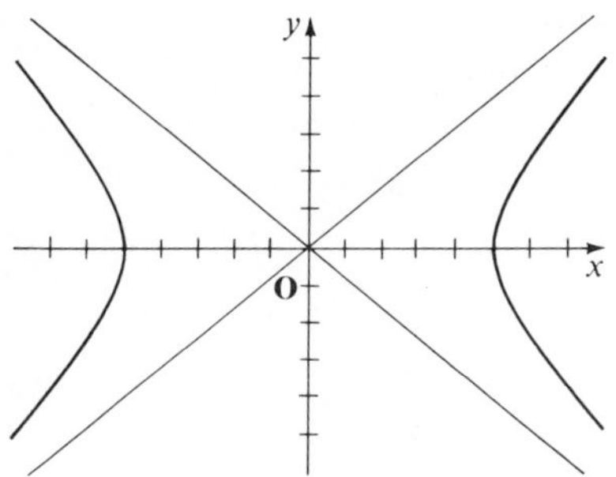

15. $V_1(-2\sqrt{2}, 0)$, $V_2(2\sqrt{2}, 0)$; $F_1(-2\sqrt{3}, 0)$, $F_2(2\sqrt{3}, 0)$; $y = \pm\frac{x}{\sqrt{2}}$

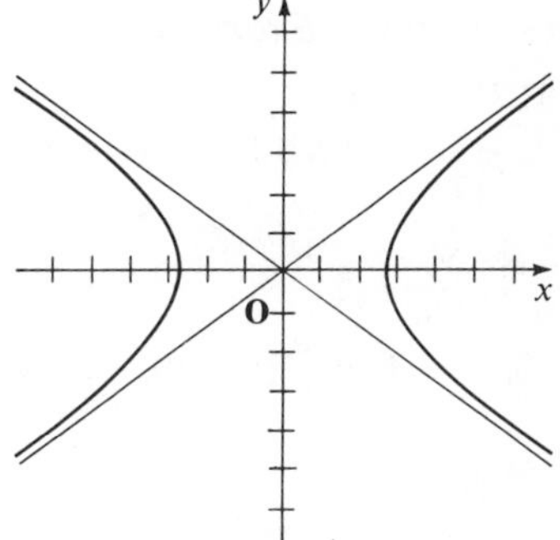

17. $V_1(0, 4)$, $V_2(0, -4)$; $F_1(0, \sqrt{41})$, $F_2(0, -\sqrt{41})$; $y = \pm\frac{4}{5}x$

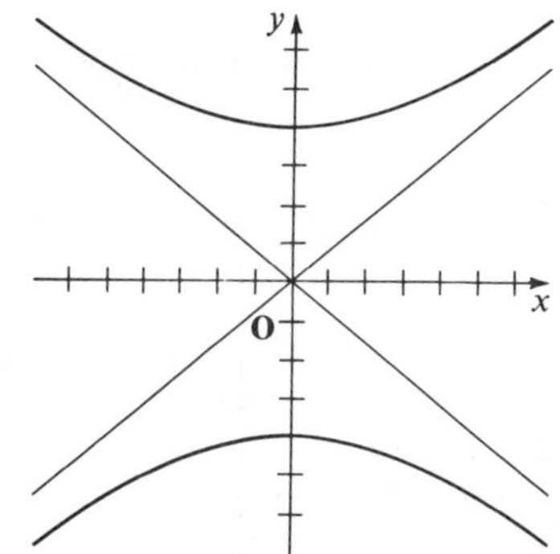

19. $V_1(0, 4)$, $V_2(0, -4)$; $F_1(0, 5)$, $F_2(0, -5)$; $y = \pm\frac{4}{3}x$

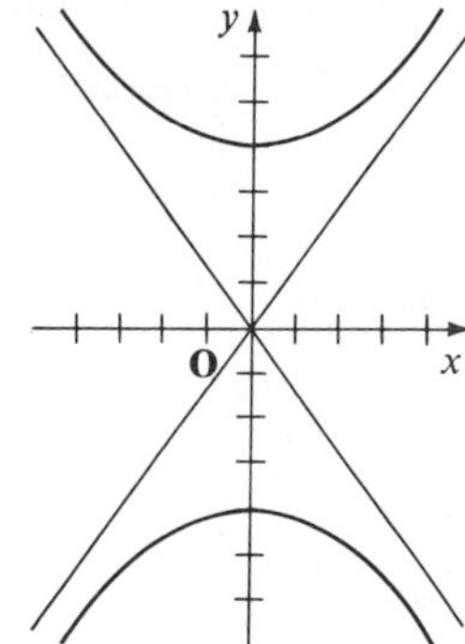

21. $\frac{x^2}{4} - \frac{y^2}{12} = 1$

23. $2x^2 - 3y^2 = 5$

25. $\frac{y^2}{3} - \frac{x^2}{9} = 1$

27. $\frac{x^2}{36} - \frac{y^2}{108} = 1$

35. $8x^2 - y^2 + 2y - 9 = 0$

37. $3x^2 - y^2 + 2y - 13 = 0$

Exercises 4–6, pages 158–160

1. $\frac{x^2}{25} + \frac{y^2}{9} = 1$

3. $\frac{x^2}{16} + \frac{y^2}{7} = 1$

5. $\frac{x^2}{225} + \frac{y^2}{200} = 1$

7. $\frac{x^2}{4} - \frac{y^2}{5} = 1$

9. $\frac{x^2}{3} - \frac{y^2}{13} = 1$

11. $\frac{x^2}{\frac{5}{2}} - \frac{y^2}{\frac{45}{2}} = 1$

13. $\sqrt{2}$ **15.** $\dfrac{1}{\sqrt{2}}$ **19.** $e = \sqrt{1 + m^2}$

23. $\dfrac{(x-1)^2}{25} - \dfrac{(y-1)^2}{\frac{125}{4}} = 1$ **25.** $\dfrac{(x-2)^2}{25} + \dfrac{(y-3)^2}{9} = 1$

27. $x = \pm 8\sqrt{5}$

Chapter Review Exercises, pages 162–163

1. $x + 2y - 2 = 0$ **3.** $xy - 13x + 8y - 8 = 0$

5. $\mathbf{C}(3, -4);\ r = \sqrt{13}$ **7.** $(x-4)^2 + (y+3)^2 = \frac{49}{2}$

9. $x^2 + 16y + 16 = 0$ **11.** $y^2 - 6y - 16x - 23 = 0$

13. $\dfrac{x^2}{169} + \dfrac{y^2}{144} = 1$

15. $\mathbf{F}_1(0, -4), \mathbf{F}_2(0, 4);\ \mathbf{V}_1(0, -5), \mathbf{V}_2(0, 5)$

17. $\dfrac{y^2}{9} - \dfrac{x^2}{16} = 1$ **19.** $\dfrac{x^2}{25} + \dfrac{y^2}{16} = 1$

21. $\dfrac{(x-2)^2}{\frac{400}{9}} + \dfrac{y^2}{\frac{175}{9}} = 1$ **23.** $\dfrac{y^2}{16} - \dfrac{x^2}{84} = 1$

Exercises 5–1, pages 177–180

1. $x' - 4y'^2 - 16y' - 10 = 0$ **3.** $x'^2 + 4y'^2 - 36 = 0$ **5.** $x'^2 + 2y'^2 = 0$

7. $x^2 - 6x - 24y - 39 = 0$ **9.** $y^2 - 8y + 12x - 44 = 0$

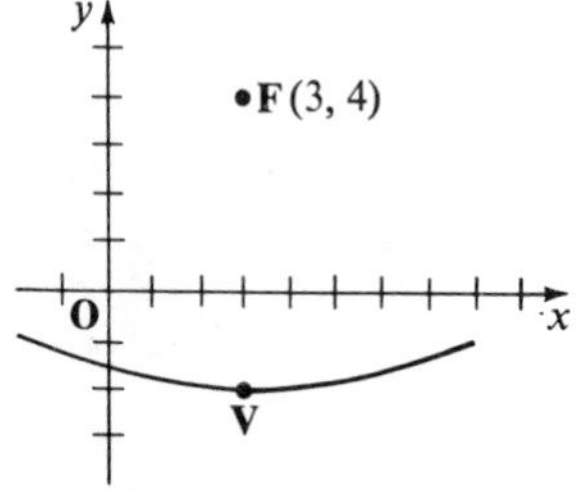

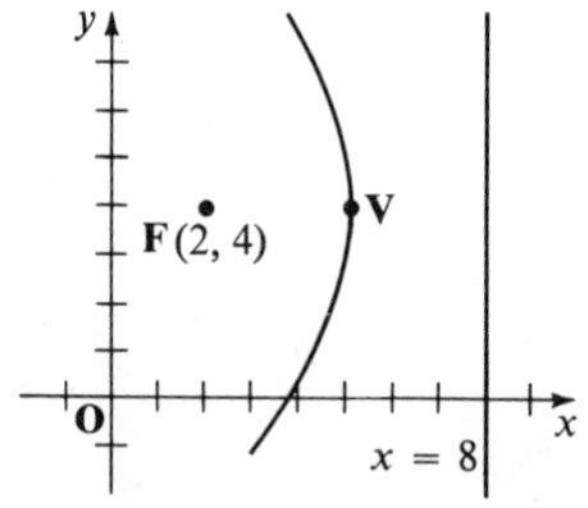

11. $9x^2 - 90x + 25y^2 - 100y + 100 = 0$

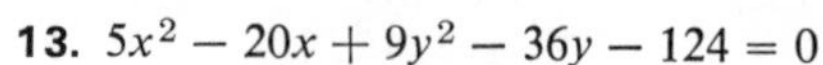
13. $5x^2 - 20x + 9y^2 - 36y - 124 = 0$

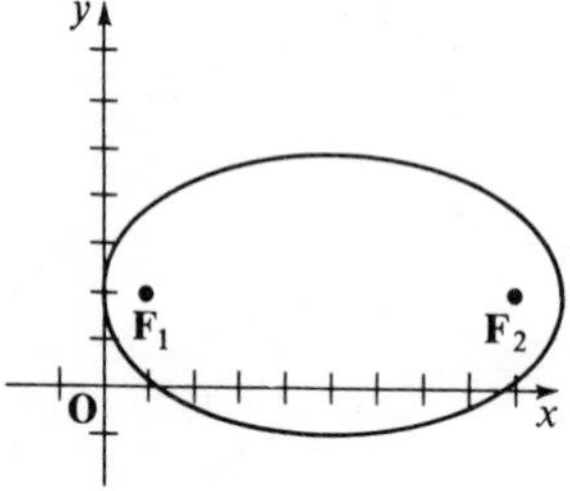

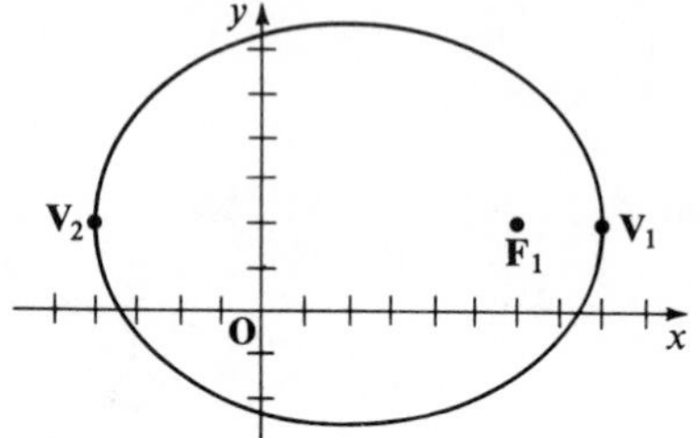

15. $3y^2 + 12y - x^2 + 6x - 9 = 0$

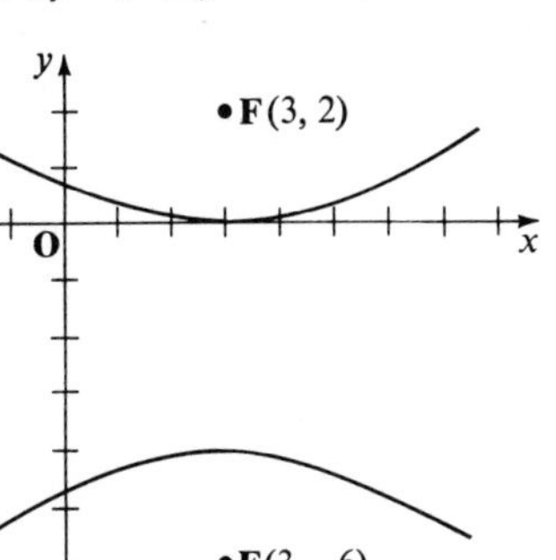

17. $24x^2 - 96x - 25y^2 + 100y - 388 = 0$

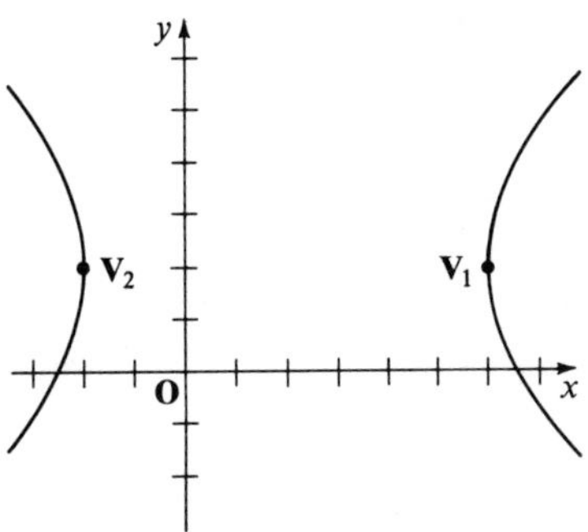

19. $x'^2 + y'^2 = 65$, circle

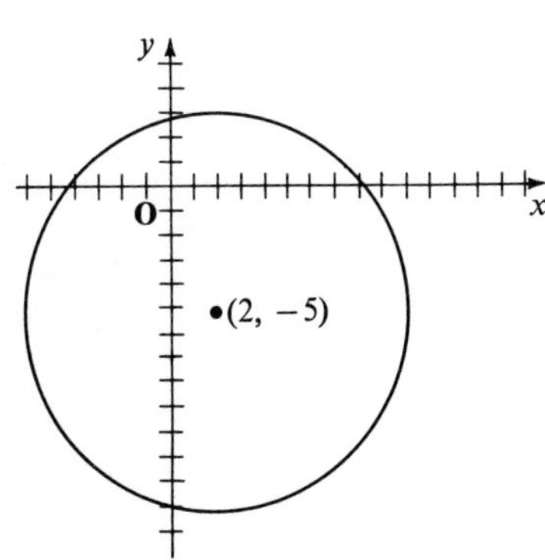

21. $\frac{x'^2}{36} + \frac{y'^2}{72} = 1$, ellipse

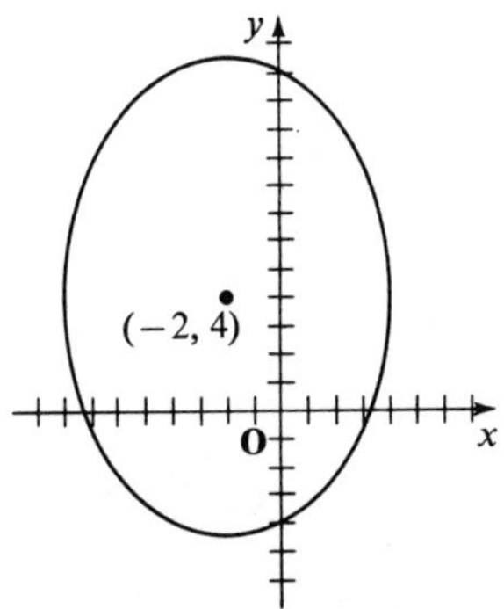

23. $y'^2 = -4x'$, parabola

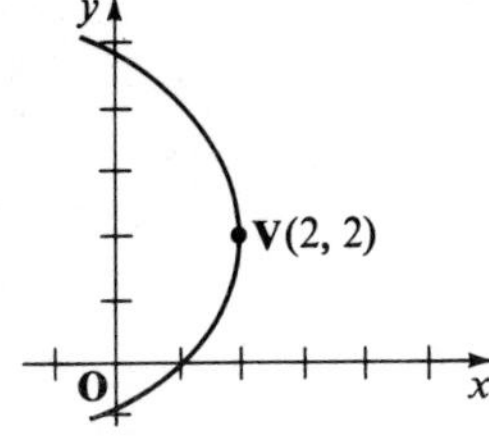

25. $x'^2 + y'^2 = 4$, circle

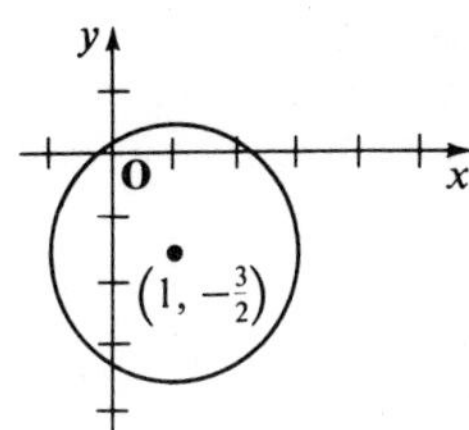

27. $x'^2 + y'^2 = 0$, point

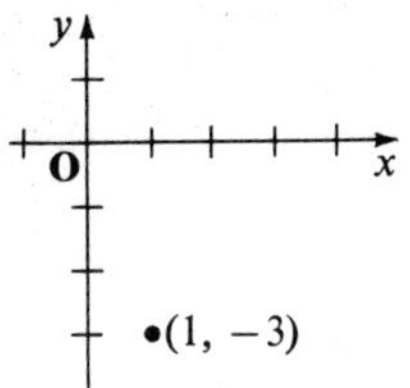

29. $\frac{2x'^2}{7} - \frac{9y'^2}{14} = 1$, hyperbola

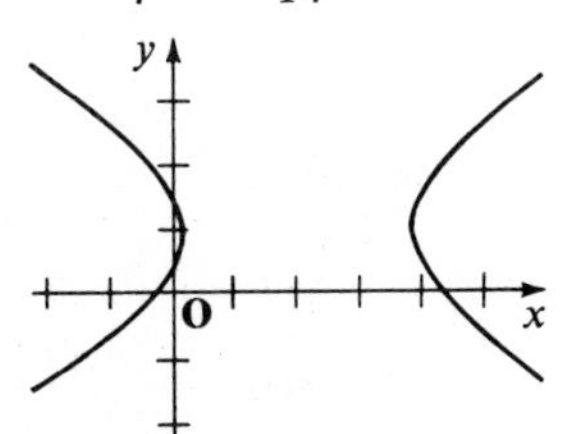

31.

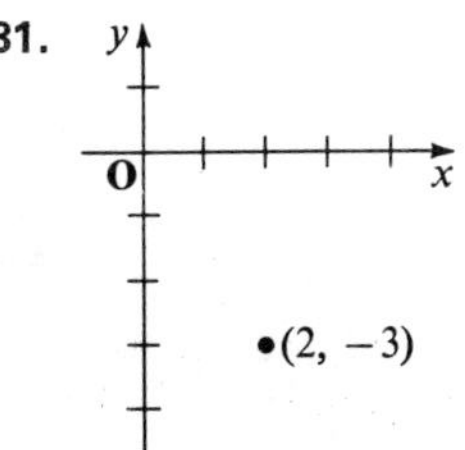

33. $\emptyset$

35.

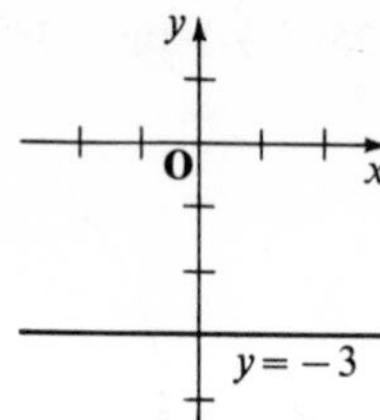

37. $x'y' + 38 = 0$

Exercises 5–2, pages 184–186

1. $\left(\dfrac{3+\sqrt{3}}{2}, \dfrac{3\sqrt{3}-1}{2}\right)$

3. $(1 - 2\sqrt{3}, -2 - \sqrt{3})$

5. $\left(-\dfrac{3}{\sqrt{2}}, -\dfrac{9}{\sqrt{2}}\right)$

7. $-5y' + 10 = 0$

9. $4x'^2 + 9y'^2 - 36 = 0$

11. $7x'^2 + y'^2 - 14 = 0$

13. $x'^2 + y'^2 = 25$

15. $-5x'^2 + 20y'^2 + 50x' - 45 = 0$

17. $y' = \frac{4}{5}\sqrt{5}, y' = -\frac{4}{5}\sqrt{5}$

19. $3x' - y' = 6, 3x' - y' = -6$

21. $x' = -\dfrac{\sqrt{10}}{5}, x' = \dfrac{\sqrt{10}}{5}$

Exercises 5–3, pages 191–192

1. $\sin\phi = \dfrac{1}{\sqrt{2}}, \cos\phi = \dfrac{1}{\sqrt{2}}$

3. $\sin\phi = \dfrac{\sqrt{2+\sqrt{3}}}{2}, \cos\phi = \dfrac{\sqrt{2-\sqrt{3}}}{2}$

5. $\sin\phi = \dfrac{1}{\sqrt{10}}, \cos\phi = \dfrac{3}{\sqrt{10}}$

7. elliptic,

$\dfrac{x'^2}{4} + \dfrac{y'^2}{8} = 1$

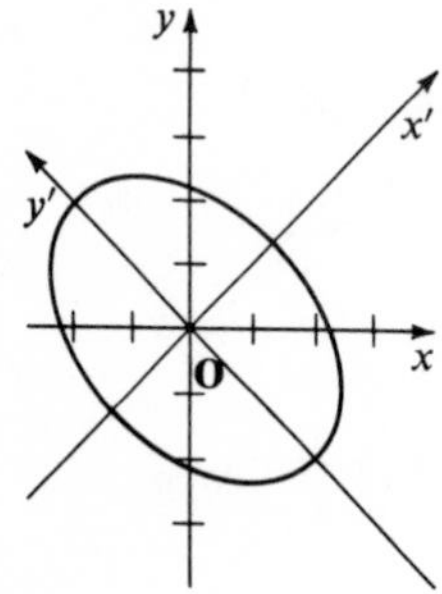

9. hyperbolic,

$\dfrac{x'^2}{3} - \dfrac{y'^2}{5} = 1$

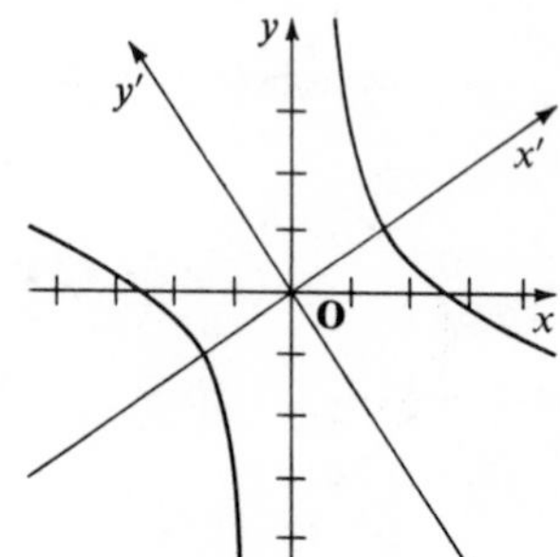

11. elliptic, $\dfrac{x'^2}{80} + \dfrac{y'^2}{30} = 1$

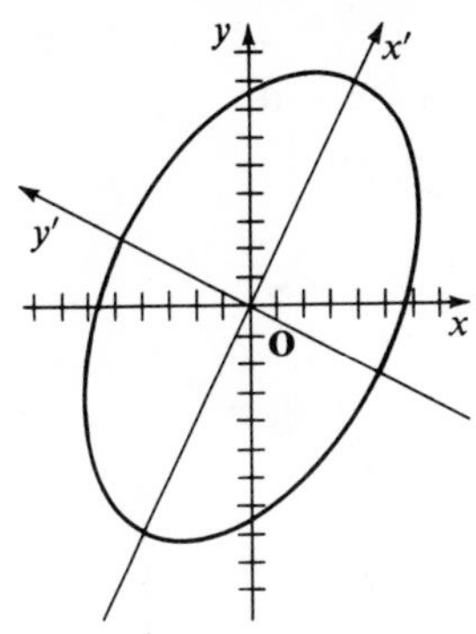

13. $4x''^2 - y''^2 = 0$

15. $\dfrac{y''^2}{4} - \dfrac{x''^2}{12} = 1$

17. $2x''^2 - 3\sqrt{2}y'' + \frac{19}{4} = 0$

Exercises 5–4, page 195

1. $-3x + 4y = 0$ **3.** $x - 2y = 0$ **5.** $2x - 3y = 0$
7. $5x + 2y = 0$ **9.** $2x - y - 1 = 0$ **11.** $2x + y + 5 = 0$
13. $2x + 3y + 12 = 0$ **15.** $2x + 3y + 5 = 0$ **17.** $x + y - 4 = 0$
19. $x - 4 = 0$ **21.** $x + y + 2 = 0$ **23.** $8x - 5y - 1 = 0$
25. $13x + 21y - 55 = 0$ **27.** $15x - 17y - 47 = 0$

Exercises 5–5, pages 201–202

1. $4x - y = 0, 4x + y = 0$
3. $x - 6y = 0, x + 6y = 0$
5. $x + 2y - 3 = 0$
7. No line through **S** is tangent to the curve.
9. $x - y + 4 = 0, x - y - 4 = 0$
11. $x - y - 1 = 0$
13. $6x - 3y + 2 = 0$
15. $2x - 3y = 0, 4x - 6y - 25 = 0$
17. $2x - y + 5 = 0, 2x + y - 5 = 0$
19. $x - 2y + 2\sqrt{2} = 0, x - 2y - 2\sqrt{2} = 0$

Chapter Review Exercises, pages 204–205

1. $4x'^2 - 9y'^2 - 60 = 0$ **3.** $x'^2 + 3y' - 5 = 0$
5. $x^2 + y^2 - 6x + 8y = 0$ **7.** $x^2 - 4x + 16y - 108 = 0$

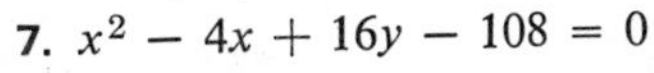

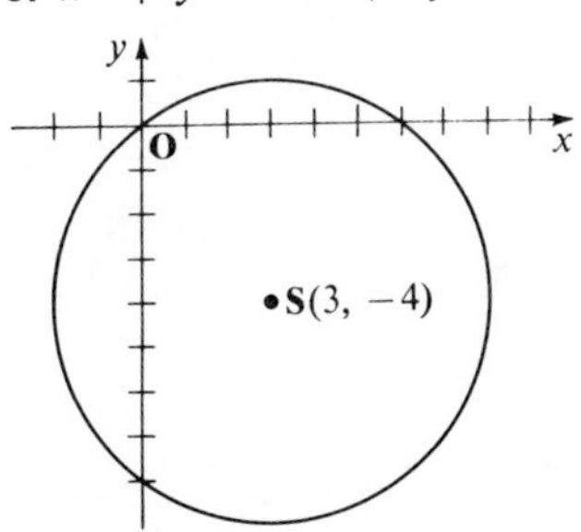

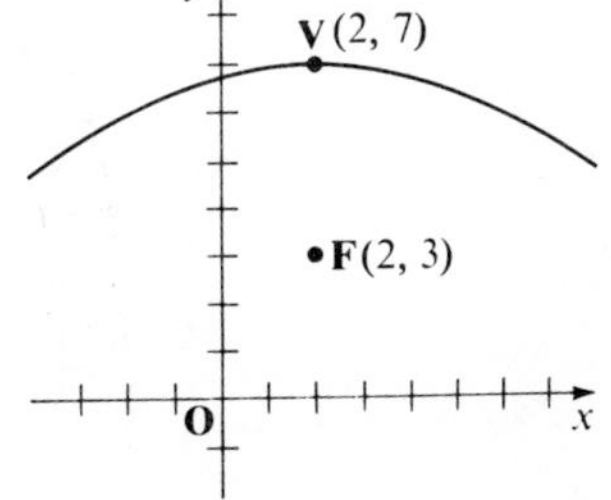

9. $\left(\frac{5\sqrt{3}-3}{2}, \frac{-5-3\sqrt{3}}{2}\right)$ **11.** $17x'^2 + 4y'^2 = 884$

13. **(a)** hyperbolic **(b)** hyperbolic **(c)** hyperbolic

15. $13x - 24y = 0$ **17.** $x - y = 0, x + 3y = 0$

Exercises 6–1, page 219

1. d **3.** a or f **5.** b or e **7.** a **9.** c

11.

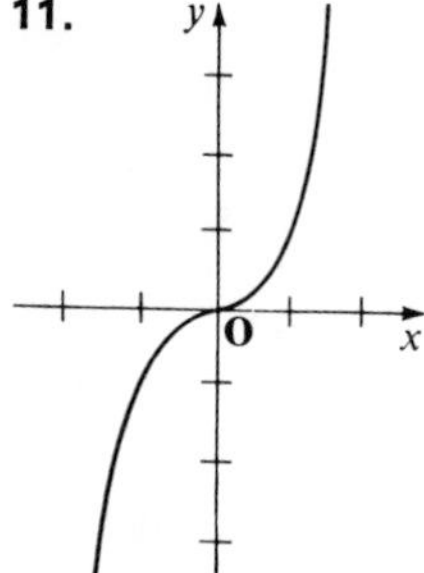

13.

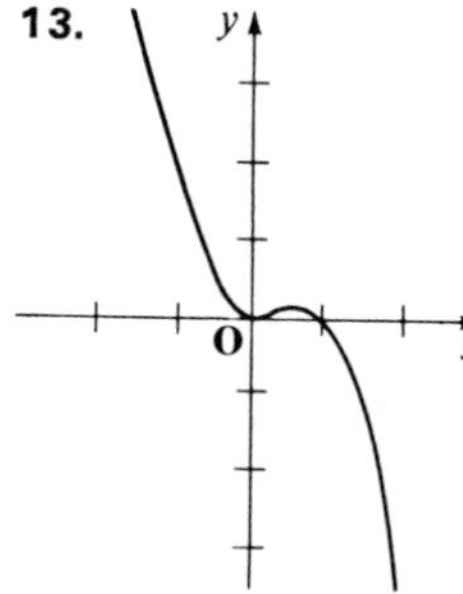

15.

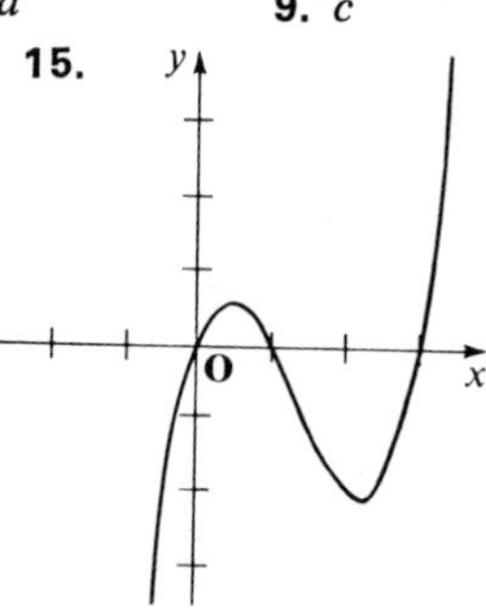

17.

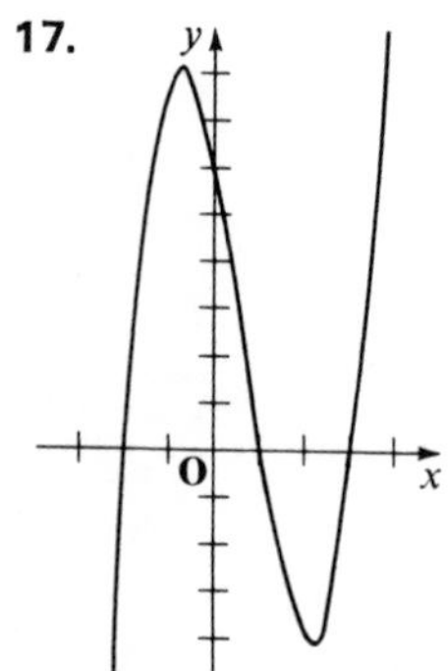

19.

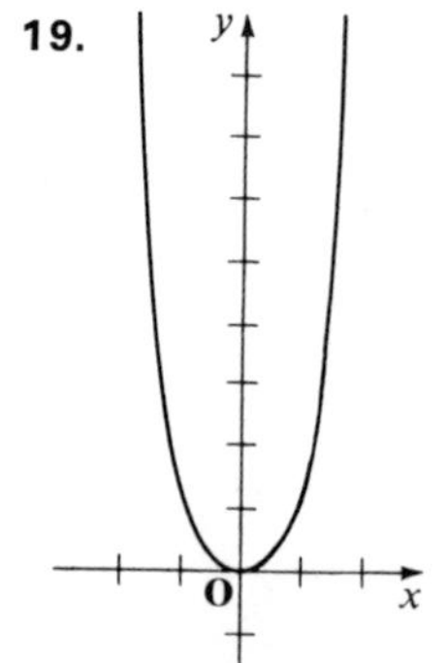

21.

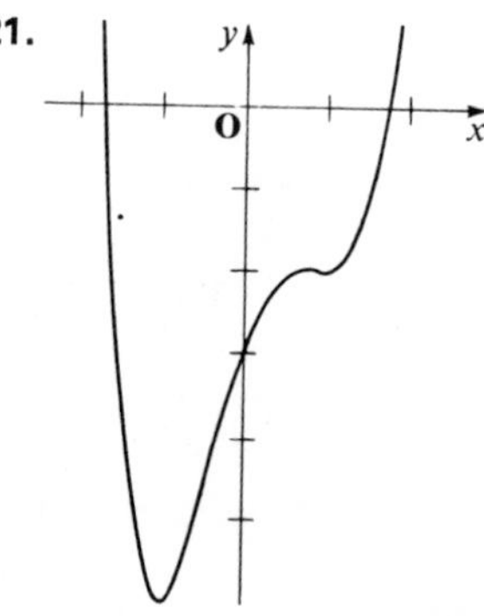

23.

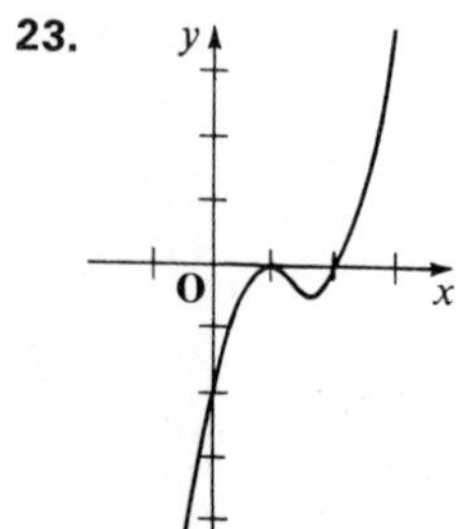

25.

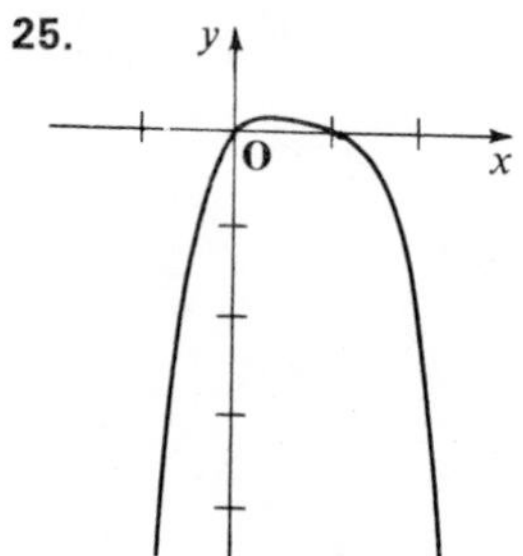

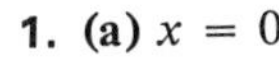

Exercises 6–2, page 223

1. **(a)** $x = 0$
(b)

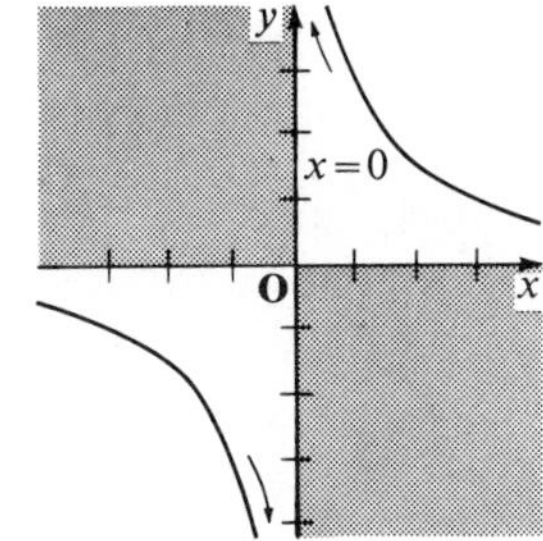

3. **(a)** $x = 2$
(b)

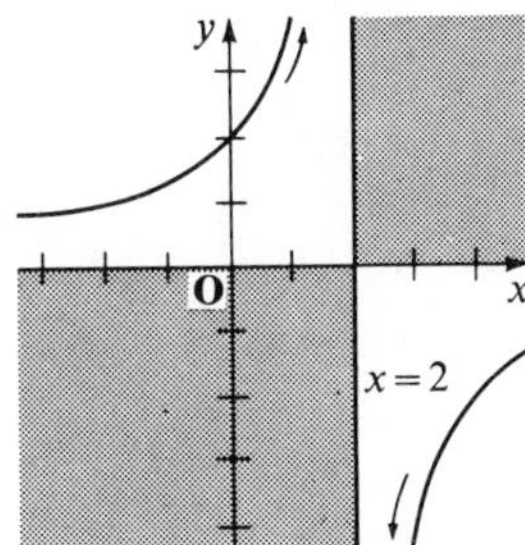

5. **(a)** $x = 0, x = -1$
(b)

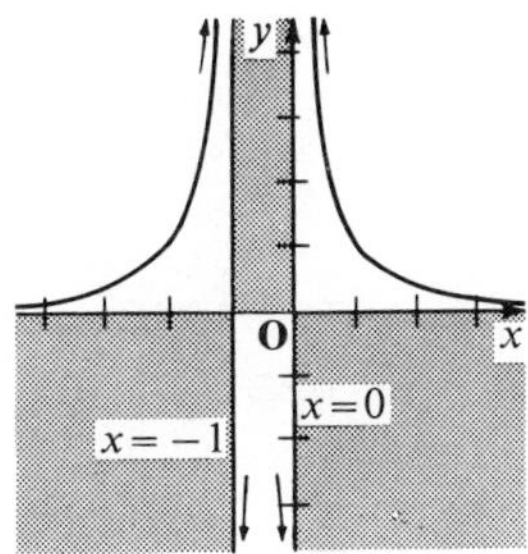

7. **(a)** $x = 2, x = -2$
(b)

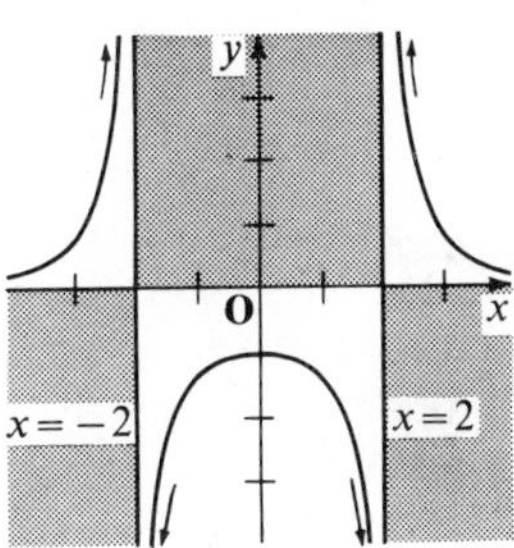

9. **(a)** $x = 1, x = -3$
(b)

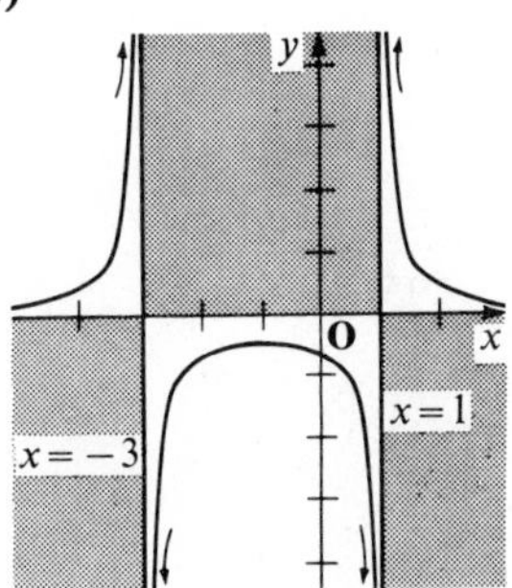

11. **(a)** $x = -2$
(b)

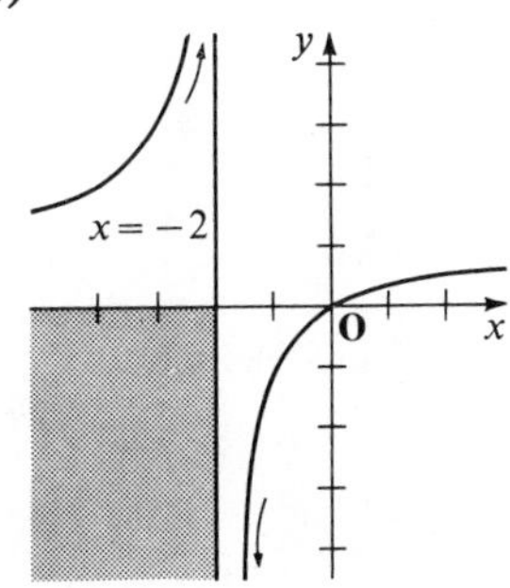

13. (a) $x = 2$
(b)

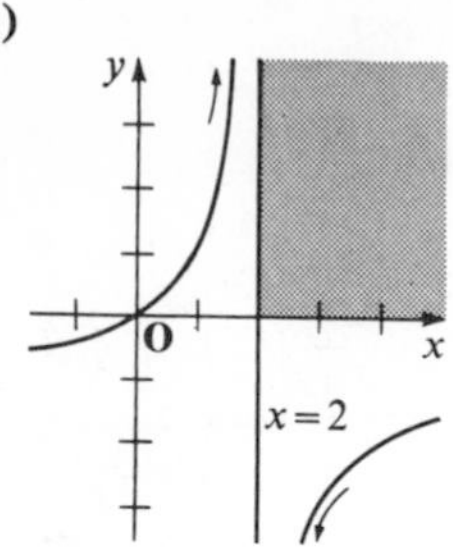

15. (a) $x = -1, x = 2$
(b)

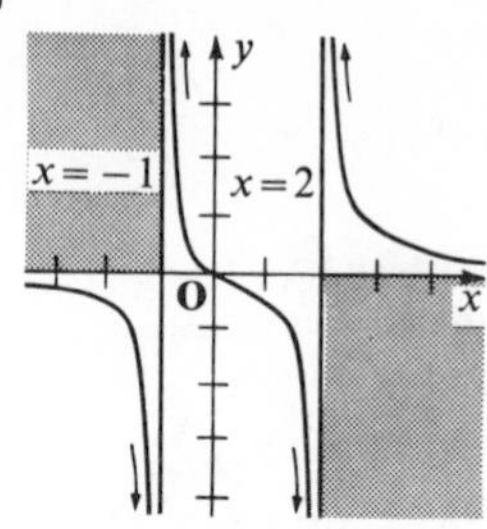

17. (a) $x = -\frac{1}{2}, x = 2$
(b)

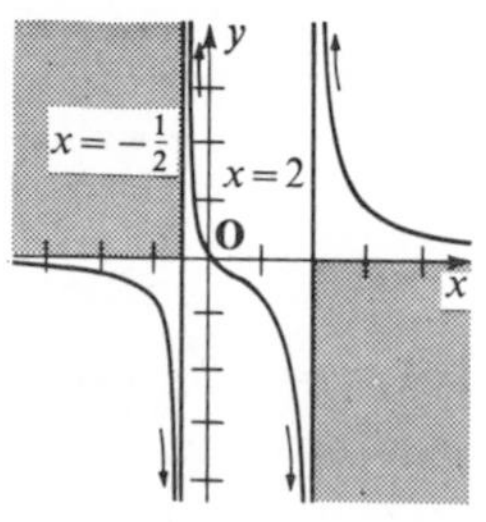

19. (a) $x = -3, x = 3$
(b)

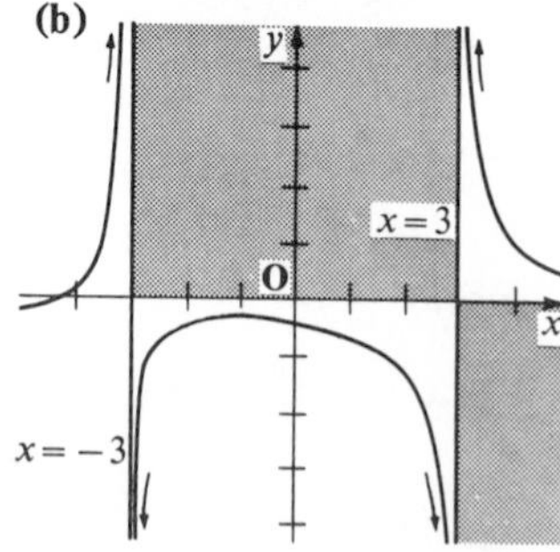

21. (a) $x = -2, x = 5$
(b)

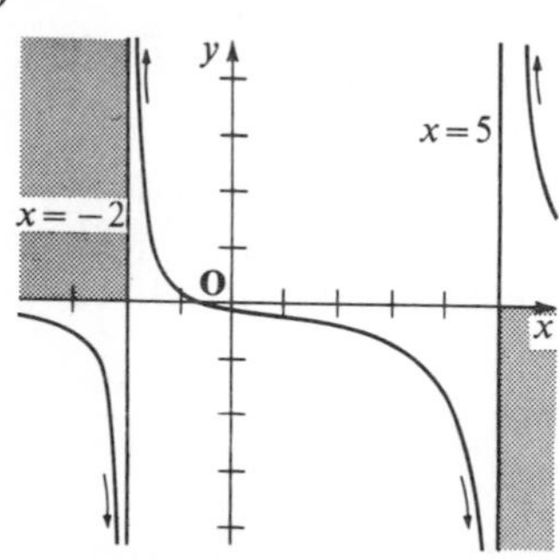

23. (a) $x = -2, x = 2$
(b)

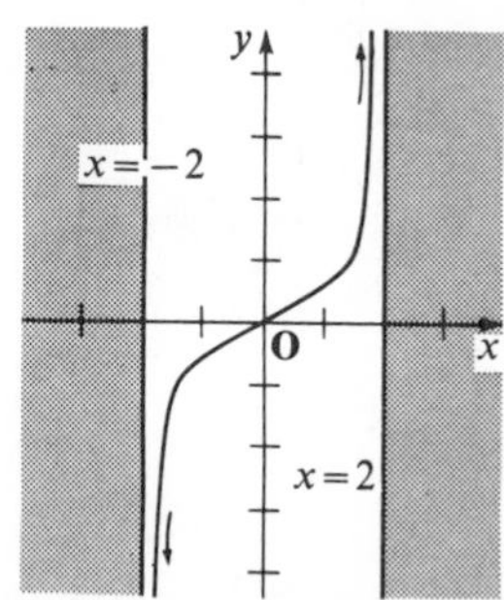

25. (a) $x = -2, x = 2$
(b)

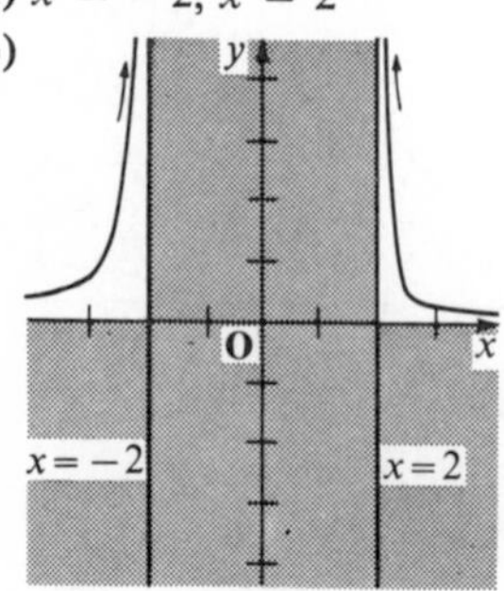

27. (a) $x = -1, x = 3$
(b)

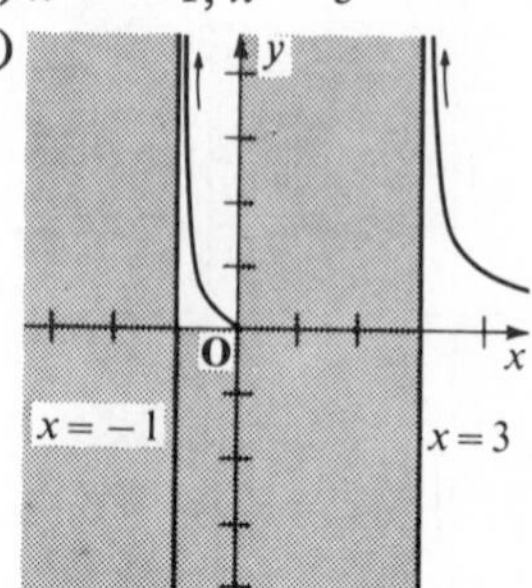

Exercises 6–3, page 228

1.

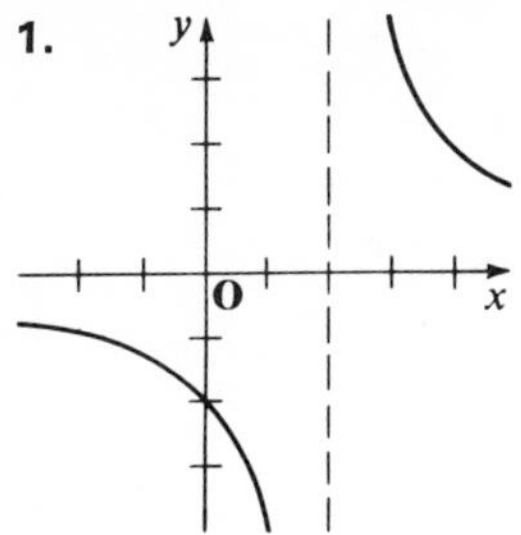

3.

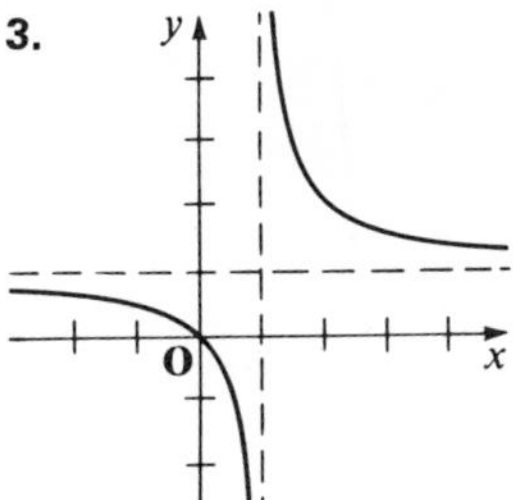

5.

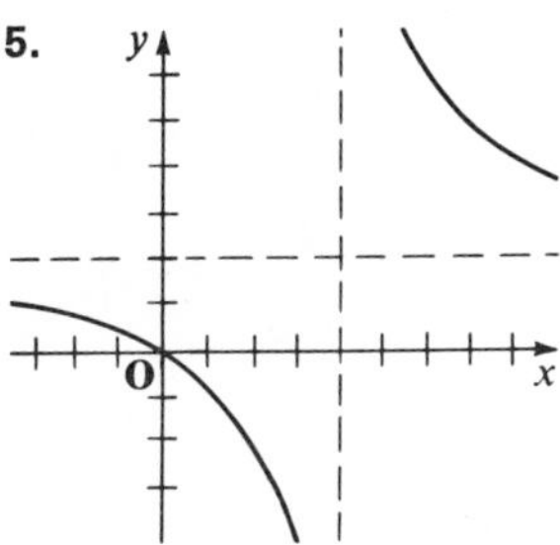

7.

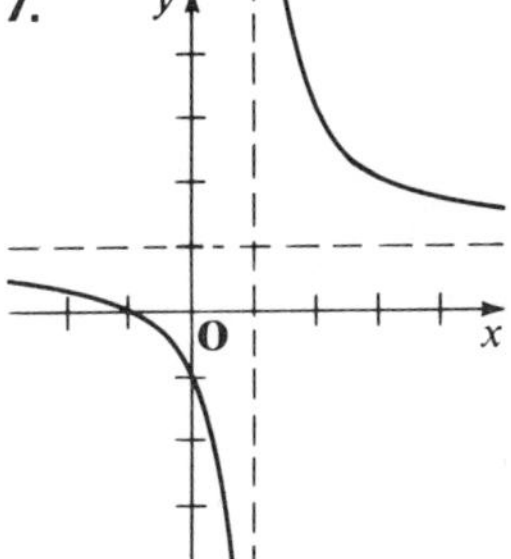

9.

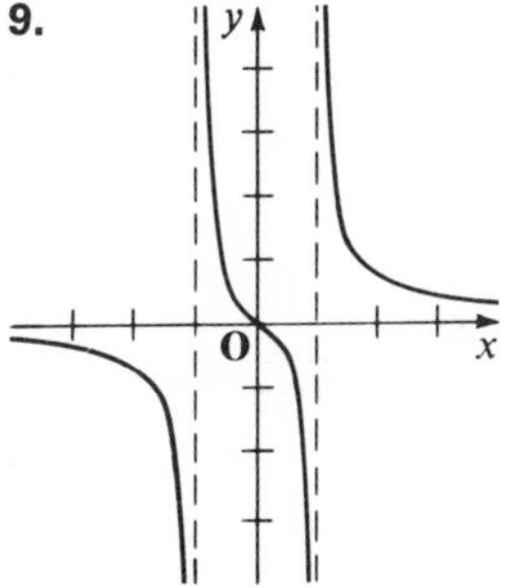

11.

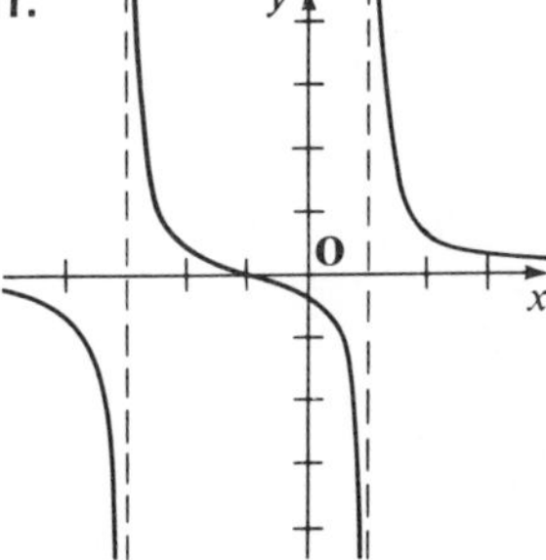

13.

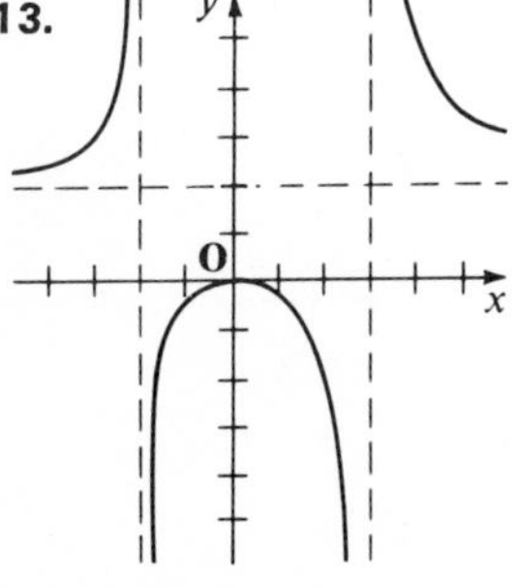

15.

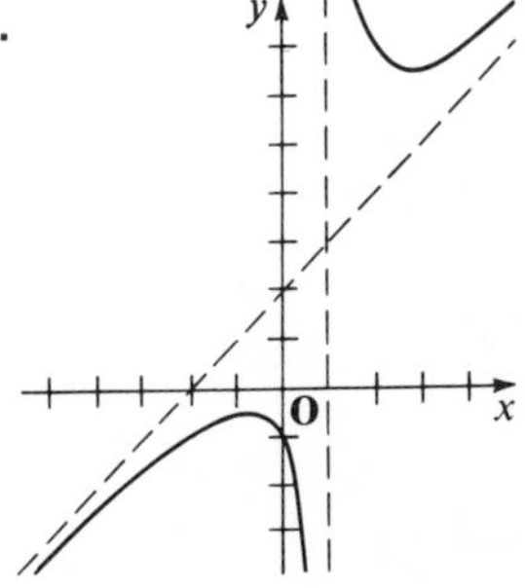

17.

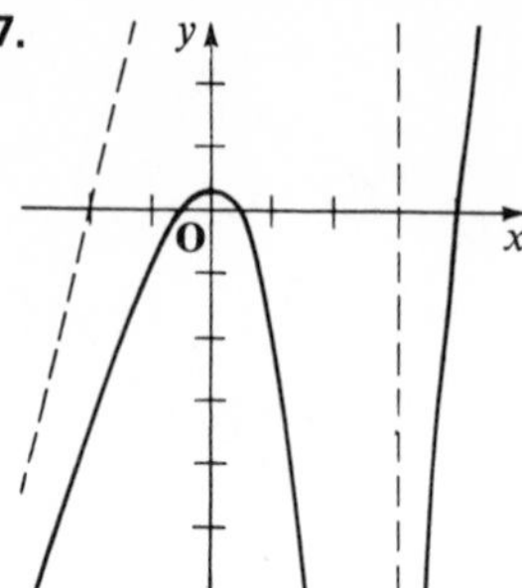

19.

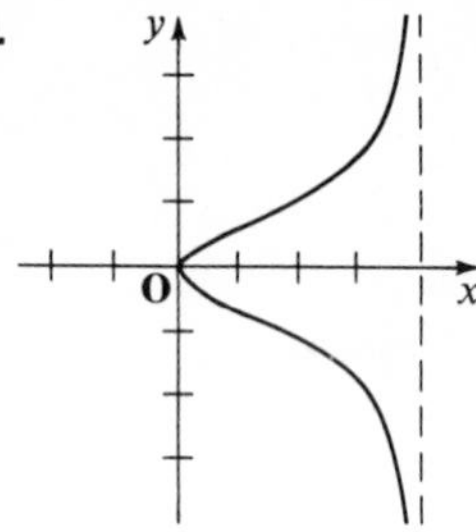

21.

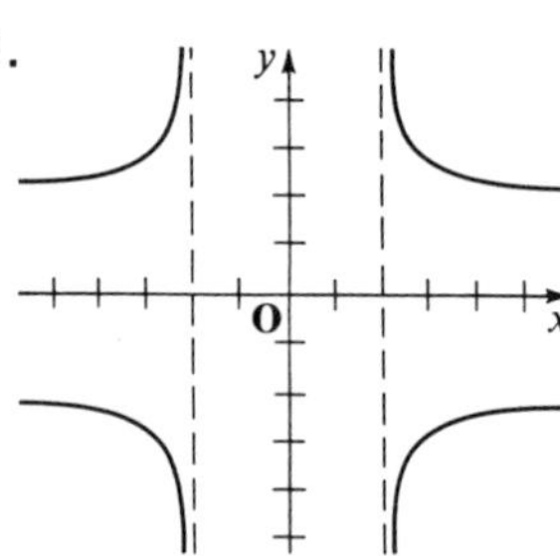

23.

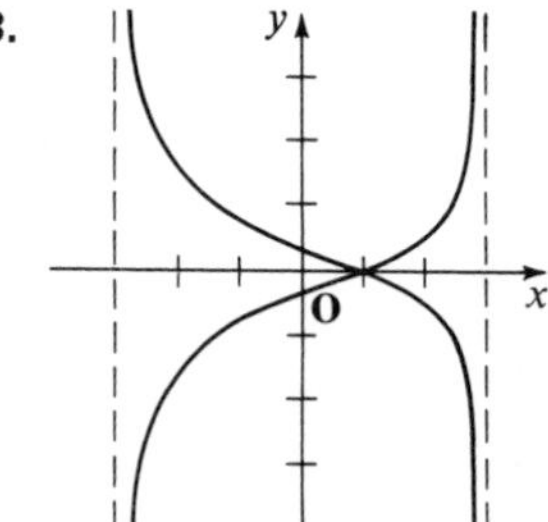

Exercises 6–4, page 230

1.

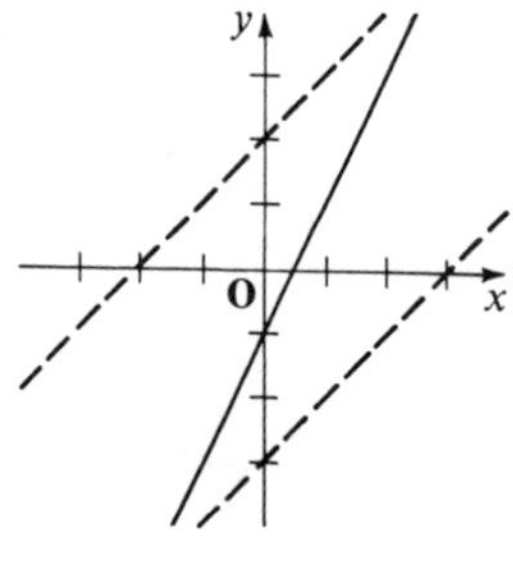

3.

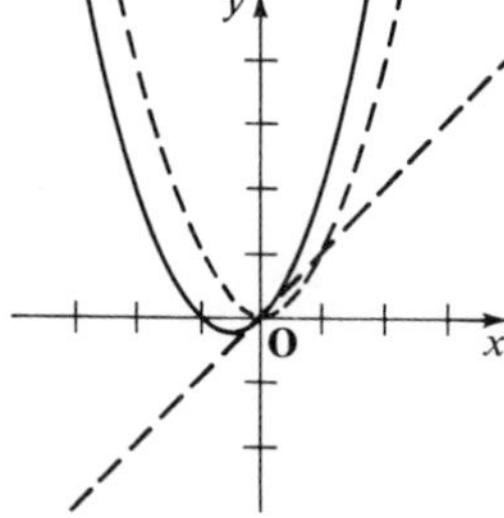

5.

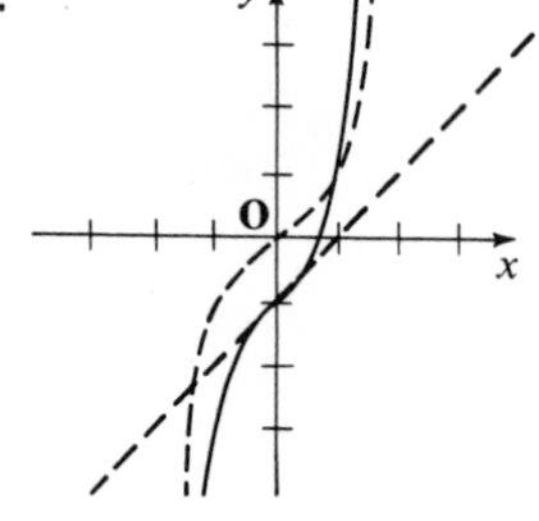

7.

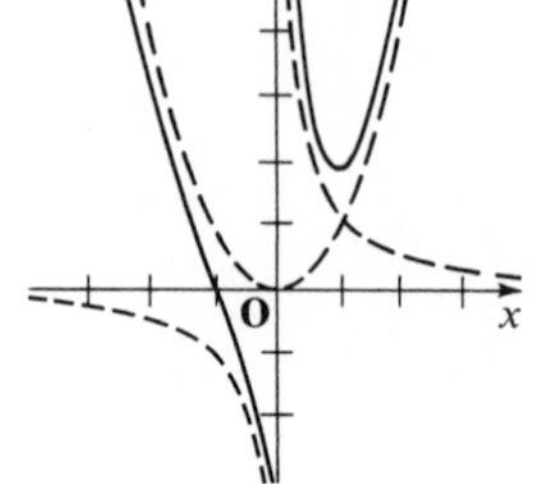

9.

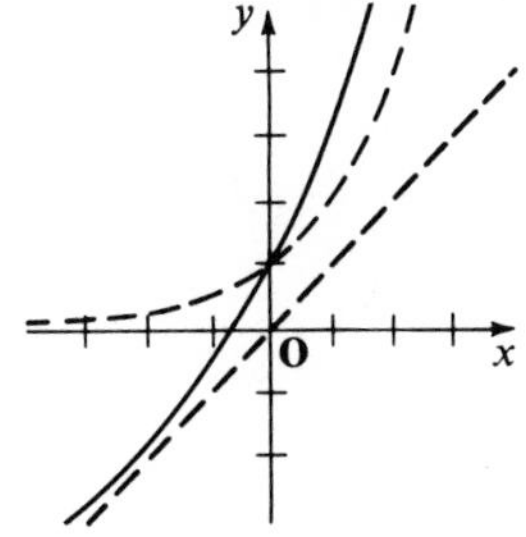

11.

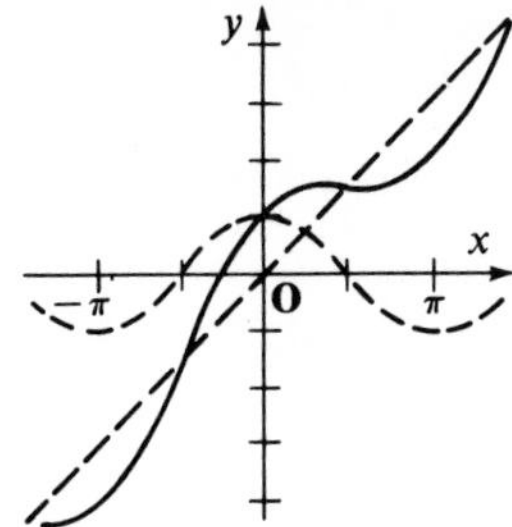

13.

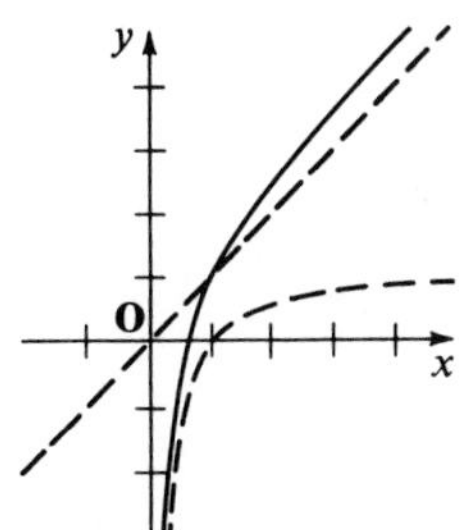

15.

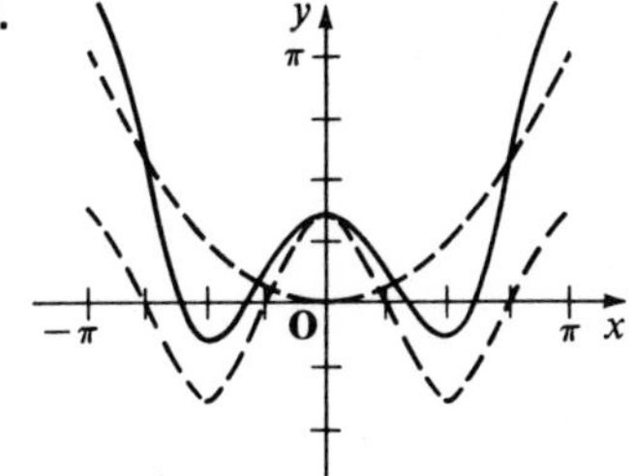

Exercises 6–5, pages 234–235

1.

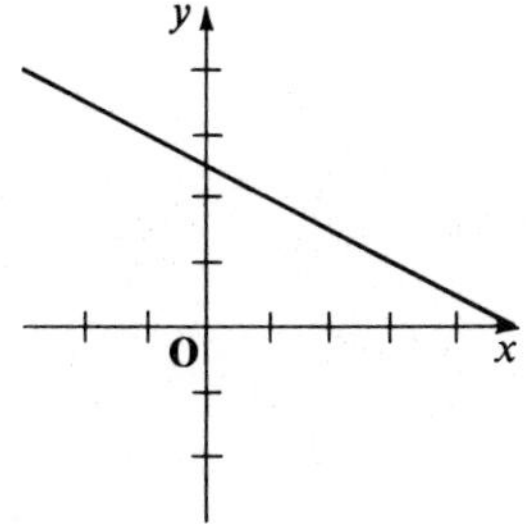

3.

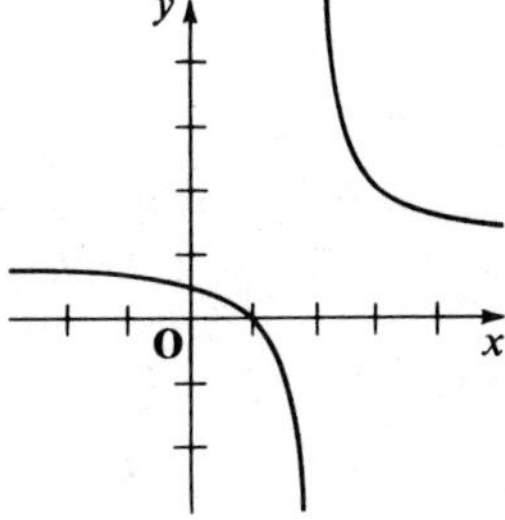

5.

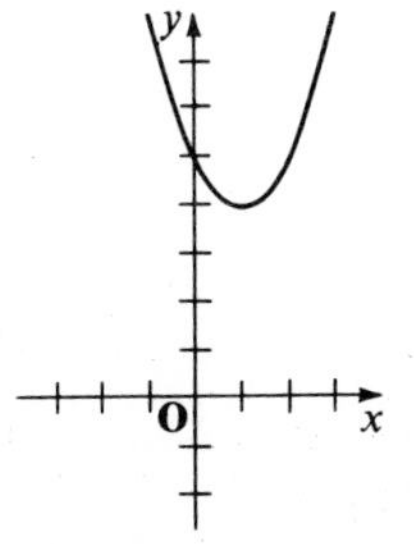

7.

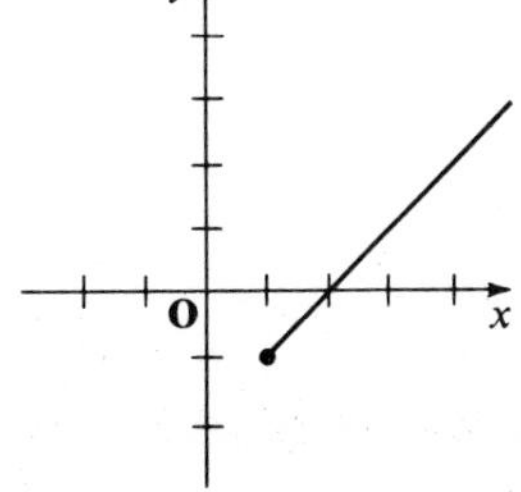

9.

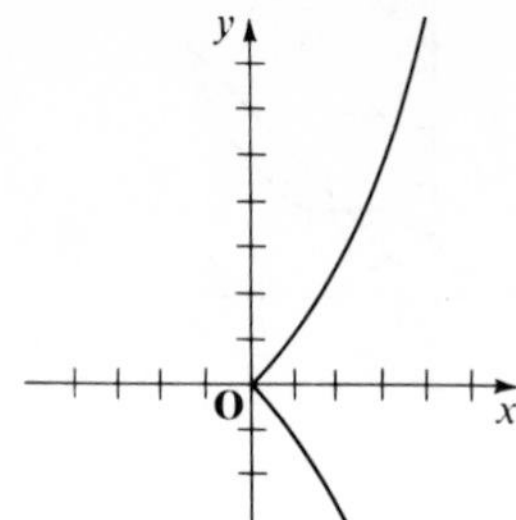

11.

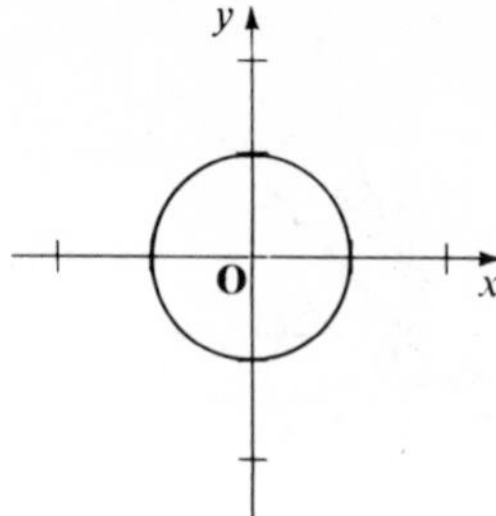

13.

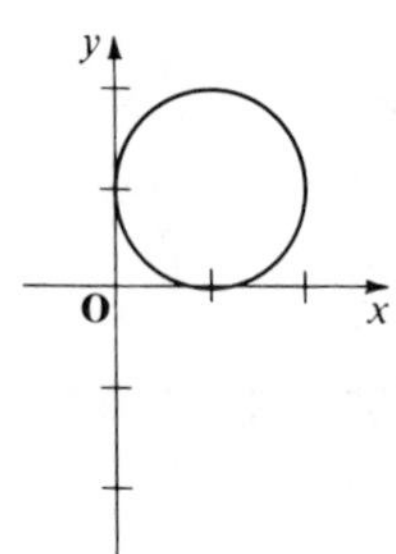

15.

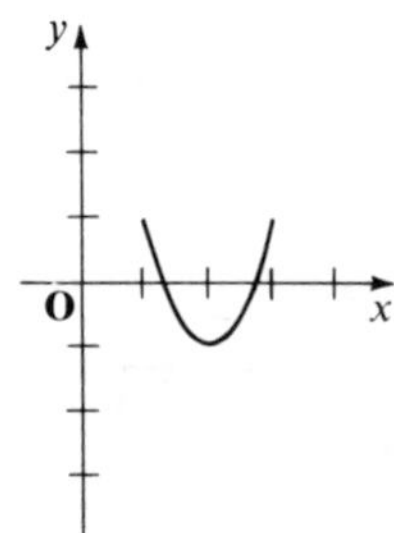

17.

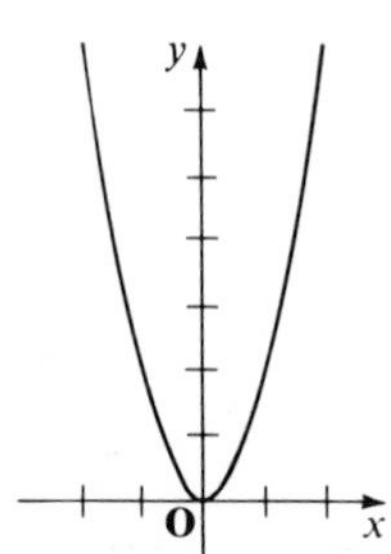

19. $y = x + 6$

21. $y = x^2 + 2x + 3$

23. $x^2 - xy + 1 = 0$

25. $x^2 - 2xy + y^2 - 2x - 2y = 0$

27. $x^2 + y^2 = 9$, same

29. $y = 1 - 2x^2$, no

31. $x^2 + y^2 = 1$, same

33. $y + 1 = 2(x - 1)^2$, no

35. $x = \dfrac{6}{2 + r}$, $y = \dfrac{6r}{2 + r}$; no

37. $x = \pm\dfrac{3}{\sqrt{1 + r^2}}$, $y = \pm\dfrac{3r}{\sqrt{1 + r^2}}$; same

39. $x = \dfrac{2(1 + r)}{1 + r^2}$, $y = \dfrac{2r(1 + r)}{1 + r^2}$; same

41. $x = \dfrac{3r}{1 + r^3}$, $y = \dfrac{3r^2}{1 + r^3}$; no

Chapter Review Exercises, page 237

1. 6, 4, 2, or 0

3. $x = 3$, $x = -1$

5.

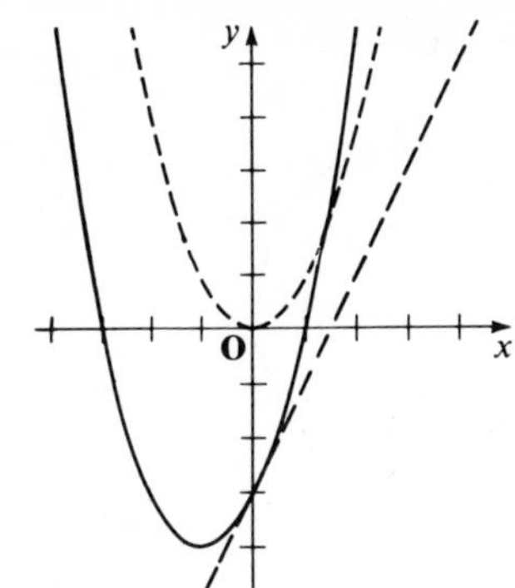

7. $y = x^2 - 4x + 3$

Exercises 7–1, page 248

1. for example, (3,385°), (3, −335°), (−3, 205°)

3. for example, (2, 480°), (2, −240°), (−2, 300°)

5. for example, $\left(2, \frac{5\pi}{2}^{R}\right), \left(2, -\frac{3\pi}{2}^{R}\right), \left(-2, \frac{3\pi}{2}^{R}\right)$

7. for example, $\left(1, \frac{19\pi}{6}^{R}\right), \left(1, \frac{-5\pi}{6}^{R}\right), \left(-1, \frac{13\pi}{6}^{R}\right)$

9. for example, (−1, 400°), (−1, −320°), (1, 220°)

11. for example, $\left(-2, \frac{17\pi}{6}^{R}\right), \left(-2, \frac{-7\pi}{6}^{R}\right), \left(2, \frac{11\pi}{6}^{R}\right)$

$\left(2, \frac{\pi}{2}^{R}\right)$
(2, 120°)
(3, 25°)
0°
O
$\left(-1, \frac{7\pi}{6}^{R}\right)$
(−1, 40°)
$\left(-2, \frac{5\pi}{6}^{R}\right)$

13. $(\sqrt{2}, \sqrt{2})$

15. (0, 6)

17. (−3, 0)

19. $(-\sqrt{2}, \sqrt{2})$

21. $(-\sqrt{3}, 1)$

23. $\left(\frac{3\sqrt{3}}{2}, -\frac{3}{2}\right)$

25. $(2\sqrt{3}, 300°), (-2\sqrt{3}, 120°)$

27. (5, 90°), (−5, 270°)

29. $(3\sqrt{2}, 315°), (-3\sqrt{2}, 135°)$

31. (8, 240°), (−8, 60°)

33. (4, 270°), (−4, 90°)

35. $(2\sqrt{2}, 315°), (-2\sqrt{2}, 135°)$

Exercises 7–2, page 250

1. $r = 5$

3. $r = -\frac{\sin\theta}{\cos^2\theta}$

5. $r = \frac{4}{\cos\theta - \sin\theta}$

7. $r = \pm\sqrt{\frac{5}{3 + \sin^2\theta}}$

9. $r = \pm\frac{4}{\sqrt{\cos 2\theta}}$

11. $r = \pm\frac{2\sqrt{2}}{\sqrt{\sin 2\theta}}$

13. $x^2 + y^2 = 16$

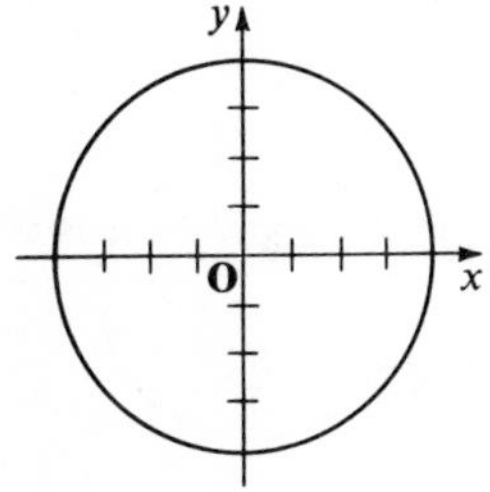

15. $x^2 + y^2 - 4y = 0$

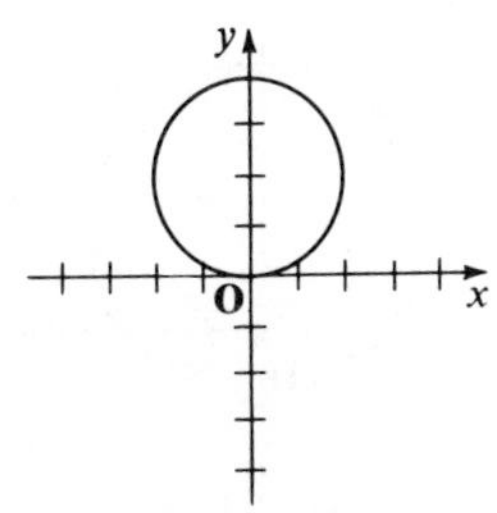

17. $y - x = 5$

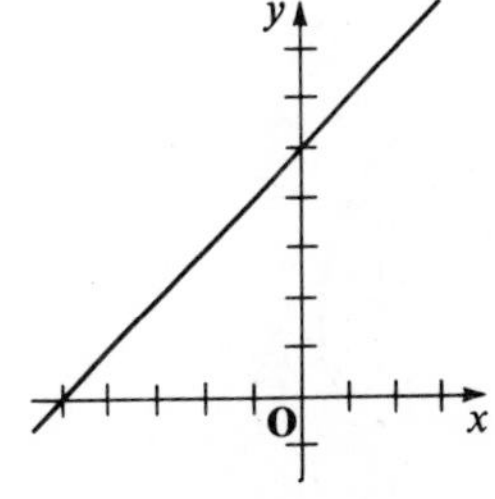

19. $8x^2 + 9y^2 - 6x - 9 = 0$

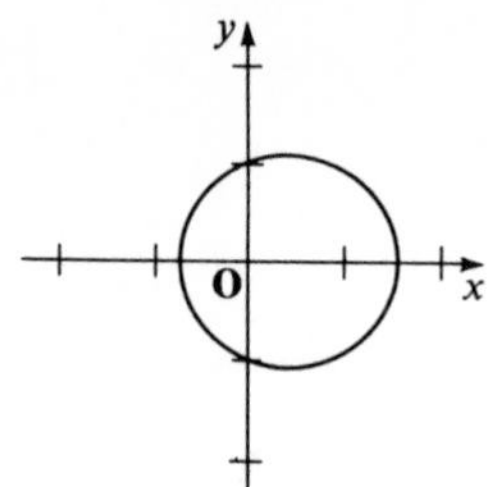

21. $3y^2 - 6y + 4x^2 = 9$

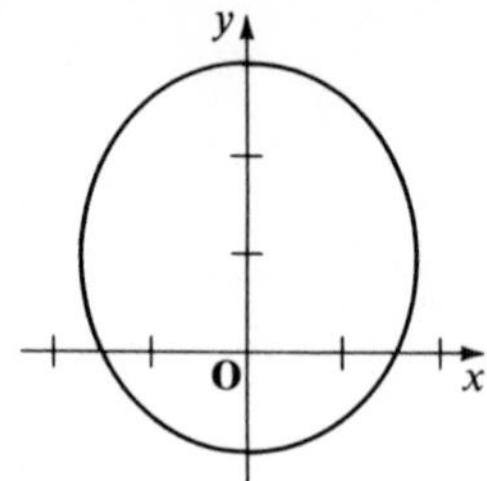

23. $24x^2 + 25y^2 - 4x - 4 = 0$

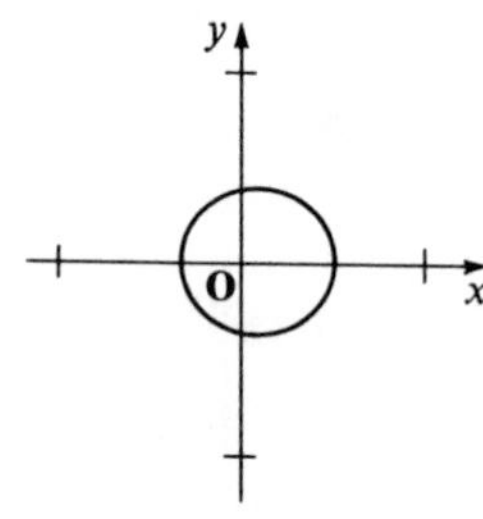

Exercises 7–3, page 256

1.

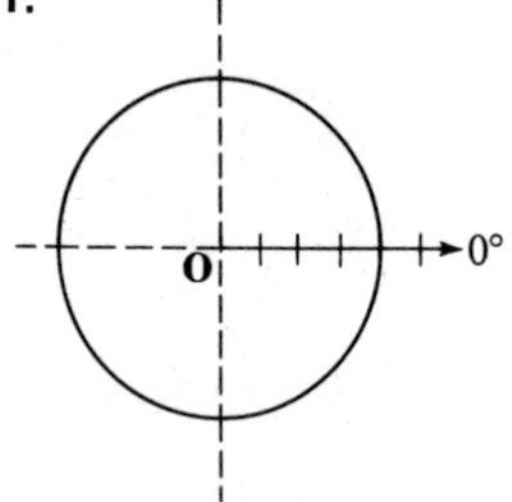

3.

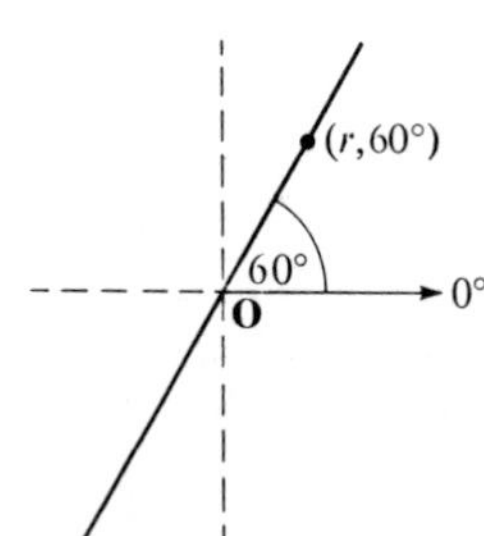

5.

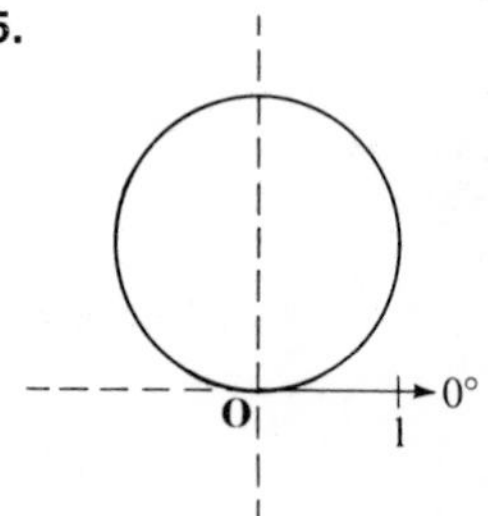

7.

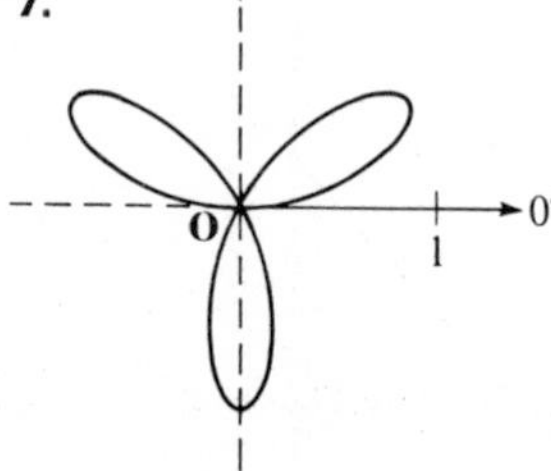

9.

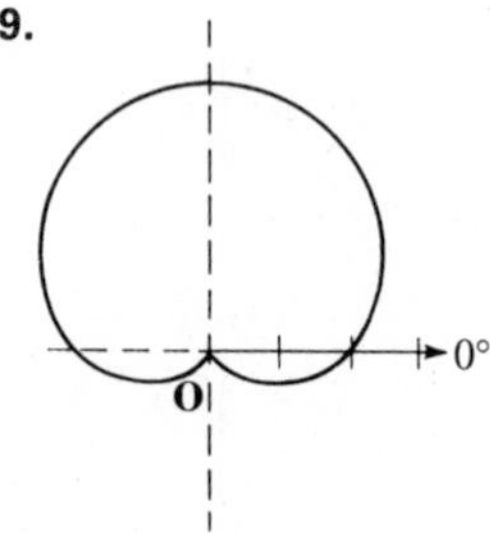

11.

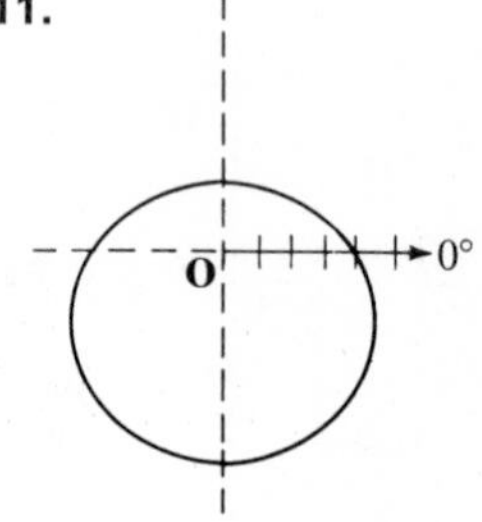

13.

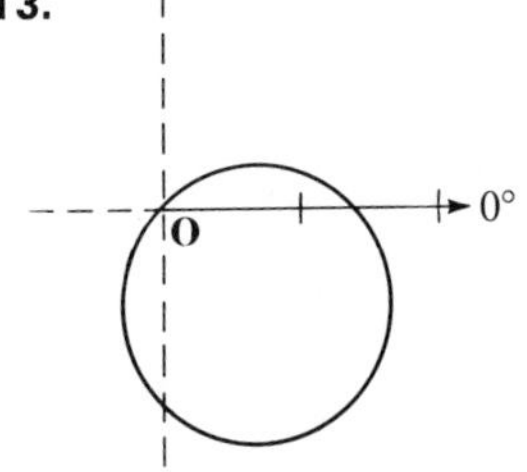

15.

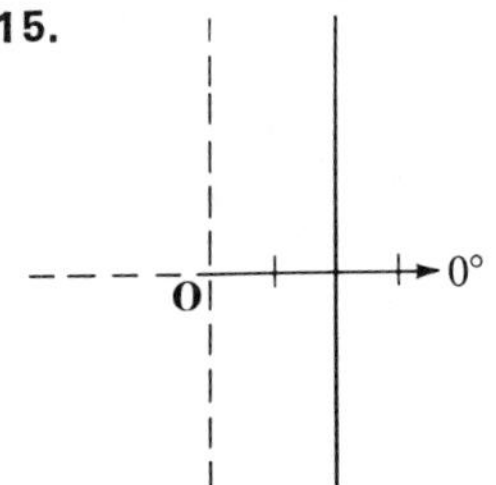

17.

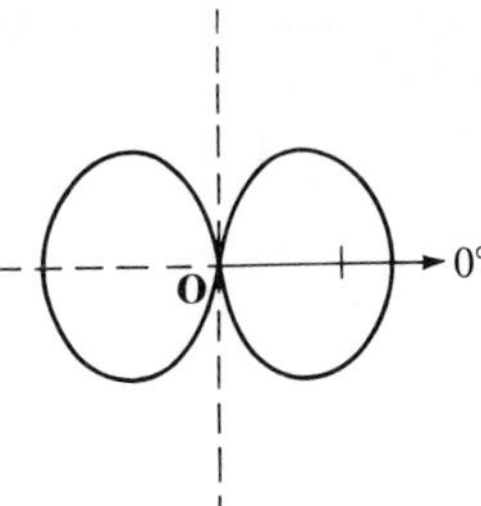

19.

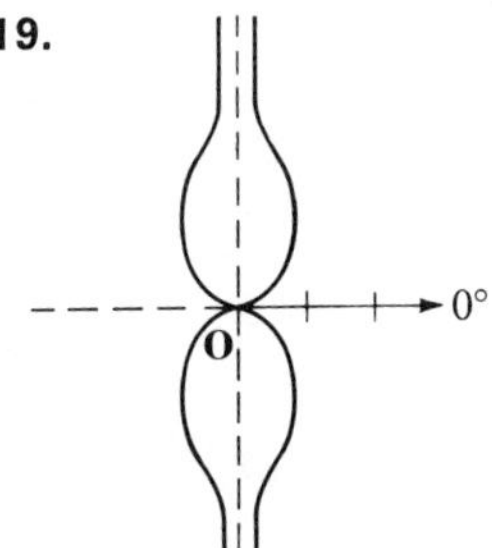

21.

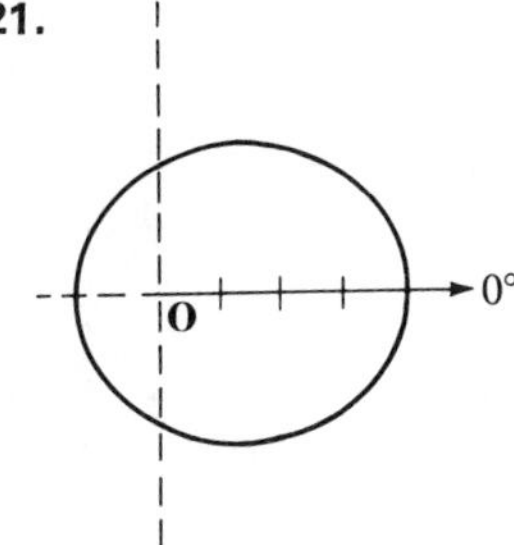

23.

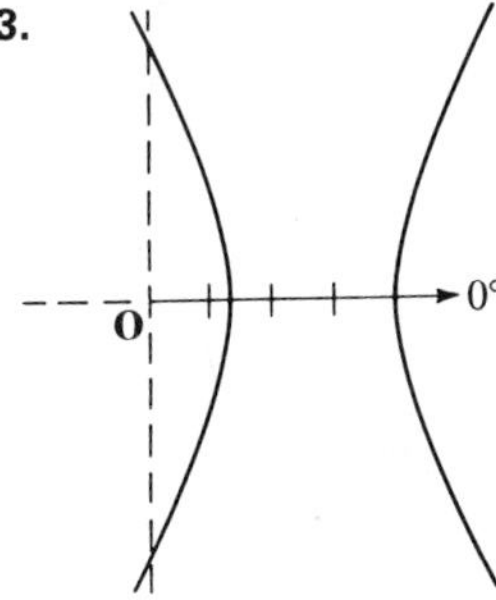

25.

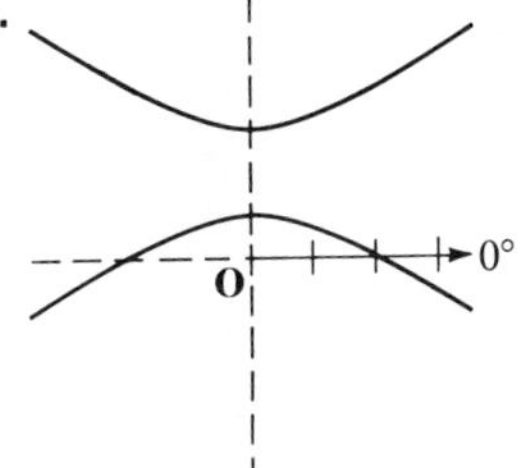

27.

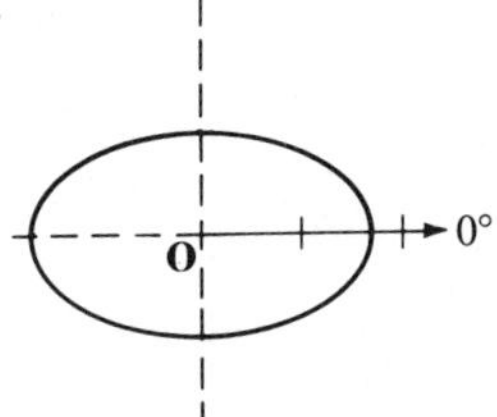

29.

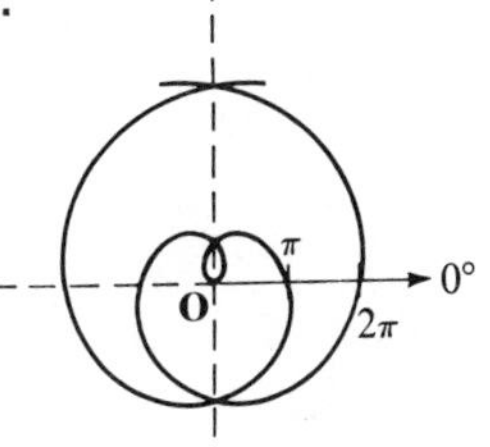

Exercises 7–4, page 258

1. $\left(\frac{1}{\sqrt{2}}, 45°\right), \left(\frac{1}{\sqrt{2}}, 225°\right)$

3. $(\frac{1}{2}, 60°), (\frac{1}{2}, 300°)$

5. (1, 45°), (1, 225°)

7. (1, 90°)

9. (4, 60°), (4, 300°)

11. (4, 90°), (4, 270°)

13. (6, 30°), (6, 150°)

15. (1, 90°), (2, 30°), (2, 150°)

17. (−1, 180°), graphically also (0, 0°)

19. $\left(\frac{1}{\sqrt{\sqrt{2}}}, 22°30'\right), \left(\frac{1}{\sqrt{\sqrt{2}}}, 112°30'\right)$; graphically also (0, 0°)

Exercises 7–5, pages 261–262

1. parabola; $x = -2$

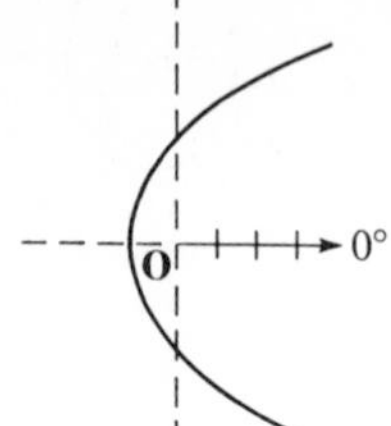

3. ellipse; $x = 6$

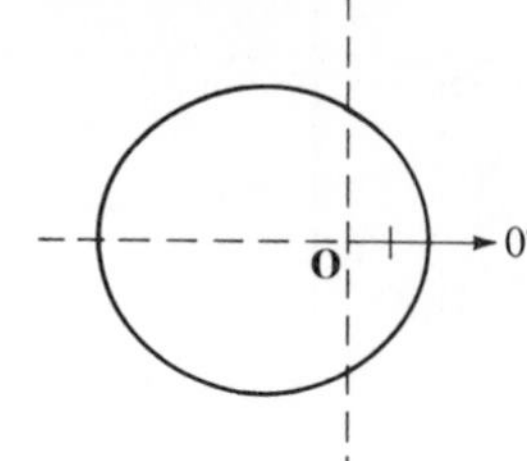

5. parabola; $y = 10$

7. hyperbola; $y = -\frac{16}{5}$

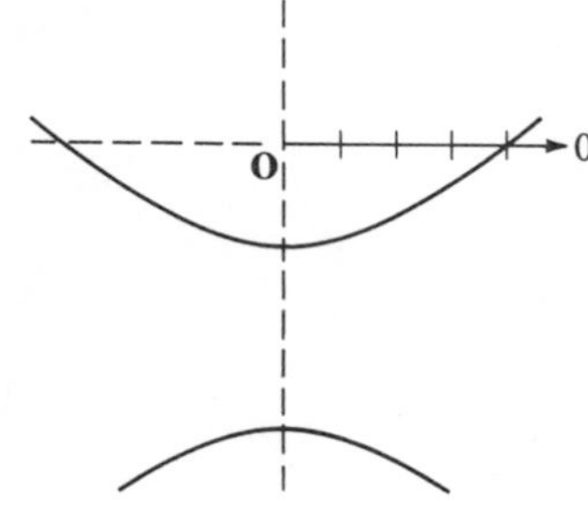

9. ellipse; $y = 2$

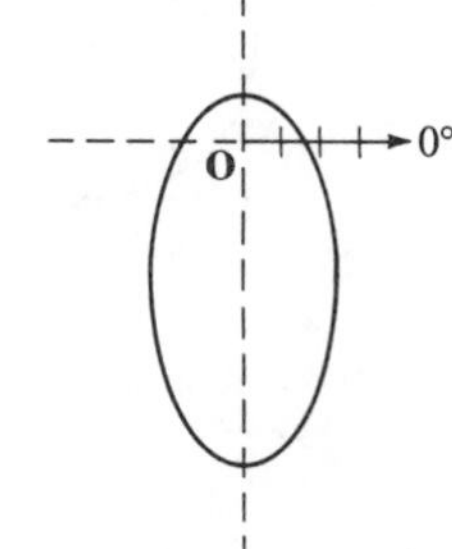

11. $r = \dfrac{4}{1 - \cos\theta}$

13. $r = \dfrac{18}{4 - 3\sin\theta}$ **15.** $r = \dfrac{8}{1 - 2\cos\theta}$

Exercises 7–6, pages 265–266

1. $r = \cos\theta$ **3.** $r = \dfrac{b}{2 + \sin\theta}$ **5.** $r = \dfrac{-3}{1 - 2\sin\theta}$

7. $r = a\cos\theta + k$ **9.** $r = a\sin^2\theta$ **11.** $r = a\cos\theta(1 - \cos\theta)$

Chapter Review Exercises, page 267

1. $\left(\dfrac{3}{2}, \dfrac{3\sqrt{3}}{2}\right)$ **3.** $r\sin^2\theta - 4\cos\theta = 0$ **5.**

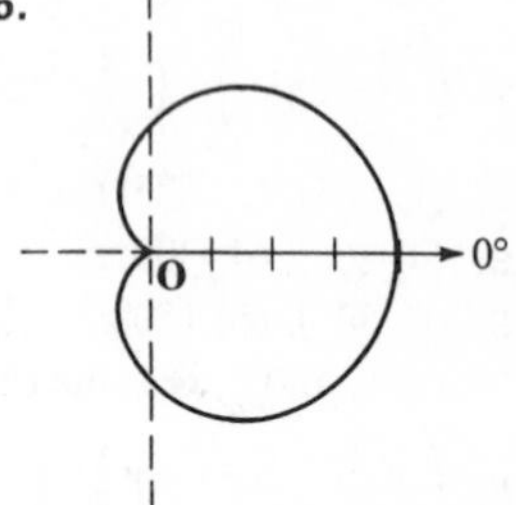

7. $(1, 90°)$ **9.** $y = -6$

Exercises 8–1, pages 277–278

1,3,5,7.

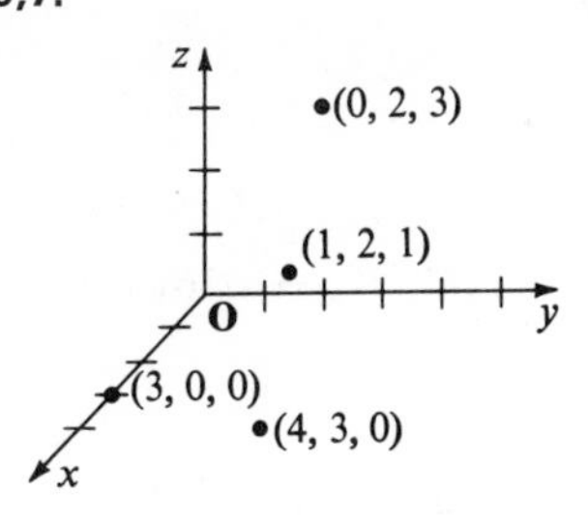

9. $z = 0$
11. $y = 0$
13. $y = 0$ and $z = 0$
15. $y = 4;\ z = 4$
17. $x = -3,\ y = 6,\ z = 2$
19. $x = 0;\ y = 0;\ z = -3$
21. 3
23. 7
25. 9
27. 3
33. $x^2 + y^2 + z^2 - 4x - 6y - 8z + 20 = 0$
35. $25x^2 + 9y^2 + 25z^2 - 225 = 0$

Exercises 8–2, pages 280–281

1. **(a)** $(2, 4, -2)$ **(b)** $\mathbf{v} = 2\mathbf{i} + 4\mathbf{j} - 2\mathbf{k}$ **(c)** $2\sqrt{6}$
3. **(a)** $(5, -5, -5)$ **(b)** $\mathbf{v} = 5\mathbf{i} - 5\mathbf{j} - 5\mathbf{k}$ **(c)** $5\sqrt{3}$
5. **(a)** $(6, 6, 6)$ **(b)** $\mathbf{v} = 6\mathbf{i} + 6\mathbf{j} + 6\mathbf{k}$ **(c)** $6\sqrt{3}$
7. **(a)** $(-5, 1, -2)$ **(b)** $\mathbf{v} = -5\mathbf{i} + \mathbf{j} - 2\mathbf{k}$ **(c)** $\sqrt{30}$
9. $(-9, 3, -6)$ **11.** $(2, 3, -2)$ **13.** $(2, -8, 8)$ **15.** $(\frac{1}{2}, -\frac{15}{2}, 7)$
17. -1 **19.** 66 **21.** -37 **23.** $\frac{19}{6}$

Exercises 8–3, page 286

1. $\cos\alpha = -\frac{1}{3},\ \cos\beta = \frac{2}{3},\ \cos\gamma = \frac{2}{3}$
3. $\cos\alpha = \frac{5}{13},\ \cos\beta = 0,\ \cos\gamma = \frac{12}{13}$
5. $\cos\alpha = -\frac{1}{2},\ \cos\beta = -\frac{5}{6},\ \cos\gamma = \dfrac{\sqrt{2}}{6}$
7. $\cos\alpha = \frac{12}{13},\ \cos\beta = -\frac{3}{13},\ \cos\gamma = \frac{4}{13}$
9. $(\frac{3}{7}, \frac{2}{7}, \frac{6}{7})$
11. $(-\frac{6}{7}, -\frac{3}{7}, \frac{2}{7})$
13. $\left(-\dfrac{7}{5\sqrt{3}}, -\dfrac{1}{\sqrt{3}}, -\dfrac{1}{5\sqrt{3}}\right)$
15. $\cos\gamma = \pm\frac{9}{11}$
17. $\left(\dfrac{8}{\sqrt{30}}, \dfrac{16}{\sqrt{30}}, \dfrac{40}{\sqrt{30}}\right)$ **19.** $(\frac{3}{14}, \frac{3}{7}, \frac{1}{7})$ **21.** $(3\sqrt{14}, 2\sqrt{14}, -\sqrt{14})$
23. $\left(-\dfrac{8}{\sqrt{30}}, -\dfrac{16}{\sqrt{30}}, -\dfrac{40}{\sqrt{30}}\right)$ **25.** $(-\frac{3}{14}, -\frac{3}{7}, -\frac{1}{7})$
27. $(-3\sqrt{14}, -2\sqrt{14}, \sqrt{14})$
29. parallel **31.** $m^\circ(\alpha) = 0$ or 180 **33.** $m^\circ(\beta) = 60$ or 120

Exercises 8–4, pages 291–292

1. $\dfrac{1}{\sqrt{2}}$ **3.** $-\frac{1}{2}$ **5.** $\frac{1}{2}$ **7.** parallel

9. perpendicular **11.** perpendicular **13.** $z = 0$ **15.** $y = 10$

17. $x = -3$ **19.** $\frac{7}{3}$ **21.** -4 **23.** $\dfrac{3}{\sqrt{14}}$

Exercises 8–5, pages 297–298

1. $(-2, 10, -6)$ **3.** $(0, -4, -4)$ **5.** $(-15, -2, 10)$
7. $(1, -17, -7)$ **9.** $(2b - a, -a - b, 3)$ **11.** $(2, -10, 6)$
23. $a = \frac{5}{3}, b = \frac{1}{3}$ **27.** $\sqrt{230}$

Chapter Review Exercises, page 300

1.

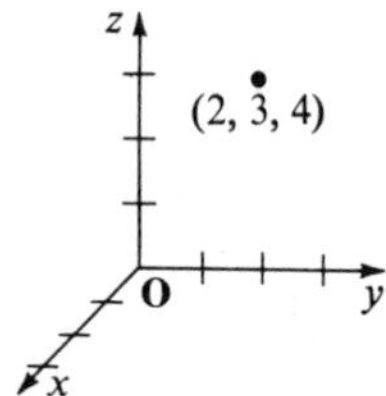

3. $2\sqrt{3}$
5. -3
7. $\left(\dfrac{5}{\sqrt{6}}, -\dfrac{5}{\sqrt{6}}, \dfrac{10}{\sqrt{6}}\right)$
9. $y = -\frac{1}{2}$
11. $(-7, 7, -14)$

Exercises 9–1, pages 311–312

1. $\mathbf{u} = (-1, 5, 7) + r(-3, -4, -4)$;
$\left(-\dfrac{3}{\sqrt{41}}, -\dfrac{4}{\sqrt{41}}, -\dfrac{4}{\sqrt{41}}\right), \left(\dfrac{3}{\sqrt{41}}, \dfrac{4}{\sqrt{41}}, \dfrac{4}{\sqrt{41}}\right)$

3. $\mathbf{u} = (1, -2, -3) + r(1, -1, 5)$;
$\left(\dfrac{1}{3\sqrt{3}}, -\dfrac{1}{3\sqrt{3}}, \dfrac{5}{3\sqrt{3}}\right), \left(-\dfrac{1}{3\sqrt{3}}, \dfrac{1}{3\sqrt{3}}, -\dfrac{5}{3\sqrt{3}}\right)$

5. $\mathbf{u} = (2, -3, 4) + r(3, 5, -5)$;
$\left(\dfrac{3}{\sqrt{59}}, \dfrac{5}{\sqrt{59}}, -\dfrac{5}{\sqrt{59}}\right), \left(-\dfrac{3}{\sqrt{59}}, -\dfrac{5}{\sqrt{59}}, \dfrac{5}{\sqrt{59}}\right)$

7. $(3, 4, 2)$ **9.** $(-3, 5, 3), (0, 9, 1)$

11. $(1, 3, -2), (3, 4, -5), (5, 5, -8)$ **13.** $\frac{19}{21}$ **15.** $\frac{8}{117}$ **17.** 0

19. $\dfrac{x}{1} = \dfrac{y - 2}{3} = \dfrac{z + 1}{4}$ **21.** $x = y = z$

23. $\dfrac{x - 3}{-1} = \dfrac{y + 1}{-2} = \dfrac{z - 4}{-2}$ **25.** $\dfrac{x - 2}{-4} = \dfrac{y + 1}{2} = \dfrac{z - 3}{-6}$

27. $\dfrac{x - 2}{3} = \dfrac{y - 1}{2}; z = -4$ **29.** $x = 5; y = -1$

31. $\dfrac{x - 1}{1} = \dfrac{y - 4}{3} = \dfrac{z + 2}{-6}$ **33.** $x + 2 = \dfrac{y + 4}{-1} = \dfrac{z + 1}{-1}$

35. $\dfrac{x - 1}{-1} = \dfrac{y - 2}{1} = \dfrac{z - 3}{6}$ **37.** $\dfrac{x - 1}{3} = \dfrac{y - 5}{2}; z = -7$

39. $x = 6; y = -1$

Exercises 9–2, pages 316–318

1. $-x + 3y + 5z - 20 = 0$ **3.** $3x + 4y - 5z + 2 = 0$
5. $x = 0$ **7.** $3x + 2y + 6z - 23 = 0$
9. $5x - 4y + 3z - 15 = 0$ **11.** $11x - 17y - 13z + 3 = 0$
13. $\dfrac{x-1}{1} = \dfrac{y+3}{-3} = \dfrac{z-4}{2}$ **15.** $\dfrac{x-1}{4} = \dfrac{y+1}{-3} = \dfrac{z-2}{2}$
17. $\dfrac{x-3}{2} = \dfrac{y+1}{2} = \dfrac{z-4}{-1}$ **19.** $x - 4y + 3z + 1 = 0$
21. $-2x + y + z + 1 = 0$, excluding point $(1, 2, -1)$
23. $2x - y - 3z - 7 = 0$ **25.** $43x + 3y - 14z - 34 = 0$
31. $-18x + 4y + 17z - 25 = 0$

Exercises 9–3, pages 324–326

1. $\mathbf{u} = (2, 1, 0) + r(-5, 1, 7)$ **3.** $\mathbf{u} = (1, 0, \frac{4}{3}) + r(-9, 6, 2)$
5. **(a)** $x = 1 - r, y = 3 + r, z = -2r$ **(b)** $\dfrac{x-1}{-1} = \dfrac{y-3}{-1} = \dfrac{z}{2}$
7. **(a)** $x = 2 + 2r, y = -1 + r, z = -r$ **(b)** $\dfrac{x-2}{2} = \dfrac{y+1}{1} = \dfrac{z}{-1}$
9. point $(1, 2, 3)$ **11.** $\emptyset$ **13.** line, $\dfrac{x-5}{-3} = \dfrac{y}{1} = \dfrac{z+3}{5}$

15. **17.** **19.**

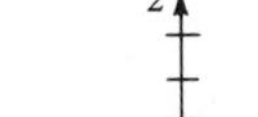
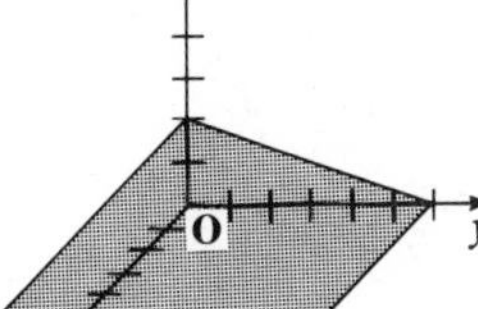
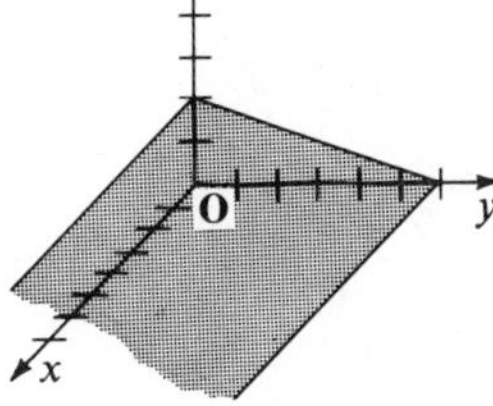

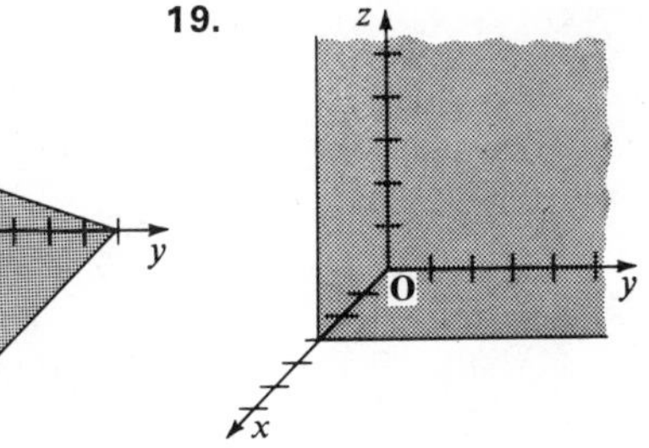

21. $2x - 6y + 3z - 67 = 0$ **23.** $\dfrac{x+3}{3} = \dfrac{y-1}{-7} = \dfrac{z-6}{-5}$
27. 0 **29.** $\dfrac{4}{\sqrt{26}}$
35. $\mathbf{u} = (0, -2, 0) + r(-1, -5, -3)$; $\mathcal{P}_1 \cap \mathcal{P}_3 = \emptyset$;
$\mathbf{u} = (0, 6, 4) + r(-1, -5, -3)$; $\mathcal{P}_1 \cap \mathcal{P}_2 \cap \mathcal{P}_3 = \emptyset$
37. $\mathcal{P}_1 \cap \mathcal{P}_2 = \emptyset$; $\mathcal{P}_1 \cap \mathcal{P}_3 = \emptyset$; $\mathcal{P}_2 \cap \mathcal{P}_3 = \emptyset$; $\mathcal{P}_1 \cap \mathcal{P}_2 \cap \mathcal{P}_3 = \emptyset$
39. $2x - y + 3z + 4 = 0$; $2x - y + 3z + 4 = 0$; $2x - y + 3z + 4 = 0$;
$2x - y + 3z + 4 = 0$
41. $x - y + 3z + 4 = 0$; $\mathcal{P}_1 \cap \mathcal{P}_3 = \emptyset$; $\mathcal{P}_2 \cap \mathcal{P}_3 = \emptyset$; $\mathcal{P}_1 \cap \mathcal{P}_2 \cap \mathcal{P}_3 = \emptyset$

Exercises 9–4, pages 328–330

1. $(-1, 2, -3)$ **3.** $(2, -6, 5)$ **5.** $(1, 11, 0)$
7. $-4x - 3y + 5z + 2 = 0$ **9.** $5x - 4y + 8z + 5 = 0$
11. $9x + 13y - 7z - 14 = 0$ **13.** $x + y + z + 8 = 0$
15. $y = -2$ **17.** $8\sqrt{3}$ **19.** $(0, -1, 2)$ **21.** $(2, 1, 1)$

Exercises 9–5, pages 333–334

1. $\frac{7}{3}$
3. 7
5. 2
7. $\frac{1}{3}x - \frac{2}{3}y + \frac{2}{3}z - 9 = 0; 9$
9. $\frac{7}{9}x + \frac{4}{9}y - \frac{4}{9}z + 5 = 0; 5$
11. $\frac{6}{7}x + \frac{3}{7}y - \frac{2}{7}z - 12 = 0; 12$
13. 2
15. 6
17. $\frac{26}{3\sqrt{14}}$
19. 8
21. $x + 4y - 3z - 2 = 0$ or $3x + z = 0$
23. $3x - 2y + 6z + 49 = 0; 3x - 2y + 6z - 49 = 0$

Exercises 9–6, page 337

1. $\sqrt{6}$ **3.** $\frac{12\sqrt{894}}{149}$ **5.** $\frac{13\sqrt{6}}{3}$ **7.** $2\sqrt{6}$
9. $\sqrt{13}$ **11.** $\sqrt{5}$ **13.** $\frac{\sqrt{10}}{7}$ **17.** $\frac{11\sqrt{19}}{57}$

Chapter Review Exercises, page 338

1. **(a)** $(x, y, z) = (3, -1, 2) + r(-2, 6, 3)$
(b) $x = 3 - 2r, y = -1 + 6r, z = 2 + 3r$
(c) $\frac{x-3}{-2} = \frac{y+1}{6} = \frac{z-2}{3}$
3. $-x + 2y + z = 0$
5. $(x, y, z) = r(1, 3, 5)$
7. $(16, 10, -18)$
9. $\frac{20}{13}$
11. $\sqrt{6}$

Exercises 10–1, pages 348–350

1. $(x - 5)^2 + y^2 + (z + 7)^2 = 4$
3. $x^2 + y^2 + (z + 1)^2 = 25$
5. $(x + 1)^2 + (y + 2)^2 + (z - 2)^2 = 49$
7. $(x - \frac{3}{2})^2 + (y - \frac{2}{3})^2 + (z + \frac{1}{2})^2 = \frac{1}{16}$
9. sphere, center $(1, -2, -3)$, $r = 3$
11. sphere, center $(-\frac{1}{2}, -\frac{3}{2}, 0)$, $r = \frac{\sqrt{10}}{2}$
13. point $(4, 3, -2)$
15. $\emptyset$
17. $(x - \frac{1}{2})^2 + (y - \frac{1}{2})^2 + (z - \frac{1}{2})^2 = \frac{3}{4}$, $C(\frac{1}{2}, \frac{1}{2}, \frac{1}{2})$, $r = \frac{\sqrt{3}}{2}$
19. $x^2 + y^2 + z^2 = 3$, $C(0, 0, 0)$, $r = \sqrt{3}$
21. $x^2 + y^2 + z^2 - 2x + 4y - 6z + 5 = 0$, $C(1, -2, 3)$, $r = 3$

23. $x^2 + y^2 + z^2 - \frac{1}{15}x + \frac{13}{3}y + \frac{23}{3}z = \frac{68}{5}$, $\mathbf{C}(\frac{1}{30}, -\frac{13}{6}, -\frac{23}{30})$, $r = \sqrt{\frac{1133}{60}}$
25. $6x + 2y - 3z = 49$
27. $-9x + 6y + 2z = 73$
29. $-x + 3z = 15$
31. $4x - 2y + z = 14$
35. plane, $3y + z = 0$
37. not on sphere
39. circle in plane $y = 3$, $r = 4$, $\mathbf{C}(0, 3, 0)$
41. point $(0, 0, 5)$

Exercises 10–2, pages 354–355

1.

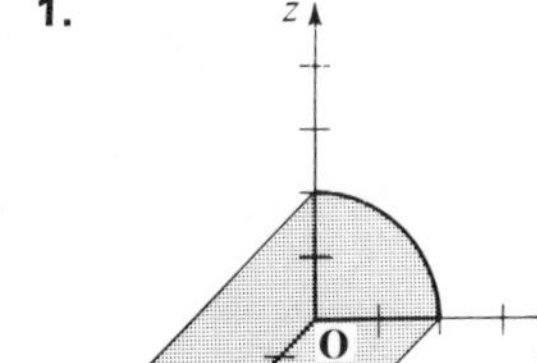

3.

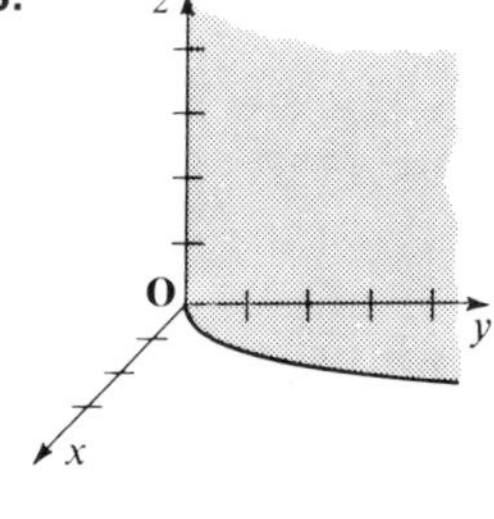

5.

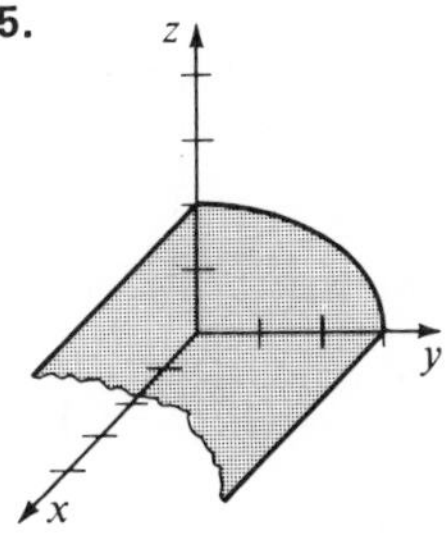

7. $x^2 + 2y^2 + 2z^2 = 1$
9. $2x^2 - y^2 - z^2 = 1$
11. $x + y^2 + z^2 = 4$
13. $x^2 - 4x + y^2 + z^2 = 21$
15. $x = 2, y = 4, z = r$; $x = -2, y = 4, z = r$
17. $2y^2 + 2z^2 = 1$

Exercises 10–3, pages 363–365

1. **(a)** elliptic paraboloid **(b)** xz-plane: $9x^2 + 4z^2 = 0$, xy-plane: $x^2 = 4y$, yz-plane: $z^2 = 9y$ **(c)** for $x = 4$: $z^2 - 9y + 9 = 0$; for $y = 4$: $\frac{x^2}{16} + \frac{z^2}{36} = 1$; for $z = 4$: $9x^2 + 64 = 36y$

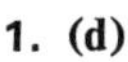

1. (d)

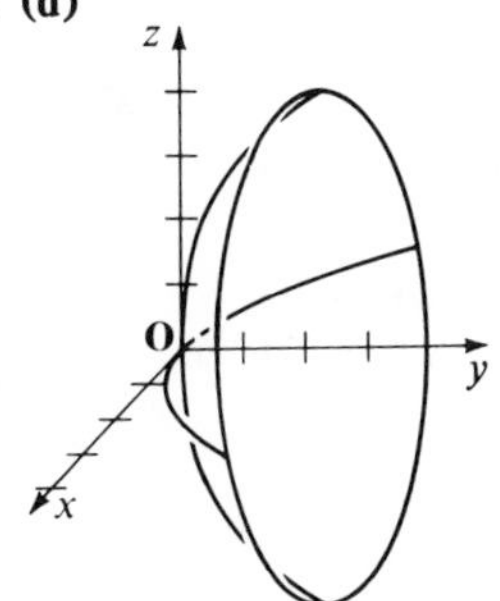

3. (d)

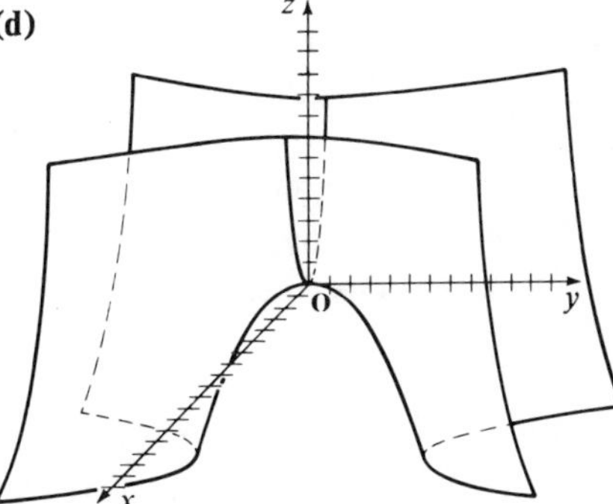

3. **(a)** hyperbolic paraboloid **(b)** xz-plane: $9x^2 = 4z$, xy-plane: $9x^2 = y^2$, yz-plane: $y^2 = -4z$ **(c)** for $x = 4$: $y^2 - 144 = -4z$; for $y = 4$: $9x^2 - 16 = 4z$; for $z = 4$: $9x^2 - y^2 = 16$

5. **(a)** ellipsoid **(b)** xz-plane: $\frac{x^2}{1} + \frac{z^2}{9} = 1$; xy-plane: $\frac{x^2}{1} + \frac{y^2}{4} = 1$, yz-plane: $\frac{y^2}{4} + \frac{z^2}{9} = 1$ **(c)** for $x = 4$: $\emptyset$; for $y = 4$: $\emptyset$; for $z = 4$: $\emptyset$.

5. (d)

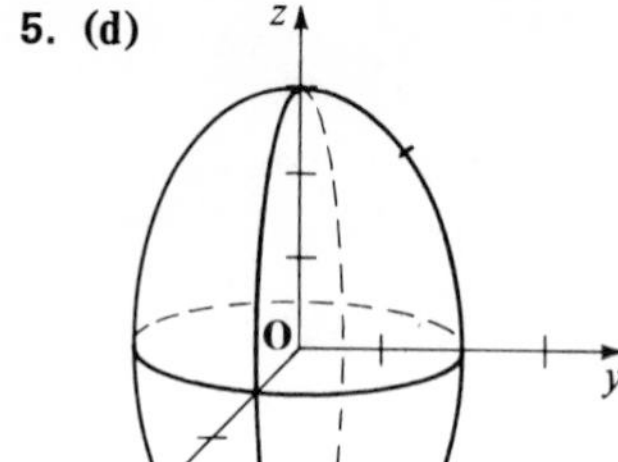

7. (d)

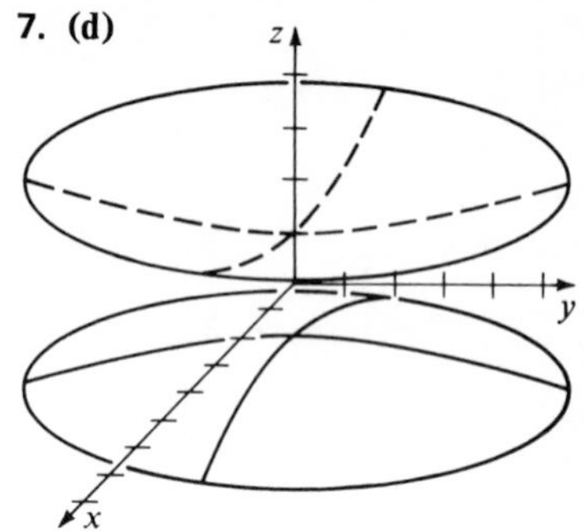

7. (a) hyperboloid of two sheets

(b) for xz-plane: $\frac{z^2}{1} - \frac{x^2}{4} = 1$, for xy-plane: $\emptyset$, for yz-plane: $\frac{z^2}{1} - \frac{y^2}{9} = 1$

(c) for $x = 4$: $\frac{z^2}{1} - \frac{y^2}{9} = 5$; for $y = 4$: $\frac{z^2}{1} - \frac{x^2}{4} = \frac{25}{9}$; for $z = 4$: $\frac{x^2}{60} + \frac{y^2}{135} = 1$.

9. $\frac{(x+2)^2}{4} - \frac{(y+4)^2}{4} - \frac{(z-1)^2}{4} = 1$, hyperboloid of two sheets

11. $\frac{x^2}{5} + \frac{(y-3)^2}{10} - \frac{(z-\frac{1}{2})^2}{\frac{5}{2}} = 1$, hyperboloid of two sheets

13. $z = (x+1)^2 - \frac{(y-1)^2}{\frac{1}{4}}$, hyperbolic paraboloid

15. $(x-2)^2 + z^2 = y + 9$, elliptic paraboloid

17. $\frac{x^2}{2c} + \frac{z^2}{2c} = y + \frac{c}{2}$, elliptic paraboloid

Exercises 10–4, pages 369–370

1. $x'^2 + 3y'^2 + 4z'^2 = 11$, ellipsoid, $\mathbf{C}(2, -3, 0)$
3. $2x'^2 - 3y'^2 - z'^2 = 5$, hyperboloid of two sheets, $\mathbf{C}(1, -2, 2)$
5. $3x'^2 - z'^2 = -3y'$, hyperbolic paraboloid, $\mathbf{V}(-2, 1, \frac{2}{3})$
7. $9x'^2 - 4y'^2 = 5z'$, hyperbolic paraboloid, $\mathbf{V}(0, -1, -2)$
9. hyperboloid of one sheet
11. hyperboloid of one sheet
15. spheroid

Exercises 10–5, pages 374–375

1. $\left(\sqrt{2}, \frac{\pi}{4}, 1\right)$; $\left(\sqrt{3}, \frac{\pi}{4}, \cos^{-1}\frac{\sqrt{3}}{3}\right)$

3. $\left(\sqrt{7}, \cos^{-1} - \frac{2}{\sqrt{7}}, 2\right)$; $\left(\sqrt{11}, \cos^{-1} - \frac{2}{\sqrt{7}}, \cos^{-1}\frac{2}{\sqrt{11}}\right)$

5. $(5, \cos^{-1}\frac{3}{5}, 5)$; $\left(5\sqrt{2}, \cos^{-1}\frac{3}{5}, \frac{\pi}{4}\right)$

7. $\left(\sqrt{7}, \cos^{-1}\frac{\sqrt{3}}{\sqrt{7}}, -1\right)$; $\left(2\sqrt{2}, \cos^{-1}\frac{\sqrt{3}}{\sqrt{7}}, \cos^{-1} - \frac{1}{2\sqrt{2}}\right)$

9. $(0, 4, 1)$ **11.** $\left(-\frac{3}{\sqrt{2}}, -\frac{3}{\sqrt{2}}, -5\right)$ **13.** $(0, 1, \sqrt{3})$

15. $\left(0, -\frac{5\sqrt{3}}{2}, \frac{5}{2}\right)$ **17.** $x^2 + y^2 + z^2 = 4$ **19.** $x^2 + y^2 = 9x$

21. $x^2 + y^2 + z^2 = 4$ **23.** $x^2 + y^2 + z^2 = 4\sqrt{x^2 + y^2}$

25. **(a)** $r^2 + z^2 = 9$ **(b)** $\rho = 9$

27. **(a)** $r^2 + 9z^2 = 36$ **(b)** $\rho^2(1 + 8\cos^2\phi) = 36$

29. **(a)** $r^2 + z^2 - 2r\cos\theta = 0$ **(b)** $\rho(\rho - 2\sin\phi\cos\theta) = 0$

31. **(a)** $r\cos\theta = 4z^2$ **(b)** $4\rho^2\cos^2\phi - \rho\sin\phi\cos\theta = 0$

33. **(a)** $r^2 + z^2 = k^2$ **(b)** $\rho = k$ **35.** **(a)** $z = 4$ **(b)** $\rho\cos\phi = 4$

Chapter Review Exercises, page 377

1. $C(4, -3, 0)$; $r = 5$ **3.** $-x + 2y + 2z = 4$

5. $4x^2 + y^2 + 4z^2 = 16$ **7.** $J = 0$

9. elliptic paraboloid, $V(0, -2, 2)$

11. $\left(\sqrt{2}, \frac{7\pi}{4}, -1\right)$; $\left(\sqrt{3}, \frac{7\pi}{4}, \cos^{-1} -\frac{1}{\sqrt{3}}\right)$